Between Two Worlds: A New Introduction to Geography

Houghton Mifflin Company   **Boston**

Atlanta   Dallas   Geneva, Illinois   Hopewell, New Jersey   Palo Alto

Between Two Worlds: A New Introduction to Geography

Robert A. Harper   *Professor of Geography, University of Maryland*

Theodore H. Schmudde   *Professor of Geography, University of Tennessee*

Cartography by Francis & Shaw, Inc.,
New York, N.Y.

Printed in the U.S.A.

Library of Congress Catalog Card Number: 72-6890

ISBN: 0-395-12075-6

Preface

THIS BOOK and its approach grew out of the continuing effort of each of the authors to introduce college students to geography. The primary stimulus came from the geography community at Southern Illinois University during the 1960s where the two of us and others on the staff carried on a running dialogue about the problems of teaching geography to undergraduates.

A parallel stimulus developed from the challenge of organizing and carrying out a series of four summer institutes for teachers funded by the U.S. Office of Education. We had to face up to the problem of introducing geography to teachers who often had little or no background in the field. We also had to make the experience one that offered them a framework from which to develop courses of study that were geographically sound as well as pertinent to the lives of students. The institute of the summer of 1968, involving a cooperative venture between Southern Illinois University at Carbondale and the University of Maryland, embodied an approach very similar to that of this book. Although the area of concern was primarily the geography of the United States, it was the interest of the teachers from that institute, in terms of both their requests for written materials and feedback from them on things they tried in using the approach, that stimulated this writing attempt.

Our thinking has been much influenced by the ideas, arguments, and experiences of faculty colleagues, particularly Frank H. Thomas, Douglas Carter, Jean Gottmann, David Sharpe, and the late Charles Colby. Nevertheless, students have been the critical ingredient stimulating our continuing experimentation and innovation. We also benefited from the climate of experimentation and change that permeated the whole campus during this period of rapid development at Southern Illinois University at Carbondale.

The audience for this book, then, consists of students enrolled in courses variously entitled Introduction to Geography, Regional Geography, Cultural Geography, and Human Geography—in short, nearly all those programs of introductory geography of the nonphysical variety. It can be used by both majors and nonmajors.

We wish to acknowledge particularly the help of Doyne Horsley, Edward J. Lombardi, and Francis Nicholas. Doyne did most of the research on the maps and tables in the book and proved to be not only a meticulous researcher but a thoughtful contributor to the concept of the book as well. Thanks go also to Ed Lombardi for his research on the map and tabular data. Frank has accomplished the difficult task of developing an instructor's manual that is consistent with the thrust of the book and that challenges each teacher to build beyond it as much as he is capable of and disposed to doing.

Robert A. Harper
Theodore H. Schmudde

Contents

Between Two Worlds: A New

Introduction to Geography

Prologue

IF THERE is a single overriding purpose in *Between Two Worlds,* it would be that of broadening the reader's scope. The range of man's activities and the places where these activities occur form a complex of interrelationships. Thus, while we may study the trees—that is, the components of individual places—we will not lose sight of the intricate design of the forest. We will view the order in human life from a "systems" perspective, in which interrelationships between components of the man-environment complex are seen as having greater significance than the specific character of the components themselves.

Our framework is based on the view that societies, or systems, fall more or less into one of two categories: (1) the closed human system that has been traditional throughout most of man's history, and (2) the modern, worldwide, interconnected human system that has developed with modern transportation, communication, and technology. An important aspect of any system is its approach to the use of environmental resources. There is much overlap of these theoretical system types in any given example one studies. No place in the world—neither Washington, D.C. (Chapter 1) nor Ramkheri, India (Chapter 2)—exemplifies completely one or the other. It is our hope that the reader will first come to some understanding of these two systems or "worlds" and then see how specific places throughout the globe partake of both these worlds. What is important here is not our particular framework, but recognition by the reader that, at the beginning at least, geography is a way of seeing—a way of identifying, organizing, and understanding the world.

After studying two contrasting places and articulating the perspective of the book in Part One, we present in Part Two the human and physical systems which fit together at any particular place to give it its distinctive character. Parts Three and Four then examine particular parts of first the industrialized world and then the underdeveloped world within the framework we have constructed.

The book is designed to be a catalyst that will stimulate and encourage the reader to go beyond the ideas and information in it. The conscious avoidance of jargon, the inclusion of photo essays and inquiry boxes which suggest further steps to be taken with the material provided in the text, the numerous maps—all this has been done to emphasize the ideas, not simply to catalog information. Our hope is that the reader will go beyond the ideas and information contained in this book to form separate opinions, decisions, and conclusions for himself. As Hamlet said in another context, "There are more things in heaven and earth, Horatio, than are dreamt of in your philosophy."

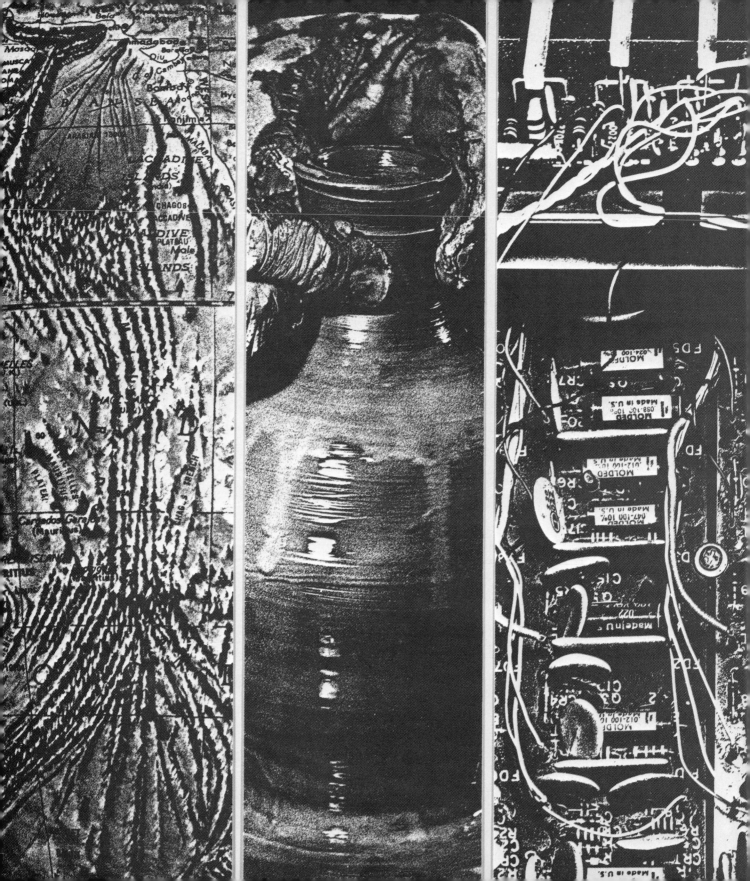

PART ONE   THE GEOGRAPHIC PROBLEM:

FINDING A FOCUS FOR

THE WORLD

Chapter 1   Metropolitan Washington: Center of an Interconnected Network

I MAGINE WASHINGTON, D.C., on almost any weekday morning, winter or summer, between 7:30 and 9:00 A.M. The view from a traffic-reporting helicopter shows bumper-to-bumper traffic backed up as the cars try to cross the bridges over the lower Potomac River to enter downtown Washington. The mass of traffic coming in from the Virginia suburbs slows to a halt in the "mixing bowl" where the major freeways come together in the area of the Pentagon. This is just part of the multitude of autos, buses, and trucks bringing more than 400,000 workers into downtown Washington each workday morning.

Life in the teeming anthill of Washington calls for a daily migration of people from their homes to jobs and back again. The bulk of this movement is into the heart of the city, but at the same time there are also crowded traffic snarls on the beltway that rims metropolitan Washington because many people now go to jobs in the suburbs as well.

At the same time, more than half a million schoolchildren make their own daily migration on foot, by bicycle, by car, or in hundreds of school buses. But for most children this movement is only a few blocks or a few miles, and very few go to schools downtown. Years of planning and development by local school boards have created a system of schools distributed as nearby as possible to where the children live. And, since more than two out of every three people live in the suburbs outside the District of Columbia, school migration forms a pattern different from that of the people going to work.

Later in the morning still another, less concentrated movement will begin to take place. Nonworking people will begin to make their daily trips to the store, to the doctor, or to visit friends. Again, some of these will be on foot or by bus, but the majority will probably also be by automobile. Like the schoolchildren, only a minority of these people will travel into downtown Washington. Most trips will be local, within one neighborhood of the city proper or within one sector of a suburb. Few Marylanders travel regularly to suburban Virginia or even to the District except to work, and the same is true of Virginians with respect to suburban Maryland.

The lower-income groups which dominate many parts of the District have few cars and limited mobility. Thirty per cent have no cars and are entirely dependent on public transportation. Their job opportunities, shopping possibilities, and social contacts are restricted to those places in the metropolitan area with bus connections. Since the bus system is designed mostly to bring suburbanites and District residents into the central city, it does not help city residents to find jobs in the suburbs or to move easily from one sector of the city to another. Inner city residents have complained bitterly about their relative inaccessibility to the job opportunities of the outer metropolitan area and about the expense of public transportation.

During the day, a massive fleet of trucks moves endlessly throughout the metropolitan area, delivering goods to offices, factories, stores, and even individual homes. The trucks pick up cleaning, trash, and special orders from those same places and maintain the appliances, equipment, and complex public utilities so essential to the operation of this complex metropolitan system.

On weekends the patterns change. There is an almost complete halt in the major migrations to work and school. In their place shopping patterns increase, and new flows appear to churches and recreational centers.

But most of the major movement in the metropolitan area occurs each weekday morning and evening as people travel to and from work. Otherwise, most Washingtonians, both suburbanites and inner city residents, live in their own local "life space." They have little regular experience in other parts of the metropolitan area and have only a vague concept of the rest of the District. Other neighborhoods and suburban communities are almost foreign territory.

Figure 1-1 shows the Washington of its residents. The daily "life space" of any individual or family occupies just a small fraction of the total metropolitan area. Each person encounters only a minute number of the almost three million inhabitants. Yet everyone identifies with Washington. Every family listens to Washington radio stations or watches Washington TV news; many read the Washington daily papers, follow the Washington Redskins, and pay taxes to municipalities in the Washington area. Each family is dependent on others who live there and who usually are personally unknown to them. All residents partake of Washington's services; all suffer from the pollution and congestion that besets the whole metropolitan area.

As we can see, metropolitan Washington is more than a location measured by latitude and longitude, more than an environmental combination of atmosphere and land resources. It is a place where people live—and it is the living which focuses the geographer's attention on the place. He is interested in understanding the how and why of human life at particular places on the earth. Each place has its unique combination of locational, environmental, and human conditions, and it is the geographer's concern to understand how they interact.

Washington: Movement on a Different Scale

Anyone who is familiar with the United States recognizes that Washington, as the capital of the country, probably ranks among the top metropolitan areas of the world in importance. This real significance cannot be seen in the morning traffic jams or in the crowds at shop-

Figure 1-1 Washington, D.C. and Its Environs

Commercial

Densely built-up residential areas with more than 10 dwelling units per acre

Less densely built-up residential areas

Industrial

U.S. Government and Defense

Institutional

Parks and permanent open space

—— Main highway

<-+-> Main railroad

✈ Commercial airport

■ Shopping center

ping centers and sports events, which are also found in other large cities. It is not the local movements within the metropolitan area that make Washington distinctive among cities. We must look at other evidence.

Washington is one of the world's busiest air centers. In a recent year more than 15.5 million passengers passed through the three public airports serving the metropolitan area (one of the three also serves the Baltimore metropolitan area).

This means that the population coming and going between Washington and other places by air was more than five and one-half times the resident population. Of course, air travelers are just a portion of the people moving between Washington and other places. Many others travel by auto, bus, and railroad.

Even these vast movements of people in and out of Washington tell only part of the story of Washington's connections with other places. Recently, the telephone company serving the Washington metropolitan area estimated there were some 76 million long-distance telephone calls leaving the area annually. If we assume an equal number of incoming calls, the total would be some 152 million telephone calls a year— more than fifty times the total resident population of the metropolitan area. These figures do not include calls over permanently leased government lines since the company does not monitor the number of calls on these lines. And there must be many other messages coming in over other privately leased lines and by telegraph, cable, and microwave connections.

In the light of these figures, the resident population of Washington may be seen as just the upper part of a largely submerged iceberg of interaction. In other words, the visible daily life of Washington is a very poor measure of what Washington really does. The three million residents of the metropolitan area operate an enterprise involving many more people than those who actually live there.

The airline passengers include tourists to be sure, but the majority are businessmen and government officials on business. The long-distance telephone calls include many personal calls, but most are undertaken as part of governmental or private business. When one considers all these contacts by air, phone, and other means, the total scope of Washington's daily communications with other places can barely be imagined.

Thus, we see a second dimension of Washington and its population: the interaction with places and people beyond the metropolitan area. Washington does not operate as a self-controlled entity. It cannot provide itself with all the required goods, services, and ideas of life. Rather, Washington lies within a much larger network which stretches into many different parts of the country and the world. The people of Washington depend on other parts of the world for food, clothing, and most other needs. Washingtonians buy foreign cars as well as those made in Detroit. They drink Florida orange juice, eat California lettuce, and broil Midwestern beef. The wealthy women of Washington wear the latest clothing designs from New York, Paris, and Rome. The rest buy the same ready-to-wear clothes sold in cities and towns throughout the country.

Washington also performs important functions for people in other places. Many of the people who commute to work through Washington traffic are on their way to jobs in government offices which deal with problems of people, local governments, and business throughout the United States. Others work in agencies involved in international relations. Some of these workers are employed by business associations, lobbying groups, and professional societies representing interests of people and businesses in other parts of the country. Still others are foreign nationals who work for embassies or consulates of their governments or for international organizations.

For most of the people in the Washington metropolitan area, the city is a place to live and a place to work. They experience the city in terms of their particular job, their school, their shopping world, and their recreational activities. Only a few move beyond Washington as part of their work or make frequent long-distance phone calls. Washingtonians worry first about the congestion of their streets, the crime rate in their part of town, the crowding in the school and their local neighborhood, and the busing arrangement that serves that school. The news of Washington affects them since many have government jobs. But the average family is probably more concerned with how things went at school, what the news is from grandma in Des Moines, or how the Redskins did in the game last night.

What, then, should we know about Washington? What is significant? Are the figures on size, population, employment structure, and income important? Is Washington more important for suburbs, inner city communities and slums, or for its problems of pollution and race relations? Should one concentrate on Washington as the capital of the United States with influence over the farm program, the highway system, the use of wilderness areas, and the welfare of the poor? Or is its role as the capital of one of the world's strongest nations most significant?

A Major Capital, but Only a Modest-sized Metropolis

Washington is the capital of one of the largest, most influential countries. In population, however, it does not stand out either among the

Table 1-1 METROPOLITAN AREAS OF THE WORLD'S LARGEST CAPITALS, 1964

CAPITAL	POPULATION	RANK
Tokyo	15,400,000	1
London	11,025,000	2
Moscow	8,450,000	3
Paris	8,000,000	4
Buenos Aires	7,700,000	5
Mexico City	6,100,000	6
Peking	4,600,000	7
Cairo	4,200,000	8
Seoul	3,200,000	9
Djakarta	3,150,000	10
Delhi–New Delhi	2,900,000	11
Manila	2,900,000	11
Madrid	2,757,000	13
Lima	2,300,000	14
Washington	2,265,000	15

SOURCE: *Encyclopaedia Britannica* Atlas, 1970.

Table 1-2 LARGEST METROPOLITAN AREAS OF THE UNITED STATES, 1970

METROPOLITAN AREA	POPULATION	RANK
New York	11,529,000	1
Los Angeles	7,032,000	2
Chicago	7,000,000	3
Philadelphia	4,818,000	4
Detroit	4,200,000	5
San Francisco–Oakland	3,110,000	6
Washington	2,861,000	7
Boston	2,754,000	8
Pittsburgh	2,401,000	9
St. Louis	2,363,000	10
Baltimore	2,071,000	11
Cleveland	2,064,000	12

SOURCE: *Statistical Abstract of the United States: 1971.*

Table 1-3 EMPLOYMENT STRUCTURE OF MAJOR METROPOLITAN AREAS OF THE UNITED STATES IN 1970 (per cent)

	TOTAL EMPLOYMENT	MANUFAC- TURING	TRADE	SERVICES	TRANSPOR- TATION AND UTILITIES	CONSTRUC- TION	FINANCE, INSURANCE, REAL ESTATE	GOVERNMENT
New York	4,861,000	20.9	20.9	20.5	7.8	3.5	10.6	15.8
Chicago	2,981,000	31.4	22.5	16.9	6.9	4.0	6.1	12.1
Los Angeles	2,897,000	28.2	22.3	18.9	6.0	3.8	5.9	14.5
Philadelphia	1,796,000	30.5	20.5	17.7	5.8	4.8	5.7	14.8
Detroit	1,483,000	37.6	19.9	14.8	5.3	3.5	4.6	14.3
Boston	1,291,000	21.5	22.7	24.9	5.9	4.0	7.3	13.7
San Francisco– Oakland	1,264,000	16.1	21.3	17.8	10.6	4.8	7.8	21.5
Washington	1,157,000	3.8	19.6	21.8	5.2	5.9	5.9	37.7
St. Louis	899,000	30.5	21.3	16.9	7.5	4.5	5.2	13.9
Pittsburgh	875,000	31.8	20.3	18.3	6.8	4.9	4.3	12.6
Cleveland	859,000	34.5	21.4	16.2	6.0	4.1	4.9	12.8
Baltimore	808,000	24.2	21.9	16.7	7.1	5.4	5.4	19.2

SOURCE: U.S. Bureau of Census, *Statistical Abstract of the United States: 1971* (92nd ed.), Washington, D.C., 1971, Sect. 33.

world's capitals or among the largest metropolitan areas of the United States. Table 1-1 shows that it ranks fifteenth among world capitals in population. It is slightly smaller than Lima, Peru, and Manila, Philippines; and those are capitals of countries with far smaller populations. Among major metropolitan areas of the world it ranks fortieth, and even in the United States it ranks seventh (Table 1-2). Why is the capital of the world's most powerful political and economic country not also the greatest metropolitan center in the world? Let us look at the evidence.

A SPECIAL-PURPOSE METROPOLIS

Table 1-3 indicates one of Washington's distinctive characters as compared with the largest metropolitan areas of the country—those with populations of more than two million. The concentration of employment in one activity, government services, is obvious. Notice that government workers in Washington constitute a larger percentage of all those employed there than the concentration in any single activity in any other metropolitan area except Detroit, where approximately 38 per cent of employment is in manufacturing. If we think of Detroit as a specialized manufacturing center, then Washington must be thought of as a specialized government center. It is the federal government's "company town."

The concentration of 30 per cent or more of all workers in one activity is not unique to Washington, as you can see from Table 1-3. Six of the other largest metropolitan areas have such concentrations; but in all six it is manufacturing that is the most important activity.

Washington, the national capital, is the center of government only; there is virtually no manufacturing there.

It is both the lack of manufacturing and the concentration of government functions that make Washington distinctive. Notice in Table 1-3 that it is relatively low not only in manufacturing but also in trade, transportation, and utilities. Washington does not fit the normal pattern of a major metropolitan area in the United States. And it is the lack of the usual urban nongovernment functions that makes it different from London, Tokyo, Moscow, Paris, and even Mexico City and Buenos Aires. They are centers of business, finance, trade, manufacturing, and culture for their countries as well as being capitals. In the United States, however, the major center for such activities is New York City, the largest metropolitan area.

CHOICE OF THE SITE OF THE CAPITAL

Washington was specifically established as the seat of national government. The idea had been to create an entirely new federal city unencumbered either by pre-existing street patterns and buildings or by existing local commercial and political interests. Land was taken from the states of Maryland and Virginia and set aside as the Federal District within which the government could develop in its own fashion.

The location was selected not primarily for reasons of topography or climate. Rather it was very close to the center of population of the existing 13 states from New Hampshire to Georgia. It was close to both the geographical center and the center of population of the new republic. Moreover, it lay along the Potomac River, which provided easy access to the Chesapeake Bay and the Atlantic. It was also the quickest route to the distant states. Like many other cities along the seaboard at the time,

Washington was at the fall line, the zone of falls and rapids that marked the edge of the coastal plain. Ships could navigate the broad lower Potomac as far as Washington, but the fall-line rapids blocked them from traveling farther upstream.

From the beginning, then, it was Washington's location with regard to other places that was most important (Figure 1-2). Its location was determined at least in part by accessibility to the capitals and large cities of the states. Through the years, Washington has continued to be a key center of transportation routes: first toll roads and sea routes, then a canal system and railroads, and now highways and airlines. Communications have been equally important. First there were courier and mail services, then telegraph systems, now telephones and radio and television networks. It has been necessary to develop systems of transport and communication that can connect not only with points throughout the United States but also with capitals and large metropolitan centers throughout the world.

WASHINGTON'S GROWTH: A MEASURE OF GOVERNMENTAL INFLUENCE

The growth of Washington as the capital city has reflected both the burgeoning of the country and the rise of the federal government in the affairs of the American people. In the period of nation-building it was a major center of population, ranking eighth among the country's cities in 1850 and ninth in 1880. But its single-purpose character was a limiting factor as business and industry developed throughout the country between 1880 and 1930. By 1920 Washington had fallen to fourteenth place, behind Milwaukee, Buffalo, and Baltimore. It was not until 1940 that the metropolitan popu-

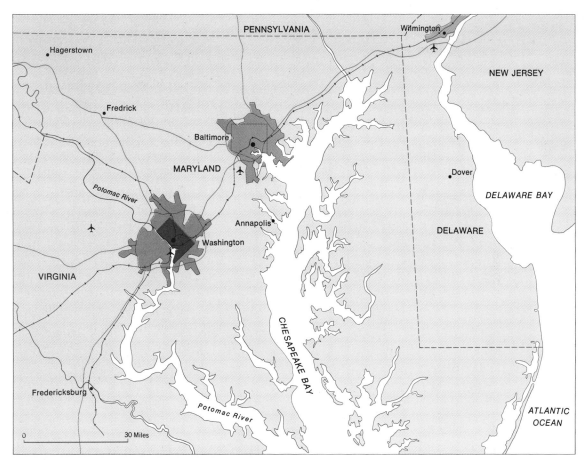

Figure 1-2 Washington, D.C.: Its Place in the Surrounding Geographic Area

Urban area

—— Interstate highway

+——+ Main railroad

✦ Major commercial airport

lation of Washington reached one million. Nine cities exceeded its population at that time.

Since 1940 the metropolitan population of the United States has grown, and Washington has been one of the most rapidly growing centers. This rapid increase seems directly related to the growing importance of the federal government in the affairs of the country. Federal expenditures in 1940 amounted to only 9 per cent of Gross National Product; by 1970 they were 30–35 per cent. Moreover, the higher percentage represented a much larger sum, $313 billion in 1970 as compared with $9 billion in 1940. (The declining value of the dollar accounts for only a small part of this increase.) Although most of this money is actually spent elsewhere, the decisions concerning its allocation are made in Washington and the funds are administered from there. With total assets of $400 billion in 1968 both in real property and intangibles, the United States government is a huge enterprise, about ten times as large as the American Telephone and Telegraph Company which is the world's largest private business corporation. Between 1940 and 1965, federal employment in Washington approximately doubled to almost 300,000 persons. In this period the total population of the metropolitan area increased two and a half times.

The Importance of Government Today

The legislative and judicial branches of the government together employ less than 10 per cent of the total government work force. Thus, most of the federal employees in the Washington area are neither making nor adjudicating laws. Rather they are in the executive branch under the President, carrying out and administering the business of government.

In addition to the government's civilian em-

ployment of 300,000 people, a large military component has been rapidly growing since 1950. An important segment of this military population works side by side with civilians in the Department of Defense, mainly at the Pentagon. This famous building complex has more than 27,000 military and civilian workers. There are also various other administrative, staff, and research centers manned by the military. But close to 85 per cent of all military personnel are attached to operational military establishments like Andrews Air Force Base in Maryland.

Apart from their administrative functions, many departments and agencies of the executive branch are also engaged in research programs on a massive scale. Some of this research is carried out at various centers over the country, but over fifty important facilities with 50,000 employees, are located in the Washington area. The majority belong to the Department of Defense, but there are important civilian research centers, including the Smithsonian Institution, the United States Geological Survey, the Bureau of Standards, the National Institutes of Health, the Goddard Space Flight Center of NASA, the Beltsville Agricultural Research Center, and the United States Weather Bureau.

The modern city is tied neither to the productive resources of its local area nor to the need to manufacture its own products. Most of its needs are supplied from elsewhere in the world. Of all major cities in the United States, Washington perhaps exemplifies best a place where very few people work with their hands or process resources, yet the decisions made there have worldwide impact.

LIMIT
4 STEAKS
PER
CUSTOMER

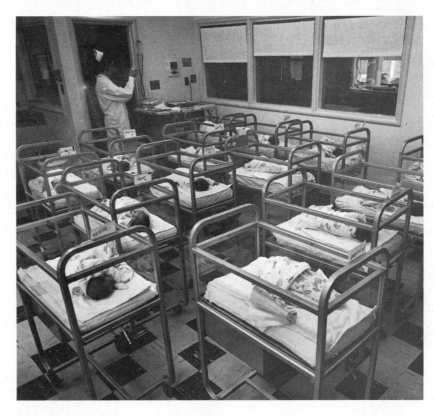

Socially and economically, Washington is a divided city. The more affluent and middle class, especially those with children, move to the suburbs to find open spaces and cleaner surroundings. Those in the lower middle class and on the bottom rung of the ladder are left in the less attractive and heavily congested pockets of the inner city, though of course there are exceptions. Many middle-class and affluent persons prefer to live in fashionable neighborhoods like Georgetown within the city limits. Conversely, some suburban towns such as Arlington have sections of lower middle class and the poor.

The McGrady family fits one of the more familiar metropolitan patterns. Their home is a large sprawling country house on five acres in a rural setting about 25 miles from the city. Seven members of the family plus an assortment of pets fill the house.

Dan McGrady works in Washington as Washington correspondent for a leading Midwestern newspaper. He has tried many ways to outwit the traffic jams he invariably runs into on his way to downtown Washington, but he can't beat the rush unless he leaves home before 6:30 in the morning or returns after 6 P.M. His wife, Laura, also a writer, regularly drives into Washington for interviews and research. She starts out later in the day so that the journey into the city goes somewhat more easily, accompanied by less frustration and headache for her.

The McGradys are deeply involved in many facets of Washington. Anything that happens in the metropolitan area is not only grist for their typewriters but makes its impact on their daily life as well. Indeed they have much less interest in strictly local matters involving the township they live in than in the gossip and events that move the metropolitan area.

The Hamilton family lives on the fringe on Silver Spring, about 18 miles from downtown Washington. Their Colonial house, situated on a wooded one-acre plot gives the family the kind of privacy it desires. Russ is a rising young executive of the telephone company whose office is but seven miles from the house. He is one of an increasing number of white collar workers who are finding jobs in the suburbs. The Hamiltons have very little contact with the central city, but they view that as an asset. Russ feels a real sense of relief when he hears the reports of traffic snarls on the radio and knows it is a quick seven miles to his own suburban office.

For the most part, life has the mark of relative isolation and comparative ease. Distances mean little to a two-car family, although Betsy is often irritated by her role as permanent chauffeur. She must use the car to get a loaf of bread, to cash a check, to take the kids to the dentist, or to visit friends. In good weather the kids can bicycle to school, but in winter that is added to her itinerary.

Like most suburban families, the Hamiltons only go through Washington's downtown district to show out-of-town friends the sights. There are movies in all of the suburbs which show the same current features once found in the "big town" and the family has little interest in museums or theaters.

The local shopping mall about five miles from the house is the focus of shopping life. There the family can find a supermarket, a beauty parlor, a variety store, and a branch of the local bank. There is a larger shopping center 12 miles away with a choice of department stores and specialty shops for major purchases.

The Hamiltons' closest contact with downtown Washington is via television, but it has no more meaning for their personal lives than Washington news has for people in Denver. If Russ moves up the company path he may well be moved to another metropolitan area, and this potential mobility keeps them from feeling rooted to the larger community.

The life of the Richardson family in the inner city is neither as comfortable nor as glamorous. They live in one of Washington's poorest areas. Raised in the deep south, Henry came to the capital with only ninth-grade education and experience as a farm worker. There is virtually no demand for his limited skills, and the best he can do is to find work as a janitor or helper. His wife works at night as a cleaning woman in one of the downtown office buildings to help make ends meet.

Home for the Richardsons is a four-room flat in an old "row" house that was once fashionable but is now completely rundown. The house could be remodeled, but landlords in neighborhoods of this type make only absolutely essential repairs and have done nothing for years. Although the family manages to hold onto a ten-year-old car, both adult Richardsons use public transportation to travel to and from their jobs and they now fear reports of a proposed 10-cent fare increase which will raise their monthly expenses by more than eight dollars.

Most of the Richardson life is confined to the neighborhood. Martha Richardson shops for the groceries at a store two blocks away, making daily trips to gather the family's supplies. The church and the children's school are another block-and-a-half distant. Young Henry, Judy, and Sam spend most of the time they are not in school playing on the sidewalks of the block or in the alley behind the house. If they want to try the public park a few streets away, they have to cross a busy thoroughfare crammed with buses and cars most of the day. Nor does the school playground offer an alternative; it is always badly overcrowded. Except for work, the Richardsons are anchored to the blocks around their flat.

A few times each year, they venture out to the suburbs in their car to shop at discount stores for the clothing and other merchandise they need. What friends they have in Washington lead pretty much the same circumscribed lives within the same neighborhood.

In many ways the Richardsons relate to their neighborhood as they might to a small town. They live in one of the most important world capitals, but the news on their local radio station seems totally divorced from their lives and their only contact with the government is through local agencies such as the police or welfare departments. But Washington is a large city, not a small town and the contrast between their lives and those of the more affluent citizens of Washington is apparent to them every time they turn on the television or take a bus through the other sections of the city. Like the poor in so many cities, they are confronted by the dichotomy between the cosmopolitan world that all the visitors to Washington see and the reality of their own poor neighborhood.

ORGANIZATIONS DEALING WITH THE GOVERNMENT

The tremendous expansion of governmental activity in Washington has brought with it the rapid growth of organizations that do business with the federal government: foreign governments, international organizations, national associations, nonprofit organizations, lobbying groups, and research and development firms. There are embassies and legations of virtually all foreign countries recognized by the United States.

International organizations headquartered in Washington include the World Bank, the Organization of American States, the International Monetary Fund, and the North American Regional Office of the United Nations. Associations and nonprofit organizations want listening posts and lobbying bases from which to influence legislation. The heading "Associations" in the Washington classified telephone directory covers six pages. They include national organizations such as labor unions, business and trade associations, charities, and professional societies; groups representing farm, transport, power, conservation, planning, and special civic interests; and veterans' associations, religious groups, and political organizations. While the Washington staffs of both foreign governments and national associations are generally small, size is not an accurate measure of their influence, for they are the spokesmen for people from all over the nation.

Most research and development firms are extensions of governmental research activities. They operate on government contracts or subcontracts. Nearly all are small operations employing fewer than 100 employees, although several have more than 2,000. They tend to be oriented toward the physical sciences, engineering, and data processing, but there are medical and social science firms as well.

The number employed in such organizations is about one-sixth that of federal civilian employment, and like federal employment, a large share of the workers are in high-paying positions. In fact, Washington has a higher proportion of workers making over $10,000 a year than any other large American city.

Washington as a Focal Point for a World Beyond

It is common to look at a metropolitan area or any urban place in terms of the number of people who live there. We have seen Washington as only seventh among the big metropolitan areas of the country. However, Washington's impact on the nation is based less on the number of inhabitants than on the influence that its governmental and other decisions have on other areas. It is not so much what Washington does for Washingtonians as what it does for the population of the United States. This is difficult to measure, but we get some idea of Washington's ties with other areas by measuring its air-passenger traffic and telephone connections. As we have seen, many more people come and go from Washington airports than live in the entire metropolitan area, and long-distance phone ties in a year are more than fifty times the resident population.

THE FLOW OF PEOPLE AND IDEAS

By examining its airline schedules we can obtain one measure of Washington's direct contacts. In 1971 Washington airports had daily direct-flight connections with 182 places in the United States. There was direct connection with all but 9 of the 100 largest metropolitan areas. But flights reached only 65 of the more than 100 places with populations between 100,000 and 500,000, six places with 50,000–100,000 population, and only forty-six of the hundreds of places with less than 50,000 people.

Washington's air connections do not reach all states. There are no direct flights to Vermont, New Hampshire, South Dakota, Wyoming, and Idaho. All are sparsely populated, and there are no big cities in those states. Nevada, which also has low population, does have air service to its large cities, Las Vegas and Reno.

The pattern of frequency of air service is quite different from that of flight connection. Only nineteen places within the United States were directly connected to Washington by twelve or more direct flights a day in 1972 (Table 1-4), yet flights to these nineteen places totaled over half of all flights from Washington. The New York metropolitan area alone accounted for nearly 20 per cent of all direct flights. Ten of the nineteen metropolitan areas with most service are among the country's twelve largest. Most other places with intensive service are regional metropolitan centers in particular parts of the country, like Atlanta, Minneapolis, and Dallas. Miami is both a large metropolitan center and a famous resort area. Norfolk, Harrisburg, and Salisbury, Maryland, represent a different sort of connection: three important centers within the Washington regional area.

From the pattern of direct-flight connections we can deduce a series of different types of airline connections that Washington has with the rest of the United States. First, there is the frequent service to the major metropolitan centers of the country, like New York and Los Angeles. Next are the frequent flights to smaller regional metropolitan centers in the South, Southwest, and Midwest. Third are the less frequent flights to metropolitan areas of 500,000 population and larger. Similar service goes to those smaller communities, many of them with less than 50,000 in population, within 200 miles of Washington or along the South Atlantic coast.

Table 1-4 Cities with Twelve or More Direct Flights per Day to Washington–Baltimore on August 15, 1972

City	Flights
New York–Newark	93
Philadelphia	64
Chicago	54
Atlanta	42
Boston	36
Dallas	29
Miami	21
Los Angeles	19
Pittsburgh	18
Baltimore	17
Detroit	17
Minneapolis	17
Cleveland	16
Salisbury, Md.	15
San Francisco	15
New Orleans	13
Harrisburg	12
Hartford	12
Norfolk	12

Source: *Official Airline Guide, North American Edition,* August 15, 1972 (semimonthly), The Reuben H. Donnelley Corp., Chicago.

Finally, there are scattered centers served by one or two flights a day.

Surprisingly, Washington has few direct-flight international air connections. It has daily flights to only seven political capitals. The only major centers linked by more than one flight a day are Toronto, Montreal, London, and Mexico City. Some other cities including resort areas of Mexico and Bermuda also have direct flights. Flights to other places are on a less-than-daily basis, apparently showing the predominance of New York as the international airport for the eastern United States. Since New York is really very close to Washington, it serves as the dispatching point for most foreign-bound passengers from Washington.

It is probable that telephone traffic would show patterns similar to those of the air routes, but connecting many more communities. Twenty-five per cent of all monitored telephone calls from Washington terminate in one of four cities—New York, Chicago, Philadelphia, or Baltimore.

Types of Places Connected

Washington's relations to the rest of the country, if not the world, can be hypothesized from airline data. Its most important links are with other large metropolitan centers, particularly the top dozen or so, and with smaller regional centers out over the country such as Atlanta, Dallas, Denver, and Seattle or Portland. Links to other communities are generally much less important, almost disappearing west of the Mississippi River. However, the Washington metropolitan area also serves in its own right as a regional center for Delaware, Maryland, Virginia, and West Virginia and even for southern Pennsylvania, North Carolina, and northern South Carolina. Finally, it has long-distance connections with resort centers, particularly in Florida and the Caribbean.

Types of Transactions

The first order of connection between Washington and the rest of the world is through people and ideas, not goods. As a center of government Washington makes decisions, not cars or television sets. Lawmakers come to Washington to establish the rules of corporate and personal conduct, business and education for the country. Through indirect measures such as price supports and direct investment in irrigation projects and flood control facilities, they determine how the country will use its agricultural land and other resources. They also set the climate for business by means of tax structures and control over the country's financial structure. Government spending programs play a large part in the structure and population of cities and rural areas as well. Federally supported urban renewal has changed the appearance of inner cities, and government-supported freeways have allowed suburbs to sprawl. Government welfare programs have contributed to the migration of poor to the cities and government aid programs have been designed to support people in depressed areas such as Appalachia. The government controls interstate freight rates, issues television and radio licenses, charters banks, and supports schools. Its laws concerning segregation and civil rights have changed the cultural patterns of the country.

As we have seen, the government is also the largest single purchaser of goods and services. In 1970 the Department of Defense alone spent $36 billion, and almost all of it was for contracts with business firms. It is not surprising that the major military contractors maintain Washington offices, and more and more research and development facilities are locating in the Washington area.

Relations with Its Own Local Earth Environment

With nearly three million people, almost a third of whom have annual family incomes of $10,000 or more, the Washington metropolitan area is also a great consumer of goods. It lies in an area with a long growing season but only moderately productive soils. Much of the land to the east of the city is low and sandy, while that to the west is hilly with only patches of good soil. There are no local deposits of oil or iron, and the coal of the Appalachians is more than 100 miles away. Clearly the city is not supported industrially or agriculturally from its own locality.

Most goods used by Washington residents, like those of any metropolitan or urban place in the United States, are drawn from specialized producing areas spread over the country or the world. An elaborate system of transport, finance, and trade handles the movement of goods between the producing areas and Washington.

In one year food sales in the metropolitan area were more than $722 million, while the total value of all farm products raised there was less than 5 per cent of that figure. In the same year retail sales of other than food were $2,744 million, whereas the total value of goods manufactured in the area was only 20 per cent of the total.

Beyond providing the basic essentials for urban life—air, water, and land—the environment of the metropolitan area is not a major supplier of food and material goods for Washington residents. Through the years, residents of the Washington area have made value judgments concerning its differing possibilities. Premium value has been placed on residential sites in the higher, more rolling sectors west of the coastal plain, particularly on the lands along the upper Potomac valley and west of Rock Creek Park (see Figure 1-1). On the other hand, the coastal-plain lands along the Anacostia River have as a rule been the site of lower-income housing and much of Washington's industrial and warehousing activities.

The water supply for most of the metropolitan area is drawn from rivers traversing the area, particularly from the Potomac and Patuxent Rivers. These rivers flow from the higher Appalachian Mountains in the west so that, even in this case, a large share of the city's needs is drawn from beyond its metropolitan area. Water storage and treatment facilities have been developed in the metropolitan area, but there is great difficulty expanding these fast enough to keep up with the growing population and its increasing use of water due to more air conditioning, extra bathrooms, and swimming pools. The Washington area has been forced to institute water rationing during recent periods of summer drought.

Waste disposal is an even greater problem. Most sewage is disposed of in the major rivers that flow into the lower Potomac and into Chesapeake Bay. While much of it is treated, some is not, and there have been problems treating detergents and phosphates.

Trash and garbage are burned or spread over waste disposal areas in sanitary landfills. Closer and closer restrictions are being placed on the burning of wastes because of air pollution; thus the demand for landfill increases greatly. But the problem of how to handle the ever-increasing sewage and garbage of the growing metropolitan area without despoiling some aspect of the local environment is not solved.

Although the offices, industrial and public buildings, and residences of the metropolitan areas to raise its children. In contrast, the conditions—by heating in winter and, increasingly, by air conditioning in summer—the population and vegetation depend on the particular local atmospheric conditions. Summer is long and hot; spring and fall are mild and pleasant; and winter is cool and wet with only one or two heavy snowstorms annually.

Metropolitan areas are designed to function on man's terms, not on those of the environ-

THE AMOUNT OF LAND NECESSARY
TO SUPPORT THE METROPOLITAN AREA

What amount of land would be necessary to support a population as large as that of the Washington metropolitan area?

1. Let us assume that all the people of the United States share in the products of the farms of the country.
   a. In 1970 there were 203,184,000 people dependent on 1,121,000,000 acres of agricultural land in the United States.
   b. Thus we have a total of 5.0 acres of agricultural land per person.

2. The population of the Washington metropolitan area in 1967 was 2,861,000 people.
   a. Thus we could assume that the amount of land needed to support the population of the Washington metropolitan area would be 5.0 × 2,861,000 or 14,305,000 acres, or about the total agricultural acreage of Maryland and Virginia.

3. But notice that there were 6,866,000 other people living in Maryland and Virginia outside of the Washington metropolitan area. We must assume that each of them also requires 5.0 acres for support.

   It is obvious, if our assumptions hold, that the agricultural land immediately surrounding the Washington metropolitan area could not support the population of the metropolitan area.
   a. What could be wrong with our assumptions?
   b. Are these important problems? If so, which ones?
   c. What conclusions can we draw about the support of Washington?
   d. What alternative assumptions might you postulate?

ment. The weekday schedule of regular movement to and from work goes on every week of the year. Interruptions occur only when there is a particularly heavy rainfall or snowstorm, and then for only a few hours. Severe hot spells create overloads in electrical systems. The power from generators outside the metropolitan area is taxed by the tremendous demands of air conditioners and breakdowns can result.

The exhausts of at least three-quarters of a million motor vehicles used in the metropolitan area produce an atmospheric haze almost every day of the year. During periods of high pressure and low wind, there is no way for air to remove the pollutants being continuously added, and pockets of smog form. As with water use, the problems of smog come not only from overloading man-made distribution systems but also from actually discharging more waste materials into the air than the atmosphere can readily absorb. At issue, then, are the technological ability and financial willingness of the people to clean up the environment.

The World of Washingtonians

Today the built-up portions of the Washington metropolitan area spread out more than twenty miles from downtown Washington, far beyond the limits of the District of Columbia. This outward sprawl is a result of both the growth of the metropolitan population and changes in commuting patterns.

When the District of Columbia was established, the area of ten miles by ten miles ceded by Maryland and Virginia seemed more than adequate for future growth. Washington was to be a "planned city," but the plan by a Frenchman named L'Enfant was designed at a time when most people walked to work and the wealthy traveled by horse or carriage. The emerging city was of necessity confined to a few square miles surrounding the Federal Mall. The plan itself covered an area only about three by six miles. Even after more than sixty years commuter transportation had not changed, and Washington had not reached the limits of L'Enfant's original design. This was the time of the two-story row house that still characterizes much of the District. Houses were tightly placed on narrow lots with no space between them.

By 1917, owing to streetcars and commuter railroad, the city of Washington had spread well beyond L'Enfant's design but not beyond the District boundaries for the most part. Distinct suburbs had been established in both Maryland and Virginia where there were towns with streetcar and commuter railroad connections. For the first time the District boundaries appeared inadequate for the functioning urban center that was Washington. Between the World Wars the coming of the automobile set the pattern for suburbia beyond the District, and today more than two of every three persons in the metropolitan area live in suburban Maryland or Virginia.

A single functioning metropolitan unit has thus sprawled across the original political boundaries established for it. What was planned as a governmental and living unit now lies within the political jurisdiction of two different states as well as the Federal District, four counties in Virginia and two in Maryland, and five incorporated cities in Virginia and twenty in Maryland. When it comes to considering the development of the metropolitan area as a whole, the result is chaos: there is no single water or sewer system; there are different operating electricity companies serving different parts of the area; phone companies charge different rates; sales, property, income, and gas-

oline taxes vary from one area to another; there are various school districts. Still, no serious discussions have been undertaken to integrate the political structure and the basic public services of the area. Planning agencies and elected officials focus attention on their own segments of the metropolitan area.

Today 38 per cent of Washingtonians live in suburban Maryland, 31 per cent in Virginia, and 31 per cent in the District. As in most large metropolitan areas in the United States, the division separates different populations: suburbanites generally have high incomes and city residents low incomes; suburbanites are predominantly white, district residents black.

Within the District of Columbia two different groups of people live in largely separate parts of the city. On the one hand, there is still a significant number of whites within the city. On the other, there are masses of blacks. Other differences are more marked than that. The whites are generally young people without families, who work for the government or businesses in the center of the city, enjoying the active life of a major metropolitan center with its restaurants, nightclubs, museums, and cultural attractions. Here, too, are elderly whites who have stayed on or perhaps returned to city life after raising their families. But essentially missing from the white population are the families. It is the family faced with problems of integrated public schools and neighborhoods who has fled to the more attractive suburban areas to raise its children. In contrast, the black population of Washington still consists mostly of families, and public schools of the District of Columbia are over 90 per cent black.

As might be expected, the white and black populations of Washington, D.C., occupy different parts of the area. The affluent whites live in the northwest close to the Potomac on the lovely, rolling terrain of the piedmont. In contrast, blacks occupy the rest of the city much of which is lowlying land on either side of the Anacostia River.

Until recently the suburban communities generally had formal and informal covenants which banned blacks from purchasing housing. These have been declared unconstitutional, and blacks have been moving to the suburbs. Naturally it is almost exclusively the better-educated, higher-paid blacks who, like their white counterparts, can afford to move to the suburbs. There have been general apathy and considerable opposition to the construction of publicly owned, low-income residential areas by the suburban political units.

Although governing the world's most affluent country is the business of Washington, the real management of government involves only a tiny fraction of its population. For most federal employees and most workers in embassies and associations dealing with government, work is an eight-hour day, a job typing, filing, and keeping accounts.

Decision-making in the federal government is largely in the hands of Congress, the President and his cabinet, the Supreme Court, and a handful of military commanders—a total of less than a thousand persons in this metropolitan area of almost three million. This is the group that functions as the corporation board and the executive headquarters for the government of the United States. As such, it is also the focus of Washington life. The families of these leaders form the nucleus of Washington's society. It is they who hold the endless cocktail parties, and it has been said that the social life of Washington is merely the after-sundown extension of the day's business—politics and government.

Yet this unrelenting single-mindedness of life in the inner circle of Washington attracts an important segment of the talent from communities throughout the country. The ambitious—both the young and those already successful in their own right—see Washington's inner circle as the opportunity to become a part of the power of American government.

Functioning Metropolitan Washington Today

With its dependence on world connections, Washington itself is highly compartmentalized. Lacking a comprehensive local government and an overall identity with the metropolitan community as a whole, and made up of people who have no contact beyond their own tiny portions of the metropolitan area, its residents often have little understanding of the way of life of other people in the many parts of the area. Governmental and business leaders within Washington are in regular contact with centers of commerce and politics throughout the United States and the world, but the majority of the Washington population does not see much beyond its own job, family, and neighborhood. Major contact with the outside world comes through the mass media—newspapers, magazines, radio, and television. And only a segment of such information is concerned with the metropolitan area or even originates within it. Thus, Washington, one of the key metropolitan centers of the world, dependent on contacts with the world beyond its limits, is populated by people whose lives are largely confined only to portions of the local area.

SELECTED REFERENCES

Alsop, Stewart J. O. *The Center: People and Power in Political Washington,* Harper & Row, New

York, 1968. Noted columnist deals with Washington's first function: politics.

Federal Writers Project. *Washington: City and Capital,* Government Printing Office, Washington, 1968. A revised version of a basic reference volume on Washington done during the Depression years.

Green, Constance M. *The Secret City,* Princeton U. Press, Princeton, N.J., 1967. The most recent book by the woman who is Washington's personal historian.

Harper, Robert A., and Ahnert, Frank. *Introduction to Metropolitan Washington,* Association of American Geographers, Washington, 1968. A look at Washington in terms of both its internal function and its role in the United States and the world.

Chapter 2   The Village of Ramkheri: A World in Itself

RAMKHERI is a small village of about nine hundred persons in central India about 350 miles northeast of the city of Bombay. It has neither a rail station nor an all-weather road. The nearest road passes three-quarters of a mile away, and entrance to the village is over a carttrack in an old stream bed which is flooded in rainy weather.

The village itself is not very important even in Madhya Pradesh, the Indian state where it is located (Figure 2-1). Yet Ramkheri is important to us because it is representative of some 750,000 other Indian villages in the world's second most populous country. It is an example of a way of life completely different from the one illustrated by Washington, D.C. Moreover, Ramkheri is much more typical of life for the majority of the world than Washington.

In using Ramkheri as an example of a traditional village, it is important to remember that it is not typical of all villages, even all villages in India. Every place has a distinctive character based on the culture, and traditions of the people and the nature of the terrain. It is only in the most general geographic terms that we can refer to Ramkheri as being typical of other villages, but by examining that one place carefully we should get some insights into the nature of life in villages in all countries.

The Village

Despite pressure for change by the government, life in Ramkheri goes on much as it has for many centuries. Families rise with the dawn, for in this agricultural community there is usually

The material in this chapter is largely based on *Caste and Kinship in Central India,* a study by Adrian C. Mayer which was published by the University of California Press, Berkeley (1960). Note that Ramkheri is a fictitious name for a real village in this part of India.

much to do. Since there is no electricity and fuel for lamps is expensive, the daylight hours are important to all. Essentially, village life takes place from dawn to dusk. When the sun goes down, most activity in Ramkheri comes to a halt.

As in Washington there is a daily journey to and from work, but in Ramkheri the traffic is mostly along the footpaths leading from the village to the surrounding fields. Most of the workers in this village are farmers. Since there is no winter, they walk from the village to their fields almost every morning of the year and return each evening.

Since Ramkheri is supported by agriculture, the village itself is mainly a place to live, not just to work. Nevertheless, there are the work-sheds of the village craftsmen—the carpenters, blacksmith, potters, tanners, and barbers—and a few tiny shops.

The village consists mostly of one-story houses made of mud reinforced by straw and cow dung. The poorer houses have palm-thatch roofs, while the better ones have roofs of locally made tiles or corrugated iron. The houses serve more as shelters, barns, and storehouses than as places to live. The few windows have no glass, only a barred frame with a wooden shutter. Houses have no chimneys. In the evening when cooking goes on inside over open fires, it appears that the whole village has been set afire and left to smolder. Smoke can be seen oozing through cracks in the roofs.

The smaller of two basic house styles has one large room separated into three compartments by earthen grain-storage bins. One compartment is used for the family's livestock, a second contains the hearth for cooking, and the central room is used for sitting and sleeping. In the more elaborate houses the stable is separated from the living quarters by a small courtyard which includes the sitting area. Such a house

has a separate storeroom, and the kitchen is also separated by a wall.

All buildings have small earthen verandas on which the men of the house can sit and talk or smoke with passers-by. On a few houses the verandas have been enclosed with a barred framework.

There is little privacy in the houses not only because of the openness of their interior structure but also because they are closely spaced. A family's life goes on in close proximity to its neighbors, and villagers are involved in each other's affairs whether they want to be or not.

The two largest buildings near the village center were formerly the residence and the office of the landlord. But he is long gone, and one building is now the village school. The other is the headquarters of the Village Committee sponsored by the state government. Nearby is a small shrine to a Hindu goddess. Across the way from the shrine and through a gateway is a courtyard and small temple built by the landlord but now in care of the village priest. In the midst of these public buildings stands the village square, a cleared, unpaved area strewn with carts and usually occupied by several cattle. This is where village festivals and meetings are held; there is little else to attract people to the square. There is no cafe, no men's club, no clustering of shops and shopkeepers. The real life of the village takes place in the neighborhoods where people live. These are connected by streets and alleys, some wide enough for bullock carts and some so narrow that only two people can pass.

Because Ramkheri is off the main road, government officials and outsiders rarely visit it. Dewas, a town of more than 25,000 people, is seven miles away, and Indore, a major city of almost 400,000, is twenty-two miles farther, but it is a long journey to either for people who must walk or go by bullock cart. Now some

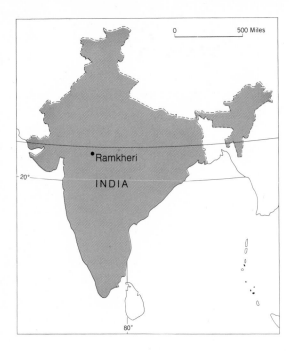

Figure 2-1 India, Showing the Location of Ramkheri

villagers have bicycles. They make the trip to sell their produce in the market at Dewas and to buy special items. In addition, young people from Ramkheri have taken jobs in Dewas and Indore. Thus, the villagers have contacts with larger places. But they feel uncomfortable there. The larger centers benefit from the profits of trade and the taxes the villagers pay, but seem unaware of them as people. Some people consider them second-class citizens—hicks or rubes from the country. Government officials assigned to teach farming techniques, animal husbandry, or health in Ramkheri hope they will not have to remain long in such a small place.

Still every year brings more government interest in the lives of the people of the villages. Government programs aim at improving farming practices and providing better health, education, and more voice for the local people in village politics. Meetings are often held in the village school. The government radio station with its farm programs, music, and news can be heard throughout the day from a battery-powered radio on the veranda of a headman's house.

Still Largely a Local World

Although the villagers have contact with the world beyond their immediate area, it is the tiny space within a few miles of the village center that provides the basic needs of the village families. The distance to neighboring villages is only two or three miles. Yet from slightly more than five square miles of the earth's surface must come basic food and material support for the nine hundred people who live there.

This area supplies most of the food eaten in the village, the sun-dried bricks and daub walls of the houses, the tile and thatch for roofs, and the dung-cake fuel used for heating and cooking. The chief sources of power are human muscles and bullocks and other local animals. Basic tools such as hoes, plows, threshing implements, and carts are locally made. Most of the manufactured products used in daily life are made by village carpenters, tailors, potters, and the blacksmith. Most of these craftsmen serve the farmers who make up the majority of the population, and their major source of income is an annual payment of grain from each. In turn the craftsmen provide basic services at no other charge to the farmers.

AN AGRICULTURAL BASE TIED TO THE LAND

The chief concern in Ramkheri is to provide food for the families of the village. Only secondary attention is paid to crops for sale outside the area, and even then the market, Dewas, is just a few miles away.

The chief crop is sorghum. From it are made thin cakes of unleavened bread, the staple food of the village. These are the base for most meals. Mixtures of chili, leafy vegetables, and seasoning cooked in water are spooned over these cakes. So important is sorghum that the villagers who speak of "mother earth" also talk of "mother sorghum." A village saying is "just as our mother feeds us, so the earth feeds us; for without sorghum we could not live." Sorghum also has many ceremonial uses in village rites.

Other major crops contribute to the food supply. The edible seeds of pod-bearing plants such as peas, beans, and lentils are used in making the daily curry dishes. Wheat, much less important in the daily diet than sorghum, is a luxury dish to villagers. Because little land can be irrigated, rice is only a minor crop in this part of India.

In recent years the government has encouraged the sale of produce in Dewas. Cotton is now the second leading crop. Groundnuts and sugar cane are both local necessities and sources of cash.

Most of the land is farmed by the landowners and their families. The average size of properties ranges from 10 to 20 acres. Farms in this area are larger than the Indian average because of the drier nature of the local terrain. This dryness results in lower production possibilities per acre. Rarely do the properties lie in a single unit; more commonly a 10-acre farm may include several separate pieces. These farms represent about as much land as a family itself can operate. Even so, it is often necessary to hire additional labor because there are virtually no imports of modern energy. Thus, a number of farmers hire villagers to work at contract labor, usually on a yearly basis for payment of some cash and a share of the harvest. Extra laborers needed at harvest time are paid a daily wage.

CHANGES WITH THE SEASONS

Life in an agricultural village such as Ramkheri varies with the season. Plowing, planting, and harvesting must be fitted into the climatic sequence of this part of the world, and since farmwork is of first importance in the village, the pattern of family and social activity follows the seasonal sequence, too.

The agricultural year starts in mid-April near the close of the hot, dry season. Religious farmers mark their bullocks with holy signs, and their wives make offerings in their homes to this new year. It is considered important to choose a day that is personally auspicious to begin work. From April through the middle of July is the first busy period of the agricultural year. This is the time to plant the crops that grow during the rainy summer monsoons. Plowing, planting, and weeding must be done. Since all work is accomplished by human and animal

The villages shown in Table 2-1 are found in parts of India very different from Ramkheri. Some of them are in Gujarat State north of Bombay and some in West Bengal State north of Calcutta. Here is evidence to test whether Ramkheri is representative of village India.

The problem you have is to compare the written description of Ramkheri with the statistical profiles of the other villages in the table. What are the limitations of the statistical profiles as compared to the written description?

What written description can you draw from the statistics? Are there any facets of village life that you can draw from the table that were not described in the written material on Ramkheri? What are the basic characteristics of the villages presented in the tables?

Would you expect that the villages would be generally similar to Ramkheri in terms of the basic system of life? If so, what are the key points of similarity?

Do you see any distinctive features of the villages in the table that indicate differences from Ramkheri?

muscle and with very primitive implements, it takes great effort and lots of time. Every moment of daylight must be utilized.

After mid-July until mid-September the workload lightens, partly because many days are lost because of rain. The heavy soils cannot be worked when they are very wet, although the crops must be weeded when possible. Sorghum, planted in rows, can be weeded using a bullock-drawn cultivator; fields planted broadcast must be weeded by hand.

In mid-September when the rains cease and the crops are ripening, the workload is not heavy but constant vigilance is needed in the fields to keep away bird and animal predators. During that season men live in small huts on stilts in the fields. Land left fallow during the first crop season is now plowed and harrowed in preparation for the cool-season crops.

Harvesting and threshing occur from October through December. Maize, a minor crop, is harvested first, then groundnuts, pod-bearing plants, small amounts of rice and hay, and, in December, cotton and sorghum. At the same time newly harvested fields must be prepared for a second crop to be grown during the cooler season from January through March. The final three months of the calendar year are the busiest of all, for there is no rain to hold up work and labor has to be expended on two different crop sequences at once. Everyone, even the craftsmen, are enlisted in this work.

By January work slackens. Only sugar cane is harvested at this time, and sugar-making takes place right in the fields. Then in March the second-crop season is completed with the harvesting of wheat and pod-bearing plants. A slack period lasts until the start of the next agricultural year in April. This is the time of village festivals and family events. Early spring is the time of betrothals and marriages and for making visits and pilgrimages. Most other festivals occur between June and November.

Table 2-1 BASIC DATA ON THE INDIAN VILLAGES, 1951

	GUJARAT STATE VILLAGES			WEST BENGAL STATE VILLAGES				
	A	B	TOTAL	A	B	C	D	TOTAL
Area (sq. km.)	4.9	1.5	6.4	1.0	2.0	2.2	0.6	5.8
Number of houses	264	49	313	n.a.	n.a.	n.a.	n.a.	n.a.
Number of households	274	60	334	117	164	62	62	405
Population	1,135	265	1,400	414	784	165	273	1,636
Males	596	134	730	n.a.	n.a.	n.a.	n.a.	n.a.
Females	539	131	670	n.a.	n.a.	n.a.	n.a.	n.a.
Density (per sq. km.)	232	177	219	n.a.	n.a.	n.a.	n.a.	n.a.
Illiterates								
Number	686	165	851	361	685	148	168	1,365
Percentage	60.5	62.2	60.8	87.5	88.2	89.4	61.5	83.5
Total agricultural employment								
Number	927	236	1,163	249	746	157	273	1,425
Percentage	81.6	89.1	83.1	60.0	96.4	93.9	100.0	87.1
Farm owners and/or tenants								
Number	650	120	770	162	434	145	161	902
Percentage	57.2	45.2	55.0	39.0	55.3	87.9	59.0	55.1
Farm laborers								
Number	249	102	351	87	312	12	112	523
Percentage	21.9	38.6	25.1	21.0	39.5	7.3	41.0	32.0
Noncultivating landlords								
Number	28	14	42	n.a.	n.a.	n.a.	n.a.	n.a.
Percentage	2.5	5.3	3.0	n.a.	n.a.	n.a.	n.a.	n.a.
Total nonagricultural population								
Number	208	29	237	165	38	8	0	211
Percentage	18.0	11.0	17.0	40.0	4.9	4.8	0	12.9

SOURCE: T. Fukutake, T. Ouch, and C. Nakane, *The Socio-Economic Structure of the Indian Village,* The Institute of Socio-Economic Affairs (Tokyo, 1964), pp. 5, 102–103.

INEFFICIENCY OF AGRICULTURE

To those of us in the modern world of machinery and inanimate energy, the amount of effort and time that goes into traditional agriculture is unbelievable. It takes two months of steady work to prepare and plant a field smaller than a city block. The hard ground must be broken by a single-bladed wooden plow with a metal tip, the product of village craftsmen. Manure collected daily from the houses in the village has been placed in individual family compost pits on the edge of the village, but then it must be spread over the field, a task that can take two men with hoelike shovels, two baskets, a bullock team, and a cart as many as thirty days during the hot season when temperatures reach 100° F. The farmer and his bullock team might take another eight days to break up the plowed clods and level the field. Finally the field can be sown in a single day by several men using a seed drill and the bullock team.

The harvests of sorghum at the beginning of the cool season and wheat at the start of the hot season require the efforts of almost every able-bodied person in the village. Harvesting of wheat is done by hand with sickles. The task requires additional hired labor and takes several days. The grain, still on the stalk, must then be transported to the threshing floor, another time-consuming task.

Threshing requires from four days to a week. The grain must be dried in the sun in the morning; then in the afternoon eight to ten bullocks or cows are driven round and round a stake in the center of the threshing floor until their hooves knock the grain off the stalks. The men thresh the grain, while the women guard the crop from stray cattle or goats and rake the threshed ears from the floor. The hay is removed with a forked stick, and the grain and chaff are piled up by the women who fill the winnowing basket. The farmer stands on a stool, and when a gust of wind passes he pours out the contents of the basket. The wind carries the lighter chaff away as the grain settles to the floor. Another worker sweeps up the grain and places it in burlap sacks.

Traditionally there have been two places where the threshing of the village has been done. Threshing time had a festival air when farm and village families worked and visited together. In recent years, however, individual farmers have developed threshing floors out in their own fields as a convenience, thus changing the social patterns of the village as well as the farming practice.

Economic and Social Dimensions of Caste

Life in Ramkheri has been altered by centuries of human life just as the people have been affected by the nature of the environment. The result is a remarkably sophisticated pattern of human relationships that form the economic, political, and social system of the village.

Like the rest of India, Ramkheri has an elaborate series of social castes based largely on occupation and ritualistic food habits. To Hindus death and body wastes are considered polluting, although all products of the cow are pure. Vegetarians usually enjoy high social standing since, in theory, they live without taking life. Those who have not had to work with their hands—landlords, priests, and warriors—are of highest rank. Farmers, around whom the whole village economy revolves, are at a middle level above the craftsmen and traders who depend on them. Lowest in the hierarchy are persons who eat meat, kill animals, or touch polluted things such as dirty clothing. Leatherworkers, whose traditional duty involves removing dead animals from village streets and using leather to make sandals and drums, and sweepers, who have had to handle animal excrement, are at the bottom of the caste system. Stoneworkers and those who eat pork also are low in the scale.

DISTINCTIONS AMONG THE CASTES

Because there are various degrees of ritual purity associated with the hierarchy of castes, persons within a given caste are careful about their associations with persons of other castes. The quality of one's purity can be lessened, or polluted, by certain contacts with castes having an inferior quality. Much of this ritual is associated with food-handling. A superior caste will not eat from the cooking vessels or hands of a caste it regards inferior, nor will its members sit next to inferior people when eating at ceremonial occasions such as weddings. There are similar rules affecting drinking and smoking. There is great rigidity and formality in the relationships between people of different castes. The people of a given caste have grown into an almost extended-family relationship. Members owe each other respect, affection, and obligations. Such friendships are important because friends lend each other equipment and money, and work in each other's fields. In Ramkheri, as in most villages, each caste occupies a particular area, or neighborhood, in the village. Persons from the highest castes—those associated with leadership in the village—live near the center of town. Farmers and others of the middle castes live farther out, and the homes of the lowest tend to be on the fringe of the village.

The caste system produces a series of socially and ritually homogeneous groups ranked along a ladder of prestige. Each of the resident caste members has a definite economic, social, and ceremonial role within the village. Moreover, there are established practices to be followed in dealing with members of each caste. Caste relations guarantee economic cooperation outside the family circle. It is good business for the farmer to be on favorable terms with basketweavers, carpenters, and blacksmiths who are sources of farm equipment. The farmer's good will and credit are important assets in survival, for he is rich only at harvest time when his needs are usually low. He may be very poor at sowing time when services are needed. Mistakes in dealing with members of other castes may cause hostility, and offended parties might withdraw their services.

Payment for services falls into three classes: those for which an annual fixed payment of grain is provided, those for which payment is made on each occasion that the service is used, and those for ceremonial duties including gifts at designated times. Craftsmen such as the car-

penter, potter, and barber belong in the first group. The second includes other craftsmen such as the tailor whose services are irregularly used and the tobacco-curer who buys from villagers and sells the products in the town market. The keepers of the two shops in town are commonly paid in amounts of grain which they, in turn, must sell in order to buy additional shop supplies. Those supported by gifts include the priest and the keepers of various village shrines. By custom gifts are given when particular rites and ceremonies are performed. In addition, there are certain days when the priest is offered other gifts.

Likewise, certain castes have specific social and religious obligations. In planning a marriage, building a new house, or arranging entertainment, the householder must call on a wide range of castes. At a wedding a drum must be played, an umbrella carried, the bride rubbed with ointment; each of these rites must be performed by an individual of the proper caste. It is said that to survive one requires the cooperation of only a few castes; to enjoy life and do things in the proper manner require the cooperation of many.

BREAKS IN THE OCCUPATIONAL HIERARCHY OF CASTE

Although one is born into the caste of one's parents and is not likely to rise above it socially, political and economic forces do allow a person of one caste to work at other things. There are landowner-cultivators in Ramkheri from more than two dozen different castes. Persons of the warrior, goatherd, tobacco-curer, barber, and carpenter castes are, in fact, farmers. As a result, members of the farmer caste own a minority of the land around Ramkheri. In the same way persons of several other castes operate carpenter shops in the village. Some castes, such as cotton-carders, oil-pressers, weavers, and tobacco-curers, find that their services have largely been replaced by products of industry which can be obtained in the village shops or the town market. Some of the members of these castes have left the village altogether to become factory workers and teachers or to join police forces and the army. But they are still closely tied to their families and thus to their castes in Ramkheri.

Among the people who farm the land around Ramkheri, there is an important distinction between landowners, tenants, and farm laborers. The landowner is well recognized since land is the fundamental source of wealth. As the manager of his own property he has a high degree of independence. Tenants work parcels too small for effective individual operation or land owned by absentee landlords from Dewas or beyond. Most tenants are sharecroppers. Their shares depend on the proportion the landlord contributes. For a daily wage farm laborers are hired either on annual contracts to assist a farmer or when needed at harvest or planting time. Like landowner-cultivators, tenants and laborers come from a variety of castes. As a rule, they are more likely to come from the lower castes.

The Headmen, Symbols of Authority

The most respected men in the village are the three headmen. They are members of the Rajput caste who traditionally have been rulers, warriors, and landlords. In Ramkheri the Rajput headmen had been designees of the maharajah who until recent times owned all the land in the village. As official representatives of the maharajah the headmen were responsible for collecting the land taxes of the village and for maintaining law and order. They were both the symbol of political power in the village and the village representatives to higher authority.

The power of the maharajah is now gone; his land has been distributed among the villagers. A village council supported by the Indian government makes important decisions about local matters. Nevertheless, the headmen still remain the most important people in the local hierarchy. They still collect the land taxes, and although they have no official power, much of their traditional authority remains. The village has no police force or court, so disputes have always been brought before the headmen for decisions. Like a mayor, the headmen have served as hosts for the village when governmental officials and other dignitaries visit. They are the most important figures during village festivals.

Actually the headmen have no real badge of office. Authority is simply passed from one generation to another. But the position of headman may not continue much longer. As the children of the headmen become educated, they commonly leave the village. Once they have tasted life in the larger world, they often have little interest in returning to the village. They prefer to work in government service as officials or teachers. Although the day when the headmen might dominate a village and its people is past, they remain very important to life in Ramkheri, and its people regret their passing.

In traditional societies like Ramkheri, the land is the source of all food and building materials, and the people have little technological equipment to aid them in their labors. The interaction between man and land here is as intense as one can find anywhere.

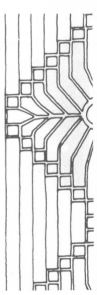

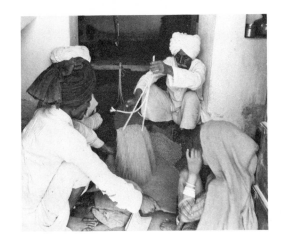

Under the traditional pattern the life of the richest and poorest people in Ramkheri is much the same. Patra, one of the headmen, lives in one of the largest houses of the village. Although there are perhaps twenty rooms in the maze of buildings in his compound, he and his immediate family occupy only two rooms. Other quarters house his brother's wife and children and his mother, who is in charge of the household by virtue of her age.

Patra rises at daylight like almost everyone else. As soon as most of the village is up, it is too noisy for anyone to sleep. Besides Patra has a lot to do.

Like the other villagers Patra's day starts with a bath carried out in Indian fashion by scrubbing and rinsing without removing his clothes. He wears a loosely wrapped garment that looks like a woman's full-length skirt; it is the standard garb of village men. When he has completed his bath, he wraps a fresh garment around him and deftly removes the damp one. Then he prepares to eat breakfast.

The morning meal is usually large, consisting of fried sorghum cakes and a curried stew of vegetables topped off with a few cups of milky tea. But during the warm season he has a cold meal, prepared the night before, and washes it down with tea.

Most days Patra goes out into the fields to supervise the operation of the several hundred acres that he owns. He must oversee all stages of the crop from planting through harvesting, threshing, and storing. He must check on the progress of his workers and deal with any problems that have arisen. It is not easy to get enough people to work for him, and he tries to get the most work he can out of his helpers.

In the dry season there is less to do. Patra may rest in the cool of his house or play cards on the temple porch. Much time is spent on his own front porch, hearing the problems of villagers. He may be called on to settle a dispute between a landlord and a tenant or between two villagers over divorce compensation. Some days he removes his village garb, changes into Western clothes, climbs on his bicycle, and rides to Dewas. At the headquarters of the government farm program he can order supplies, get information on crops, and perhaps place an order for special seed or fertilizers. He may also visit the many cloth and merchandise shops, or just stop for tea and sweets and conversation.

The evening meal comes just before going to bed. It is large and centers around hot dishes prepared that day. He eats alone. His wife eats after he finishes, and the children eat when it is convenient.

Patra's wife is rarely seen outside the confines of the house. She leaves only for latrine duties in the fields or to go visiting in other villages with her mother-in-law and to show off her young boys. She also regularly returns to her own village to visit her parents. She has numerous household duties, but the management of the house is still in the hands of her mother-in-law.

Patra's children live like other Ramkheri youngsters. Since there are few children of their social status, they play with the other village children even though none of the parents are always pleased. They attend the three-grade village primary school. Patra can also afford to have his children tutored at home in the evening. They will later go on to secondary school elsewhere, and the boys will go to college if they are able.

Motiram, a farmer who owns three acres of land, lives in a family compound with the families of his two brothers. But their homes are just crude huts. He follows much the same daily routine as Patra, but his meals are more austere, he has simpler clothes and home furnishings, and he rarely travels to Dewas. His children get no tutoring and probably will go no further than the village school in their education.

Like Patra his day begins at dawn with a morning bath. His bath, however, is taken at a village pond, and he uses no soap or scented oil.

Motiram works in the fields every day during the growing season. Most days he works for Patra, for he had to mortgage two of his three acres of land for money to marry off his eldest daughter. The remaining acre will not feed his family for the year, so he and his wife and two children must spend most of the day, six days a week, working for Patra. He makes twenty cents a day for this work while his wife makes ten cents and each son seven cents. On Friday Patra gives his workers a holiday. This is the day Motiram must work in his own fields.

Motiram's meals vary greatly over the year. During the dry season there is little available. He cannot afford to buy vegetables, and his own supply could not be stored for use at this season. The sorghum cakes will be accompanied by a chili pepper or two. But during the best times of the year he, too, will eat curried vegetables.

Motiram's leisure time is spent discussing crops with other villagers. Occasionally during the dry season he visits relatives in a distant village. The life of Motiram's wife is much less restrained than that of Patra's wife. Her daily duties bring her into contact with the other women of the village. They gather to gossip as they bathe in their corner of the pond and draw water for their families' daily needs. Of course, she works in the fields with her husband and children.

Motiram's children have the run of the village when they are young, and in addition to field work the boys go to school. The girls, on the other hand, are taught womanly duties instead of going to school. All Motiram's children show the effects of their meager diet. They do not get either enough calories or proteins and vitamins, and are susceptible to colds and childhood diseases. Motiram has lost three young children to either typhoid or dysentery.

The Beginnings of Change

Today much of the traditional power of the headmen rests in the new Village Committee and an associated Comprehensive Committee. Both are elected at village meetings. The former, created by authority from the Indian Government, has eight members including two from a much smaller village nearby. The Village Committee had begun to take over the collection of land taxes from the headmen and is now the official agent of the government in the village. As such, it is the local contact for new national social and economic programs aimed at producing change at the village level. The six members of the Committee from Ramkheri are important people politically, but, unlike the headmen, they do not hold arbitrary and continuing authority. A new election is held every five years. Membership on the Committee surprisingly comes from various classes. There are two members from the Rajput caste, but neither is one of the village headmen. The other four from Ramkheri are each from a different caste.

The Ramkheri villagers found that the small Village Committee was not representative enough and so have come to depend on the Comprehensive Committee consisting of 43 of the 250 adults in the village. This Committee has no formal authority but has begun to take over the establishing of village rules and the adjudication of disputes. Since the people of Ramkheri like to reach a consensus on issues, the meetings of the Comprehensive Committee are used as village assemblies where major questions can be discussed and consensus established. Several meetings of the Comprehensive Committee were devoted to the issue of a new road to the village. Deciding how to build it and where to lay it were issues the whole village was concerned about.

Change is slowly coming to Ramkheri. Caste

has been outlawed. The elected Village Committee has replaced the long-recognized headmen. But as civil-rights legislation in the United States has indicated, it is easier to change a structure than the ways of people, and many in Ramkheri continue to honor practices of the past.

The Risks of Change for a Subsistence Economy

This problem of change can be seen in Ramkheri's agriculture, too. The basic methods of food production have not changed in hundreds, perhaps thousands, of years.

Farming practices among the small farms of Ramkheri remain woefully inefficient by modern standards. Farmers scatter wasteful quantities of unselected, untested seed. Young plants are lost to birds, insects, and wild animals. Manure and compost are unprotected from sun and rain. Crops are stored in the open in the farmer's house where rats, worms, and weevils reduce them to powder.

In the future it may be possible to increase productivity tremendously by applying modern technology. New varieties of sorghum could double or triple production. Insecticides, adequate grain storage facilities, chemical fertilizers, rat poison, and decent fences would all contribute greatly to Ramkheri's productivity.

There are reasons why the farmers in Ramkheri have not found a better way of doing things. Change incurs great risk. Farmers unskilled in the use of fertilizer may destroy their crops by applying it more liberally than required. An improperly mixed insecticide can ruin a crop. If seed is purchased through government channels rather than from traditional village sources, who will provide the loan for a daughter's wedding? Modern agricultural equipment would have to come from outside the village, jeopardizing the farmer's traditional

ties with the blacksmith and carpenter—and such ties are more than economic. These craftsmen are neighbors and friends, but they also have religious functions that make their presence essential at births, marriages, and deaths. At the same time the blacksmith and carpenter hesitate to improve their products; there is no assurance that their farmer-customers will accept the new implements. Besides, they are likely to cost more than farmers are willing to pay.

An Agricultural System Which Is Not Economically Viable in a Money Economy

It should be remembered that agriculture in villages like Ramkheri is not a profitable industry if costed out. The farmer produces food for his family and sells only a small part of his production. If he paid wages to the members of his family or paid cash for the fodder consumed by his animals, he would lose money every year. But the crop must not only feed his family, it must also pay for seed and other farm needs. It must provide clothing for the farmer's family, and pay for birth, wedding, and funeral ceremonies. The farmer must try to make a profit in cash, but the price he receives for his produce is far below its true value in terms of labor costs and capital invested.

This deficiency results in part from the nature of agriculture: harvests occur at roughly the same time over most of India, and because storage facilities are poor, the farmer must sell at whatever price he can or lose much of his crop through deterioration. His plight is also the result of government pricing: with many new city laborers and low-paid government workers who even now spend most of their income on food, food prices must be kept low if workers are not to starve. As a consequence, controlled prices force the farmer, already the poorest man in India, to support not only the urban poor but also the wealthy and middle classes in faraway cities.

Change comes most easily to the large landholders in the Ramkheri region. They have acreages large enough to yield a profit even at low prices. The wealthy can use chemical fertilizers or American-made moldboard plows so big that they must be pulled by three pairs of bullocks. Those who regularly visit the larger towns, such as Dewas, where government offices are located, have access to seed information and agricultural advice on new crops and techniques. Moreover, these changes favor relatively low-yield grain crops that occupy large acreages and require little labor. The moldboard plow requires only half as much labor as the single wooden plow; chemical fertilizers can be applied with much less labor than cow dung and compost. Adequate fertilizer, good seed, and healthy plants reduce the amount of labor needed for weeding, thinning, killing insects, and even for harvesting.

Ramkheri villagers would be affected by any decrease in the demand for labor on the lands of large landowners since many villagers depend on contract farm labor or harvest work. Even slight changes in the demand for labor may drastically reduce the needs of the large farms, further lowering the living standard of the poorest farmers.

GOVERNMENT PROGRAMS AND SOCIAL CHANGE

Government programs also continue to put pressure for change on such traditional practices as animal sacrifice, arranged marriages, and the caste system. Villagers view themselves as very moral, but government officials with a different set of standards regard matters differently. The real problem is finding substitutes for traditional practices. People in Ramkheri do not know what should replace their traditional

ways, nor do they see any advantage in renouncing them.

In the same way, the government considers the institution of the Village Committee as a step toward democracy. But it upsets the hierarchy of political power in the village and contributes to a disruption of traditional social and ceremonial relations. Who will replace the headmen as the key figures in ceremonies, and who will entertain visiting dignitaries in a manner that the villagers judge proper?

The government desires to eliminate the menial castes and the less remunerative occupations, but this will also produce problems. Where missionaries have converted large numbers of people in the lower-ranking castes, the new converts refuse to play drums at ceremonies and to remove dead animals from village streets. People are told that they must stop discriminating against those in low-ranking castes, but they are not told what to do in order to obtain the services formerly performed by them.

The winds of change blow fitfully in the Indian village. Government officials tend to blame the villagers for lack of progress, and the villagers in turn tend to blame the government officials. New problems arise for every one solved. Changes come piecemeal. They attack pieces of the problem, but they are not responsive to disruption created in the total system.

Ramkheri within Its Region

Ramkheri has links with the world around it. With the concern of the new Indian government for national economic and social progress that is to reach every village, outside pressures are increasing. Even so, it should be clear that the differences between Ramkheri and Washington, D.C., are more than differences in size or culture. Essentially Ramkheri's dependence on the environmental base is fundamentally different from Washington's. The American capital has vital ties with a myriad of other places, even other continents. Via these tie-lines come the basic food and other necessary resources. In the same way, the influence of Washington spreads to all communities in the United States, to capitals in all countries of the world. In very small measure it even extends to rural India.

In contrast, Ramkheri's basic ties are with the environment within several miles of the village. From the land within sight of the village comes most of the food and other needs of the people. It is within this limited sphere that the essential cultural ties of the people of Ramkheri exist; it is here that they feel comfortable. At the core of their lives is the farmers' dependence on the land they work.

The village of Ramkheri exists because protection and control of both people and livestock require the clustering of houses. Unlike the United States, individual farm residences are not located out on farmed land.

Village life offers many advantages: the services of carpenter, blacksmith, barber, potter, basketweaver, and other artisans; the protection provided by the headmen; the pleasantries of talks with neighbors at night, of ceremonies; shrines essential for the practice of one's religion, so important to peace of mind and to crop production. It is good to be a part of a village and one takes real pride in it; but it is the family that is the basic unit of social and economic life.

Variations on the Village Theme

The region in which Ramkheri is located is made up of villages like Ramkheri which are essentially the residences of basically independent family socio-economic units. Because some villages are in better environmental situations, have more industrious inhabitants, or both, villages vary in size from a handful of

residences to several thousand families. Not all can produce the complete range of foodstuffs needed; not all can support a complete staff of specialists and technicians to produce things considered to be essential to economic and social life. Thus, there is some exchange of goods and services between them as different villages concentrate on the production of specialized goods and services. This interchange of goods is, however, secondary to the primary dependence of most families on their own tiny farms within walking distance of their homes.

In most of India four different levels of villages can be distinguished. There are tiny hamlets of less than 300 people on poorer land, and since hamlets usually are found in poor environmental situations, everyone is poor. These villages are made up of people mostly from only one caste—often either farmers or shepherds. They have few artisans and must depend on larger centers for supporting services. Because of their poverty the hamlets are of little importance to the larger villages nearby and are largely ignored by regional governmental officials. They have little social life, few ceremonies, no theaters, and no fairs. No wealthy landowner lives there.

Villages with between 300 and 700 people form the second level. They have a range of castes and thus provide some supporting services for themselves. Such villages vary sharply. Some are supported by herding, others by different crop combinations. Although there is a mixture of castes in each village, most people are middle-class farmers, and social life features parties, entertainments, and elaborate ceremonials. Commonly one wealthy landlord has been the central figure in the community.

Medium-sized villages like Ramkheri contain 700 to 1200 people. They may be the residence of several large landowners as well as many middle-class farmers and laborers. A large number of different castes are present. Such a village approaches self-sufficiency; its wide range of workers provides the basic goods and services, and it is the center of many fairs and ceremonies. A village of this size is the focus of attention for government officials and government programs. It has a good school and receives considerable government assistance.

The large village which may have 6000 or 7000 people is as much a central place for other villages as a unit in itself. It often occupies a key agricultural location and thus generates its own wealth, but it also serves other functions. It is a base of operation for minor government officials such as village-level workers with jurisdiction over smaller villages in the area. In fact, the large village may include a number of hamlets and smaller villages for administrative purposes. It often has the equivalent of a junior high school and is a market for surrounding communities.

The Interconnection between the Two Worlds

This is the way the people of India and other traditional cultures have spread over the earth's surface. Each group utilizes the landscape within the limits of its own environment. They tend to be successful where the environment is suited to their traditional farming practices, and less successful in other environments. In every traditional culture the basic ties are to the land directly surrounding the community, and success and failure are determined by the geographic and cultural parameters.

India, like most other places, has a hierarchy of towns and cities which has emerged above the level of the villages. Dewas, only seven miles from Ramkheri, represents that "other world," which is so different and so impersonal. The road and railroad from Dewas lead to Indore, a metropolitan area of more than half a million people. Railroads and airlines in turn connect Indore to Hyderbad, Bombay, Calcutta, and New Delhi, the capital of India. From these large centers India has air and communication links to Washington, London, and other capitals of the world.

The dichotomy between the life of people in Washington and those of Ramkheri is striking. They are extreme examples of two totally different ways of life. Most people in the world live in communities that lie somewhere between the cosmopolitan world of a political capital and an isolated village in a poor country. But if we can understand the way in which the two extremes of the spectrum interact, then it is easier to understand the relationships among all the millions of places that lie between them.

SELECTED REFERENCES

Beals, Alan R. *Gopalpur, A South Indian Village*, Holt, Rinehart and Winston, New York, 1962. An anthropologist looks at a village in another part of India.

Lewis, Oscar. *Village Life in Northern India*, U. of Illinois Press, Urbana, Ill., 1958. Village life in yet another part of India viewed by one of the most lucid anthropologists.

Maver, Adrian. *Caste and Kinship in Central India*, U. of California Press, Berkeley, 1966. The full study of Ramkheri and its people upon which this chapter is based.

Nair, Kusum. *Blossoms in the Dust*, Praeger, New York, 1962. A novel in a village setting by an Indian author.

Wiser, Charlotte V., and William H. *Behind Mud Walls, 1930–1960*, U. of California Press, Berkeley, 1963. One of the first studies of an Indian village providing a comparison over time.

Chapter 3   A Framework for Comprehending the Worlds of Man

THE WORLD of mankind is tremendously varied—much more so than the examples of Washington, D.C., and Ramkheri indicate. It is a world of ice caps and hot equatorial lands, of mountains and lowlands, of wet and dry climates, of crowded places and vast empty stretches. It is a world of many different peoples and cultures, social and political systems, human values, and ways of making a living. It is as real as a view through a window or a chat with a friend, yet as abstract as the concept of the earth as a sphere or the thought of the nearly four billion people who inhabit it.

With so many people of different cultural origins living on earth, with a seemingly infinite range of variations in environmental conditions, understanding man's world is a complex task. Yet it is the concern of geographers to deal with this range of variations from place to place, to find order in the arrangement of people and places, and to bridge the gap between the concreteness of the local environs and the abstractness of the globe.

The Need for a Conceptual Framework

What we need if we are to gain an understanding of man's place in the world is a frame of reference that encompasses the multitudinous data and theories about specific situations and places. Washington, D.C., Ramkheri, or any other place seems to have little meaning when examined only as an entity in itself. A place is not only significant to the people who live there but becomes meaningful to students of geography when seen in the perspective of the global scene. In the same way, overpopulation, adequacy of resources, and pollution are world issues as well as national or local problems.

In this book we develop a conceptual framework for studying man's life in various places in the world. In Chapters 1 and 2 the framework was not specifically stated. Now it seems important to state it explicitly and in some detail. As we view particular regions in terms of this framework, we should recognize that it is merely a first approximation. We may want to modify and refine it, just as physicists moved from the idea of molecules to that of smaller atoms, or from the simple atomic structure involving neutrons, protons, and electrons to a more complex explanation.

Finding a Perspective for Understanding Man's Life in the World

From the beginning of life, the human problem has been and will remain how best to live on the planet Earth. Each segment of mankind discovered that it had to wrest the necessities of life from the particular environment in which it found itself. This problem has been more than an individual matter, for mankind is essentially social. In examining how human life varies from place to place, geography has focused on the group rather than on the individual. That focus has become easier with the advent of censuses, other sources of group data, and the development of quantitative techniques for analysis of group behavior.

In studying the world of man, geographers have tended to view the world as a patchwork globe and to analyze the man-environment system of each piece separately, as well as to compare the systems of different pieces. In Washington and Ramkheri we saw not only two different environments and two different human groups with distinctive cultures but two very different ways of utilizing the environment. Using these examples as a base, we can examine the characteristics of each of the two systems they represent.

The Traditional Model

In the early days of human development, man's capacity for understanding and working with the earth environment was so limited that the attention of both the individual and the group was centered on figuring out how to sustain life on that small part of the earth the group occupied. The only human instrument available for developing the local environment was the collective capacity of the group for achieving results through trial and error. Having no means of rapid transportation, the group could not have even a continental range, much less a global one. The environmental base from which all the necessities of life had to come was within walking distance. There was little exchange or interaction with other places.

Of course, regions have never been entirely separate. Despite the difficulty primitive man faced in moving over the earth, remarkable journeys were made. Individuals and groups moved back and forth, and although they could not carry much in the way of goods, there was traffic in ideas and particularly valuable commodities, such as seeds, good-luck symbols, and medicinal potions. Through this limited travel, new techniques, new crops, new inventions were introduced, but many of them were at first rejected because they did not fit the cultural values of the local group. Some innovations had little or no effect in sustaining life, while others caused fundamental changes in how the group wrested its livelihood from the surrounding environment.

In the traditional world, each village and tribe is distinct. For example, in India there are more than 750,000 villages, and more than 150 languages are recognized, not counting the hundreds of dialects. Such cultural variation is also found among tribal groups in Africa and among the native Indian populations of Latin America.

Separation of groups by vast distances and primitive means of transport and communication limited them to a local focus and limited most early contacts to peoples in one part of the world. Thus, although ideas could move from tribe to tribe in Africa or from village to village in India, the chances of contact between the African continent and the Indian subcontinent were very limited. Following ocean currents and using sails to harness the wind, man did have some early intercontinental contacts, but these remained infrequent until a few hundred years ago. Therefore, large sections of the world developed broad cultural patterns in isolation from other sections, so that the similarities in the villages or tribes within a region were more pronounced than their differences.

Culture

As we look back at Washington and Ramkheri, the people and their cultures seem to have dominated the scene. First, there is the nature of the people themselves. The people in Ramkheri are very different from those in Washington. They differ in physical appearance, but there are much greater contrasts in their ways of living—in social institutions, values, and, particularly, in their perception of the possibilities of their environment. The way a group of people in a particular part of the world lives is known as its culture, and the people who make up that group are identified as a distinctive culture group. The particular culture group (or set of groups) occupying a place at a particular time constitutes the most fundamental component in the geographic interrelationships of that place.

The ideas of a culture group about how to live rest upon the traditions and values built up by many generations before them. Like the an-

nual growth rings of a tree, each generation's life rests on folkways and artifacts left by past generations: ideas of what to eat, what to wear, what is good and bad, etc., come from the past, as do field patterns, streets and buildings, and political boundaries. Even Washington's streets still follow the routes established in the original city plan, and the boundaries of the District of Columbia, long obsolete, still hold.

All culture groups are under some pressure of change. Traditional societies have experienced fewer outside pressures for change and fewer contacts with other groups than have modern societies. For them cultural change comes very slowly. In the modern world great value is placed on progress and change. Modern transportation and communications expose the cultures of many parts of the globe to ideas and objects from other areas, and change may occur rapidly. But traditional or modern, the particular culture is the filter through which all new ideas and developments must pass. As a result, the culture group is always a product of both the pressure of new ideas and the structure of culture inherited from the past.

This traditional model for man's life assumes that the world is divided into separate regions, each with its own culture group. It focuses on the fundamental problem of each group, which is to live in its particular earth environment and use the knowledge that has evolved within its culture. The analysis of different groups reveals many variations in the fundamental model: different degrees of know-how, different group values, different earth environments.

Properly applied, this model fits life in Ramkheri, in much of rural Asia, Latin America, and Africa, and even in portions of Anglo-America, Europe, and the Soviet Union. The basic problem is always the same: the earth environment of the local area, often within sight of home, must provide the basic needs of life.

Washington and the Modern, Interconnected Model

The traditional scene does not enable us to understand Washington, New York, London, Paris, Tokyo, Peking or any of the other major metropolitan centers of the world, or even of most of the smaller cities and towns. None of these centers expects to support its population from local environmental resources. Their stores and marketplaces abound with products from all over the earth. In turn, goods, money, and ideas from these centers move throughout the world.

Most farmers of the United States Corn Belt, the Argentine pampas, Latin-American and African coffee plantations, and Soviet collective farms do not expect to produce all the essentials needed to support their own lives either. Rather, their plan is to mass-produce as much of a particular commodity as they can sell in the cities and to buy whatever they need with the money received from the sale. Thus, they have transcended their local environments. Their resource base stretches out to encompass many different environments, and they sell to people many miles away.

The mechanism that makes all this possible is an extensive network of transportation and communication facilities which enables man to contact different parts of the world rapidly and cheaply. In the early stages of technology caravan routes, rivers, and oceans offered alternatives to local self-sufficiency. Over these routes goods, people, and ideas could move from one particular earth environment to another. With primitive technology, however, the quantity that could actually be carried between locally based communities by human porters, animals, wagons, or sail-powered vessels was limited.

The transformation of transportation which began with the Industrial Revolution in eighteenth-century Europe and the subsequent development of telegraph, telephone, radio, and television not only have reduced the travel and communication time between places but have also tremendously increased the quantity of goods, people, and ideas that can be moved over long distances. More and more, this interconnected model assumes mass consumption by persons throughout the system and specialized production at particular points within it. Thus, each producing area has the possibility of supplying consumers throughout the system, and the producers, in turn, of consuming goods from any other point. The supermarket in any community can stock goods from almost any place in the world. The important barriers to trade are commonly political.

The modern, interconnected system radically changes man's relation to his environment. No longer is the population of a particular point in the system bound by the limitations of either the character of the earth environment at that point or the thought and technology of the particular culture group. Theoretically all peoples within the system can draw the best goods and ideas from any other part of the network. Thus, points throughout the varied earth environment may form the production base for the system, and the collective knowledge of all persons in the system provides the capacity to increase the understanding of the environment and its possible use. Technology developing within the United States and the Soviet Union, in particular, even offer the possibility of drawing upon the resources of the moon and other bodies in space.

As we have said, this new interconnected system for supporting human life did not spring forth full-blown in the twentieth century. From early times sea travel, and later overland travel, allowed some cultural interaction. The Chinese, Greeks, Phoenicians, Norse, Venetians, and

Arabs all expanded their regular contacts well beyond their own culture groups. The Roman Empire was an early attempt at an interconnected system, but the first effective example of the truly interconnected system was the British Empire. That empire connected producing areas on all continents, but it really organized only a small fraction of the possible earth resources. Other European countries developed their own intercontinental networks on a smaller scale, and the United States developed its version of the interconnected system by organizing the varied resources of a continent-wide country into a functioning whole and by strengthening its ties with Europe. Today, other examples of the interconnected model on different scales can be seen in the Soviet Union and Japan. To an important degree, each country in the contemporary world is trying to follow the interconnected model by developing its own national resources while at the same time tying into a growing network of world-wide connections. Consequently, just as there are many variations on the locally based model, there are numerous forms of the interconnected system.

It could be argued that the interconnected system is just a large-scale version of the traditional model; perhaps modern transportation has simply extended the range of the human-resource base. Instead of limits within walking distance from home, that range is now global. But it is not quite that simple; there are other differences as well. We usually think of the locally based culture in terms of the organization of a limited contiguous territory—the area occupied by a particular culture group, the state or territory, the economic region. But the British Empire, with its connections across the seas, began a series of point-to-point linkages across areas that were themselves part of the system. The interconnected system with its global scale is increasingly like that. Activity is located more and more in giant metropolitan centers, and modern airplanes and telephones, which theoretically can connect any two places, primarily link widely separated metropolises. The lands between these points have only secondary links to the system.

The nature of the interconnected world is illustrated most dramatically in jet air travel. The traveler moves from one metropolitan center to another, from one modern airport to another, regardless of whether his destination is a nearby city or a metropolis on another continent. Moreover, since the plane often flies above a cloud cover which obscures the rest of the world, the traveler might never see the world between his origin and destination. The time on the plane is a necessary wait to be endured by eating, drinking, watching movies, reading, and sleeping. Although his destination may be another country or another continent, it seems very similar to life in the metropolitan area he has left. Modern metropolitan life provides remarkable similarities at points within traditionally different culture areas. The new metropolitan culture shows components from many different traditional culture areas. Women's fashions may be drawn from Africa or Asia as well as from Europe; restaurants specialize in foods from all over the world; gift shops sell exotic goods from many places.

Functional Aspects of the Models

To understand the workings of any culture, we must shift our attention from examining how things appear in the landscape to seeing how things operate. In doing this we can identify three components of a functioning system: deciding, producing, and consuming. In modern business parlance, *decision-making* is the function of management: deciding what to do and then managing production and distribution to consumers. *Production* includes making things of all sorts—not just the products drawn directly from the environment by agriculture, fishing, mining, and forestry but the products of manufacturing and construction as well. *Consumption* is the use of what is produced.

Although we have expressed the concepts of management, production, and consumption in economic terms, they can be applied to all aspects of a culture, to such things as religion and education and politics.

From the discussion of the traditional, locally based society in Chapter 2 on the village of Ramkheri, we have seen that the portion of the earth environment with which the group must work is small and that its possibilities are therefore limited. Any isolated group is comparatively limited in both tools and energy sources. Its ability to produce is limited by the small number of workers available, their limited technology, and their narrow cultural perspective. Amounts of produce needed for consumption are predictable. In areas with large harvests there may be no way of storing the extra produce, and much of a good crop may go to waste. In bad times, on the other hand, many will go hungry.

To a large degree most people in the traditional group are producers, consumers, and decision-makers all at once. The individual family decides what it needs for life, organizes itself to produce it, and consumes virtually all of it. Economic decisions are limited by the narrow range of resources and by the small number of consumer-producers. There is little labor specialization and exchange of goods. In Ramkheri there were only a few craftsmen who were not also farmers. The leaders of a tribe or village may make certain decisions and carry them out. Sometimes a representative from each family might be involved in making a decision, passing it on to the people, and carrying it out.

In the modern system there is a characteristic separation of the functional components of decision-making, production, and consumption. Specialization is the rule. Even in a modern system, each primary producer (person, family) is also a consumer, but primary production accounts for only a tiny fraction of all production and for only a fraction of all consumption.

We have seen that the modern system reaches out over a large area to place specialized production facilities wherever the environmental resources make it most economically feasible within the limits of the governing unit. Each producing area, or unit, produces only the things it can sell competitively in a particular marketplace. The dairy farm, the cattle ranch, the orchard, the cotton or coffee plantation, the lobster fishing fleet, the coal mine, the iron-ore mine, the oil well—each supplies a particular commodity that no other unit could supply as easily.

Most consumption occurs in urban areas, particularly the large metropolitan centers. That is where the most important production takes place in terms of manufacturing and services. Most decisions are also made in urban centers, especially the largest ones. The occupants of office buildings in New York City and Washington do not produce anything in the usual sense; they are not manufacturing anything. But they are managing goods and people—making decisions for vast business empires which include widely spread producing and consuming units, some of which may be far away.

Consider a major internationally-based oil company. It is engaged in production, it has a system of marketing outlets to consumers, and it must manage the whole operation. It pumps oil from the ground from fields in Texas, California, Canada, Alaska, Venezuela, Saudi Arabia, and Indonesia. It refines oil in cities in Texas, England, and the Netherlands and in the New York and Philadelphia metropolitan areas. It makes petrochemicals in most of the same places. It distributes its products throughout most of the United States, Europe, Latin America, and parts of Africa and Asia. It manages this vast operation from headquarters in New York City and subcenters in Houston, Caracas, and London. This company, like most modern business firms, produces where it is best to produce, markets where it has many potential customers, and manages from business centers where it has ready access to banking and business services and to communication to all points in its vast network of operations.

A giant insurance company, whose salesmen call on people and service policyholders throughout the country, operates on the same principle. The money collected is used to pay for routine operations, but also to finance commercial and residential building in cities and towns scattered over the United States and to capitalize national and international corporations. Yet management headquarters are in downtown New York City, the business and financial capital of the United States.

Or consider an agency such as the U.S. Forest Service. It operates under a mandate from Congress to manage the public forest lands of the country. It manages National Forests in all except seven states and has individual rangers and offices across the country, but its operational headquarters are in Washington, D.C., the political capital of the country.

Unlike primary production, manufacturing, supplying services, wholesaling, retailing, decision-making, and management are activities best carried on in urban places, and within limits, the larger the center the better. Such centers have more consumers, better transportation and communication facilities, more supporting business and technical services, and a larger labor pool than rural settings. It is not surprising that the huge metropolitan centers of today have appeared with the development of the modern, interconnected system and that the largest of them are located in the industrial countries.

Moreover, although in a closed society ideas stem largely from individuals in the group, in modern society the ideas that circulate throughout the system come from the large units of the system—the corporations, the financial organizations, big government, and even large universities rather than small colleges. The individual inventor has been replaced by the research and development corporation and the local philosopher by university and corporation "think tanks." Ideas flow through the system just as goods do, but the ideas that flow most easily seem to be those sponsored by institutions rather than individuals.

With mass transportation and mass communication, cultures become more similar. Clothing styles, hair styles, musical tastes in countries all over the world begin to look and sound alike. The modern system, through its very interconnectedness, tends to have a homogenizing effect on the world's natural diversity.

A Frame of Reference for Viewing the World

In this book we maintain that the two models of human life that we have presented provide a useful basis for understanding the variations in life on earth. Man's life in any place partakes of both the traditional, locally based system and the modern, interconnected system. We must consider exactly how these two basic systems confront and affect each other at any place. Sometimes one system predominates, and sometimes the other.

Nowhere does either of the two systems occur in pure form. Ramkheri, our example of the traditional, locally based model, has some contacts with the outside world. Virtually anywhere on earth someone from an interconnected system—doctor, missionary, scientist, or tax collector—has entered a traditional community. Eskimos drink tea and carry guns. Some Africans raise cotton to sell; others leave their villages to work in mines. All peoples are located on territory that is claimed by some sovereign state.

All centers of an interconnected system have contacts with some locally based communities. In any part of the world, one can find elements of both systems in juxtaposition. In Zaire most people still live in locally based communities, but Kinshasa, the capital and largest city, has regular communication with the world-wide network, as have the Katanga copper region and other specialized production areas. Mexico is both Mexico City, one of the largest metropolitan centers of the modern, interconnected world, and a native Oaxacan village with few contacts beyond walking distance.

At first glance it might seem that the whole of the United States is part of an interconnected network. But what about the American Indians on reservations, or people in "depressed areas" such as Appalachia or the Ozarks? A recent survey found not only large-scale poverty but malnutrition and starvation in parts of the United States. Surely a case can be made that the inner-city ghettos of the country have a large population that is really not a part of the modern, interconnected system at all. Likewise, there are large numbers of people in cities such as Calcutta, Rio de Janeiro, and Lagos who live in a metropolitan environment in much the same way that their forebears lived in traditional, locally based communities. They do not participate in the interconnected systems of which their cities are a part. This is another way in which the traditional and interconnected systems coexist.

The two systems have been described largely in economic terms, but in reality they also represent strikingly different ways of life. This is easy to see in a locally based system. There the

isolated culture group has had to work out the dilemmas of life for itself. The group's resolutions of these dilemmas make up a distinctive way of life.

Language is a good index to the problem of isolation, as we saw in India. Language is the medium for transmitting ideas within the group, and each isolated group develops its own language or dialect, much as it develops its own political system. The fact that a group speaks a distinctive language indicates that there has been little need for most of its peoples to communicate beyond the geographic limits of its local area.

The modern, interconnected system is also more than an economic one. For the first time a way of life is beginning to emerge that can be seen not just in a single locality or even in one country or on a continent. Metropolitan life throughout the world is taking on certain common characteristics. There is common reliance on many of the material goods of the modern world and also adoption of increasingly similar ideas. Music and art styles become international in the metropolitan centers of the world. Common problems and political movements are found in metropolitan communities throughout the world. They seem equally natural in New York, London, Rome, Mexico City, Rio de Janeiro, and Tokyo.

The two-model framework of human society gives us a basis not only for comparing one place with another but, more important, for comprehending how pieces fit into the larger whole of man's universe. Some places function with little involvement beyond the local area. Others depend a great deal on interaction. In any case, we cannot expect to understand a place just in terms of itself. We must see it within a larger setting. A geography of any area must take into account the contacts between it and the rest of the world. The examination of the larger network may seem less significant in regard to an underdeveloped country such as Afghanistan where much of the population still lives within the locally based system. The degree of isolation is, however, as significant as the degree of interaction. And, although many places are largely isolated, there are always some outside links.

In today's world, it is important to keep the global scale in mind. Modern technology makes it possible for mankind to interact economically, politically, and culturally on a world-wide scale. Thus, we must constantly evaluate both the degree to which particular places do function that way and the implications of this globalism for modern society.

As participants of the modern, interconnected system we know many of its strengths and weaknesses from firsthand experience. We may not have had any direct involvement with the traditional, locally based system, but we come to it with certain preconceived notions. Some see it as an inferior system and thus the work of inferior beings. Others look at it romantically as "the good life" close to nature. As scholars concerned with a search for better explanations of the world, we must try to overcome such prejudices and be objective.

There is little evidence to indicate in which system the majority of people are better off spiritually and psychologically. We have no measure of the degree of "happiness" of the participants in either system. We certainly might question whether the people of India or Ghana are better off today, even in a materialistic sense, than they were before contact with the modern system. We may wonder if the frustration of "rising expectations" brought by contact with the modern world has diminished their peace of mind. At the same time, many people in our giant metropolitan centers wonder whether they are living as satisfying a life as their ancestors did. The modern system has brought world wars, colonial exploitations, and

the very real prospect that all human life will be wiped out by an atomic holocaust. Although such value judgments are not to be ignored, our task is rather to understand the two systems, how they work, and the rationales behind them.

An Approach to the Study of Place

Whether we are examining the traditional system or the modern, interconnected system at a place, we must deal with a set of interrelationships involving people and the earth environment. Thus, in dealing with a particular human system, we must examine the interactions of a series of variables that will provide significant insights into the workings of life in that place. Four variables should provide us with a basic understanding of life at any place, if we know how to relate them. They are (1) the operation of the earth environment, (2) the culture or cultures of the people living there, (3) the technological know-how possessed by the group, and (4) the ties between the people at that place and those in other areas. These are the components, or variables, with which geographers commonly work. If we are to learn to think geographically, we need to analyze each of the components and understand why they vary from place to place.

It is equally important to recognize that knowledge of each of the individual components in isolation is not the end we are seeking in geography. We want to understand how a place "works." Rather than looking at a place as something exotic, we are seeking an understanding of the rationale of life and the interaction of the four components that give a place its character.

We might think of a place as being described by a mathematical expression having four factors. The combination of these factors characterizes that place and differentiates it from all other places. We must examine each of the factors separately, but such an examination is just the means to a more important end: to understand how the factors fit together. We must not be diverted by the intricacies of any one component. Our attention should always be on how the components interrelate to characterize life at a place.

In considering the components separately, we should recall the two different models of human life that we illustrated with Washington and Ramkheri. The four components that we now will be examining have already been used loosely in Chapters 1 and 2 and earlier in this chapter.

THE EARTH ENVIRONMENT

The fundamental characteristic of a place is the physical environment within which all life occurs. It is the source of all the essentials upon which life depends—air, water, energy, food—and of the materials from which we have developed clothing, shelter, and machines. The environmental characteristics of a place—its roughness or flatness, temperature regime, moisture availability, character of soil, and mineral occurrences—determine the resources available. From these the people must obtain the basic essentials of life, assuming there are no contacts with other supporting environments. Lives are dominated by the earth environment. Dry seasons, heat, and meager supplies of wood all have an impact on a society. Nor do Washingtonians escape the need to adjust to the year's cycle of seasons, the conditions of the terrain and soils, and the particular mineral-resource base.

The earth environment is not merely a passive stage on which life moves. A large number of events take place in cycles that even the most primitive cultures can read. For example, Ramkheri anticipates the coming of the rainy and dry seasons each year and plans accordingly. Washingtonians have an elaborate system of heating and cooling devices which counteract the contrasting temperatures of winter and summer. Only the unpredictable extremes of the environment, to which man cannot adjust, wreak havoc on life. Too much rainfall in too short a time brings flooding; a severe shortage of moisture is always a problem; tornadoes, earthquakes, and volcanic eruptions bring disaster to any culture within their path.

CULTURAL RESOURCES

The components of the environment include the materials that sustain life, but each natural resource is actually perceived in cultural terms. There were no signs in nature saying: this is corn, its ears should be roasted; this is coal, light it and it will provide warmth or energy to do work; this is pure water that can be drunk; this is polluted water, it should be avoided. Each culture group has had to experiment and discover the resources it can use in various ways or to adopt ways learned from other groups. Thus, what are resources to one group may not be to another. There was a time when men had only stone tools. Then some learned of bronze, and later others learned of iron and steel. Today all but a small number of peoples have the capability of using iron-bearing rock to make iron and even steel implements, but only people with sufficient knowledge, capital, and a highly organized market system can build steel mills and make machines from the steel.

TECHNOLOGY

Among the contrasting cultural elements of Ramkheri and Washington, one stands out very prominently: the technological capacity of the culture. This is the stock of experience that determines the capabilities of people to manipulate their material existence. Anyone from the industrial world is struck by the tremendous amount of physical effort expended for the small amount of product gained per individual in Ramkheri. The great efforts of humans and animals there accomplish little, we would say, when compared to Washington, where for a century men have known how to harness electricity, gas, and other energy sources to drive laborsaving machines.

The separation of the world into high- and low-productivity areas, wealthy and poverty-stricken, involves many human values and desires and social organizations, but it is first of all a matter of technology. The large metropolitan areas could not be supported today without the elaborate system of production and supply that makes use of a technology far more advanced and sophisticated than that of Ramkheri.

Present-day technology at any place is the result of the accumulation of technological know-how within the culture through the years. In the traditional society this knowledge is made up of hard-won individual discovery through trial and error and of word-of-mouth instruction in knowledge so gained. In modern society, contacts with other places allow any modern center anywhere in the world to draw on the accumulated technology of the rest of the modern world—either the production knowledge itself is applied or the manufactured products made possible by that knowledge are imported.

At any place, then, there is an accumulation of technological ideas and output. Each successive generation finds itself with a body of technical know-how with which to work and with an accumulated stock of facilities and tools. New investment, whether in tools or ideas, can build upon the supply inherited from the past. It must be remembered, however, that such accumulations include many products—goods or ideas—that are obsolete in addition to those that work successfully.

TIES BETWEEN PEOPLE IN DIFFERENT PLACES

The final factor in understanding a place is the extent of contacts which it has with other places. In Washington these contacts were shown to be more important than the ties to the local environment. They included dependence on other areas all over the world for food, resources, and business and political connections. The ease of transportation and communication between places depends in part upon other factors: accessibility in terms of natural barriers or avenues in the earth environment, cultural interest in trade or other interactions, and the extent of the technological development of transportation and communication systems. There is a constant flow of people, goods, and communications between Washington and hundreds of other places throughout the United States and the world. And although such is hardly true of Ramkheri, even there goods are sold outside the village, and some contact is maintained with the larger world.

Two Models: Combination and Modification

The ideas developed in this chapter should be useful in examining the places in Parts Three and Four of this book. It is important to consider them together, not simply as separate entities. Technology must be viewed with the resources of the place in mind and within the framework of the existing culture.

The two human systems, the traditional, locally based and the modern, interconnected, should be regarded as models of very different ways of human life with very different problems in terms of environmental management and cultural development. But in the real world we will not find these two models. Instead we will see a variety of modifications of the two. In each particular part of the world there will be variations in the geographic components that will result in modifications in the system. Thus, traditional life in Latin America is different from life in Black Africa or Asia. And traditional life in the highlands of Latin America will show differences from that in the lowlands. In the same way we know that the interconnected system that has evolved in the United States is different from that of Europe or the Soviet Union.

Moreover, the two systems in their various modifications are found in different combinations in each place. It is easy to see the dual nature of life in a largely underdeveloped country like Brazil where some of the people in Rio de Janeiro or São Paulo are obviously part of the modern, interconnected system, in contrast to people in the rural areas of that country. But the two different systems are also in juxtaposition within the United States or any of the industrialized areas of the world and are present even within individual metropolitan areas such as Washington.

In looking at any place as a living, functioning entity, we must first consider the production that supports development, examine the consumption pattern, and discover who makes decisions and how the system is managed.

Before we proceed to an examination of some of the regions of the world as examples of the traditional and the modern human systems found in particular parts of the world, let us examine each of the geographic factors in greater depth and then consider their geographic distributions over the earth. This is the purpose of Part Two.

SELECTED REFERENCES

Bacon, Phillip. *Focus on Geography*, 40th Yearbook, National Council for the Social Studies, Washington, 1970. Essays outlining present-day thinking in geography.

Black, Cyril Edwin. *The Dynamics of Modernization, A Study in Comparative History*, Harper & Row, New York, 1966. In terms of the political scientist and historian this study uses much the same framework proposed here. The focus is on the modernization process as it has evolved in different parts of the world.

Broek, J. O. M. *Geography: Its Scope and Purpose*, Charles Merrill, Columbus, O., 1965. A distinguished geographer commissioned to prepare a statement on geography and its viewpoint writes for teachers.

Haggett, Peter. *Geography: A Modern Synthesis*, Harper & Row, New York, 1972. One of the leading thinkers among the new wave of geographers presents his view of an introductory course in geography.

Haggett, Peter. *Locational Analysis in Human Geography*, St. Martin's Press, New York, 1966. An advanced summary of methods of analyzing the components of human settlement.

James, Preston. *All Possible Worlds, A History of Geographical Ideas*, Bobbs-Merrill, Indianapolis, Ind., 1972. One of American geography's elder statesmen interprets the development of geographic thinking.

MacKaye, Benton. *The New Exploration*, Illini Books, Urbana, Ill., 1962. The conflict between the modern interconnected system and traditional life as seen by a regional planner in the 1920s.

Philbrick, Allen. *Man's Human Domain*, John Wiley, New York, 1962. An introductory college text organized around the theme of the organization of space by man.

PART TWO   THE GEOGRAPHIC COMPONENTS

OF PLACE

Chapter 4   Man
and His Culture

THE CULTURAL heritages of mankind can be examined separately from the physical environment. In examining their components it is important to keep in mind the two microcosms we studied in Part One. This chapter identifies specific characteristics of human culture in various places and evaluates their importance to both of the proposed geographic models.

The Modern, Interconnected System as a Veneer on Top of the Traditional, Locally Based Society

As we look at the human component of different places in the world, we must remember that almost everywhere the long-range, interconnected system has emerged out of the locally based system. But at our particular point in time, the transformation from old to new is not complete. It is as if we are looking at a frog at some stage in its transition from tadpole to frog: parts of the creature resemble what has been and parts imply what is to be. The long-range, interconnected system may never encompass the whole world, but if the trends of the past several hundred years are projected, it just might.

Everywhere the interconnected system rests upon vestiges of traditional man-made structures and remnants of ideas from the past. The ideas and traditions are often more important obstacles to change than the structures. When the centers of many western European cities with their 500-year-old buildings and street patterns were destroyed during World War II, there seemed to be an excellent opportunity to rebuild them to fit modern needs. Although the buildings were gone, the titles to property and traditions remained; there were existing sewers, utility lines, and other artifacts of the past. Thus, while buildings were of new architectural design, in almost all cases the rebuilt centers bore the pattern of the past with little regard for the movement of modern street traffic and the need for auto parking. Culture, race, and the present political pattern of separate states are all fundamental facets of the twentieth century inherited from the traditional, locally based world of the past.

The Newness of the Modern System

The modern, interconnected world is a very recent development. Important interaction between peoples on different continents dates largely from the European Age of Exploration. This age began with Columbus's voyages at the end of the fifteenth century, but explorations did not become really extensive until development of the steel-hulled, steam-powered ships in the nineteenth century. Efficient trains, planes, pipelines, telegraph, telephone, and television have all developed since 1880. Until then man was mainly dependent on animal power, wind, or his own muscles, all of which are inefficient and slow. For most people in the industrial world, the interconnected system is really a product of the twentieth century.

Race: By-product of the Isolation of the Traditional Society

Anthropologists present evidence of the existence of separate human groups on different continents at least 40,000 years ago. For thousands of years primitive means of transportation and communication prevented people in one area from interbreeding with those from another. Each group remained in its own part of the world, married, and had children. It is not surprising that each of these groups, remaining largely separate and interbreeding only within its own population, developed its own set of inherited characteristics. It is the external biological variation among these once-isolated groups of people that we use to distinguish races.

During the thousands of years of isolation from one another, human groups have taken on distinctive physical appearances. People of one part of the world resemble one another closely, yet are different from those in other places. Thus, men have tended to identify with people who look like them and to be suspicious of people from other areas who look different. Only recently have significant numbers of one traditionally isolated racial group moved into the territories of another. The movement of Caucasians into the Americas, South Africa, and Australia is a phenomenon of the last few hundred years. So was the removal of blacks to the Americas and the migration of Asians to the United States and Africa.

The physical isolation of groups in particular parts of the world has not only produced racial variations through time but has also led us to associate cultural differences with the various races. European cultures are equated with the Caucasian race, African cultures with the Negro race, and Asiatic cultures with the Mongoloid race. Some people assume that variations in technological, political, and economic development are related to racial characteristics. In fact, these variations are more closely related to cultural and geographic differences.

Transportation and communication which have produced the modern, interconnected system have increased contacts among races, although most people still do not have contact with people of other races. Even where they do meet, there are usually strong cultural taboos against intermarriage. But as the world-wide, interconnected system brings peoples of different races into closer contact, these cultural in-

Figure 4-1 presents perhaps the most readily apparent index of race: skin color. Since mankind is a single, intrabreeding species and each individual is distinct, any racial classification is arbitrary, and any racial map is therefore crude. Moreover, the significance of biological differences is meaningful only in cultural terms. Within a given culture group there are great ranges in size, shape, and even skin color of individuals, but these are generally ignored. Of all biological differences, skin color and associated facial characteristics and hair offer the most immediate signal of race to us because, as the map shows, there are great regional differences in these human characteristics from one traditionally isolated area of the world to another.

What relationships are there between variations in skin color over the earth and variations in the earth environment? Are any correlations between the two significant to the story of racial differences?

hibitions are being challenged. Many North and South Americans have European, African, and Indian roots. United States soldiers stationed overseas often marry local girls. Continued location of American forces in Japan, Korea, and Indochina has increased the number of children with mixed racial heritages.

Racial differences become important when particular groups point them out as part of their social consciousness. Caucasians in the United States and South Africa have made sharp distinctions with regard to Negroes, but they are less conscious of Asian Mongoloids, a tiny minority. People of mixed Caucasian-Negroid blood are considered "black" even if Negroid features are almost nonexistent. These individuals could as well be called "white" unless completely black, but cultural conventions in North America have decreed otherwise. By contrast, a dominantly Caucasian society in most of Latin America takes little account of the Negro but is very sensitive to the Mongoloid Indian, whose distinctive way of living is looked down on.

Culture: A Result of the Isolation of Traditional Society

Many of the characteristics that we associate with race are really differences in ways of life. Decisions are made by a group about what to eat, what clothing to wear, what is "right" and "wrong" on social, political, and ethical matters. Not all of these decisions are made at one time. Neither are they unchanging once established. But each group communicates its way of life and its value system from one generation to the next. Each group evolves its own system of living, indeed its own way of perceiving the environment. It has its own culture.

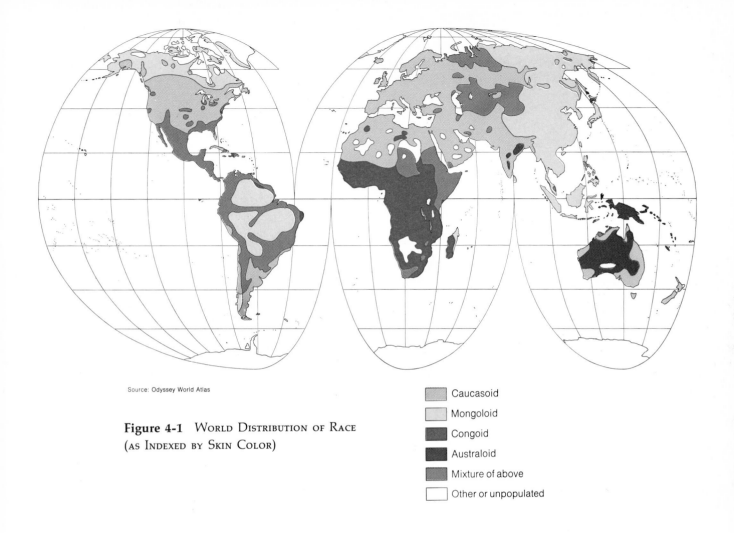

Source: Odyssey World Atlas

Figure 4-1 World Distribution of Race (as Indexed by Skin Color)

Caucasoid
Mongoloid
Congoid
Australoid
Mixture of above
Other or unpopulated

BASES OF CULTURAL DIFFERENCES

Variations in ways of life from one culture group to another may have been affected by a number of variables. The particular earth environment presents its own set of obstacles and possibilities. For example, mid-latitude Europe and tropical Africa present very different problems of survival. From time to time particular groups have also been inspired by their creative leaders. Aristotle, Jesus, Julius Caesar, Luther, Marx, and Darwin all have had tremendous impact on European culture. But it has been the group, rather than the individual, that has gradually evolved its accepted patterns of life and its conventional wisdom about how life should be lived.

Some individuals within the group may not accept its "conventional wisdom" and demand change. The group may adjust its ways to suit protesters within the community, or it may reject the proposed changes. Life styles may change as the result of individual inventiveness, but only to the degree that the group accepts new ideas.

ADVANTAGES OF CULTURE

It is generally recognized that the culture group has been one of man's great assets. Drawing upon the abilities of particular individuals, the group provides not only the frame of reference for everyone within it at a particular time but also a continuity of life from generation to generation. The individual learns the lessons of living from the group; he does not have to start from the beginning. Each generation has a basic body of knowledge about life with which to start. It can then add on to the sum of men's experience. In this way basic cultural continuity extends from one generation to another, but it is also modified by the additions of each generation.

CULTURAL LIABILITIES

Culture can be a great human asset, but it is also a tremendous liability. A culture's preconceived way of living greatly limits its possibilities for using any environment. People develop persistent convictions about what is right or wrong, what can or cannot be done. These cultural "mind-sets" often present obstacles to the development of the earth that are as great, or greater, than environmental limitations. An outsider often can quickly see a group's cultural "blinders." It is not hard for Washingtonians to provide solutions to some of Ramkheri's problems. It is less easy for them to see the mind-sets affecting Washington.

An Example of Cultural Mind-sets

Perhaps an example of cultural mind-set within the United States is in order. The Middle West of the United States has one of the world's finest environments for agriculture. There Americans have developed one of the most productive agricultural systems mankind has ever known. Yet, although the middle western environment permits the production of more than 200 different crops, most of the land is used to raise less than half a dozen: corn, soybeans, wheat, oats, and hay. Why? Because these are the crops our culture uses as a food base for eating and for feeding livestock.

The fact that this cultural outlook is limited has become obvious in recent years because of the very productiveness of middle western agriculture. We have produced so much of some of these crops that we have had great surpluses. To prevent a glut on the market that would drive prices down and hurt the farmers, the government established a program for limiting the acreage that they could plant of a particular crop. In return, the government has guaranteed them a base price for what they do produce.

When too much wheat was produced, wheat

acreage was limited by the government. Farmers had to decide what to plant in former wheat fields. Faced with such a problem, they planted more of the other established crops—soybeans, corn, and hay. The land could be used to produce dozens of other crops, but the farmers did not even consider most possibilities. Moreover, they probably could not have found a market for them anyway, since it is not generally acceptable in our culture to eat millet or use sunflower seed or some other exotic crops. Limitations are not only self-imposed by the farmer; they are also imposed by the culture that has a narrow perspective of what is good to eat.

Language, Religion, and Political States: Further Results of Isolation

Today, when daily contacts throughout the world seem to make world-wide communication essential, there are some 3,000 different languages in use. Thirty of them are spoken by at least 20 million people. In Europe alone there are about 120 different languages in current use. One of the problems of the United Nations and other international bodies is communication because of the many language differences. Many people are trained in foreign languages, but at the United Nations it is still necessary to provide simultaneous translation for the five major languages. The business of the world is usually conducted in the languages of the major industrial powers, but these are the native languages of only a tiny minority of the world's people. The presence of so many languages in the modern world stands as a measure of diversity. Each group evolved its own word designations for the same physical phenomena or the same ideas. The multitude of different languages in the world is an indication of the separateness of different groups in the past, and language is one of the key factors used by anthropologists to distinguish separate culture groups.

If man were emerging in today's modern, interconnected world without a past history of separateness, George Bernard Shaw's dream of a single world-wide language might be possible. But the diverse history of man has allowed language to become one of the cherished cultural values. Second languages are taught in schools, but one's mother tongue will continue to be the primary one.

As the need for international communication has increased, there has been a simultaneous rise in interest in individual local languages. Each culture group has shown interest in its past and considers language an important aspect of that past. India, for example, rejected English as its national language when freed from Britain and sought a language of its own. Further, the political subdivisions of India were drawn largely to distinguish individual language groups. Newly emerging countries have encouraged the teaching of the languages of their minority groups, and some are developing written languages for the first time. Even minorities within modern industrial countries have sought roots in their own local languages. Welsh is taught over British television, and Gaelic has been reintroduced in Ireland. In the United States blacks with renewed interest in their African heritage are learning Swahili.

RELIGION: VARIATIONS IN MAN'S VIEW OF THE MEANING OF LIFE

More fundamental than the differences among cultural areas are the views held by various culture groups about the meaning of life. Each geographic area has its own religious beliefs to answer the basic questions of man's existence. Religions have formed the basis for different

Language, the basic medium of communication, is probably the most distinctive and important cultural index. A given language indicates the limits of easy communication within a group. Notice that the major language groups shown in Figure 4-2 follow closely the separation of biological groups by skin color (Figure 4-1).

The subdivisions of language within any group present further insight into cultural divisions; they show the importance of tribal or national divisions on a more local level. Even the map in Figure 4-2 does not indicate individual languages. Notice that it lumps French, Italian, and Spanish in the Romance subgroup and German, English, and the Scandinavian languages in another subgroup. All Indian languages in Latin America and the Canadian-Alaskan area are included together. The large number of different languages in the world—about 3000—prohibits rendering them on a world map of the scale used in this book.

Would you say that there is a difference between the patterns of languages of the predominantly modern regions of the world and those of regions with dominantly traditional ways of living? Why does Europe have a variety of languages whereas Anglo-America, Latin America, and the Soviet Union are each dominated by a single language group?

rules of conduct, for distinctive perspectives of life, for speculation about the place of the individual in the world, and even for particular uses of the earth environment. These are a major factor inhibiting communication between cultures.

Figure 4-3 shows the distribution of religious groups throughout the world. Animism, a primitive religion which attaches spiritual power to earth, atmosphere, and all living things, has been dominant among the peoples of Africa, the polar areas, and the Pacific islands. In South Asia, Hinduism has dominated. In East Asia, the religions have been Buddhism, Confucianism, Taoism, and, in Japan, Shinto. Buddhism embraces the idea of the submergence of the individual to a non-worldly state of infinite being. In Confucianism man is seen as related to the supreme deity, Heaven, through his ethical actions on earth. Disasters are viewed as the result of the chastisement of Heaven for misdeeds, particularly of the Emperor, the "Son of Heaven." Shinto stresses the descent of the Japanese people and islands from the gods. An East Asian is likely to hold a combination of tenets from all of these religious teachings.

Southwest Asia–North Africa is the realm of the Islamic religion; Allah, the one God, is the supreme ruler, and men's lives are subject to God's will. There religion is a particularly strong force in the daily life of the people. Europe, the center of the Judeo-Christian religious heritage, has still other emphases, such as the personal relationship between God and the individual, and man's stewardship of God's resources on earth.

The importance of religion in shaping the practices and perspectives of a culture can scarcely be overstated. In today's world Hinduism in India is perhaps the best-known example, and its impact is graphically illustrated by

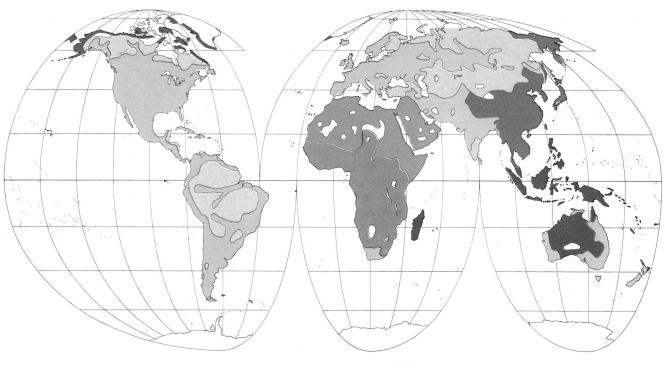

Figure 4-2 WORLD DISTRIBUTION OF
LANGUAGE GROUPS

☐ Indo-European

☐ Sino-Tibetan and other Asian

☐ Niger-Congo, Sudanic and Khoin

☐ Malayo-Polynesian, Papuan and Australian

☐ Semito-Hamitic

☐ Other

☐ None

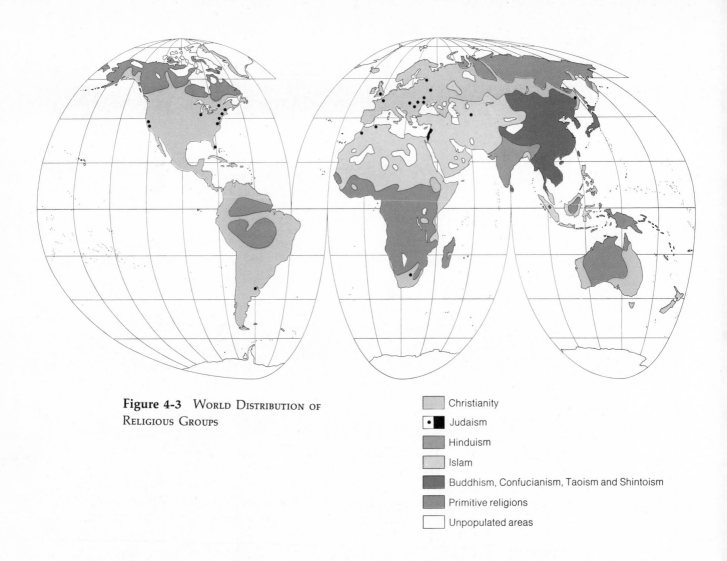

Figure 4-3 WORLD DISTRIBUTION OF
RELIGIOUS GROUPS

	Christianity
	Judaism
	Hinduism
	Islam
	Buddhism, Confucianism, Taoism and Shintoism
	Primitive religions
	Unpopulated areas

the fact that cattle compete with the human population for foodstuffs, even though famine is an ever-present danger. India has the largest cattle population of any country in the world, yet religious practices forbid the killing of cattle for meat. In fact, there is a general reluctance to kill any animal. Monkeys and other pests destroy an important share of the harvests and complicate the problems of sanitation and health.

The Hindu religion teaches that all created things are one and that a person's soul may have been previously born in animal form or may be so born at another future time. This belief is fundamental to the Hindu view of human existence. Thus, obviously, control of the animal population is far more than a matter of proper governmental decree or education. Change is not going to take place without basic modifications in the Hindu concept of man's life in the world. Moreover, the average Westerner's perception of the Indian cattle predicament stems from his particular view of man's place in the world and what seems right to him.

Hinduism has also provided scriptual sanction to the rigid system of social caste which is recognized as one of the great cultural obstacles to Indian economic progress. The caste system is a continuing force in the Indian village, despite its official abolition by the government. It is too deeply rooted in the ways of village life.

The impact of Hinduism on Indian life is but a single illustration of the importance of religion. In the Moslem world there are taboos on eating pork and drinking liquor, but religious doctrine teaches that all running water is good to drink, even though modern scientific research has shown that such water may, in fact, be polluted.

The Christian religion has given distinctive character to human values as well. Like the Moslem and Jewish religions, but in contrast to other religious groups, it places particularly high value on the individual. It asserts that each human being is of special concern to God. Some have argued that this stress on the individual has been important to the development of democratic political tradition, even though democracy was practiced among the Greeks before the birth of Christ. Many dictators have also ruled in the name of Jesus Christ—and many battles fought under the sign of the cross.

Some people also believe that the stress on the individual and on what has to be termed the Protestant Ethic is essential to the free enterprise system of western economies. It would probably be more accurate to say that individualism has become an important aspect of western culture as a whole and is no longer related to any religious tenet. Although the church has lost some of its influence as the political and economic leader of men, it still plays an important role. The effect of Roman Catholic opposition to birth control and abortion legislation is a prime example of this.

As might be expected, contrasting religious views have presented some of the greatest difficulties to intercultural contacts. Aggressive religious groups have been driven by evangelistic zeal to convert the misguided to the "true" faith. This has been a particular characteristic of European Christians in their expansion. One of the drives in the Spanish colonization of the Americas was to convert the Indians to Christianity. And, in the time of empire building during the nineteenth century, many people joined the Christian crusade to bring the gospel to the heathen of Africa and Asia. Missionary effort was an integral part of the colonial domination of Africa and Asia, and as a result Christianity has been identified as an aspect of European colonial exploitation.

In the modern, interconnected world where

people are more affluent and better educated and where other diversions compete with the church, religion has become a less important force in society. Many of the competing secular forms are, however, virtually religions in themselves. Communism in the Soviet Union could be viewed as a state religion. Lenin and Marx are revered as prophets or saints, and their teachings are as essential to Soviet life as the Bible is to practicing Christians and Jews.

Although the United States has not evolved an official state religion, there are homes in which the American flag is as important as the crucifix, and a great many families have become more ritualistic about their television-watching than their church-going.

In Western Europe and the United States a large proportion of the population no longer participates in any organized religious activity. Social critics have pointed out that a concern for material goods and scientific progress have replaced religion in establishing the value system of these societies. The "good life" is said to come from attaining material goods and position in society. In other words, belief in the modern, interconnected system as the source of satisfaction has replaced traditional Judeo-Christian values. It is significant that recent protests against the modern industrial complex and the affluent society have fostered a return to various forms of religion including Oriental mystic beliefs and fundamental Christianity.

POLITICAL STATES: FUNDAMENTAL DECISION-MAKING UNITS

Political fragmentation is another result of the separation of people into separate culture groups. Just as distinctive religious values have emerged in different cultures, so have varying political beliefs and political systems. Thus, there have been different forms of tribal rule, feudal fiefs, kingdoms, dictatorships, and de-

mocracies and a range of other forms of government. No matter what the *form* of rule, the authority of the political regime has had a geographic spread. And in early days, like culture itself, that authority had a very limited reach.

Today there are more than 140 different sovereign political states in the world, most of them with populations the size of some individual states of the United States. Such tiny nations seem to be anachronisms in a world where emphasis is on power on a large scale and where modern businesses function on a worldwide basis. Following the general pattern of the fragmentation of other cultural factors such as language and religion, the nations of today stem from a different era and represent a totally different scale of interaction.

But the division of the world into many separate political nation-states is even more significant than the fragmentation of language and religion. Nations are man's political organizing units. They demand the primary loyalty of people living within their borders and are probably the most important decision-making bodies in the world today. A nation-state not only establishes the basic economic and social rules of life for people within its borders, it is also the primary agent for its citizens in dealing with the rest of the world. Political states negotiate peace and make war.

Perhaps the nation-states, with their origins in cultural diversity, best characterize the fragmented nature of the world today. Moreover, as decision-making entities they are perpetuating man's parochial view of his world. If modern civilization had originally emerged with the present system of world-wide transportation and communication, there might well be world government now. But, like language, nation-states are deeply rooted in man's cultural patterns and represent a very important part of present-day cultural perspectives. Thus, they are likely to remain with us for a long time.

Economic organizations such as corporations can spread out to embrace operations on all continents. By contrast, governments are more limited today than a hundred years ago. In the early days of industrialization, the powerful European countries were able to establish empires by the political domination of underdeveloped areas. But by the middle of the twentieth century change had begun. A basic principle of the United Nations is the "self-determination of peoples," and since World War II more than forty countries have achieved independence from former empires.

Technology: The Basic Difference between the Two Models

Perhaps the most obvious and important difference between farmers in Indian villages and farmers in the United States lies in technology. The villager is limited to simple tools—hoes, wooden plows, and simple threshing floors—and human and animal energy sources. By contrast, American farmers employ modern tractors, self-propelled combines, and other machines using gasoline and electrical energy. Technological differences of this type form the most apparent and economically and politically significant variation between cultures in the world today. This fact is widely recognized in the terms *industrial* and *nonindustrial* or *technologically advanced* and *technologically less advanced.* There is no doubt that different areas have different technological capabilities and that individual affluence and national economic and political power lie in the hands of the technologically advanced areas. In contrast, areas that have lagged technologically are weak politically and economically, and their populations are largely poverty-stricken.

The outward thrust of Europe and its success in colonizing and dominating other cultures both politically and economically were, to a high degree, outgrowths of the technological advantage held by the Europeans. And the struggle of non-European cultures against European colonialism has not been merely a campaign to retain political freedom. It has been an effort to develop economic independence and acquire the kind of living standard that European culture has achieved through the perfection of modern technology.

THE MOST IMPORTANT THREAD OF CULTURAL PROGRESS

Technological progress has been one of the most important threads in man's cultural development. If he depends only on his own muscle power, a man has little control over the natural environment. But he has been able to use his intellect to accomplish work far beyond his limited physical capacity. He has developed tools and machines which utilize other energy sources.

Technological knowledge offers material well-being for those people who can apply it. Furthermore, the capacity to increase the fund of technological know-how gives man a sense of progress. The primary drawback to this progress is that change occurs so rapidly that society is continually having to adjust.

It is important to recognize that technological development has depended on far more than individual inventions and discoveries. People must be receptive to innovation and must be able to adapt their living patterns to the changes brought about by new products.

Equally important in the development of technology is the organizational structure it demands. More sophisticated machines have created large-scale production, and modern transportation allows for mass-marketing. Factories call for subdividing work assignments and tremendous capital investments. Although a craftsman in his own shop can manage his business, handle its sales, and still take an active part in the production process, a factory requires not only a foreman to supervise the job but also an office full of support personnel—persons who keep the books, make the sales, and determine long-range company plans. Besides manufacturing, the modern economic system also depends on giant transportation systems, great financial organizations, and huge wholesale and retail complexes.

THE EVOLUTION OF TECHNOLOGY

The first tools were designed specifically to make human muscles more efficient. Shovels, knives, bows, baskets, and carrying poles were among the tools created to be powered and controlled by human muscle. With the discovery of other energy sources such as animals, water, wind, and fire, new tools were designed, including plows, water wheels, sails, and cooking devices. With more powerful and dependable sources of energy, man shifted from simple tools to machines that could do the work of the men who had formerly operated the hand tools. Wagons, grinding mills, cannon, spinning wheels, clocks, and hundreds of other machines were made with moving parts. Finally, with the discovery of inanimate energy sources such as coal, oil, natural gas, electricity, and atomic energy, there was a further technological revolution. Tools and machines can now be created that operate on almost any scale and at any speed. They have tremendous flexibility. Today computer-controlled machines produce automobile engines and load coal trains without any direct use of human energy or control.

Man's capability for creating material wealth has been "revolutionized" by the use of tools. From a small producer using just his own mus-cles, he has become the manager of an amazing array of machines that operate day and night, year after year, in his presence or absence. But the technology of tools and machines has left a further heritage to any civilization. It has bequeathed a complex of artifacts constructed with that equipment: buildings, transportation routes, clothing, books and phonograph records, and all the rest. As a result every generation has a tremendous supply of man-made goods with which to work. These represent tremendous capital which the present generation can utilize. When one considers the time and effort necessary to build a house, complete a road, develop a water or sewer system, the great value of this technological legacy becomes apparent.

But like other inheritances from the past, the capital accumulation of material, tools, and machines is also a liability. No generation can wipe out its past; it must build on what is there. Houses built in the era when coal was used to heat have to be converted to gas or oil and then air-conditioned. Roads built for horses and buggies must be adapted to automobile commuters. The cost of wiping out all the capital accumulation from the past seems too great. Nowhere is this problem more sharply apparent than in the great metropolitan centers of the world. Urban change calls for removing outmoded parts of a city; but these things cannot be done cheaply or quickly. As a result, people live in inadequate housing and drive on inadequate streets. Moreover, who is to say that, if by some special effort all of a metropolitan area could be renewed today, the renewed area would fit the needs of tomorrow?

Men are gregarious by nature. They organize themselves into groups for political, economic, religious, and social functions. In the course of time the groups within a culture develop similar aspirations and values. This combination of common attitudes and experiences determines the culture of a people and distinguishes one culture from another.

DEVELOPMENT OF ENERGY SOURCES

Since tools and machines depend on energy, man's greatest technological achievement has been the development and control of energy sources. Manpower has been mobilized on a large scale throughout history. Big families have been used to increase production from the family farm. Slaves have been captured in battle or bought to do heavy physical labor. Masses have been organized in China and elsewhere to build roads, dams, and factories.

But other sources of energy have become far more important than the human one. More than 10,000 years ago the first domestication of animals began, and since then, horses, water buffaloes, camels, and other animals have been put to work as beasts of burden. They have not only transported goods, they have also pulled plows and run machines such as the Indian threshing machine. The discovery of fire enabled man to derive energy from timber, grasses, crop wastes, animal manure, coal, oil, and other combustible objects. The force of running water, winds, steam and hot water from the earth, and even the tides has been used. Now we are beginning to harness the forces resulting from the release of nuclear energy and from direct radiation from the sun. Through most of man's history, however, he had little energy at his command and was not always effective in producing enough, even for his own domestic needs.

Until some 200 years ago the energy available for human use was largely limited to that derived from humans, animals, trees, and other plants and from wind and water flows. Generation and use occurred at the same time. Very little energy could be stored. At best wood could be cut and stored for consumption at some other time and water dammed for later use.

THE ENERGY REVOLUTION: TAPPING THE EARTH'S RESOURCES

In the eighteenth century the Industrial Revolution in England opened a whole new storehouse of energy. This revolution followed the invention of new machines and the organization of the factory system which replaced traditional handicraft. But, most of all, it was an energy revolution which has transformed man's life.

The Industrial Revolution started with harnessing the force of running water in the small streams on the slopes of the Pennine mountains in England. It advanced through the discovery of ways to utilize coal, oil, natural gas, and radioactive materials. Energy from the sun, absorbed by living plants, had been captured and stored in rocks as coal and oil or gas. The discovery of ways to utilize this energy in engines and power plants released new wealth of power. Now in our automobiles we burn in an instant the gasoline refined from petroleum which accumulated in the earth over a million years ago. Our power plants make electricity from coal that also took centuries to form.

THE SIGNIFICANCE OF MODERN ENERGY USE

The geography of energy consumption over the earth is fundamental to an understanding of "industrial" and "nonindustrial" worlds. (See Figure 4-4.) Today's traditional society depends in large part on the same sources of energy that man has used throughout recorded history. These sources are comparatively poor in terms of producing material wealth. On the other hand, modern society has harnessed sources stored in the earth from past geological eras to develop a degree of material productivity unbelievable 200 years ago. In Figure 4-5 it can be seen that a few industrialized nations consume most of the world's energy.

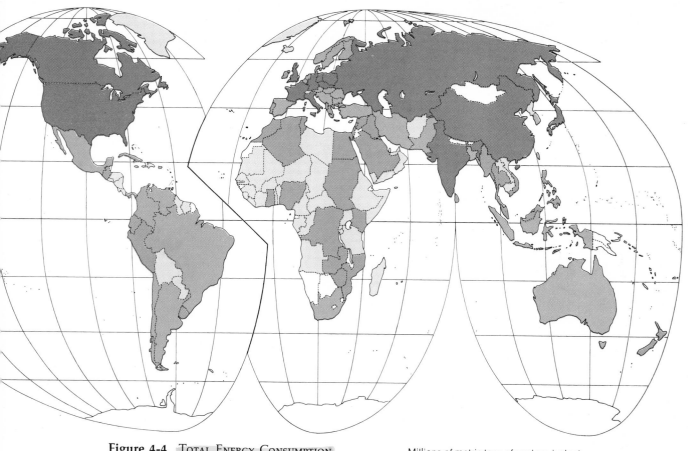

Figure 4-4 TOTAL ENERGY CONSUMPTION OF COUNTRIES

Millions of metric tons of coal equivalent

Less than 1
1-9.9
10-65
80-1,000
More than 2,000
No data available

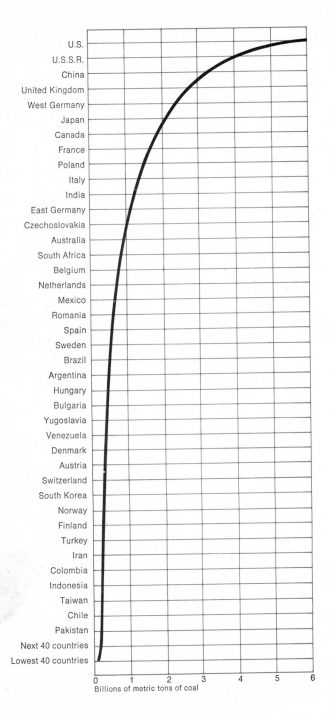

U.S.
U.S.S.R.
China
United Kingdom
West Germany
Japan
Canada
France
Poland
Italy
India
East Germany
Czechoslovakia
Australia
South Africa
Belgium
Netherlands
Mexico
Romania
Spain
Sweden
Brazil
Argentina
Hungary
Bulgaria
Yugoslavia
Venezuela
Denmark
Austria
Switzerland
South Korea
Norway
Finland
Turkey
Iran
Colombia
Indonesia
Taiwan
Chile
Pakistan
Next 40 countries
Lowest 40 countries

0   1   2   3   4   5   6
Billions of metric tons of coal

Figure 4-5 Total Energy Consumption by Country, Shown Cumulatively

The difference between the energy sources of the industrial and nonindustrial societies is more than just the relative quantity of energy available to the industrial world. Inanimate energy can be transported long distances and concentrated in a particular place to do prodigious jobs impossible in previous times. One thinks of the Egyptian pyramids built by human labor as an amazing feat calling for the assemblage of tens of thousands of workers. But the concentration of power in one nuclear explosion is far greater than the power that could ever have been assembled in the pre-industrial world. In addition, modern energy can be stored until needed, used in measured quantities, and turned off and on. In contrast to human slaves or animal power, the "energy slaves" of inanimate power are completely at man's bidding.

In the traditional society the individual producer simply is not efficient enough to accumulate wealth. A family is able to provide its basic needs only with the expenditure of tremendous effort. In such a society wealth comes from being able to command a share of the productivity of many workers. This is why slaves have been important. They were the only supplement or substitute for one's own work. Before the Industrial Revolution only a tiny fraction of any group had wealth. As a rule the wealthy were those who controlled large tracts of productive land and drew a share of the output in exchange for the privilege of using the land. Thus, the wealthy in feudal times were the landlords. The affluent in the American South and in Spanish Latin America were the plantation owners whose land was worked by slaves. It is the same today where traditional society exists. The headmen of Ramkheri are the traditional landowners who have both affluence and power.

But in modern society the vast quantities of energy slaves which power the technological equipment increase productivity per person to such a tremendous degree that it is possible not only for the factory owner to be wealthy but for the workers to be paid at such a rate that their standard of living is greater than that of feudal landlords of the past and is certainly more than that of the village headmen. The affluence that spreads down to workers in today's society comes not from the superior productivity of the workers themselves but from the economy's utilization of quantities of coal, oil, natural gas, and other energy slaves. The system is so productive that there is wealth unknown in the pre-industrial world, wealth that may be widely spread among workers.

To illustrate the point, consider a worker on a modern auto assembly line in the United States. He is paid $3 to $5 an hour even though his particular task may be only to attach door handles by using a pneumatic screwdriver. Such a man cannot be considered a highly skilled worker; he can be trained to do his job efficiently in a few hours. He is highly paid, not because of his personal skill, but rather because the assembly line on which he works can produce a $3,000 automobile every minute, thanks to a multitude of power-driven tools and machines. Since the power-driven equipment is redesigned through the years to operate more and more efficiently, the assembly process has become more productive. The benefits of this greater productivity are increased profits for management and higher wages for the workers.

The worker on the assembly line is so well paid that he can afford to buy the $3,000 automobile and many other machine-made products. Industry benefits as well from his being a consumer of the products of this automation. It is the ever-increasing utilization of cheap, controlled energy slaves that supports the affluent industrial society.

In contrast to the assembly-line operator,

consider the highly skilled craftsman in India or Iran doing intricate metalwork for which he had to train for years. He receives only a few cents a day. Without the availability of machines it takes days or weeks for his deft, trained hands to produce a single brass tray. Poorly paid, this worker can scarcely afford the basic essentials of life. He and most of his countrymen are poor and can buy very little to stimulate economic growth.

Differences in technology have resulted in more than different levels of ease and affluence. The whole outlook of decision-makers, producers, and consumers is very different. Recently much has been said about the dehumanizing nature of assembly-line work. One worker may attach the same three bolts on car frames as the cars move by him each minute, eight hours a day, forty hours a week. Such work, day after day, year after year, offers little personal satisfaction and causes physiological strain for these workers. Weekends for recreation may be essential therapy.

Automation using computer programming to provide control over complex machines offers a solution to the dehumanization of the assembly line. Such automatically controlled machines will no longer need human workers to guide them. On fully automated lines, workers will be needed only to check, maintain equipment, and repair breakdowns. Thus, where hundreds of persons have worked on an assembly line, perhaps with automation only a handful will be needed.

Conservatism of Traditional Society

Traditional society with its limited technology has low productivity and tends to conserve the status quo. Old ways are revered, and traditional practices are followed with meticulous

Table 4-1 shows the relative importance of traditional and modern energy sources. Even in the underdeveloped countries where traditional society dominates, modern energy sources are of major importance. How can we explain that fact? Is there an industrial-nonindustrial dichotomy with respect to sources of energy used?

Table 4-1 WORLD ENERGY CONSUMPTION BY PERCENTAGE OF SOURCE (1963)

TRADITIONAL SOURCES	PER CENT	INDUSTRIAL SOURCES	PER CENT
Manpower	1	Coal and lignite	38
Work animals	4	Petroleum	28
Fuel wood, manure, waste	13	Natural gas	14
		Waterpower, nuclear power, geothermal power	2
Total	18	Total	82

SOURCE: R. A. Harper, "The Geography of World Energy Consumption," *Journal of Geography*, October 1966.

care. As we saw in Ramkheri, farmers and village craftsmen show strong resistance to any change in their ways and oppose new practices that affect their basic economic support.

Such views are not hard to understand when considered from the perspective of rural people. Some of them are living on the margin of life and death. Others have enough for the essentials of life and a few luxuries. They know that past practices work, and they expect them to continue to work in the future. Only a natural calamity—a great flood, a severe drought, or an insect plague—presents real danger to their way of life. Especially favorable environmental conditions during a growing season may ease the strain by providing a surplus.

Such a system allows little margin for error if physical sufferings of human beings is to be avoided. Crops may be stored from harvest to harvest, but they are usually completely consumed in daily living. There is commonly little or no reserve accumulated from year to year to serve as a source of support if a harvest season is bad. Even if there is a surplus, storage is difficult without modern facilities; much of a crop is lost to rodents or rot. Under these conditions the risk of change is tremendous. How can one afford to try something new—a different seed, a plow, a new fertilizer? What if it should somehow fail?

Added to the dilemma of the poor subsistence dweller is his lack of education and a cultural structure built for preserving traditional ways. In such a society all accepted practices call for maintenance of the status quo. It is a brave man who attempts to make a change. Often it is only the wealthy one who can afford to take risks or the one whose educational experiences extend beyond those of the local society.

Technological development generally calls for a great deal of reorganization, too. It requires a change in institutions, in styles of living, and, most of all, in attitudes. Such changes are among the hardest of all to accomplish. Often the blinders of cultural heritage stand as greater obstacles than resource limitations.

Change in the Modern World

In contrast to the locally based system, the interconnected system is oriented toward change. A recent book on California, perhaps the most modern part of the United States, asserted that if the state had a single constant, it was change, and if it had a symbol, it was the bulldozer. The problem of people in the modern society is not how to conform to tradition but how to adapt to continuing change.

Why is the modern world committed to change when the traditional world fears it so? The vital element seems to be the technology that has produced the modern world's material well-being. Of course, virtually all the factors that made traditional society so conservative are reversed in the modern world. An affluent society has a large margin for taking risks; if the gamble fails, all is not lost; it can try again or try something else.

In the modern business world, new products mean new markets, and new markets mean new profits. Since the emphasis is on improvement and new models, much of the profit is reinvested in research and the development of new products or new designs. The major operations of this type in Washington, D.C., for example, are the research and development programs funded by government grants for new military equipment and studies for social as well as technological change.

Social scientists speak of the human problems involved in living in the modern world with its emphasis on change: the loneliness, the uncertainty, and the frustration. They emphasize the problems of political and other institutional adjustment to the frantic pace of change. It appears to be much easier to accomplish technological change than to carry out the cultural changes necessary for proper accommodation to the new technology. As we have seen, the nation-state in the world had its roots in pre-industrial times. The small size of most countries is a disadvantage in today's economic world. Yet, the political structure cannot be adjusted to the possibilities offered by the larger-range economy under the modern, interconnected system. In the same way, city boundaries established before the automobile have not always been extended to take in the new and larger functioning city beyond the municipal limits. So we have the present problems of the separation of city and suburb.

Population and Resources: The Issue of the "Population Explosion"

Much is said about the population explosion in both of the situations we have described. But the results of this explosion are very different in the two systems. In the traditional society population has been growing rapidly while the use of the environment has remained almost unchanged. Thus, with each generation the pressure of population increases.

In the modern world, however, the population explosion takes on a different character. The problems are those of congestion and consumption. People are crowded into urban centers with populations in the millions. There are thousands of people per square mile. The movement of people to and from work in such congested areas produces the problems we saw in Washington.

Even worse is the twofold problem presented

by increasing affluence that goes along with increasing population. This has led to a rapid rise in consumption, which in turn places greater pressure on the world's resource base and multiplies the problems of waste disposal. A single stream may be the source of water for drinking and bathing; at the same time it may serve as a sewer. With low-population densities and a swift-flowing, high-volume stream, such multiple use may present no severe problems. Most wastes, including food, clothing, and tools, could be thrown away, and they would gradually decay. But in a modern metropolitan area the problems of waste disposal are overwhelming. City life depends on materials gathered from all over the world. Most of these come carefully packaged to protect them. So both the product and its package must be disposed of. Moreover, metropolitan areas are the major foci of energy consumption. Although some of this is electricity produced from fuels elsewhere and transmitted as electricity, much is gas, oil, and coal which is burned in city buildings and in city vehicles. The problem is: how to dispose of wastes without overwhelming the environment. Yet, as with traditional society, the problems of overpopulation in the modern world are culturally derived as much as they are the result of simple limitations of the earth environment. Here is another example of human adjustment lagging behind technological change. The increasing problem of waste disposal, so simple in primitive society, had not been recognized as a major difficulty of metropolitan life until recent years.

At present, population is rapidly expanding with no limit in sight, while the earth environment is obviously finite. All that we have to work with is contained within the relatively small planet Earth. Not only is it limited in size, its resources are likewise finite.

THE MALTHUSIAN PRINCIPLE

A long-established principle of social science has been reintroduced to describe the situation of overpopulation. It was originally put forward by the English economist Thomas Robert Malthus before the Industrial Revolution. He believed that population increased in geometric progression but that agricultural production, the primary support of the population, could increase only in a strict arithmetic progression. In the eighteenth century agricultural production could be increased only by clearing new land for additional crops. The ultimate results of this situation were obvious. Population would soon overtake man's ability to clear more land for agriculture in England or anywhere else. Then the competition for available food would intensify. Famine and warfare would inevitably follow, which would, in turn, control population growth. But the predictions of Malthusian principle did not really come to pass in Europe. Today there are four times as many people in Europe as there were in Malthus's day, and most of them live at a much higher standard than was common in the eighteenth century.

THE CASE OF EUROPE: AN UNEXPECTED OUTCOME

What happened to produce results so different from Malthus's predictions? The Industrial Revolution changed his formula completely, at least in the short run. Contrary to his equation, the total amount of production increased even more rapidly than population. There was more and more production for the growing population to share, thanks to the new technology. The combination of machinery, fertilizer, new seeds, better animal breeds, better harvesting methods, and insecticides expanded output rapidly. Moreover, transport ties to food-producing areas on different continents made other food available to Europeans. The number of those who worked in factories, trade, and shipping grew larger as the modern, interconnected system spread over the world, with Europeans supplying manufactured products and managing the commercial networks. Today, the population problem in Europe is one of congestion, not starvation.

RESOURCE QUESTIONS INVOLVE CONSUMPTION AND TECHNOLOGY

Can the earth environment continue to absorb the increased amount of waste being transmitted to it without being so changed that it will threaten the delicate balance of life? An additional question is: can technology continue to move at such a pace that it will outdistance population growth and improve material existence for the entire world? If the earth is finite, its resources must be limited. Sooner or later the limits of the earth's resources will be reached—or will they? Can man achieve a balance with nature by recycling, reusing, and creating synthetically? At present no one really knows. Surely the growth of population in the modern, interconnected world, its increasing affluence and the resulting consumption of resources, and its growing concentration in the great metropolitan centers of the world will bring the need for answers ever closer.

THE TRADITIONAL MODEL

The results of the population explosion are quite different in the traditional, locally based society. Here we find the Malthusian equation still at work. The past 200 years have brought a tremendous increase in population but little basic change in the system of resource use. In Ramkheri population growth is almost outrunning the ability of the economy to produce; without the importation of food supplies from the outside world, famine would be rampant. In fact, parts of India have suffered from famines in this century.

Population in India has more than doubled in the past one hundred years. Yet traditional agriculture and the Indian farm village remain the standard of most of the Indian people. It is not hard to see that rural living a hundred years ago might have been more affluent than today's. Each year more and more mouths have had to be fed by traditional agricultural output.

Today population densities are much higher in India and China than in most parts of Latin America and Africa, although the latter have some of the highest growth rates in the world. They too may face famine in the near future unless the population is checked or a major change occurs in the economic base.

These countries need far more than a way of stabilizing population. They require some means of dramatically increasing production. One answer would be for the traditional model to reap the technological benefits of industrialization and the modern system. They need to channel modern energy sources and scientific knowledge. In Ramkheri and in other traditional societies, people utilize their environment intensely and carefully, but modern technology could significantly increase that production. Hybrid seed, fertilizers, steel-tipped plows, and insecticides would all improve the crop. There would also be less waste if the harvest could be stored in rodent-proof granaries. These communities need first to change the management of their resource base and then to increase the technological capabilities for its utilization. There are widespread ramifications to any

The following generalizations about the order of trade connections between major world areas can be made.

1. First-order trade occurs between the industrialized areas of the world. These are the major producing areas in terms of manufacturing and the major consuming areas for manufactured products and the resources needed to support manufacturing. The importance of Europe is partly a matter of the number of international boundaries, but there are other peculiarities as well. For example, why is trade between the United States and the Soviet Union so small?

2. The next order of trade occurs between the industrial and nonindustrial areas. Each major industrial area has its own particular links with the nonindustrial areas. The ties of Anglo-America, Western Europe, and the Soviet Union with nonindustrial areas vary substantially.

3. The trade between one nonindustrial country and another is the least important. In general the trade between these countries, even neighboring ones, is small compared to their trade with industrial countries.

change that takes place within the tightly structured life of the village. Undoubtedly economic changes must be made if the people of the traditional world are to survive even at present levels. The basic question is not only whether changes can take place but whether the society can change economically and survive culturally. Any change threatens the basic structure of the culture.

Moreover, if the society becomes part of the modern industrial system, it will have other problems to face. Whether the earth environment can support a completely modernized world is indeed questionable. Many people believe it is the people in the modern, interconnected system who pose the real resource problems because in their affluence they require more land, utilize more energy and minerals, consume more food, and accumulate more waste. Moreover, the very industrial processes used to convert these resources into more utilitarian products create enormous amounts of wastes. These are the actually dangerous pollutants.

The Organization of Area by Human Activity: Patterns of Interconnection

So far in this chapter, we have been examining cultural and technological factors as if the world were a mosaic of cultures. As we saw earlier, although that may be a satisfactory way of examining traditional societies, the long-range, interconnected system has links to all parts of the world. To get a full picture of the modern system, we must add still another factor to our geographic components: spatial interaction. We must not only see the extent of the transport links that hold the modern system together but also get some idea of the volume of traffic that flows through the system.

First Approximation of Interconnection: Movement of Goods in World Trade

From the value of goods traded from one culture area to another we can learn both the value of different kinds of goods and the relative importance of countries in commercial activity. The system of trade is indeed world-wide. All the population centers and the different culture areas participate.

As a map of shipping shows (Figure 4-7), the heaviest trade is between Western Europe and Anglo-America. Although the Soviet Union figures less prominently, its trade ties with Western Europe are among the largest in the world. In contrast, the relatively small trade flow to and from Asia, where more than half the world lives, indicates that it plays only a minor role in the international trade system. Notice that imports to Japan, the one industrial country of Asia, are as great as those to the rest of Asia. Similarly, Australia and New Zealand, with less than 1 per cent of world population, appear as much a part of the commercial flow as Subsaharan Africa, which has more than seven times as many people.

This pattern of trade emphasizes the importance of Anglo-America, Western Europe, the Soviet Union, and Japan in another way. Trade does not simply occur from continent to continent. Very little trade moves between Latin America and Subsaharan Africa or between Southeast Asia and the Middle East. For each of these culture areas the first-order connections are with the industrial areas of Anglo-America, Western Europe, the Soviet Union, and Japan.

The Rich Who Buy and the Poor Who Can't

The peculiar pattern of international trade supports the ideas that have been emerging concerning the modern, long-range, interconnected model and the traditional, locally based model. Much more is produced in the industrial areas, and their people are much more affluent. These areas make and buy more goods. Although their interests in goods cause them to reach anywhere in the world for the things that they want and need, the greatest share of those things is produced in their own particular culture area or in one of the other industrial areas. They draw upon the nonindustrial world for raw materials and fuels that may not be in adequate supply at home, for tropical products that cannot be grown at home in the mid-latitude parts of the world, and even for supplies of staple goods if there is not enough production at home, as is often the case in small, crowded Europe. But the larger share of their money is spent neither on raw materials nor foods but on manufactured goods made in their own or other industrial areas. This is a matter of personal experience for all of us. We know that a person planning to buy an automobile in the United States expects a choice between American and European cars, and now a choice of Japanese imports, too.

On the other hand, the areas with only partial industrialization and with a large proportion of their populations still in the traditional system produce little for trade. Most of their people are poor and can buy little. If such areas produce goods, they are likely to be agricultural products or raw materials which have important markets only in the industrial areas of the world, not in other developing countries.

International Air Traffic: Movement of People

The passenger flows on the major international air routes reinforce the generalizations made about international trade. As Figure 4-6 shows, the heaviest air traffic is across the North At-

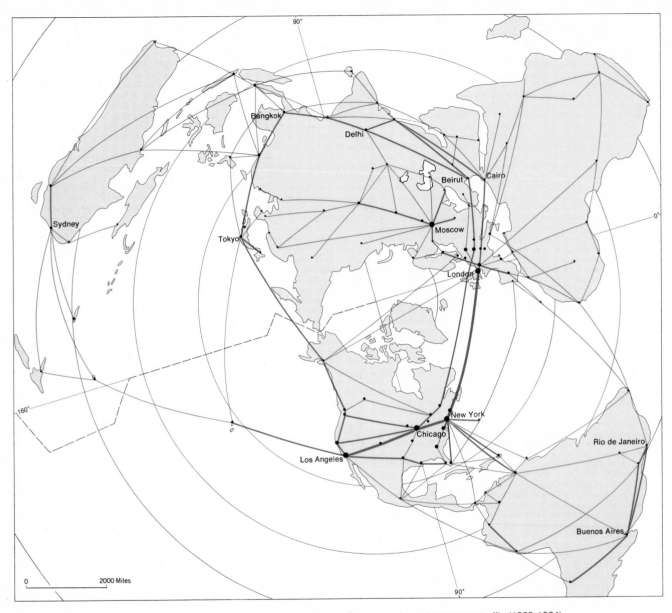

Source: Pergamon World Atlas

Figure 4-6 PATTERNS OF INTERNATIONAL AIR TRAFFIC

Annual passenger traffic (1963-1964)

—— Less than 500,000 persons

—— 500,000-2,000,000 persons

━━ More than 2,000,000 persons

Annual passenger traffic at airports

· Less than 1,000,000 persons

· 1,000,000-3,000,000 persons

• 3,000,000-10,000,000 persons

● More than 10,000,000 persons

lantic between two key industrial centers, Anglo-America and Western Europe. All other important routes connect these industrial centers with individual nonindustrial areas. All these areas are important subcenters of the interconnected system whose major axis connects Europe and Anglo-America.

Arteries of Business and Government Connection

Except on the North Atlantic the largest number of airline passengers are not tourists, but government officials and businessmen. The air routes shown in Figure 4-6 give a close approximation of the major arteries of the modern, long-range, interconnected system in the world. The figure shows that the air network does not simply connect countries but does also connect particular centers within countries. In most countries the connection is with a single major metropolitan center. Thus, the air network outlines the links between a series of large metropolitan centers which have developed around commercial centers in the world-wide, interconnected system.

Like world-trade centers in Anglo-America, Western Europe, and the Soviet Bloc, the airline system focuses on the largest metropolitan districts of those areas—New York, London, Paris, Rome, Moscow. These centers are then linked with other metropolitan points in the nonindustrialized parts of the world—Mexico City, Rio de Janeiro, Johannesburg, Cairo, Calcutta.

Linking Centers of Modern Metropolitan Life

The traveler on world airlines finds conditions much the same wherever he travels. The planes are all made in Anglo-America, Western Europe, or the Soviet Union. The service is basically alike. Airports have similar modern terminals. Taxis and buses make connections with city hotels, which tend to provide similar modern service. Even the downtown areas of cities are not all that different. All have modern office buildings, traffic-congested streets, and modern restaurants and shops. Clerks, taxi drivers, and airport attendants often speak the international languages of business and government. These great metropolitan centers, closely tied commercially and in constant political contact, have developed similarly so that they can function efficiently in the modern business and governmental world.

International jet travelers from New York often find it easier and quicker to reach London than to get to Plattsburgh in upstate New York, and surely connections between São Paulo and metropolitan centers in the United States and Europe are far better than those with most parts of their own country. Here, then, is the juxtaposition of modern and traditional life within the same world and even the same country.

MODERN SURFACE TRANSPORT: GOODS AND PEOPLE MOVING WITHIN COUNTRIES

A further measure of the extent of the modern, interconnected system can be taken from the facilities for modern surface transportation over land (see Figures 4-7 and 4-8). Virtually the entire area of Western Europe, the United States except Alaska, and Japan, is within 25 miles of a road. Most of eastern United States, Western Europe, the western part of the Soviet Union, and Japan are covered with dense railroad networks. But even in these industrialized areas, the networks thin out in outlying areas and along the frontiers.

Contrast these areas with other parts of the world. Few parts of Latin America, Africa, and Asia are so closely connected by roads for motor vehicles. And the railroad patterns can scarcely be called networks. Even in densely populated China and India the modern transport system is far from complete.

Within countries in the nonindustrial parts of the world, land transport systems tend to

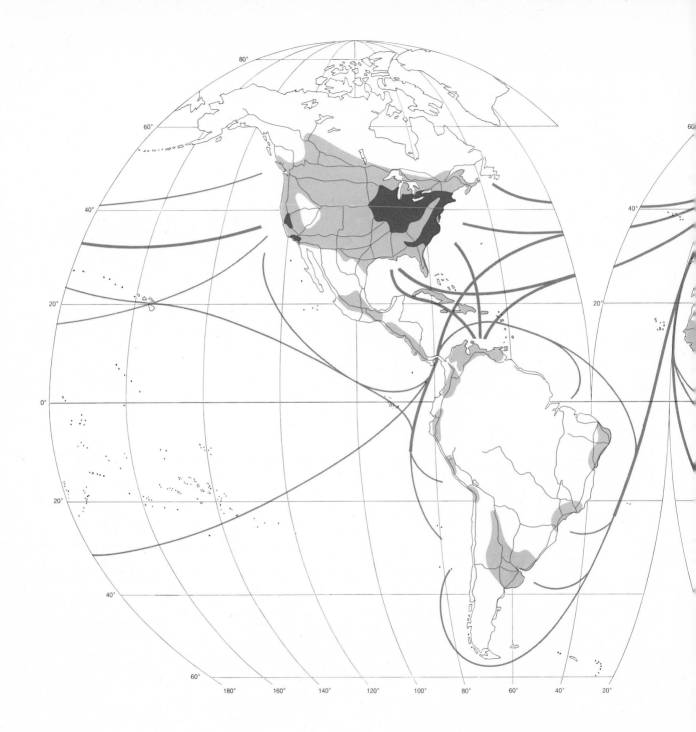

Figure 4-7 WORLD SURFACE TRANSPORTATION:
SHIPPING ROUTES AND ROAD DENSITIES

Source: Pergamon World Atlas

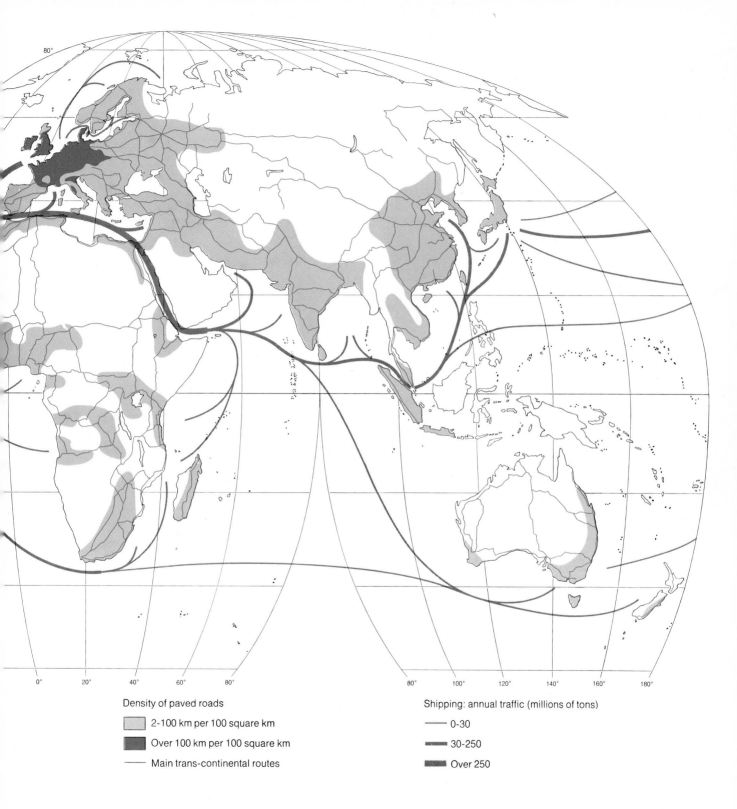

Density of paved roads

2-100 km per 100 square km

Over 100 km per 100 square km

Main trans-continental routes

Shipping: annual traffic (millions of tons)

0-30

30-250

Over 250

Figure 4-8 World Surface Transportation: Railroad Densities

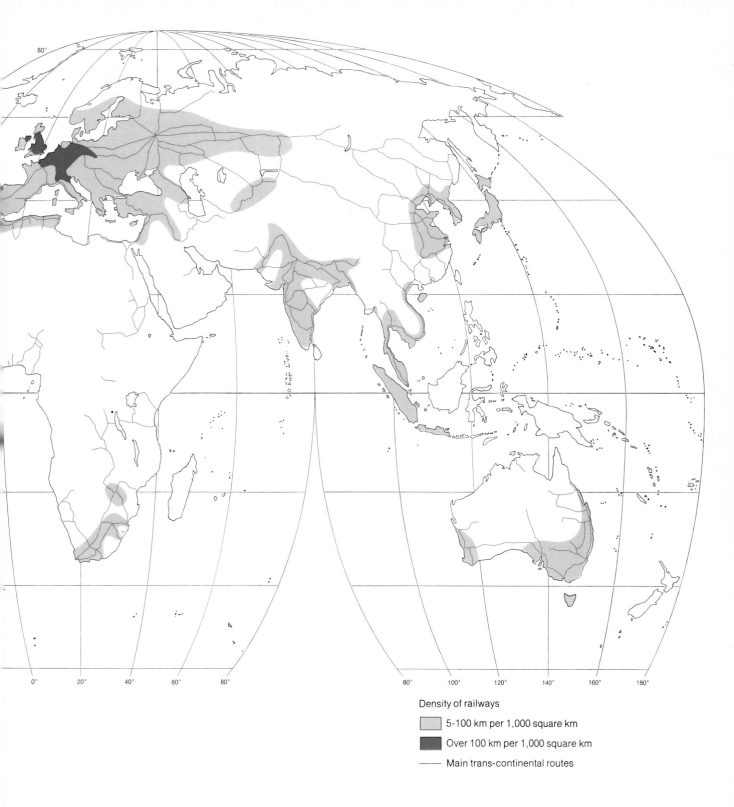

Density of railways

5-100 km per 1,000 square km

Over 100 km per 1,000 square km

——— Main trans-continental routes

develop first, and there are few international connections except for ports. Railroads commonly extend only from the coast inland. It is not hard to understand why such railroads were built. Most of them were designed to connect some mining or plantation area with a seaport where goods could be shipped to Europe or the United States, not to provide service to the country itself.

CULTURAL VARIATIONS WITHIN THE MODEL SYSTEMS

After even the rather cursory examination of human cultural variables in this chapter, the importance of culture should be clear. The two models for human society are modified by variations in their cultural heritage. The United States, Western Europe, the Soviet Union, and Japan may all be examples of the modern, interconnected system, but there are significant differences among them that result in modifications of the system. Each has its own language, history, government, customs, laws, and technology also varies from country to country. Cultural differences are striking in the nonindustrialized world as well. Africa, Latin America, and Asia may have the same general problems in terms of attempting to take their places in the modern world, but many specific cultural facets will also affect their plans and the implementation of them. The distinctive characteristics of the human culture of any place must be examined as part of the total geographic picture.

SELECTED REFERENCES

Barnett, Harold, and Morse, Chandler. *Scarcity and Growth,* Resources for the Future, Johns Hopkins Press, Baltimore, 1963. Chapters 11 and 12 present an argument counter to that of Malthus. The idea is that resources are not lost, only changed in usefulness, hence scarcity is a function of technology's ability to renew their usefulness.

Broek, J. O. M., and Webb, J. *A Geography of Mankind,* John Wiley, New York, 1968. The most popular of introductory, human geography textbooks develops the concepts of culture and race, livelihood, settlement, and population change.

Cipolla, Carlo. *The Economic History of World Population,* Penguin Books, Baltimore, 1962. A historical perspective on the use of energy and world population is the basic purpose of the analysis. The importance of energy control in the development of the world of today is argued, and the ultimate problems of population growth are explored.

Hall, Edward T. *The Silent Language,* Doubleday, Garden City, N.Y., 1959. Also in paper by Fawcett, Greenwich, Conn., 1966. This book explores how a culture establishes behavioral and perceptual patterns that are used by its individual members and are unrecognized by members of another culture.

Jones, Emrys. *Human Geography,* Praeger, New York, 1966. An English geographer provides another perspective of human geography.

Manners, Gerald. *The Geography of Energy,* Aldine, Chicago, 1964. The implications of energy consumption and production examined in greater detail.

Morrill, Richard. *The Spatial Organization of Society,* Wadsworth, Belmont, Calif., 1970. The organization of human settlement seen from the perspective of spatial economics.

Shepard, Paul, and McKinley, D., eds. *The Subversive Science: Essays toward an Ecology of Man,* Houghton Mifflin, Boston, 1969. A collection of essays by persons working in human ecology provides a provocative introduction to population, environment, and ecosystems.

Simoons, Fredrick S. *Eat Not This Flesh: Food Avoidance in the Old World*, U. of Wisconsin Press, Madison, Wisc. 1961. Reviews how food taboos are related to religious and other cultural beliefs.

Sopher, David E. *Geography of Religions*, Foundations of Cultural Geography Series, Prentice-Hall, Englewood Cliffs, N.J., 1967. This book explores the many facets of religious beliefs and associated practices that are integral to the ways humans have organized their lives and that have significance to the character of places. It also includes useful discussion of cultural traits other than religion.

Zelinsky, Wilbur. *A Prologue to Population Geography*, Prentice-Hall, Englewood Cliffs, N.J., 1966. This thin volume introduces not only the idea of population geography but its cultural implications as well.

Chapter 5   The Earth Environment

IN PART ONE we posed the idea that places can be viewed in terms of a spectrum ranging from the traditional, locally based village to the modern, interconnected city. This view implies that life at particular places is a product of human will and that the behavioral patterns of those who live there influence the world around them. The assumption is appropriate: people do dream up and invent the world they live in. Furthermore, creation of this world has been a slow process of individual and group development. Man has evolved human institutions and useful artifacts. By trial and error he has learned how to use the physical environment. Thus, besides the human system, there is another system: the physical environment.

Man in a Physical Environment

Man must sustain his biological and material existence from the raw products and the energy of the physical environment. Survival ultimately rests on a continuous consumption of materials that supply the energy and the chemical components of life. Hence, man clearly cannot live independently of his physical environment.

On the other hand, can the physical environment exist independently of man? In one important sense, it can. If mankind were to disappear from the face of the earth, the physical environment (that is, the system of processing material and energy) would continue. But in another and probably more meaningful sense, the physical environment devoid of human beings is not the same as the environment with them. Man's activities are an integral part of the environment, and in this sense human beings are an integral part of the environment as they mold it.

Changes men make in physical environments consist of piecemeal alterations designed to attain specific ends. Initially each change may seem to modify only the local environment, but with the increasing use of technology, the variety and magnitude of such changes have increased to the point at which modifications of environments go well beyond local limits and become world-wide in scope. A recent example is the localized use of low concentrations of pesticides such as DDT. These concentrations have spread, via the processes of the environment, to all parts of the earth's surface. DDT, like many man-made pesticides, is resistant to breakdown in nature, and once it gets into the food chain, it can become highly concentrated in some animals. Some of these animals live far from the areas where DDT is actually used. Neither the stability nor wide diffusion of these pesticides was anticipated by the people using them, which indicates how imprecise is our understanding of the environment.

Because of our biological inheritance, however, the configuration of the environment must not deviate radically if the species is to endure. DDT contamination of animal food sources, for example, may prove disastrous for people too. This example documents that man's modifications of environments for his own ends cause continuous emergence of new conditions which are often unanticipated and which can become undesirable. It is not inconceivable that the imprecise and compulsive use of new technology may cause states of environment to emerge which are not compatible with the biological needs of human beings. This seems to be the crux of current concern over environmental quality.

During the first decades of this century, geographers showed wide interest in the role of environment in human activity, but this took the form of simple generalizations about human responses to environmental endowments. For the most part the world was envisaged as the stock of resources which largely determined human activity and material wealth. This concept suggested that environmental conditions greatly influenced the material livelihood that men could attain; indeed, there was some belief that the standard of living was determined by the available resources of the environment of the particular place.

This view was perhaps understandable before the impact of rapidly expanding technology. Most people were still living off the land as agriculturalists of one type or another. Management of the environment was almost entirely dependent on human and animal power (biological energy). The scope for altering the environment seemed narrow indeed, and the limitations imposed by local land endowments appeared to dominate.

The notion of environment as a determinant would appear at odds with the advances made by human technology over the past fifty years. The modernized technological system has tremendously increased man's use of power and his potential to exploit the physical environments of the world. Thus the opposite position, often taken by persons awed by technological accomplishments, is that there are practically no environmental constraints. This view asserts that there can be unlimited manipulation and control over environments through technology. The validity of such a position can be seriously questioned in the light of current pollution problems, however, and many knowledgeable people feel that we can no longer be unconcerned about what is happening to our environment. Some, in fact, see this as one of our greatest contemporary problems. Nevertheless, few would again accept the earlier position that the environment dominates man.

Our perspective is that it is more appropriate to see the human-environment relationship as a system with mutual feedback. People have

A good example of the feedback system is the extensive use of chemicals for fertilizers, herbicides, and insecticides in agricultural management. The immediate benefits have been tremendous increases in crop yields in the United States since 1940. Within the past decade, however, the benefits of abundant food and feeds have begun to extract the longer-term costs of changing the quality of surface and ground water, accumulating undesirable chemicals in foods and feeds, and selectively destroying various plants and animals. It is now much clearer that management of the environment for the purpose of increased crop production has had repercussions throughout the environmental system. Many changes were unexpected and some were undesired. If the undesirable ones are not to continue indefinitely, some management practices in agriculture and in all related human systems—ranging from the fertilizer industry to eating habits—will have to be altered. Changing the human system to mitigate the undesirable feedback from the environment could take one of two general forms. Either men could cease to do those things that brought undesirable consequences, or they could make further modifications in the environment. As we attempt to manage the environmental resources of agriculture, change will inevitably occur in the environment, and these changes will sooner or later demand others because the environment will have changed.

both material requirements for their physical existence and other demands on the environmental resource base. They must manipulate their environments in order to survive. As long as people exist, they will exploit environments. In recent times their manipulations have often been in the grand proportions of huge dams, long-distance water diversions, chemical farming, mineral exploitation, and the outpouring of large quantities of combustion wastes by power-generating facilities and automobiles. There have also been the less grand, though often pervasive, alterations of vegetation, soils, and water by traditional societies. All such changes are accompanied by overall changes in the environment and have side effects requiring further adjustments.

In the modernized world, successes in meeting the material wants and desires of people are being offset by the rising problems of waste disposal, pollution of environments, and hazards to health and well-being. These problems arise from our recently acquired wealth. We have overlooked, or been ignorant of, the detrimental consequences to the environment of our increasing application of technology in pursuit of greater material wealth. Moreover, the many immediate successes of newfound technology have made us increasingly accustomed to quick solutions. But this is becoming counterproductive as our short-term gain results in a long-term loss or heavy costs. Thus, the attainment of more acceptable states of the human-environment system will require increasing our understanding of how environments work and anticipating the long-term consequences of human management schemes.

Environments as Physical Systems

Becoming more responsible about the uses of environments requires an understanding of how they work. Any environment is seen as a system of processes by which materials are constantly being transformed into new combinations and moved about. As a consequence, environments are continuously changing. It is within this context that man attempts to exploit and manipulate natural systems to serve his needs and wants. By so doing, of course, he alters the system of processes that compose each environment. And, unless the workings of environments are precisely understood and considered, he may unwittingly create undesirable or even disastrous environmental conditions.

We can view the dynamic environment as a series of machines, each working to create change. Among these machines are biotic life, which produces change as it grows, and the erosion of wind, water, heating, and freezing that breaks up earth solids and transports them to other places. Like all machines environmental processes are powered by energy; in this case the principal source of energy is solar radiation. In summer when biotic activity is high, erosion is usually low, whereas in winter when vegetation lies dormant, much erosion occurs.

Every place on earth receives its own endowment of solar energy which is absorbed and reradiated from the solid earth, the atmosphere, and the water surfaces. One might better grasp this idea by visualizing the earth's surface, or interface, as having an energy economy with inputs, outputs, and some temporary storage. An energy economy provides only a means to estimate the potential of its environment to process materials. Energy alone accomplishes very little; it is intercepted at the earth's surface, and most of it is radiated back into the atmosphere and outer space. The most important phase of the energy exchange takes place at the surface. What is more, the proportion that becomes involved with evaporating water is the same component essential for the functioning of the *hydrologic* cycle. Most of the important

processes carried out in environments, such as plants growing, materials weathering, and soil eroding, require water. Without moisture most of the activity of the environment would cease.

THE SEQUENCE OF WORK IN ENVIRONMENTAL SYSTEMS

The interaction of endowed energy and moisture is the driving force for environments. Knowing how energy and moisture interact gives useful clues about the performance of such things as the growth of plants, chemical activity, and the likelihood of earth materials' being moved. There are varying intensities and particular sequences within the annual cycle of environmental activity, however. It is not a steady condition but a series of processes that vary in kind and intensity. Because of these fluctuations, an environment gains its identity from the repetitive rhythms that change throughout the year. Therefore, the variety of environments can be seen as a matter of different intensities and sequences of work performed.

Thus, an analysis of the sequencing of a particular environment may be helpful in studying energy, moisture, and other factors that interact. While, in theory, every place can be thought of as having its own unique environmental system, in fact we can identify certain generalized conditions over wide areas, at least in terms of how the principal components of energy and moisture function. To illustrate, we will now turn to the moist mid-latitudes and examine the changing environmental processes throughout a year's sequence.

AN ANALYSIS OF THE MOIST MID-LATITUDE ENVIRONMENTS

The idea of sequence can be understood by analyzing the environmental rhythm common

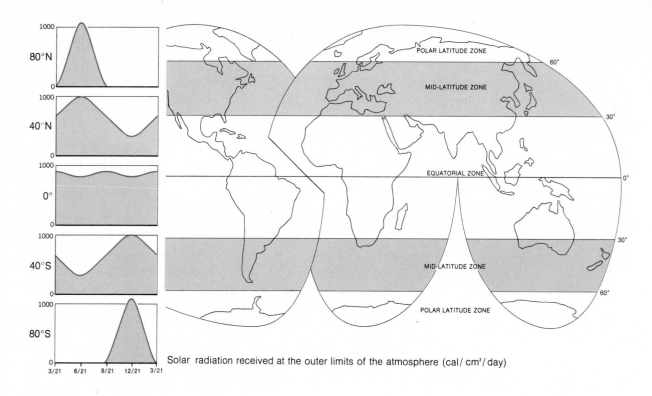

Solar radiation received at the outer limits of the atmosphere (cal/cm²/day)

Figure 5-1 LOCATION OF THE MID-LATITUDE ZONES OF THE WORLD. *Also shown are schematic energy-distribution curves for all five zones.*

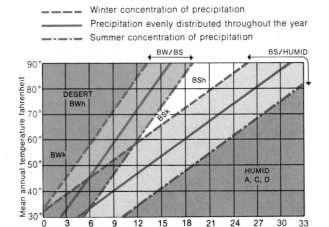

LIMITS OF THE REGIONS OF DRY CLIMATE

- - - - Winter concentration of precipitation
───── Precipitation evenly distributed throughout the year
-·-·- Summer concentration of precipitation

Figure 5-2 A Highly Schematic Summary of Temperature and Moisture Conditions for the Humid Mid-latitudes. *See accompanying text for a discussion. As you study the graph, remember that while it is not represented directly here the mid-latitudes are to a large extent characterized by an overlap of the conditions found in the zones adjacent to them—the arctic in winter, or the antarctic and the tropical in summer.*

to much of the mid-latitudes. In reading through this analysis, we should concentrate on how changes between variable energy endowment and moisture supply affect the processes of environmental work.

The choice of the mid-latitudes for analysis is arbitrary, though relevant. Much of the world's land area and a large share of its population are located there. These areas also experience annual cycles in their energy endowments and provide an opportunity to diagnose the effects of variable energy inputs. Furthermore, be restricting the analysis to those mid-latitude areas where there is a year-round supply of moisture, the significance of adequacy and inadequacy of moisture can be clarified.

For this analysis the mid-latitudes can be arbitrarily defined as 30° to 60° north or south latitude (Figure 5-1). Across these 30° of latitude, annual energy endowments decline as latitude increases. Energy-distribution curves for the year remain essentially similar, however, with one high and one low period. Moisture supply for places within this broad zone can vary in amount, timing, or both. Annual amounts range from a few to over 150 centimeters. There may be a year-round distribution or a concentration of rain in one season.

Figure 5-2 summarizes the basic energy (temperature) and moisture conditions for the humid mid-latitudes. Similar types of data on energy and moisture also exist for most parts of the world. The analysis that follows is built around energy and moisture information shown in the graph. It can also be considered a rough model for analyzing environments of other places from appropriate data on their energy and moisture endowments.

Winter

A convenient time to begin an analysis is at the low-energy season of the year, which is at the left of the graph. This would be December or January in the northern hemisphere, or June or July in the southern hemisphere. With the energy income (heat from the sun) at its lowest level, it can be deduced that all processes directly dependent upon energy conditions would also be at their lowest levels of activity. This implies that material transformation or chemical processing would be slowest at this season. This is largely true of the weathering of materials but is especially true of the decay of organic matter. Vegetative production in particular and biotic activity in general would be at their lowest levels because most living matter is in a dormant state. Evaporation of water and transpiration by plants would be at minimal levels, and when the temperature drops below freezing, they essentially cease. From this evidence it is apparent that the demand for moisture by evapo-transpiration processes is at its lowest level.

These conditions give additional clues to the performances of other aspects of environmental work. For example, what happens to the income of moisture if the withdrawal by evapotranspiration is small? Under these circumstances might not surface run-off and soil-leaching increase?

The surplus of moisture beyond evapotranspiration demand would indicate that winter is when moisture is available for soil storage. But this is also the time when the capacity of the soil for storage has been reached. Once the soil is saturated, additional incoming moisture must either move deeper through the soil into the earth or else run off over the surface. Water moving downward through the soil works on that soil, transporting the smaller soil particles physically or reducing them by chemical action. In the same way, water running over the soil erodes it physically. Translocational processes can be rather active from the moment of soil saturation. Nevertheless, midwinter can also be a period of freeze for many mid-latitude places, a condition that deters leaching and surface erosion. Winter is clearly a time of low overall environmental work.

Spring

With the approach of the spring season, energy endowment increases and so normally does precipitation. Increased precipitation is a function of warmer air with its greater potential capacity for water-vapor storage—the source of all precipitation. Increases in energy levels eventually thaw frozen land and release moisture stored as snow or ice. Early spring is a time of maximum moisture available for leaching and run-off; it is the season of maximum flood hazard. This is the time of greatest erosion from the surface and removal by solution. Ground exposure is at a maximum; vegetation is still dormant and without leaves; and erosion from rainfall on the soil surface increases. Moreover, the absence of vegetative cover permits a large proportion of solar radiation to reach the ground surface to warm and dry it. As a result of the drying of organic litter, spring is normally the period of greatest fire hazard in the mid-latitude.

A major change in environmental activity begins when energy levels are sufficient to stimulate renewed vegetative growth. Energy endowments by midspring increase and accumulate to the point of being nearly double the winter lows. Evaporation and transpiration losses become significant so that some withdrawal of soil moisture from storage may occur. A part of the moisture income is lost by evapotranspiration processes, and less is available for run-off and leaching. The increase in soil temperatures accelerates chemical activity which renews the vigor of mineral weathering and organic decay. Increased chemical activity can also be seen as a necessary accompaniment to increased biotic activity, which is dramatically displayed in the leafing-out of plants. By intercepting much of the solar energy and a significant portion of the moisture, the vegetative layer becomes the interface and the ground surface is insulated from the direct effects of these active factors.

As energy levels continue to rise, plant transpiration increases the water demand on the moisture stored in the soil. The physiological demand for water from growing plants increases as photosynthesis increases. With this demand on soil moisture, some of the new moisture supply is entrapped in the vacated soil-moisture storage, leaving less as surplus for run-off. By late spring or early summer the volume of moisture withdrawn from the soil may approach the volume received from precipitation. This may be as much as 12 to 15 centimeters of water per month. It is probable that moisture receipt at this time of year will scarcely be in excess of what can go into soil-moisture storage, and moisture available for leaching and stream flow will approach zero. Under these circumstances there is little likelihood of erosion.

Summer

The steady progression toward higher-energy conditions and biotic activity of spring culminates in summer. This is a time of active mineral-weathering and organic decay—the *transformational* process. It is also a time of decreased erosion, ground-water recharge, stream flow, flood hazards, and leaching of solubles—the *translocational* processes. Such conditions prevail through most of the summer. In fact, during late summer the potential demand for

moisture by evapo-transpiration normally exceeds moisture income. Hence, summer is often marked by frequent droughts, dry streams, low ground-water tables, and general water shortages, even though the actual amount of precipitation received at the interface may be at or near its peak. Moisture deficits in the soil arrest plant growth and, if they prevail for long, may cause some plants to wilt and die.

Fall

By late summer and early fall the moisture stored in the soil is at minimum levels, and the opportunity to take into storage any income of moisture is at a maximum. During the fall season energy endowment drops from the midsummer maximum, and the demand for moisture decreases. These conditions make the fall season a period when soil-moisture recharge will occur because of the declining energy demand for moisture by evapo-transpiration. Demand drops below moisture income, and soil-moisture storage is gradually replenished. For a month or two the availability of storage opportunities prompts the pre-emption of the moisture supply for storage, leaving little if any surplus for run-off and ground-water recharge. Thus, moisture-powered processes such as leaching and erosion are minimal. And, the lowest stream flow is likely to be recorded as a result of combined effects of depleted ground-water stocks and soil-moisture recharge.

With the return of minimal levels of energy endowment in winter, there is a cessation of most plant growth, and evapo-transpiration demand drops to its lowest level. Further, moisture losses through these processes are at a minimum. Soil moisture is usually completely restocked by this time, and nearly all income of moisture is surplus that must go either to ground water or surface run-off.

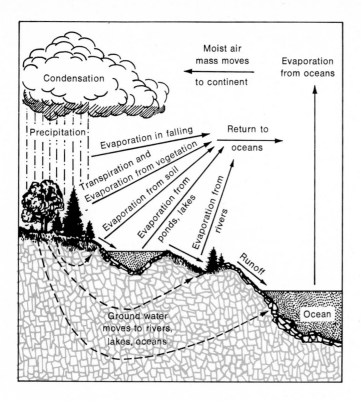

Figure 5-3 of the Hydrologic Cycle

Figure 5-3 A Generalized Representation of the Hydrologic Cycle

A General Scheme for Evaluating the Work of Environments

The description of the sequence of activity for the mid-latitude environments offers a basic approach for evaluating any environment. The evaluation rests upon determining the way moisture received at the interface is allocated, since this is also an allocation of energy potential. Where the moisture goes is an indication of which of the various processing systems will be most active. What happens to moisture at the interface can give a reasonably good estimate of the functioning of the several major aspects of environmental work.

Two general allocations for moisture arriving at the interface can be established. First, there is the moisture that infiltrates the soil and is retained there by capillary attraction. Second, there is water that is not retained by the soil because it either passes through the soil or runs off over the surface.

The first allocation is a storage of moisture more or less at the point of reception. It provides the moisture available for *evaporation losses* from the soil surface and *transpiration losses* through plants. These losses are in response to energy demands of the atmosphere for water at the interface.

The second allocation is a residual that results after the demands of evapo-transpiration are satisfied.[1] It leaves the point of reception under

---

[1] The assumption has to be made that all water infiltrates the soil so that capillary spaces (soil-moisture storage) are satisfied before water can run off or move through the soil. The logic of this assumption is acceptable in reality, as long as the intensity of moisture reception does not exceed the infiltration rate. However, in actuality infiltration rates are sometimes exceeded, especially during intense rainfalls or where soils have very low infiltration rates. Thus, this general assumption can introduce some error in reality under certain circumstances, but a more accurate alternative has yet to be developed.

the pull of gravity and is available for leaching, addition to ground-water stocks, or erosion.

These two primary allocations, one to evapo-transpiration and the other to water that must leave the point of reception, describe the basic character of hydrologic cycles of land surfaces (Figure 5-3). Evapo-transpiration forces moisture back into the atmosphere. Any moisture in excess of this loss eventually must leave the land and return to the ocean reservoirs. This latter allocation is *surplus* in the sense that it is more than is needed at the point of reception. Comparisons of actual run-off from continents with estimates of surpluses indicate that the two are in close accord, although there are obviously small losses as the moisture makes its way back to the oceans.

The two allocations are the basis for estimating the work of environments. Knowing the amount of moisture used in evapo-transpiration provides an approximation of the energy making its way through the biosystem at any given time. It is indicative of the activity of that system and also of the various chemical transactions involved with bioactivity. The amount of surplus, on the other hand, gives an estimate of the potential for translocating materials.

A simple budgeting procedure can be used to make the allocations once the evapo-transpiration demand has been determined. Such a budget is generally referred to as a *water budget,* but it is also a simple means for estimating the energy available to the environment. This technique requires a determination of the evapo-transpiration demand. This involves converting measured energy (usually expressed as temperature) into an amount of water equivalent to the amount of water it will evaporate or cause plants to transpire. The higher the energy level, the greater the *potential* demand for evapo-transpiration. Hence, the energy endowment of a place can be converted to a statement of *potential evapo-transpiration* (PE) demand.

ESTIMATES OF WATER BUDGETS FOR WORLD LAND AREAS

Basic data for evaluating environmental activity on the land areas of the world are presented in Figures 5-4, 5-5, and 5-6. These data represent one set of estimates. Other estimates differ somewhat in values because of different choices in the selection of observed data, the way the data were generalized, the method of calculating potential evapo-transpiration, and the assumed values of soil-moisture storage. The figures used here are probably as reasonable as any. Moreover, the various estimates yield comparable pictures of the basic characteristics of the water budgets of continents. Thus the absolute values are not to be emphasized, but rather the comparative values between continents and the broad patterns of PE deficit and surplus.

From Table 5-1 we can see that actual evapo-transpiration and deficit add up to the potential evapo-transpiration for each continent. Similarly, the total of surplus and actual evapo-transpiration equals precipitation. It may seem inconsistent that there can be both a deficit and a surplus for many parts of the world. However, if the annual environment is visualized as a sequence of seasons, then one season may have a deficit and another a surplus, because the deficit in one season cannot be erased by a later surplus and the surplus of one cannot be made available at a later deficit season.

Surpluses, however, offer man water to manage for use during periods of drought. Yet the figures in Table 5-1 make it apparent that, except for South America, deficits exceed surpluses, often by a wide margin. Thus, most continents lack sufficient precipitation to supply the demand of PE even with the most elaborate human management. Actual plant growth (AE) falls well below the potential (PE) of sun radiation.

A comparison of potential and actual evapo-

Table 5-1 ANNUAL WATER BALANCE OF THE CONTINENTS (in centimeters)

P = A + D

	RAINFALL R =	POTENTIAL EVAPO- TRANSPIRATION	ACTUAL EVAPO- TRANSPIRATION A +	DEFICIT	SURPLUS S
Africa	74	188	65	123	9
Asia	54	104	43	61	11
Australia	49	167	41	126	8
Europe	60	70	47	23	13
North and Central					
America	70	120	63	57	7
United States	76	134	71	63	5
South America	151	146	111	35	40
All land	75	133	61	72	14

SOURCE: J. Chang and G. Okimoto, "Global Water Balance According to the Penman Approach," *Geographical Analysis*, Vol. II, No. 1 (Jan. 1970), pp. 55–67.

transpiration in Africa and Australia shows the extent to which the potential is not attainable. Both of these continents are within the latitudes of high solar insolation year-round, hence high PE values; but because of extensive areas with low precipitation, either seasonal or year-round, they experience large deficits and have extensive areas of desert and semiarid environments.

In contrast, the potential in South America is nearly the same, but the moisture supply is much greater. Deficits are small, and the surplus is the largest of any continent. Nevertheless, the maps show that even in this continent extensive areas experience deficits as well as surpluses. Most surplus, in fact, is restricted to the Amazon basin.

In the Northern Hemisphere with its large land masses in the middle and high latitudes, little area escapes a deficit season. Areas with a surplus are also limited. The rather favorable circumstances of the eastern United States and western Europe in this regard are worth noting.

The areas with deficits of 100 centimeters or more within latitudes 40° north and south approximate environments of severe and extended dry seasons. Few of these areas have periods of even one month's duration with a surplus (see Figure 5-6). North of approximately 40°, areas of either substantial deficit or surplus do not exist except along western coasts. During what season could we expect deficits to occur in these higher latitudes? Does the time of deficit present a potential hazard to crop production?

Perhaps the most striking thing about the map of water surpluses (Figure 5-6) is the limited areas where they occur in substantial amounts. Note that the greatest concentrations of surplus occur in the tropics with their high potential for evapo-transpiration. This suggests that precipitation must be very heavy during the wet season to offset heavy evapo-transpiration losses. Even so, most of these areas, with the exception of a few areas right on the equator, also experience deficits.

Environments and Human Management

The different work budgets of the environment at particular places present different challenges to people. In places where the PE is great and there is adequate moisture available in the soil, plant growth is also great. Seasons of low PE mean that the problems of winter, whether it is wet or dry, present other problems. Drought inhibits plant growth, and periods of surplus bring the possibilities of flood and heavy erosion. Thus, human life in any place involves adjusting to environmental conditions and to distinctive problems in the management of environmental resources.

For our purposes the great variety of environmental activities over the earth can be reduced to three basic economies:

1. *Very limited environmental work.* These are the environments where either energy, moisture, or both are low throughout the year. The low-work environment is characterized by low levels of biotic and chemical work and by little translocational movement as well. Neither production nor decay can be very active. Figure 5-4 shows that a substantial portion of the land surface of the world has environments in this general category.

2. *Continuously high environmental work.* At the opposite extreme from the low-work environments are environments with high-energy and high-moisture endowments throughout the year. The hot, humid tropics, for example, are processing materials at consistently high rates throughout the year. These areas are essentially along the equator in portions of the Congo basin of Africa and the Amazon basin of South America, and on the islands and peninsulas of Southeast Asia.

3. *Seasonal environmental work.* The environments in this category have well-defined seasonal periodicities in their budgets of energy and moisture. The water budget of Chicago and the humid mid-latitude climate described previously point out characteristics of a seasonal environment. There are areas in the tropics and subtropics where environmental processing is also primarily seasonal, but they are regulated more by the timing of moisture supply than energy changes.

Let's examine each of these environmental economies in more detail.

Low-work Environments and Human Management

The low-work environments might be called the storage environments of the world because changes happen very slowly. No wonder that places with these conditions have great resources for the archaeologist or that World War II aircraft stored in the Arizona desert could quickly be reactivated during the Korean conflict because there was little deterioration. These environments are surely not the places to dispose of most things. However, deserts might be reasonable places to dispose of nuclear wastes, certainly better than the oceans where circulation of contaminants is a constant hazard.

The low activity of these environments also implies that they have limited capacity to recover quickly from disturbances. Plant growth will not re-cover an exposed area, and materials are not readily transported from one place to another. Roads across the fragile alpine and arctic tundra and wagon trails across deserts leave their marks long after their last use. Undoubtedly the recent exploration of the north slope of Alaska for oil will remain an indelible artifact in that environment for generations to come, even if exploitation of the oil never takes place, because the fragile surface has been cut by truck and tractor routes, and patterns of vegetation have been disrupted.

Figure 5-4 ESTIMATED EVAPO-TRANSPIRATION
RATES FOR AREAS OF THE WORLD

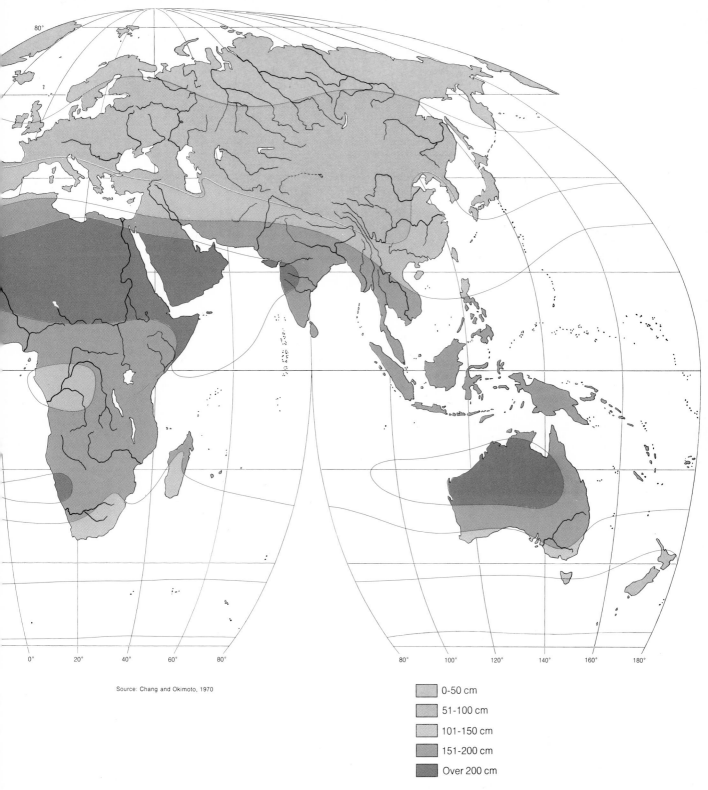

Source: Chang and Okimoto, 1970

0-50 cm

51-100 cm

101-150 cm

151-200 cm

Over 200 cm

111

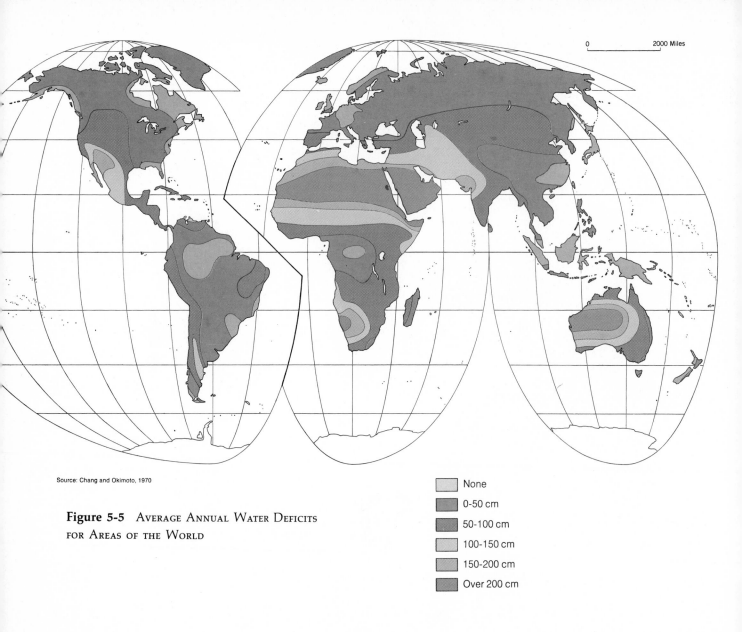

Figure 5-5 Average Annual Water Deficits for Areas of the World

None
0-50 cm
50-100 cm
100-150 cm
150-200 cm
Over 200 cm

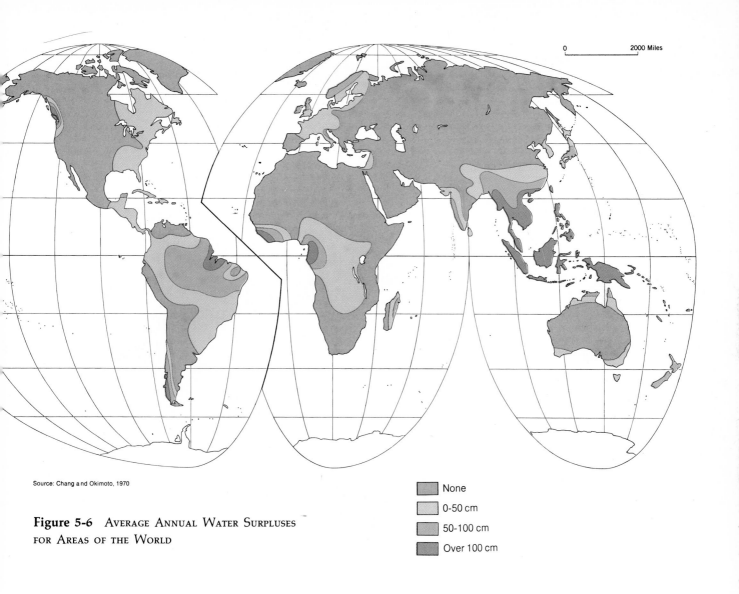

Figure 5-6 Average Annual Water Surpluses for Areas of the World

None

0-50 cm

50-100 cm

Over 100 cm

0   2000 Miles

A water budget involves three basic values: (1) potential evapo-transpiration[2] (2) precipitation, and (3) soil-moisture storage capacity.[3] Potential evapo-transpiration and precipitation are directly comparable since both are expressed as units of water. Soil-moisture storage is expressed in the same units and enters into the accounting as either a carry-over of moisture or a sink for excess moisture, depending on whether or not the soil is saturated. As an example, consider the water budget for Chicago (see Table 5-2 and Figure 5-7). Note that the budget provides values for actual evapo-transpiration, surplus and deficit. The meaning of these derived values can be briefly summarized.

Actual evapo-transpiration (AE) is the amount of water actually removed by evapo-transpiration. AE is not always as great as potential evapo-transpiration (PE) because at times precipitation (P) and soil-moisture storage (SMS) are insufficient to satisfy the demands of potential energy. The difference between AE and PE is called the deficit (D).

Studies have shown that AE relates positively to overall plant activity and correlates very well with productivity of vegetation. The generalization can be broadened to imply that AE is indicative of overall activity of the bio system—both productivity and decay. The implication is that AE seems to provide a means to estimate the actual intensity of biological activity in an environment.

Surplus (S) is the amount of water in excess of that needed to satisfy PE and SMS capacity. It cannot be retained at the place of reception and is removed by the pull of gravity. The occurrence of surplus has a logical relationship to the functioning of the processes involving water moving in response to hydraulic gradients, either across the surface or downward into the pore spaces of the subsurface. Its work in the environmental system is therefore that of translocating earth materials either as particles or in solution.

Deficit (D) is the amount of water short of fulfilling the requirements of PE (PE − AE = D). It occurs when the combination of precipitation and moisture in the soil storage does not equal PE.

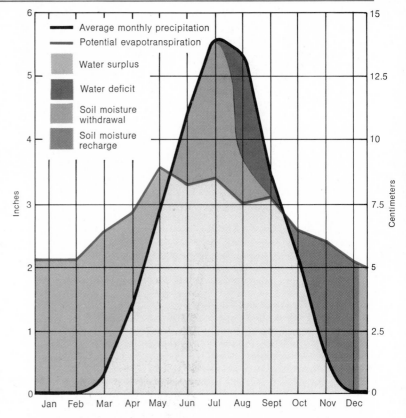

Figure 5-7 A Water Budget for Chicago, Illinois. *A graph of the data in Table 5-2.*

Deficits are generally indicative of constraints on biological growth and chemical activity. The decrease in productivity is more or less in proportion to the length and accumulation of deficits.

[2]Monthly and annual values are available for most places with some continuity of climatic record. For a rather complete source of PE values, see the series of articles by Thornthwaite and Mather listed in the references for this chapter on p. 132.

[3]Soil-moisture storage capacity varies from soil to soil. It is possible to obtain specific estimates from local soils, but it has also been found that using a general value of 10 centimeters (4 inches) is a useful approximation for large areas.

Table 5-2 A WATER BUDGET FOR CHICAGO, ILLINOIS

	J	F	M	A	M	J	J	A	S	O	N	D
Temperature												
Centigrade	−4	−3	2	8	14	20	23	22	19	12	5	−1
Fahrenheit	25	27	36	48	57	68	73	72	66	54	41	30
PE												
Inches	0	0	0.3	1.4	3.0	4.5	5.6	5.3	3.4	2.1	0.6	0
Centimeters	0	0	0.8	3.6	7.6	11.4	14.2	13.5	8.6	5.3	1.5	0
P												
Inches	2.1	2.1	2.6	2.9	3.6	3.3	3.4	3.0	3.1	2.6	2.4	2.1
Centimeters	5.3	5.3	6.5	7.4	9.1	8.4	8.6	7.6	7.9	6.5	6.1	5.3
AE												
Inches	0	0	0.3	1.4	3.0	4.5	5.6	3.6	3.1	2.1	0.6	0
Centimeters	0	0	0.8	3.6	7.6	11.4	14.2	9.1	7.9	5.3	1.5	0
D												
Inches	0	0	0	0	0	0	0	1.7	0.3	0	0	0
Centimeters	0	0	0	0	0	0	0	4.3	0.8	0	0	0
S												
Inches	2.1	2.1	2.3	1.5	0.6	0	0	0	0	0	0	0.4
Centimeters	5.3	5.3	5.7	3.8	1.5	0	0	0	0	0	0	1.0
SMS*												
Inches	4.0	4.0	4.0	4.0	4.0	2.8	0.6	0	0	0.5	2.3	4.0
Centimeters	10.0	10.0	10.0	10.0	10.0	7.0	1.4	0	0	1.2	5.6	10.0

*The storage values used a conversion of 2.5 centimeters = 1 inch in order to keep the assumed soil-storage values in round figures of 10 centimeters and 4 inches. The calculated storage, however, is consistent with the budgetary accounting of the table in which the conversion was based upon 2.54 centimeters = 1 inch. Hence, some inconsistency does occur between centimeters and inches for the months of June, July, October, and November.

KEY: PE = potential evapo-transpiration. P = precipitation. AE = actual evapo-transpiration. D = deficit. S = surplus. SMS = soil moisture storage.

It is obvious that such environments could never sustain large populations if the people had to depend on local resources. Bringing in outside resources or reorganizing available ones can, however, provide opportunities for larger numbers of people to thrive in these locations.

Irrigation represents one such reorganization of resources. It depends on using water from natural streams or on man-made schemes for relocating water. By adding water to the soil, the generally high potential energy of the environment is put to use by increasing evapo-transpiration. Greater biological growth is usually the goal. Nevertheless, irrigation management is not a simple matter of putting water into the soil. The chemical reactions activated by the added moisture set in motion a sequence of changes in the soil system. Weathering increases and alters the chemical and physical character of the soil. The accumulation of alkaline ions can harm crops under cultivation and require additional manipulation of the soil. A common practice is to add more water than needed for evapo-transpiration so that the accumulating products of weathering and solubles leach away. Often, however, this practice causes water tables to rise and induces further undesirable consequences. Because of these problems, irrigation of the desert has generally taken place on alluvial soils transported into the arid regions from more humid environments. Thus, flood plains, deltas, and alluvial fans in dry areas are the most suitable sites for agriculture. Their soils are more stable than desert soils under the moist conditions of irrigation and are the easiest lands to manage.

In spite of the problems of irrigation, current successes in some dry and cold environments seem to suggest that these environments offer opportunities for the settling of urbanized populations. An example is the spectacular growth and development in the dry southwestern part of the United States. There also have been rather vigorous efforts to develop the resources of low-energy environments of arctic Alaska, Canada, Europe, and the Soviet Union. Development of them by modernized spatial organization is largely a matter of creating facilities for importing most of the things needed to support human existence. Thus, just as a large metropolitan area can exist by drawing essential food and material resources from other producing areas, so communities can be supported in the polar areas. The United States, for example, has maintained a major air base and supporting personnel on the edge of the ice cap in Greenland since World War II.

What problems can we foresee from the large-scale settlement in these low-work environments? To make intelligent estimates we need to know more about the rates at which various wastes decay, for example. The strong seasonal fluctuations of environmental activity are known to be significant in building design. They also create specific problems for road and runway maintenance in the Arctic. The controversy associated with the construction of the oil pipeline in Alaska epitomizes the uncertainties of our knowledge and use of this environment.

High-work Environments and Human Management

The lack of biological resources is an obvious constraint to human settlement in cold and dry environments. It would seem reasonable to assume that the abundance of biological resources in environments with high-energy and high-moisture endowments (humid, tropical areas) would be an attraction for human settlement. In most such environments, however, population densities are low or even sparse. Does this mean that mankind has overlooked an obvious opportunity, or do these environments pose major problems for sustaining human population?

If one looks at the distribution of population in tropical areas (see Figure 6-4), it is apparent that the higher densities are largely limited to areas with seasonal, not year-round, moisture incomes. Where moisture income equals or exceeds potential evapo-transpiration for all or nearly all months of the year, population densities are low. The Amazon and Congo basins are outstanding examples of this. The biomass in these areas seems large, and the turnover of biomaterial has been found to be high. Yet successful use by human beings of this seemingly rich environment has generally not been realized. Grandiose efforts, even backed with modern technology, have often resulted in equally grand failures. Is there any logical reason for the problems in these high-work environments?

Some researchers have laid most of the blame on poor soils. The high-work performance of the environment has caused very active chemical decay of the surface material, and the surpluses of moisture have assured that most of the released solubles are leached away. Thus, much of the soil in these environmental areas has had its potential and actual stock of nutrients for plants removed. How, then, can the abundance of biomass exist? The answer is in the continuous recycling of those nutrients tied up in standing and decaying vegetation. Essentially, the vegetation lives off its own debris. Dead bioresidue is rapidly broken down into basic constituent minerals under high-energy and high-moisture conditions, but as rapidly as these constituents are released, the living biomass takes them up again. As a result, the standing vegetation is the source of nutrients. The highly weathered soils play a minor role in supplying plant nutrients under these circumstances.

In this case biomass production has come to depend almost entirely on its own standing stock from which it draws stored nutrients needed for growth. Removing the existing vegetation removes the source and the means of nutrient storages in the soil-plant system. The existing stock of nutrients is released by decay and removed by leaching so that production drops. Within one to three years following the clearing of a forest for farming, plant nutrients in the soil are exhausted.

One might then conclude that the way to overcome this condition would be to add nutrients obtained from other sources, such as fertilizers. Although technically possible, this procedure poses special management problems because the soils have little capacity for retaining the nutrients. Highly weathered soils are unable to hold the readily ionized form of mineral nutrients needed by plants and provided by fertilizer. Leaching removes fertilizer nearly as fast as it is added! By comparison, less weathered soils, which are much more common in the mid-latitudes and where weathering is less severe, have much greater capacities for retaining nutrient ions and reducing their losses from leaching. Under these conditions a logical design for fertilizing in the tropics would be to add a little fertilizer every few days, instead of larger amounts once a season or every few years. But the cumbersome and costly logistics of frequent applications of fertilizer is rather obvious.

Another possible reason for the seemingly poor opportunities for human existence in the humid tropics is that decay and production are continuously occurring at about equal rates. There is no dormant period following the production phase in the biocycle. In order to obtain production men must quickly gather the constant flow of products before the competition from other organisms and chemical decay consumes it. Moreover, the management of the preferred plants is complicated by the constant presence of predators which may attack crops before they mature. Finally, the intense and rapid activity of most processes means that human management often cannot respond quickly enough to mitigate the harmful effects.

ENVIRONMENTS OF SEASONAL WORK AND HUMAN MANAGEMENT

By comparison with year-round, high-work environments, management of mid-latitude environments has proved easier for agriculture. In environments with seasonal sequences there is a growth phase in spring and summer, a dormant or storage phase in winter, and a decay phase following the next spring and summer. Little can happen during the winter storage phase which is why decay is delayed to the next year. Such a sequence also means that products of the growing season mature more or less simultaneously, usually in midsummer or by late summer or early fall before the dormant season sets in.

The convenience of such a timetable is that plant production becomes available at a time when competition from other organisms or chemical decay is minimized. Most grains, vegetables, and many fruits are easily kept over the winter even without elaborate technology. Meat too can be stored, although not with equal success. So man gets to consume much of this production without competition from decay during the winter. This gives him first access to the energy stored in the production phase. Waste of consumption is then recycled by the environment during the following growing season when decay processes are again active.

Further, the sequence in environmental work makes management easier. Man can adapt his activities to those of the season; there is a time to plant, a time to cultivate and protect crops, a time to harvest, and a time to rest and get renewed for the next cycle. In humid environments, although productivity may be equal to or greater than that in the humid mid-latitudes, there is no such sequencing. All of these activities occur simultaneously. Thus, the sequencing of the work of the environment may be one of the reasons for the success of populations in seasonal environments.

That the sequencing environments offer human beings an easier setting in which to attain increased biological output is only suggested, not confirmed. However, it is intriguing to consider that the timing and the level of work of environments may be fundamental to the potential for certain forms of human management, particularly agriculture. It also suggests that organizations promoting successful resource management which have functioned under one environmental economy may be quite inappropriate for other environmental economies.

The earth environment is one of the two fundamental components of geography. The combination of earth resources, water, and radiation from the sun distinguishes one place from another. Some earth environments are highly productive; others have been made productive by technology. Man finds different opportunities and limitations in the environment around him.

124

Human Spatial Organization and Environmental Work

Chapter 6 will explore explicitly the spatial association of world population and certain environmental factors that might reasonably be expected to influence the distribution of people. Recognizing that certain characteristics of environments can influence the population in an area, we expect, for example, that extremely dry or predominantly cold areas will have low densities. And they do. Such relationships come about because some environments offer less potential than others. Historically, human existence has been closely related to the management of biological resources. In this section we shall consider the two examples, developed in Part I, of the modern, interconnected system and the traditional, locally based society, as they relate to the environmental processes analyzed earlier in this chapter. We need to consider to what extent environment is an integral part of human spatial organization.

In traditional societies, like Ramkheri, technology is largely confined to exploiting and modifying the existing biological endowment. Modern societies, on the other hand, go beyond the bounds of just their biological endowments to involve other chemical and physical endowments. They also extend beyond local areas, reaching out for material resources wherever they may be found. That is why modern societies have the capability of producing so much more material wealth than traditional societies, but it is also why they exploit so much of the world's resources.

Modernized, Metropolitan Spatial Organizations and Environments

Perhaps the most obvious and outstanding attribute of the modern world is the concentration of people in metropolitan centers. Viewed in the context of the work potential of the local environment, how can so many people sustain themselves in so little space? Obviously, they obtain little of their food or other material needs locally.

This seeming detachment from the local environment is made possible in the modernized spatial organization by transportation links and trade arrangements with other producing areas. These facilities make it possible to modify the seasonal rhythms of food supply for the consumer in a modernized organization, for example. Winter in one place is the same as summer in another, and some environments, like the wet tropics, are continuous producers of biological material. Hence, by tapping a variety of places with growing seasons at different times, a supply of fresh produce can be obtained year-round.

Similarly, as long as other people with such products are willing to sell, those in the modern, interconnected system are immune from local failures and food shortages in general. This immunity is reinforced by the technology of food storage which makes it possible to overproduce in good years and to keep a supply available at other times. Thus, the food habits of modernized societies are no longer as seasonally adjusted as those of people who depend on their local environment for foodstuffs.

In addition to breaking out of the constraints of local environment, the modernized system has developed the capability of tapping a wide spectrum of the earth's physical and chemical systems. Technological developments have opened up means for utilizing an infinite variety of physical and chemical processes. Metals and minerals derive from these sources and so does the energy obtained from nuclear reactions. These resources contribute to the creation of the manufactured goods and machines demanded by consumers and needed to support the modern system. In addition, they permit man to manage agricultural productivity better. For example, phosphorus, calcium, and nitrogen can be added as fertilizers to the soil. All told, such developments seem to give modernized societies unlimited prospects for exploiting the physical environment.

It is obvious that highly agglomerated populations depend on efficient transportation and communication because modern cities cannot depend only on the local environments for their resources. But can this generalization be extended to suggest that modernized spatial organization makes people independent of their local environments? The answer may become clear after we explore the effect of the local environment on human beings.

Local Environments as Disposal Systems

How does a city get rid of its smoke, noxious gases, excess heat, and the like? Most cities depend on the atmosphere to dissipate these wastes. Because of the compactness of cities, however, there is a limit to the amount of wastes that can be so dissipated without changing the local characteristics of the atmosphere. Eventually, if the limit is exceeded, conditions can threaten healthful existence.

Sewage disposal is another problem for congested areas. Effluent is often dumped into streams with the hope that it will be carried away. Although streams can provide low-cost sewage systems, the real cost is passed on to other places downstream in the form of purification costs and a loss of aesthetics.

The local environment, even in congested areas, is also important for the decay of solid wastes. Biowastes, in particular, decompose rather readily, and what remains can be easily washed away or become part of the soil. But there is a limit to the amount of biomaterial the local environment can process. Furthermore, most of the solid wastes of cities are made of materials much more resistant to decay. The rate of decay is usually slower than the rate of accumulation, and it is no surprise that cities are being smothered by their wastes.

These examples point to a basic principle: Concentration of population at a location means a concentration there of wastes. Production of any material object is only a temporary stage which must give way to deterioration and decay, similar to the decay of plants.

Modernized societies have developed enormous capabilities for producing and consuming material wealth, especially in the metropolitan centers. Our material affluence testifies to the efficiency of the interconnected system. The essentially unidirectional flow of raw materials to the places of human concentration, however, imposes new conditions on the local environment. Resource areas have been altered by mining, lumbering, erosion, alteration of vegetation, and other changes in the shape of the earth's surface.

Cities have acquired most of the relocated materials, either as wealth now in use, waste from production and processing, or discarded artifacts. Can such a system, geared to providing a greater and greater abundance of material wealth, be sustained without the development of a means for handling waste? Do cities have means to process wastes similar to the way biomass residues are recycled by the soil-plant system into new biomass? If the modernized city could follow the biological example and reuse the materials, it could get rid of wastes and reduce the demand for new raw materials as well.

Herein lies a problem of pollution and waste disposal that inevitably accompanies increasing productivity. The material production of the interconnected system stands in stark contrast

You may wish to investigate your food sources to determine the variety of places—that is, other environments—involved in supplying it and the extent to which your diet involves foods that are seasonal in production although available to you year-round. Obtaining information on sources of your foods may present some problems. One way to develop approximate information is to list foods that you know are not grown in your general area and to identify the environments appropriate for their production. Locate these environments on a map, and then, using whatever additional information you may be able to find, develop a geography of your food supply based on the most reasonable sources. If others in the class have done this, you might compare results. No doubt there will be conflicting opinions. To what extent are you able to resolve these conflicts?

To investigate the seasonality of production of many foods, list seasonal items and the months when each is harvested. Compare this list with your consumption pattern of these foods. You might also identify the foods that need special preparation to preserve them and those that can be stored easily. How does drying, salting and pickling retard decay that might otherwise occur. What environmental processes do they arrest?

to the material scarcity of traditional societies. The challenge of making environments more productive has been met, but at a price. Its very success carries with it unanticipated changes to physical environments that threaten their stability.

The technology of the modernized system with its exploiting of physical resources is expanding rapidly. For traditional societies technological developments have come slowly and are limited largely to the manipulation of biological production. Any side effects to the environment from this limited technology are minor. By contrast, the new technology of modernized societies is being employed so rapidly that its ramifications cannot be adequately assessed. Does this mean that modern societies will have to use less or more technology? Perhaps material production should be limited, and we should reconsider our material desires and reorder our needs to reduce environmental changes.

TRADITIONAL SOCIETY AND THE MANAGEMENT OF LOCAL ENVIRONMENTS

Since traditional society is primarily dependent on the production of its own local environment, it is popularly assumed that the more productive the environment, the more successful the local population. In this regard there is no question that if environmental work is low—as in dry and cold environments—biological production will also be low. Such an environment will therefore support relatively low densities of people.

But biological output is not the only determinant of a traditional society's well-being. The number of people that the environment must support is always an important factor. Thus, many people in an environment of high biological potential such as India's may fare less well than a few people might in an environment of low potential.

Nevertheless, the productiveness of the local environment does place limits on the density of local population that can be supported. Environments of low biological productivity have low population-carrying capacity, while those of high productivity can support larger populations.

Use of Environments by Gatherers and Herders

As indicated, the technologies for managing the resources of the biological system vary tremendously among those societies labeled traditional.

Those few societies that remain based on gathering and hunting scarcely manage the environment at all. They are scavengers and opportunists, gathering whatever the untended, wild-plant community produces at any given time and bagging whatever the cultural hunting skills and technology can manage to bring down. Such people survive much the same way that other animals do. In fact, they often compete with the animal population for the environment's output. For them survival depends on gathering fruits before they decay or on the harvest of hardier roots, tubers, and nuts.

The herder is less vulnerable because he maintains a stock of food and other necessities in the form of tended animals. But he must provide for his animals, and that keeps him on the move. In environments of low output he learns to move from place to place in order to harvest the output of natural vegetation which may vary locally. Typically he moves into moist, mountain pasturelands in summer when lowlands are parched, then goes down into lowlands where temperatures are warmer to take advantage of fall, and possibly winter, plant growth. Sometimes he tries to save some forage such as dry hay for winter.

Cultivators and Use of Environments

Cultivators, who are by far the most numerous group in traditional societies, have developed practices that combine managing the type of plants grown with redesigning the environment. Management involves reducing the competition from unwanted plants by clearing them away by means of fire and cutting, and then cultivating wanted plants and selectively destroying unwanted ones. Even the chosen crop is thinned so that the most desirable plants are favored. In this fashion many plants have become domesticated and man-dependent. Without men to protect them they probably could not survive the threat from other plants, diseases, and insects. Managing the environment usually involves modifying deficiencies in it. Most often, as for the Indian farmer, these are deficiencies of water and plant nutrients.

Water management may also involve drainage or irrigation of soil. Drainage is necessary where moisture prevents air from filling some of the pores in soil. Air is essential to the respiration of plant roots and to the overall productivity of the plant. Drainage management attempts to attain a balance of air and water. Irrigation, on the other hand, provides additional moisture for plant growth and can be accomplished either by storing excess water or by obtaining additional water from other places. Of these two the management of transferred water has proved the more complicated, not only because of the problems of transferring, but also because suspended particles tend to accumulate and create a pollution problem.

The management of plant nutrients can be difficult for most traditional cultivators. Generally, it involves conserving vegetation residue and animal manures, and returning them to the soil for decay. Vegetative cover may be cut or burned or both in another common practice to release nutrients. In most cases these practices

THE SUBSTITUTION OF MACHINES FOR ANIMAL POWER

If we recognize that in traditional societies arable land must provide not only the food for human survival but also the feed for work animals and a large share of the fuel, what might we expect if coal and oil were substituted for work animals and fuel, as they are in farming in modern societies? Horses, oxen, and water buffaloes compete for the food stocks grown on arable land, tractors fueled with gasoline do not. What are the implications for human food supply in this substitution? Would it be logical to increase world food supplies by using some form of tractor power instead of an equal amount of human and animal power? If so, what are the practical drawbacks to such a substitution in the near future? Does every traditional village or farmer have a local source of fuel for a machine if the machine was available? Is there any easy way of making the substitution of machines for animal power?

depend on recycling the local stock of nutrients. Sometimes farmers collect vegetative material from unused land for compost on cultivated land. At its best this form of maintenance conserves the available supply but does little to increase it or overcome any local deficiencies. Many traditional cultivators, however, fail to practice even these methods of management, and the nutrient supply is literally depleted.

As we have seen in Chapter 2 on the village of Ramkheri, in traditional societies nearly all the energy available for human manipulation of the physical environment is derived from current biological production of the local environment.

POPULATION AND THE ENVIRONMENT

In neither the traditional nor the modernized society is man immune from problems of his own numbers. In the Indian village, for example, life can be secure if the number of people is not too great. But there is a point at which the number to be sustained can exceed the number the existing system can provide for. This rule also applies to modernized societies in spite of their seeming capacity for producing. It is the very exploitation and pollution of environments by production technology that eventually limits the quality of life.

Places where much of the population lives on the margin of existence are often considered overpopulated. It is important to recognize that overpopulation in those areas is not a simple matter of density or of limited local resource endowments. It involves all of these in the context of the total physical environment. In traditional societies the local environment and local technology for using that environment are finite. Extending the resource base by reaching out to other places may not be appropriate for all parts of the world either. It is obviously impossible for everyone to emulate the modernized system by using large volumes of resources drawn from other places because there would be no other places to exploit.

No one can say with any precision what population the world might sustain at the levels of affluence now enjoyed by the modernized societies. Some hold that the present world population cannot be sustained at those high levels of affluence and still avoid environmental disaster. Others think that those levels could be sustained for the number of people currently populating the world, and some even predict that a larger population could be supported at levels comparable to those of affluent societies of today. For the foreseeable future, however, the evidence at hand suggests that a more realistic probability is for the world to remain split with the greater proportion of world population living in materially poor societies and the smaller proportion enjoying material affluence. Under these circumstances the prospects are not too hopeful that socio-economic order will alter its goals to be more consistent with the limits of the world's environments. It is more likely that they will proceed as they have in the past toward greater and greater exploitation of environments.

SELECTED REFERENCES

Carter, Douglas B., Schmudde, Theodore H., and Sharpe, David M. *The Interface as a Working Environment: A Purpose for Physical Geography*, Technical Paper No. 7, Commission on College Geography, Association of American Geographers, Washington, 1972. Develops the case for approaching the environment at the earth's surface as a working system. Examples of how this working system is regulated by energy and moisture and how it interrelates with the activities of man are explored.

Chorley, Richard J., ed. *Introduction to Geographical Hydrology*, Barnes & Noble, New York, 1971. A collection of articles dealing with specific aspects of water resources and their management by man.

Commoner, Barry. *The Closing Circle*, Knopf, New York, 1971. An evaluation of the role of the technology of modern production systems in environmental pollution. The case is presented in nontechnical language that is easy for a reader to follow.

Darling, F. Fraser, and Milton, John P., eds. *Future Environments of North America*, Natural History Press, Garden City, N. Y., 1966. Presents papers and discussion given at a conference of future environments of North America sponsored by the Conservation Foundation. A very broad range of topics and issues concerning environments and society are treated. Some papers are very specific whereas others are of a general nature. Collectively the material touches on a wide spectrum of problems concerning man and his use of environments.

Miller, David H. *A Survey Course: The Energy and Mass Budget at the Surface of the Earth*, Publication No. 7, Commission on College Geography, Association of American Geographers, Washington, 1968. Presents eight units to illustrate that the principles of mass and energy exchange and transformation; can be used in understanding the functioning of urban, rural, and wildland parts of the earth's surface. Each unit consists of a brief introductory overview or frame of reference and a comprehensive annotated bibliography. An excellent reference source for problems and research concerning the environment and man's involvement with it.

Thornthwaite, C. W., and Mather, J. R., eds.
Average Climatic Water Balance Data of the
Continents, *Publications in Climatology,* Drexel
Institute of Technology, Centerton, N. J.:
Part 1, Africa, Vol. XV, No. 2, 1962; Part 2,
Asia (excluding the USSR), Vol. XVI, No. 1,
1963; Part 3, USSR, Vol. XVI, No. 2, 1963;
Part 4, Australia, New Zealand, and Oceania,
Vol. XVI, No. 3, 1963; Part 5, Europe, Vol.
XVII, No. 1, 1964; Part 6, North America
(except United States), Vol. XVII, No. 2, 1964;
Part 7, United States, Vol. XVII, No. 3, 1964;
Part 8, South America, Vol. XVIII, No. 2,
1965.

The Biosphere, A Scientific American Book, W. H.
Freeman and Company, San Francisco, 1970.
A series of articles addressed to the cycles
of the biosphere and man's dependence and
use of the processes of the biosphere.

Man and the Ecosphere, Readings from *Scientific
American* with commentaries by Paul R.
Ehrlich, John P. Holdren, and Richard W.
Holm, W. H. Freeman and Company, San
Francisco, 1971. A collection of articles from
Scientific American covering a wide range of
subjects. Some primarily concern aspects of
environment and some are mostly con-
cerned with questions of human manage-
ment of resources.

Scientific American, Vol. 224, No. 3, Sept. 1971.
On the subject of energy and power, an
issue devoted to the flows and uses of
energy in the universe, on earth, and by different societies and technologies.

Chapter 6   Integrating the Geographic Components

THE PRIMARY purpose of this chapter is to compare the patterns of world population to the patterns of a selection of components that make up the distinctive geographic character of a place. The number of components included is limited, but even so, they can provide a useful basis for comparing variations of the earth. We shall evaluate the functional relationships among different environmental components and population, in particular.

Developing a "Double Exposure" View of the World

We are used to the world in which we live our own lives: it is our "life space." We have a feeling for a farm, a village, or even a metropolitan area such as Washington because we have been on the spot and have seen people living there. Although we may have to adjust to different scenery and different cultural ways when we travel to another part of the world, the scale of the things we see there is still the one we are accustomed to. The world we observe is at most only a few square miles, and the people in it loom large because we tend to focus on them. Figure 6-1 shows how one's local sphere is related in scale to some of the larger units one may be concerned with.

The world seen in maps and aerial photographs, however, is very different. Travelers looking down from jet aircraft see only patterns of terrain. The outlines of fields, roads, cities, and towns can be seen along with mountains, rivers, and forests, but the presence of people can only be inferred from the patterns on the ground; they cannot be directly observed. People, who are so prominent in our everyday experience, fade out of sight at 26,000 feet.

None of us flying in a commercial jet has seen all of Ohio or Florida, much less the entire

(a) Carbondale, Illinois

0 _____ 1 Mile

1/63,360; 1 in = 1 mile

(c) Southern Illinois

0 _____ 100 Miles

1/6,336,000; 1 inch = 100 miles

(b) Jackson County, Illinois

0 _____ 10 Miles

1/633,600; 1 inch = 10 miles

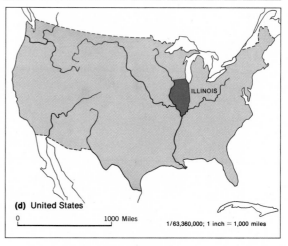

(d) United States

0 _____ 1000 Miles

1/63,360,000; 1 inch = 1,000 miles

(e) The World

0 _____ 10 000 Miles

1/633,600,000; 1 inch = 10,000 miles

Figure 6-1 SCALE. *Notice the change in scale as one thinks of his campus, city, county, state, country, and finally the world.*

United States. Yet in this book our major concern will be with even larger areas. Only rarely will we concentrate on individual countries. More often our attention will focus on places like Western Europe or all of Latin America. In Chapter 15 we shall examine those areas which contain more than one-third of the world's people.

Such areas are abstract in the sense that one must merge thoughts of places seen in photographs or recalled from memory. The abstract world of maps conforms to the sum of individual measurements taken from many places on the earth but has no real existence. Only recently has photography from space vehicles given us direct observations of areas as large as countries and continental segments. The evening weather report on TV may show a satellite photograph covering the whole of the continental United States.

One aspect of geographic understanding requires a "double-exposure" view of the world that simultaneously sees the local world of one small place such as Ramkheri and the whole abstract world of mankind or the particular part we might be working with.

The Use of Maps

This chapter is concerned with the whole world of mankind. It is built around a series of maps buttressed by supporting text and might be thought of as an annotated atlas. The maps are the major focus; they are of primary importance. The text is designed to assist in comparing the ideas on one map with those on the next. Evaluating relationships between components requires thinking and should help to develop a basic fund of knowledge about the variations of the world. The object is not to memorize the data on each map but rather to identify, compare, and associate the similarities and contradictions found between the maps. The text should help, but the significance of these relationships should be determined by each of us individually. In particular, we should attempt to categorize and group sections of the world according to the patterns of life that emerge from the maps.

Population Distribution

Figure 6-2 presents population as a distribution of small and large dots. Each small dot on the map represents 1,000,000 people outside the largest cities of the world—those rural people living out on the land on farms, in tribes, and in small settlements, or in towns and cities of less than a million people. Each metropolitan area of 8 million or more people is shown by a large dot. Because city people crowd so close together, it is difficult to represent them on a map that also shows rural people. In looking at the map, we should remember that the metropolitan concentrations of population are even larger than they appear.

We can see that the distribution of population has an uneven and rather peculiar pattern. Some places seem tremendously crowded, while most of the land area of the world is empty or sparsely settled.

But we must bear in mind that the distribution we see is not the result of a single set of cumulative decisions by mankind acting as a single group. Instead, it is the sum of the decisions of all the numerous ethnic groups with their different technological capabilities and different value systems developing through time in their own localities. The map is a composite of the many and varied perceptions and responses to numerous environments.

In the rural areas most people make direct use of the surrounding land to support them-

Figure 6-2 World Population Represented by Dots

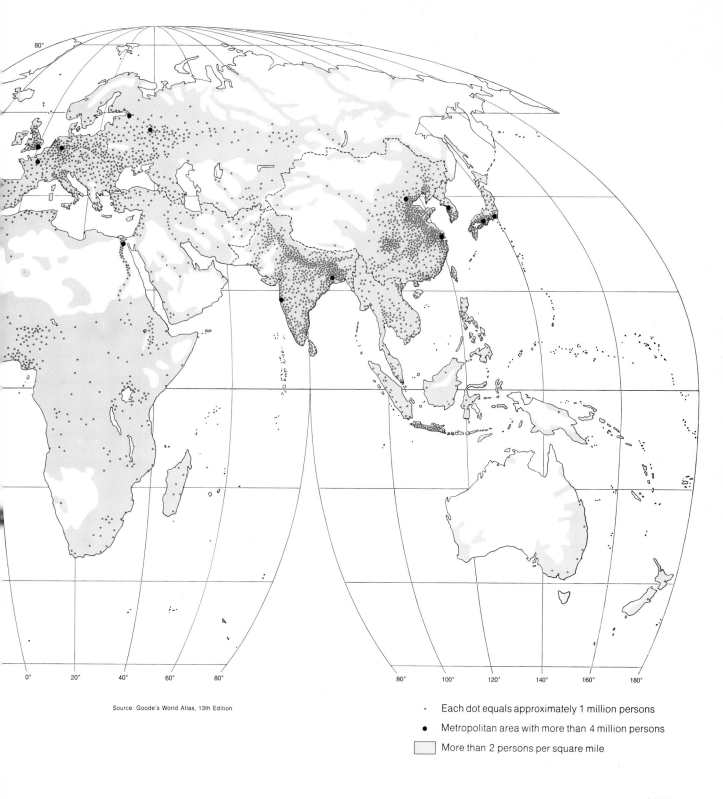

Source: Goode's World Atlas, 13th Edition

- Each dot equals approximately 1 million persons
- Metropolitan area with more than 4 million persons

More than 2 persons per square mile

selves through agriculture, hunting, and mining, for example. In big cities like Washington, on the other hand, people do not use the land to feed themselves or to provide for their basic needs and can crowd much more closely together. They are even stacked on top of one another in high-rise apartments, hotels, and office buildings.

In Washington there are great variations in density. People are more thinly spread out in the suburbs, for example, than in the city where row houses are closely placed and several families may live in a single building. Certain parts of the city are reserved for park land, to be used after school or on weekends but not for residence. Other places contain office buildings or shops where no one lives but where thousands of people work or visit during the day.

The pattern of Ramkheri is different. Here people crowd into the hamlet for family living, but members of individual families go out each day to work fields that spread for miles over the surrounding countryside. Even though every possible piece of land is put to use, tracts of poorly drained soil and rough land remain unusable for any purpose.

Throughout history separate groups have had to make decisions on the use of the land at their disposal. Decisions about land use in Asia, for example, are determined by factors different from those that influence land use in Europe. Only recently have people been able to base their decisions on the fact that they could live in one area while depending primarily on products from other places.

By looking at Figure 6-2 and observing the distribution of population, it is difficult to tell from the dots the share of world population that lives in Asia or Africa. We can say that most people live in Asia and that there are distinct major population centers: Europe, India, China,

Japan and Indonesia. But there is no clear way of seeing their relative shares of population.

Figure 6-3 gives a picture of the uneven distribution of population over the world. We see that Asia contains over half of the world's population and that no other continent has as much as 15 per cent of the total. Moreover, more than three out of every four people on earth live on the single landmass of Eurasia, even though it has only a little more than one-third of the world's total land area. On the other hand, Oceania and Antarctica combined have 15 per cent of the land area of the world but less than 1 per cent of its population.

Returning to Figure 6-2, we see that even in Asia, with over half the population of the world, almost half of the land area is empty or almost empty of people. This absence of people over large land areas is even more prevalent on most other continents. Only in Europe does population seem to be spread over almost all the land area. It is estimated that one-third of the population is located on only about 3 per cent of all the land area of the world. More than two-thirds of all people in the world live on 7 per cent of the land area.

DENSITIES

Figure 6-4 shows world population from a different perspective. Its patterns indicate the densities of population in particular places, in contrast to the dot map, which represents actual numbers of people. The emptiness of most of the world may be seen quite dramatically on the density map. Over half the area of the world has a density of less than two persons per square mile, and large additional tracts have between two and eight persons per square mile.

The crowded areas show up most strikingly in Figure 6-4. Some areas have densities of more

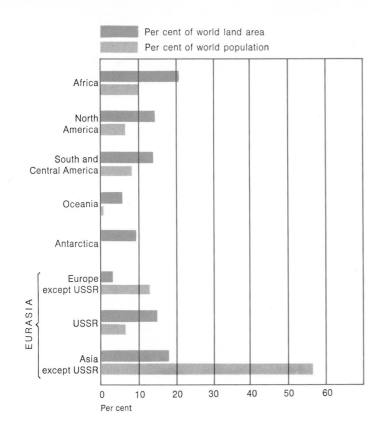

Per cent of world land area

Per cent of world population

EURASIA

Africa

North America

South and Central America

Oceania

Antarctica

Europe except USSR

USSR

Asia except USSR

Per cent
0 10 20 30 40 50 60

Figure 6-3 WHERE THE WORLD'S PEOPLE LIVE

than 500 persons per square mile. The most populous areas on the map generally have densities of more than 128 persons per square mile. We should remember that in Figure 6-4, as in Figure 6-2, the large metropolitan population concentrations are not adequately shown. Such centers exist on every continent.

RELATION TO ENVIRONMENTAL MEASURES

Even though different groups of people have made value judgments about the possibilities of the earth's environment, it may be useful to compare the population maps with certain standard environmental measures to see if population over the world has tended to distribute itself in close correlation with environmental variation.

Rainfall

Because of the importance of moisture to human life, we might confirm the old expression "It rains people." That is to say, the more rainfall a place has, the more people one finds. Figure 6-5 shows average annual rainfall in parts of the world. If you compare this with the population maps (Figure 6-2 or 6-4), you can see that people seem to avoid areas of low rainfall. But what about Egypt and parts of the Middle East? How can such population densities be supported in such dry areas? There are great concentrations of people in areas having over 80 inches of rainfall per year in Southeast Asia and East India, but lands with similar totals in West Africa and South America are virtually empty of people. The highest correlation seems to be between areas of intermediate rainfall, 20 to 80 inches, and high population, but many such areas in Latin America, Africa, and Australia are only sparsely populated.

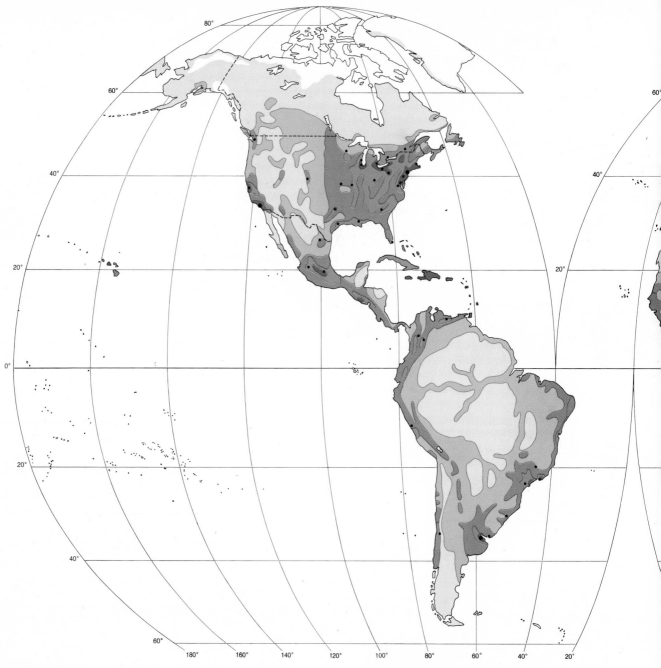

Figure 6-4 Population Densities of Areas of the World

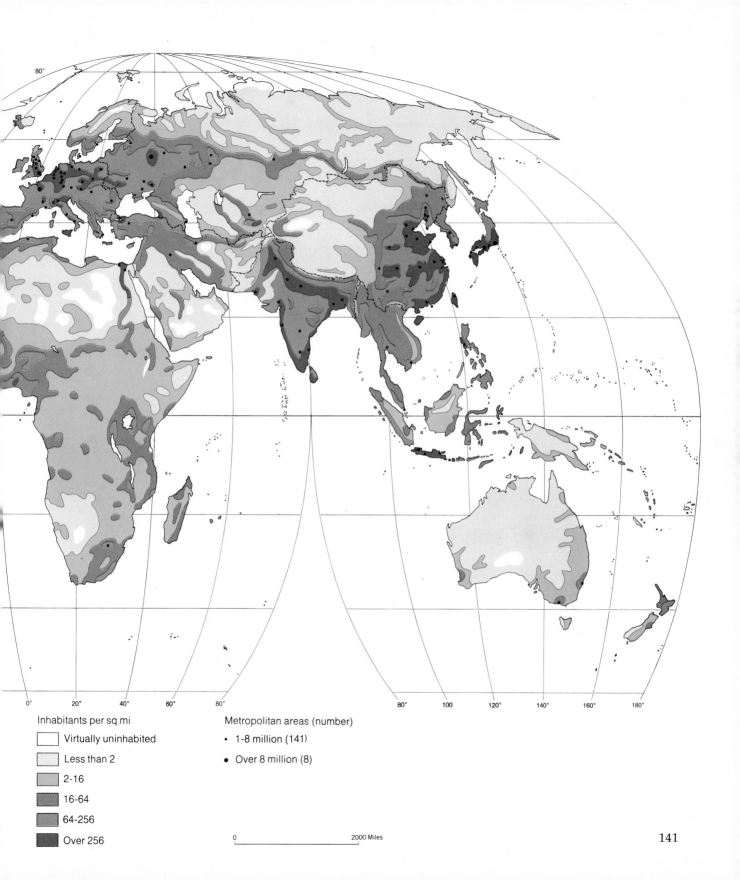

Inhabitants per sq mi

☐ Virtually uninhabited

☐ Less than 2

☐ 2-16

☐ 16-64

☐ 64-256

☐ Over 256

Metropolitan areas (number)

• 1-8 million (141)

● Over 8 million (8)

80°

0° 20° 40° 60° 80°   80° 100° 120° 140° 160° 180°

0   2000 Miles

141

Figure 6-5 AVERAGE ANNUAL RAINFALL IN AREAS OF THE WORLD

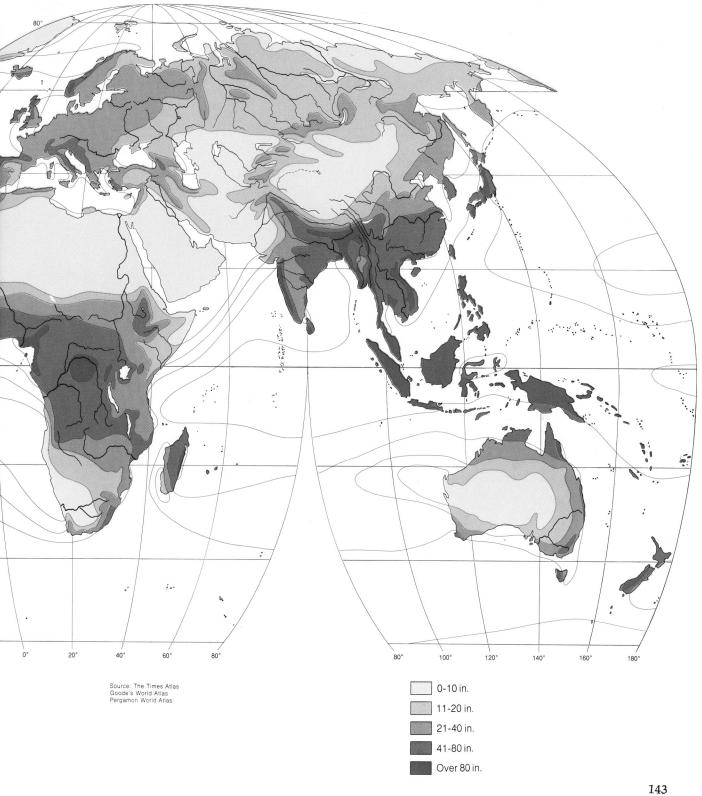

	0-10 in.
	11-20 in.
	21-40 in.
	41-80 in.
	Over 80 in.

Let's examine the densities shown in Figure 6-4 in terms of the real crowding in local areas. For example, Figure 6-6 shows local densities, rural and urban, for a number of places. First consider rural areas.

1. Density of less than 2 persons per square mile (640 acres) would mean there would be less than one farm family per square mile (assuming a family as having 4 members). The farm would be at least 640 acres—like those in many parts of the Middle West of the United States.

2. Density of 10 to 20 people per square mile would allow about 3 farms of 200 acres each.

3. Density of 20 to 75 per square mile would allow 5 to 20 farms of 30 to 150 acres. Such farms approximate the size of farms in western Europe.

4. Densities of 100 to 250 would permit about 25 to 60 farms per square mile—like the larger Asian farms.

5. Densities of 500 persons would mean that 125 farms of 5 acres each would squeeze into one square mile. This is approximately the farm size in Ramkheri.

We get a different view of densities if we consider land from an urban point of view. In the city of Chicago, streets are measured off twelve to the mile in an east-west direction, and eight to the mile in a north-south direction. There are 96 blocks to a square mile.

1. Density of less than 2 persons per square mile would mean less than 2 people lived in 96 blocks.

2. Densities of up to 250 persons place $2\frac{1}{2}$ persons per block at the maximum density. Since this

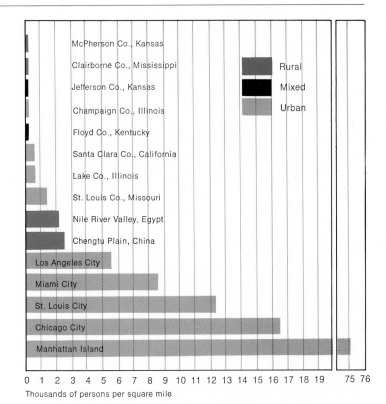

Figure 6-6   POPULATION DENSITIES OF SELECTED LOCALITIES

is smaller than urban family size, there would be less than one family per block.

3. In densities of up to 500 persons per square mile, over 5 persons at the maximum would live in a block.

These simple calculations suggest that from a rural or agricultural standpoint the densities on the map are indeed high. They approach those of Ramkheri. It takes good farmland and very productive agriculture to support such populations locally. Even so, the living standard is likely to be low. The lower densities can produce much more than a marginal living for farmers if they use machinery and inanimate energy. Low densities may also be a measure of poor terrain. In some places one square mile of land does not provide a subsistent return. Western ranches in the United States can be 100,000 acres in size. Nomadic herders might traverse even larger tracts in their migration through the year.

The densities have different meaning when viewed from an urban perspective. In the United States in 1966 the average density for a large city and its surrounding suburbia was almost 450 per square mile. In the Washington area, it was more than 1,100, and in Jersey City reached over 13,000 people per square mile.

In the central city, residential densities can be even greater. Not only are dwellings close together, but in high-rise buildings, people can be stacked on top of one another. In Chicago large public housing units run fifteen stories high and as many as 1,200 persons live in each unit. But the people who live in that apartment building go elsewhere to work or to school. Conversely, in the industrial, commercial, and institutional parts of the city, very few people make their homes there, but such areas can be very crowded during working hours.

What is the significance of population density? Compare its value to life in Washington, D.C., and Ramkheri.

Since, as we saw in Chapter 5, rainfall is most needed during warm seasons when evapotranspiration demands on the environment are greatest, we might expect population to be located in those areas where warm season rainfall is large: between November 1 and April 30 in the southern hemisphere (Figure 6-7), between May 1 and October 31 in the northern hemisphere (Figure 6-8), and throughout the year in the tropical parts of the world. Do you think this further refinement of the relation of population to rainfall provides a closer "fit" between the two distributions? How about the Amazon Valley as compared to Southeastern Asia?

Frost-free Seasons

One might generalize that more people live where the growing season (the season between frosts) is longest and rainfall is adequate (Figure 6-9). Surprisingly we see that most areas that are frost-free year-round are almost empty of people. Most of the largest population concentrations are found in the mid-latitudes, where the frost-free period is interrupted for as long as six months. On the other hand, it is also true that few people live in high latitudes where the frost-free season is very short.

Land Surface

What about the proposition that people concentrate in the lowlands and avoid mountains and other rugged land? As Figure 6-10 shows, this does not correlate very highly either. Empty lowlands can be found in South America, in the northern latitudes, and in interior Australia, while considerable populations appear in the highlands of Latin America and in parts of China and Japan. Areas of high density in Asia, Europe, and the United States seem to sprawl over several types of land surface.

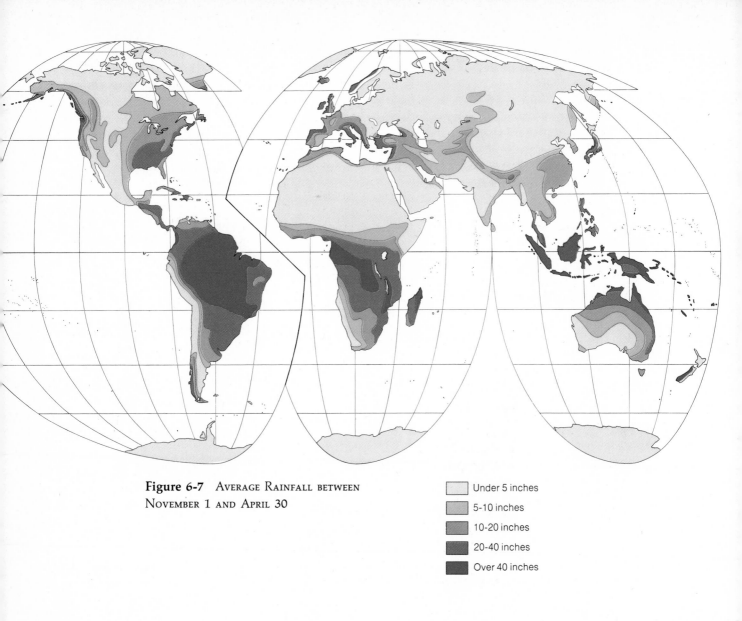

Figure 6-7 Average Rainfall between November 1 and April 30

Under 5 inches
5-10 inches
10-20 inches
20-40 inches
Over 40 inches

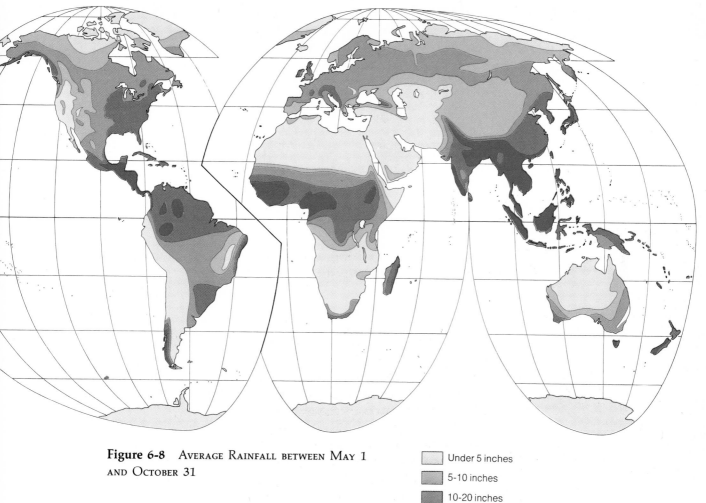

Figure 6-8 AVERAGE RAINFALL BETWEEN MAY 1 AND OCTOBER 31

	Under 5 inches
	5-10 inches
	10-20 inches
	20-40 inches
	Over 40 inches

Figure 6-9 Length of Frost-free Season in
Areas of the World

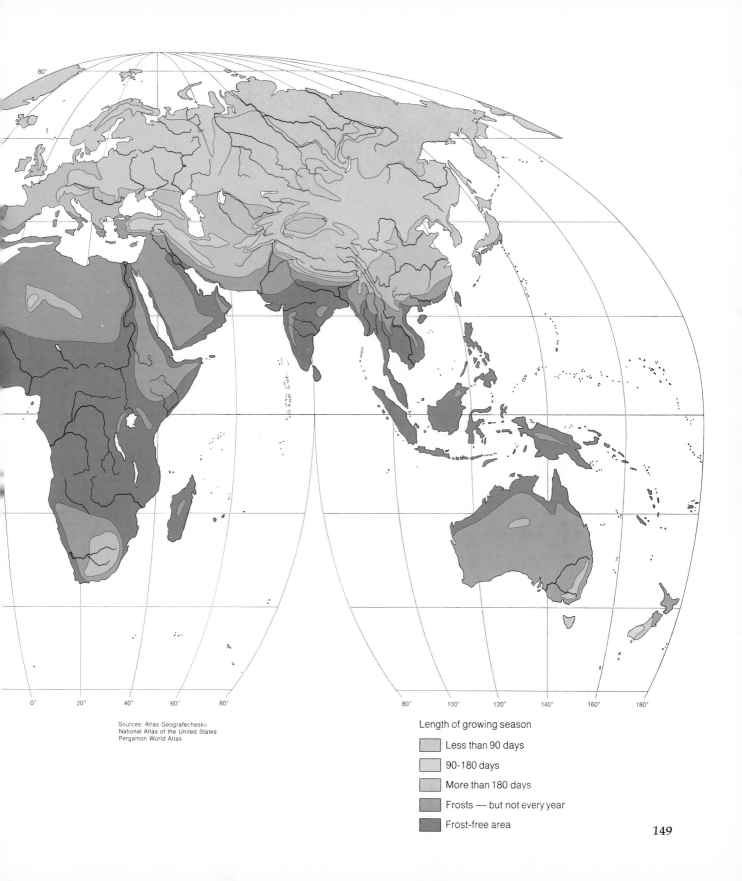

Sources: Atlas Geografecheskii
National Atlas of the United States
Pergamon World Atlas

Length of growing season

Less than 90 days

90-180 days

More than 180 days

Frosts — but not every year

Frost-free area

149

Figure 6-10 The Major Landforms of the World. *Outlines, elevations, and water systems.*

0   2000 Miles

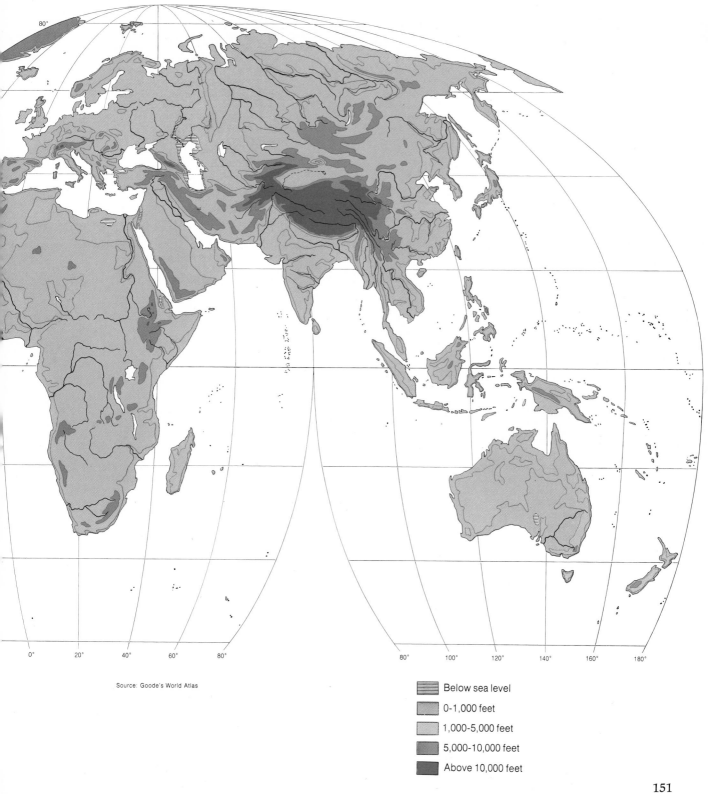

Source: Goode's World Atlas

Below sea level

0-1,000 feet

1,000-5,000 feet

5,000-10,000 feet

Above 10,000 feet

151

Does the relative emptiness of the world's land area belie the immediate threat of a "population explosion"? There seems to be plenty of unused and little-used land. But the usability of that land is another matter. Man must see considerable problems in using the world's empty areas if, after being on earth 100,000 or more years, he has left great tracts of land empty or almost so.

But the population explosion is surely something to be reckoned with, as Table 6-1 shows. World population more than doubled between 1850 and 1950, and it increased more than 50 per cent from 1950 to 1970. This twenty-year increase is greater in numbers of people than the total population of the world in 1850. Between 1970 and 2000 it may well increase by almost 69 per cent—a total equal to total world population in 1950.

However, as Figure 6-11 shows, the amount of land area newly occupied since 1800 has been relatively small, considering that the number of people on earth has increased fourfold. Also, in the face of rapid population growth since 1800, man has chosen to crowd more densely the areas already occupied rather than to move to the empty areas of the world. Why is this so?

Table 6-1 POPULATION GROWTH 1650–2000 (in millions)

	1650	1750	1850	1900	1950	1970	2000*
North America	1	1	26	81	166	228	354
Latin America	12	11	33	63	162	283	638
Europe	100	140	266	401	559	462	521
USSR	....	....	(60)†	(111)†	(193)†	243	353
Asia	330	479	749	1,186	1,302	2,056	3,458
Africa	100	95	95	120	198	344	768
Oceania	2	2	2	6	13	19	32
World	545	728	1,171	1,857	2,400	3,635	6,124

*United Nations medium growth projection.
†USSR not counted in totals for 1850, 1900, and 1950; figures are included in European totals.
SOURCE: Woytinsky and Woytinsky, *World Population and Production: Trends and Outlooks*, Twentieth Century Fund, New York, 1953; and *1970 World Population Data Sheet*, Population Reference Bureau, Inc., Washington, 1970.

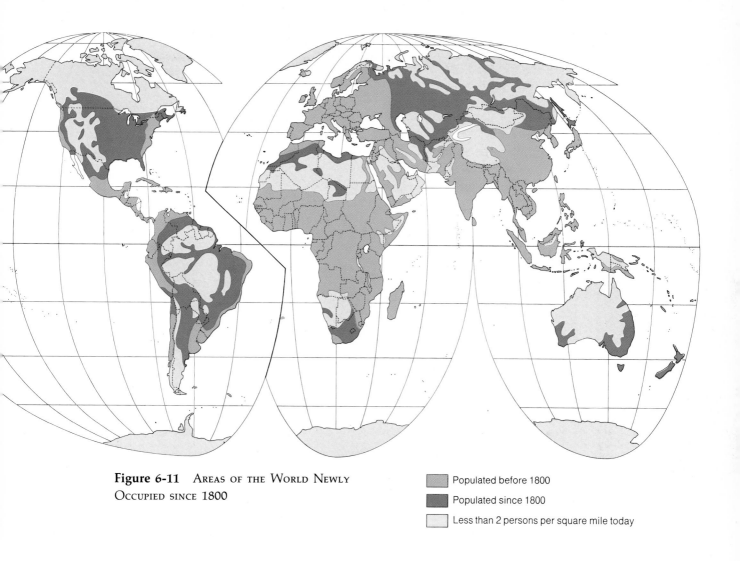

Figure 6-11 AREAS OF THE WORLD NEWLY
OCCUPIED SINCE 1800

Populated before 1800

Populated since 1800

Less than 2 persons per square mile today

RELATION TO PRIMARY PRODUCTION

If population does not seem to show close correlation to any of the major facets of the earth environment, we might expect it to have a close relationship with the human use of environmental resources. Whereas certain human activities, particularly those carried on in cities, might be only indirectly related to the environment, the primary occupations require direct use of natural resources such as soil, vegetation, animals, and minerals.

Fishing, Hunting, Herding, Forestry, Mining

The relationship between population distribution and the location of most primary activities is surprisingly not very close. Major fishing grounds are found off both densely populated areas and sparsely inhabited ones. Hunting and herding are primarily occupations of the sparsely populated parts of the world. Forest exploitation follows a similar pattern, except for managed forests and tree farms which are found in populated parts of Asia, Europe, and North America. Mining, perhaps the most widely spread distribution of all primary activities, is found in all sorts of areas ranging from the most populated to the emptiest.

Agriculture

The pattern of population does correlate with agriculture as indicated by the acreage of cultivated land (Figure 6-12). No major centers of population fall outside the important zones of cultivated land. This may not seem surprising since man depends to such a great extent on agricultural production. As we have seen, however, a city like Washington draws primary products from producing areas over the entire earth and has few ties to the land in its immediate area. In terms of primary production, Washington could as well be located in a desert or on an ice cap. Yet, the traditional ties to major agricultural regions remain significant in the pattern of population distribution.

Centers of Modern and Traditional Life

The distribution of metropolitan areas with a million or more people shows a much lower correlation with total population than the distribution of cultivated land does. Asia leads both in numbers of metropolitan centers and in total population, but the proportion of people living in large metropolitan centers is relatively low (Table 6-2). Oceania with a total population of less than 20 million people has two metropolitan centers with 2 million people each. On the other hand, Africa has only four large metropolitan centers, only one south of the Sahara.

Before turning to a consideration of the meaning of the disparity between total population and large metropolitan centers, it is useful to note that metropolitan areas are a new phenomenon in the world. Less than 200 years ago there were no cities with as many as a million people. And, as Table 6-3 shows, only 36 cities in the world were even one-tenth that size. Moreover, all but two were in the established population centers of Eurasia. The only cities of this size in the rest of the world were Mexico City and Cairo, Egypt.

The newness of large metropolitan centers in the world is an indication that they are a phenomenon of the modern world. Like Washington, D.C., they are not tied to the resource base around them. There are, however, few really new cities in the world. Most modern metropolitan centers are the result of the growth of smaller centers. They were originally political capitals, ports, or trading centers.

Because large metropolitan centers are particularly identified with the development of the

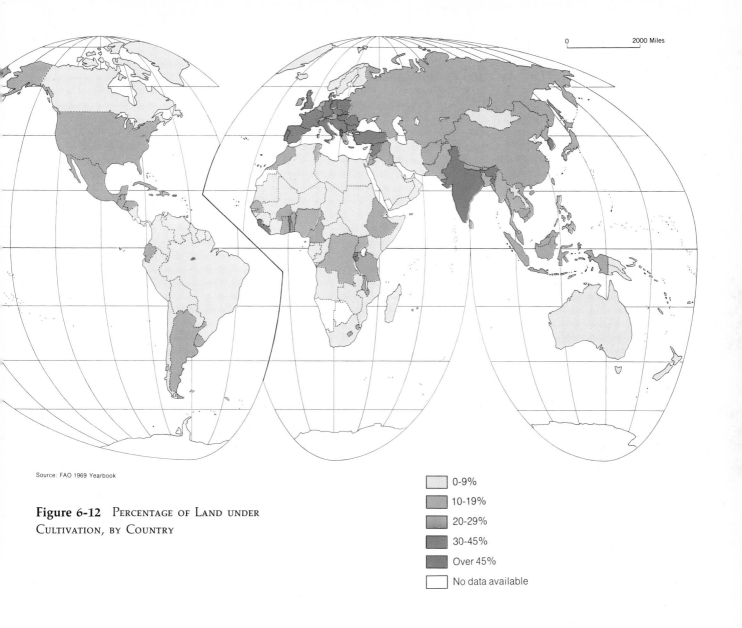

Source: FAO 1969 Yearbook

Figure 6-12 PERCENTAGE OF LAND UNDER
CULTIVATION, BY COUNTRY

0 — 2000 Miles

0-9%
10-19%
20-29%
30-45%
Over 45%
No data available

Table 6-2 POPULATION IN METROPOLITAN AREAS

	NUMBER OF METROPOLITAN AREAS OVER 1 MILLION POPULATION		POPULATION (millions)			TOTAL POPULATION	
	PER REGION	PER CENT OF TOTAL	IN AREAS OVER 5 MILLION	IN AREAS OF 2–5 MILLION	IN AREAS OF 1–2 MILLION	IN METROPOLITAN AREAS (millions)	PER CENT OF WORLD METROPOLITAN AREAS
Asia	46	33	38	52	31	121	34
Europe	38	28	24	32	31	87	24
Anglo-America	26	19	31	29	17	77	21
Latin America	11	8	24	5	7	36	10
USSR	11	8	9	4	11	24	7
Africa	4	3	5	2	3	10	3
Oceania	2	1	4	. . . .	. . . .	4	1
Total	138					359	

SOURCE: *Encyclopaedia Britannica International Atlas*, 1970.

Table 6-3 POPULATION IN METROPOLITAN AREAS OVER 100,000 IN 1800

		POPULATION (millions)			TOTAL METROPOLITAN POPULATION (millions)
	NUMBER OF METROPOLITAN AREAS	IN AREAS OVER 1,000,000	IN AREAS 500,000–1,000,000	IN AREAS 100,000–500,000	
Asia	15	. . . .	4.4*	2.3	6.7
Europe	17	. . . .	1.5	2.4	3.9
Latin America	1	. . . .	. . . .	0.1	0.1
Anglo-America	0	. . . .	. . . .	. . . .	0.0
USSR	2	. . . .	. . . .	0.5	0.5
Africa	1	. . . .	. . . .	0.3	0.3
Oceania	0	. . . .	. . . .	. . . .	0.0
Total	36				11.5

*Turkey is included in the figure for Asia.
SOURCE: Woytinsky and Woytinsky, *World Population and Production: Trends and Outlooks*, Twentieth Century Fund, New York, 1953, Table 55, p. 120.

modern, interconnected system and that system is unevenly distributed over the world, metropolitan areas follow a pattern rather different from that of world population in general. In large part, the growth of metropolitan centers has taken place in modernized areas where much of the activity of the new, interconnected system does in fact focus. Metropolitan areas are centers not only of transportation and communication but of urban-centered activities such as wholesale and retail trade, manufacturing, finance and business, and even government and medical services.

Since large metropolitan centers reflect modern systems of living, an absence of such centers would seem to indicate areas that are still primarily operating on the smaller, more localized scale of life represented by Ramkheri.

From Figure 6-2 and Table 6-2 we can see that areas having the largest metropolitan areas—those over 8 million population—which presumably are modernized to a great extent, can easily be distinguished from those areas with little metropolitan development where traditional, locally oriented life predominates. North America, Europe, Japan, and even Australia contain important areas of modern metropolitan life. There also seems to be considerable metropolitan development in parts of the Soviet Union, in Latin America, and in Asia (although not in terms of the continent's total population). Only in Africa is the absence of numerous large metropolitan areas striking.

The small number of metropolitan areas in Africa indicates the dominance of traditional, locally centered life. This seems true in most of Latin America, too, although such very large metropolitan centers as Buenos Aires, in Argentina, and Mexico City might point to important islands of the modern system in what is otherwise a largely traditional area.

The large number of metropolitan areas in Asia raises more questions. Some of these areas are in industrial Japan. But what of the large number in China and India: do they represent something different from Ramkheri, or are they simply oversized examples of that model?

In a way, the large metropolitan centers of India are a part of the traditional system typified by the Indian village. Even though most of the population and production are locally based and do not move from place to place, some movement does occur. Goods move from the village to the town market, and from there a small proportion is transported to larger regional trade centers. The big cities play a role important to the smaller places around them. They are especially important in very populous, rurally based areas such as India and China.

The large Asian cities are also part of the modern, interconnected system. They are both ports through which imports and exports pass and railroad centers. Most have international airports with world-wide links, and all have some modern factories and are the headquarters of modern businesses. Thus, they function for both old and new worlds.

PERCENTAGE OF POPULATION EMPLOYED IN AGRICULTURE

The contrast between the areas of modern metropolitan life and those of locally based rural life are obvious from another type of index. Since traditional life is based on primary activities, it would seem that a measure of the relative importance of agriculture to an area or country can provide such an index.

Figure 6-13 shows that the percentage of a country's work force employed in agriculture varies greatly. Farming occupies almost 90 per

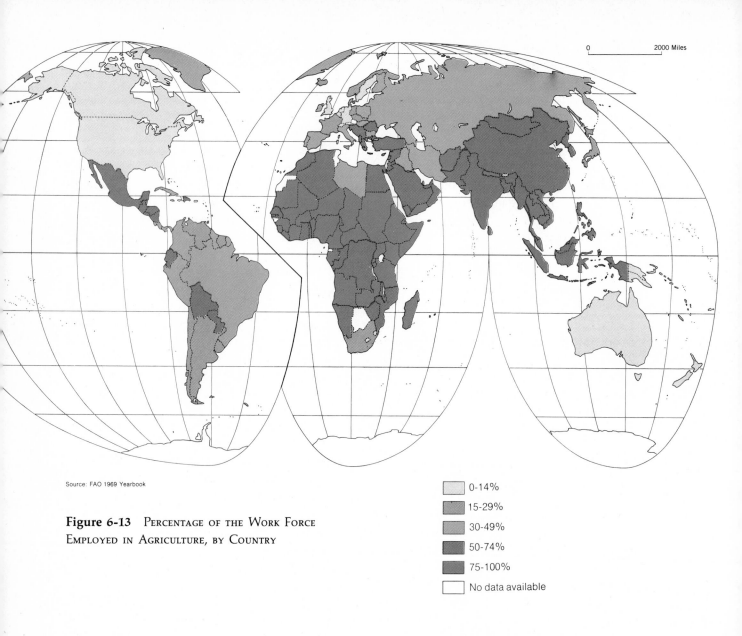

Figure 6-13 Percentage of the Work Force
Employed in Agriculture, by Country

	0-14%
	15-29%
	30-49%
	50-74%
	75-100%
	No data available

cent of the people in countries in Africa and more than 50 per cent of the people in most of Asia and much of Latin America. By contrast, less than 25 per cent of Europe works in agriculture, and the figure is below 10 per cent in the Netherlands, Great Britain, and the United States. As you might expect, the percentages are generally lowest in those countries where metropolitan centers are most prominent.

The maps we have looked at in this chapter suggest a division of the world that is familiar from magazines, books, and television documentaries: the distinction between what is commonly called the "developed" or "industrialized" world and the "underdeveloped" or "nonindustrialized" world. The developed world comprises those parts of the world that have benefited most from the Industrial Revolution—Europe, the United States and Canada, the Soviet Union, Japan, Australia and New Zealand, Israel, and, perhaps the Republic of South Africa. The rest of the world is largely nonindustrial.

PER CAPITA ENERGY CONSUMPTION:
MEASURE OF PRODUCTION

Perhaps the most direct index of the two contrasting worlds is energy consumption per capita in the various countries (Figure 6-14). The industrial world depends on the employment of machines using coal, oil, natural gas, electricity, and other sources of industrial power. The traditional world, in contrast, relies on manpower and animal power and on the use of wood and cow dung for fuel. Vast differences in energy consumption are evident from the map. Per capita energy consumption in the United States is almost 30 per cent greater than in Canada; more than twice the consumption in West Germany and in Australia; almost 10 times that of Mexico; more than 60 times that

of the Ivory Coast; and close to 1,000 times that of the lowest ranking countries on the map. This means that theoretically each individual in the United States has that much more energy available than has the man in a tradition country. Notice that, with the exception of Israel, the countries with above-average energy consumption are the industrial world: North America, Europe, the Soviet Union, Japan, Australia, New Zealand, and South Africa.

What is the significance of measuring energy consumption? What are its implications for people's lives? Where energy consumption is high, productivity is high because energy provides man with power and gives him considerable control over his environment. Think of the uses of energy in a household, in traveling from one place to another, and in keeping in touch with other people—not to mention its uses for industry. Compare life supported by such energy supplies with what it might be without them. Imagine what it would be like to live without nine of the ten energy units we rely on per person in the United States. Such is the situation in Mexico. The Ivory Coast subsists with less than 2 per cent of the energy we in the United States have available per person.

PER CAPITA GNP: MEASURE OF
PRODUCTION AND WEALTH

Economists commonly use gross national product (GNP) rather than energy as a measure of a country's productivity. GNP measures the total production of goods and services in a country during a given period of time, usually a year. It is difficult enough to collect all the data on production and the many services performed in a country. It is even more difficult to gather such data in areas where a large share of production is used locally by people to feed

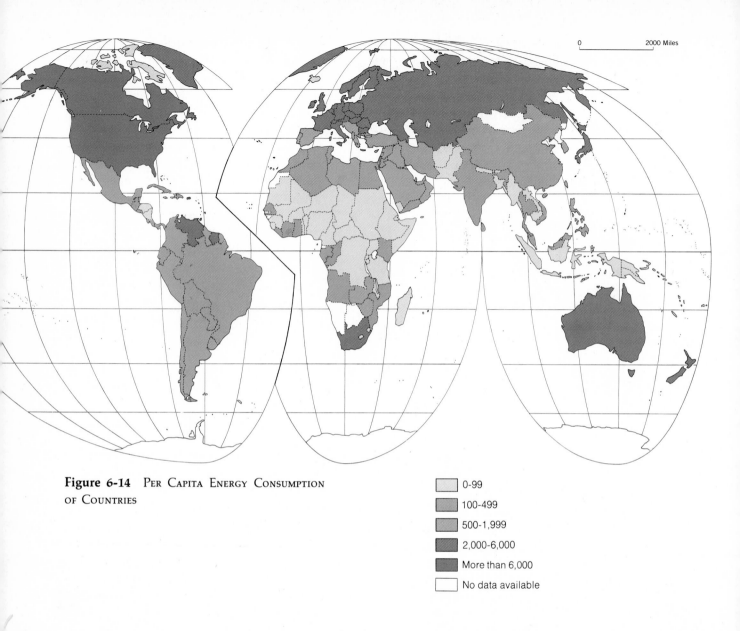

Figure 6-14 PER CAPITA ENERGY CONSUMPTION
OF COUNTRIES

- 0-99
- 100-499
- 500-1,999
- 2,000-6,000
- More than 6,000
- No data available

themselves or to trade with others in a local market. Moreover, production must be measured in monetary terms, and each country has its own currency. Bread is not only priced in dollars, shillings, yen, and lire, for example, but it also has a different value in comparison with fruit or clothing in various economies. More important, GNP does not adequately account for subsistence-type production. It is impossible to imagine how people survive in places where GNP is very low; $50 to $100 a year is just not sufficient for life. Much of the output of the subsistence producers, however, is simply not counted; actual production is greater than GNP indicators for countries where those in a subsistence economy are responsible for a large share of production. GNP is only a crude measuring stick for comparing productivity among countries. Still, it is the best we have, and since productivity varies so greatly among countries, it gives a picture of the extremes of the world. Or perhaps more accurately, since it is primarily a measure of modern economic activity, it is a good indication of the extent to which each country's population is a part of the modern system.

In Figure 6-15 GNP is converted into per capita figures for easy comparison. Here, as with energy consumption, the range between the most productive and the least productive countries is great. In the United States, production of goods and services per person is more than 10 times that of Nicaragua, Cuba, Guatemala, or Peru; more than 20 times that of Morocco or Bolivia; and almost 100 times that of Rwanda.

When we compare Figures 6-11 and 6-12 with Figure 6-2 (population) and Table A-1 in the Appendix, we find that the great population concentration of Asia, with more than half the population of the world, has failed, with the exception of Japan, to achieve significant pro-

ductivity. This failure is also characteristic of most of Africa and Latin America. The people of the world who enjoy the benefits of modern productivity in Europe, North America, the Soviet Union, Japan, Australia, New Zealand, and South Africa are still a small minority of the total—surely, no more than one-third of world population.

GNP is more than a measure of productivity, however. Since goods and services produced really add up to wealth, the map of GNP (Figure 6-15) is an index of the rich and the poor countries of the world. The industrial countries are the wealthy countries, and those less industrialized are the poor ones. As a result, individuals in the industrialized world show per capita GNP figures many times larger than those in the nonindustrial countries. This new interpretation of GNP as wealth means that the majority of the people of the world are very poor, and the wealthy living in rich, industrialized areas are only a relative few.

Per Capita Calorie and Protein Intake: Measure of Living

Productivity and wealth are often thought of in terms of the production of the goods people use, but food consumption is perhaps the most fundamental and crucial of all. Having enough to eat is the minimum necessity of life. To live, either food must be produced by the family, as in Ramkheri, or something of equal value must be offered in exchange.

Figure 6-16 shows the basic per capita food calorie intake of people in different countries of the world. Like GNP, it is only a crude measure. The human body is a living machine that runs on the energy provided by food; the actual amount of energy (measured in calories) varies with particular circumstances. People in the tropics require less energy than do those in cold

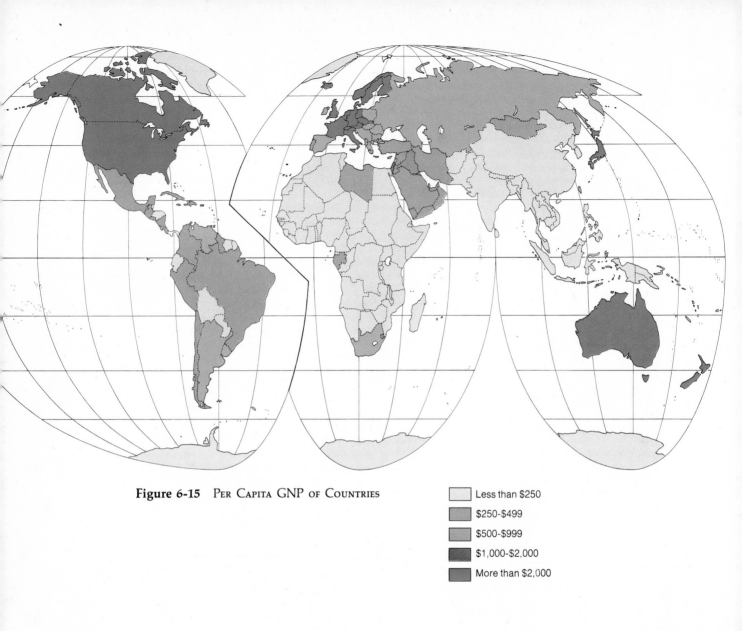

Figure 6-15 PER CAPITA GNP OF COUNTRIES

	Less than $250
	$250–$499
	$500–$999
	$1,000–$2,000
	More than $2,000

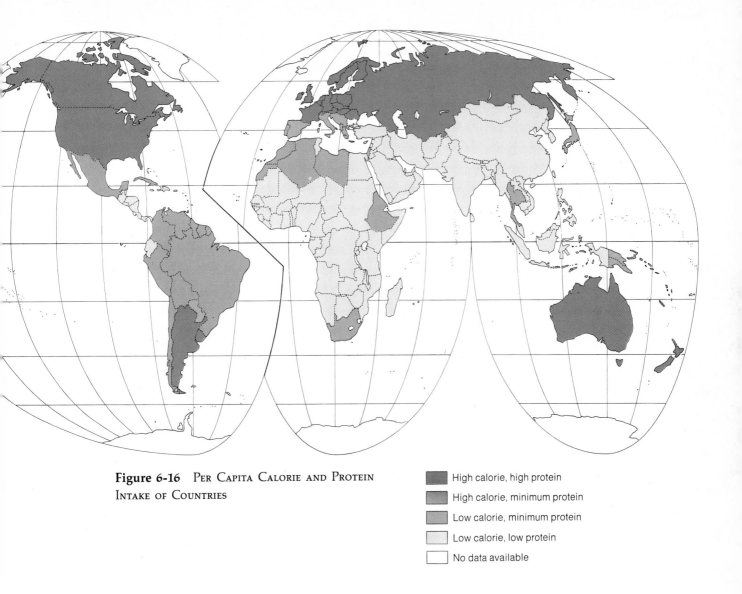

Figure 6-16 PER CAPITA CALORIE AND PROTEIN
INTAKE OF COUNTRIES

- High calorie, high protein
- High calorie, minimum protein
- Low calorie, minimum protein
- Low calorie, low protein
- No data available

climates where the body must use energy to provide additional body heat. Persons engaged in heavy work need more calories than do those who only sit and think. Small people need less food than large ones do. Growing children's needs are higher in terms of their body weights than adults'. Pregnant women need more than other women.

Moreover, the conversion of caloric energy is not a true measure of the quality of an individual's diet. A balance of essential nutrients is needed in building and maintaining body tissues. These nutrients include proteins, carbohydrates, fats, mineral elements, and vitamins. But caloric intake stands as a fair measure of diet quality as well as quantity because when the caloric content is low, nutrient quality also tends to be low.

As Figure 6-16 shows, a high caloric level is considered to be a minimum of 2,900 calories a day per person, and a low level, less than 2,500 calories. Several countries have caloric averages of less than 2,000 per day.

Just as industrial energy supply varies significantly from place to place in the world, so does the energy fuel for human activity. Parts of the world suffer severe shortages of human energy fuel in the same way that some areas are short of modern industrial fuels. Does dietary variation follow the pattern of energy consumption and GNP with a division of the world into rich and poor areas?

NEWSPAPERS AND RADIOS PER 1,000 PERSONS: MEASURE OF ISOLATION

Of course, life is more than food, production, and wealth. In today's world we think that it is important that people be informed about what is going on in their own locality and the larger world. Rapid dissemination of information is accomplished by modern means of communication, particularly the newspaper, radio, and television. Since these modern inventions make use of industrialized techniques in their production and operation, we would expect them to correlate closely with modern, industrialized areas. As Figures 6-17, 6-18, and 6-19 show, this is indeed the case.

Modern techniques of communication are more than just another measure of wealth. They offer ways to control and influence peoples' thoughts and actions in order to produce change and break traditional ways. Conversely, lack of information is a measure of cultural isolation.

In small villages like Ramkheri news must be spread by word of mouth. There are no newspapers, and much of the population is illiterate. A few radios owned or provided by the government offer the only regular contact with the world beyond. The maps show areas where similar problems of contact may exist in the world.

MODERN TRANSPORT SYSTEMS

Figures 4-6, 4-7, and 4-8 together could be considered a map of the distribution of the modern metropolitan-oriented system. They show modern means of transportation—roads, ocean-shipping routes, railroads, and major air routes. The modern world depends on them to make the long-range connections that are essential to the regional specialization needed to make this way of life possible.

Roads, railroads, air routes, and sea routes have been developed to move goods and people from one part of the world to another. The maps indicate the transport network over which goods move from one place to another. Each route can be seen to connect smaller routes to the modern interregional flows. Obviously, some links are much more important and far busier than others, but there is some flow

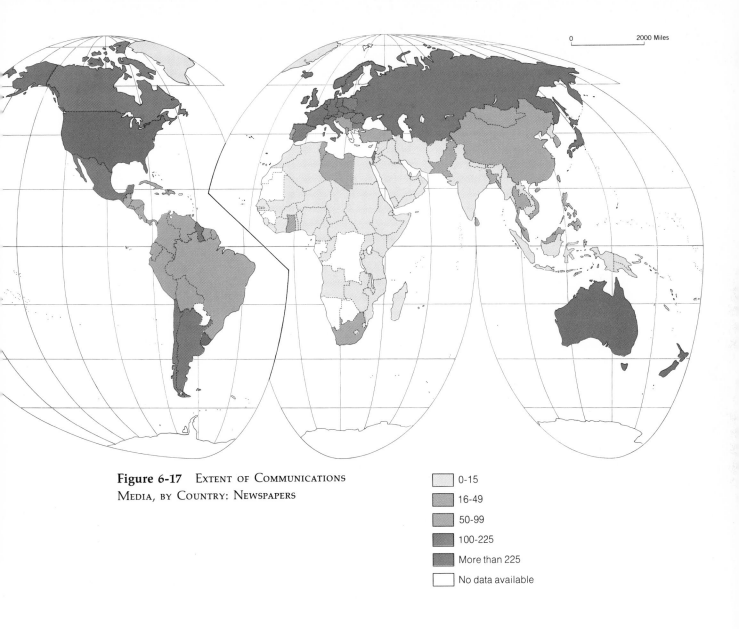

Figure 6-17 EXTENT OF COMMUNICATIONS
MEDIA, BY COUNTRY: NEWSPAPERS

	0-15
	16-49
	50-99
	100-225
	More than 225
	No data available

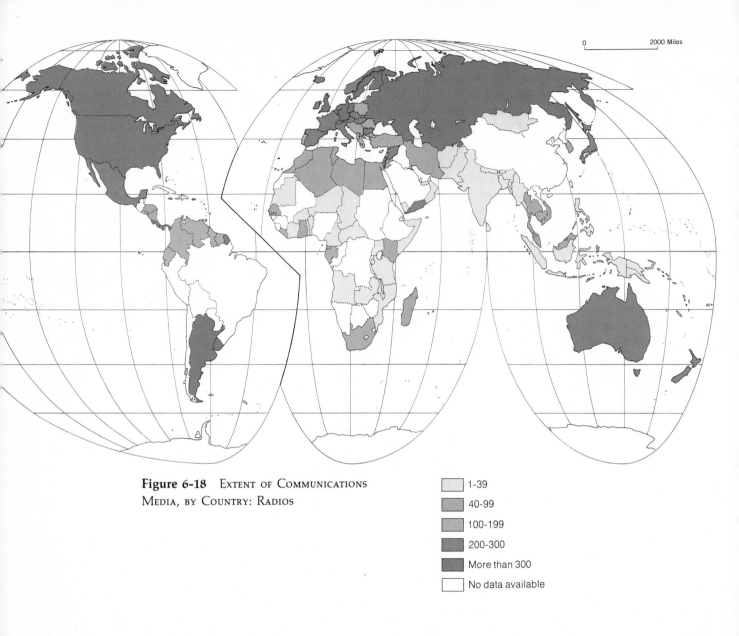

Figure 6-18 Extent of Communications
Media, by Country: Radios

	1-39
	40-99
	100-199
	200-300
	More than 300
	No data available

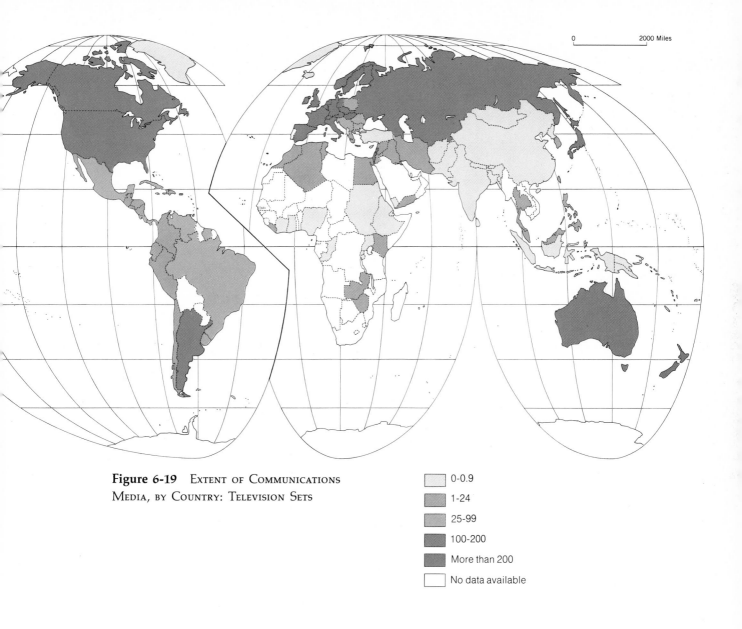

Figure 6-19 EXTENT OF COMMUNICATIONS
MEDIA, BY COUNTRY: TELEVISION SETS

0   2000 Miles

- 0-0.9
- 1-24
- 25-99
- 100-200
- More than 200
- No data available

Economic productivity, measured by energy consumption, and wealth, measured by gross national product, offer ways of separating countries into industrial and nonindustrial groupings.

1. Table A-1 lists countries of the world in terms of their per capita energy consumption and GNP.

   a. Separate this list into industrial and nonindustrial countries.

   b. Then group each of your lists regionally by continent. Do the per capita listings provide an adequate measure of political power?

2. The same table lists countries of the world in terms of their total energy consumption and GNP.

   a. Now rank the countries in your industrial and nonindustrial groupings by aggregate energy consumption and GNP, and compare the two lists.

What are your conclusions about political power? Is population a factor? How? Does population make a difference between industrial countries and nonindustrial ones? Are there underdeveloped countries with more energy consumption and total GNP than industrial countries? Are they stronger politically?

The Soviet Union and the United States are commonly considered to be the great political powers of the world, yet figures indicate that the Soviet Union's production is far below that of the United States. What other factors contribute to the apparently greater political strength of the Soviet Union when its strength is considered in relation to its energy consumption and GNP? Does the importance of the Soviet Union give a clue as to the relative importance of per capita GNP?

through all of them. Even in the wilderness frontier which lacks enough flow to justify the building of rail tracks or a road, airplanes are used to bring supplies and workers into an area and to carry out specialized resources.

The contrasts between industrial and nonindustrial parts of the world are obvious, and yet even in the United States and Europe there are gaps in the network, and such gaps are evident in the Soviet Union. Modernization in Canada, South Africa, and Australia seems to have occurred only in certain areas.

There are, however, surprisingly well developed transport networks in parts of the underdeveloped world. Note the dense patterns in parts of Argentina and Brazil. Even India has a transport network that appears more highly developed than the Soviet Union's. A great deal of money has been spent in these areas developing modern interregional transport networks.

To see the full implications of the transport maps, one must superimpose them on both the political map (Figure 6-20) and the population map (Figure 6-4). Then the dense networks and the large gaps take on significant meaning.

Analysis of the maps has suggested more than an uneven distribution of people in the world. It shows a world that can be divided into industrial nations and underdeveloped nations which have not reaped the benefits of industrialization. In this context Washington and Ramkheri appear as models applicable to large areas of the world. But even industrialized countries have less developed areas and underdeveloped countries have developed areas.

Countries: Decision-makers and Controllers

Although the data on some of the maps in this chapter have been given in terms of countries, the focus has been on continental regions of the world. But countries are the basic operational units of government today and are much more important in the functional world than continents or any other convenient divisions.

National governments are decision-makers and controllers. They set the basic rules for individuals and organizations within their boundaries. They also represent the interests of their people in relations with other countries, control the movement of trade, and set the climate of war or peace.

Since almost the whole land surface of the earth has been incorporated into one man-created political unit or another, the political map (Figure 6-20) seems to suggest the dominance of man over environment. Only the icy wilderness of the continent of Antarctica has not been brought under the political control of some government—and even there countries have made political claims.

The political map points up the compartmentalized nature of the human use of the earth. Scientists tell us that all human beings belong to the same species and that similarities among peoples should be more important than differences. Yet the basic political divisions indicate that today, thousands of years after man's first appearance on earth, he has yet to create a single world-wide organizational system. The contemporary world is made up of more than 140 separate political entities, each claiming legal power over the people who live there. Each government expects the primary loyalty of its people. The people see themselves as citizens of a particular political state, not of the world, and patriotism is considered one of mankind's primary ideals.

Like the population map, the political map reveals an uneven arrangement of countries. The vast Soviet Union encloses 15 per cent of the land surface of the earth. China, Canada, the United States, Australia, and Brazil are also huge areas. At the other extreme, Singapore is a single metropolitan center which is politically independent. There are a few vestiges of feudal kingdoms such as Liechtenstein, Monaco, and Andorra, which are too small to show up on a world map. The area of Monaco is less than one square mile.

COUNTRIES BY SIZE AND POLITICAL POWER

Table 6-4 lists the countries of the world by land area. We might expect size to be a good measure of economic strength and political power. Even though we know that growth potential and other resources are not evenly spread over the world, we still might expect that large countries are more likely to have adequate resources for development than small countries.

The table further points up the variations between countries. How much larger is the Soviet Union than Canada or China, the next largest countries? How many times larger is it than Mexico? Notice that it is more than 100 times larger than each of the 67 smallest countries or territories. Most countries are, in fact, the size of individual states of the United States.

Today, in sharp contrast to the world of 30 years ago, almost all political units are independent. The only large dependent territories are Angola and Mozambique, Portuguese colonies in Africa. But there are still a number of dependent smaller places, particularly islands and port areas, that do not show up easily on a world map. Does the size of a country correlate with its political power today?

A number of the patterns already examined might well be keys to political power in the world. Imagine the political map (Figure 6-20) superimposed on the population map of the world (Figure 6-2). Some of the largest countries in the world are in areas of low population density; some of the smallest are in areas of

Figure 6-20 Political Map of the World

Table 6-4 Countries Listed by Land Area (square miles)

Country	Area	Country	Area
USSR	8,600,000	Somalia	246,000
Canada	3,852,000	Central African Republic	241,000
China	3,692,000	Botswana	232,000
United States and possessions	3,681,000	Malagasy	227,000
Brazil	3,286,000	Kenya	225,000
Australia	2,968,000	France	210,000
India	1,226,000	Thailand	199,000
Argentina	1,072,000	Spain	195,000
Sudan	968,000	Cameroun	184,000
Algeria	920,000	Sweden	174,000
Zaire	906,000	Morocco	172,000
Greenland	840,000	Iraq	168,000
Saudi Arabia	830,000	Paraguay	157,000
Mexico	762,000	Rhodesia	150,000
Indonesia	753,000	Japan	143,000
Libya	679,000	Congo (Brazzaville)	132,000
Iran	636,000	Finland	130,000
Mongolia	604,000	Malaysia	128,000
Alaska	586,000	Norway	125,000
Peru	496,000	Ivory Coast	125,000
Chad	496,000	Poland	121,000
Niger	489,000	Italy	116,000
Angola	481,000	Philippines	116,000
Mali	479,000	Southern Yemen	111,000
South Africa	472,000	Ecuador	109,000
Ethiopia	472,000	Upper Volta	106,000
Colombia	440,000	New Zealand	104,000
Bolivia	424,000	Gabon	103,000
Mauritania	398,000	Spanish Sahara	103,000
United Arab Republic	387,000	Yugoslavia	99,000
Tanzania	363,000	West Germany	96,000
Nigeria	357,000	Guinea	95,000
Venezuela	352,000	United Kingdom	94,000
Pakistan	345,000	New Guinea Territory	92,000
South-West Africa (Namibia)	318,000	Ghana	92,000
Mozambique	304,000	Rumania	92,000
Turkey	301,000	Laos	91,000
Chile	292,000	Uganda	91,000
Zambia	291,000	Papua Territory	86,000
Texas	267,000	Guyana	83,000
Burma	262,000	Oman	82,000
Afghanistan	250,000	Senegal	76,000

Table 6-4 *(Continued)*

Country	Area	Country	Area
Yemen	75,000	Netherlands	14,000
Uruguay	72,000	Portuguese Guinea	14,000
Syria	71,000	Taiwan	14,000
Cambodia	69,000	Belgium	12,000
South Vietnam	67,000	Lesotho	12,000
Tunisia	63,000	Albania	11,000
Surinam	63,000	Equatorial Guinea	11,000
North Vietnam	61,000	Burundi	11,000
ILLINOIS	56,000	Haiti	11,000
Bangladesh*	55,000	MARYLAND	11,000
Nepal	54,000	Rwanda	10,000
Greece	51,000	British Solomons	10,000
Nicaragua	50,000	Afars and Issas	9,000
Czechoslovakia	49,000	British Honduras	9,000
North Korea	47,000	Qatar	9,000
Malawi	46,000	El Salvador	8,000
Cuba	44,000	Israel	8,000
Dahomey	43,000	New Caledonia	7,000
Honduras	43,000	Fiji	7,000
Liberia	43,000	Swaziland	7,000
Bulgaria	43,000	Kuwait	6,000
Guatemala	42,000	Portuguese Timor	6,000
East Germany	42,000	New Hebrides	5,000
Iceland	40,000	Falkland Islands	5,000
South Korea	38,000	Bahamas	4,000
Jordan	38,000	Gambia	4,000
Hungary	36,000	Jamaica	4,000
Portugal	36,000	Lebanon	4,000
French Guiana	35,000	Cyprus	4,000
Austria	32,000	Puerto Rico	3,000
Panama	29,000	Canary Islands	3,000
Sierra Leone	28,000	Sikkim	3,000
Ireland	27,000	Brunei	2,000
Ceylon	25,000	Trinidad and Tobago	2,000
Togo	22,000	Cape Verde	2,000
Costa Rica	20,000	French Polynesia	2,000
Dominican Republic	19,000	Western Samoa	1,000
Bhutan	18,000	Reunion	1,000
Denmark	17,000	Luxembourg	1,000
Switzerland	16,000		

*Figure for Bangladesh is that given for East Pakistan in *Collier's World Atlas and Gazetteer,*
Crowell Collier and Macmillan, Inc., New York, 1955, p. 131.

SOURCE: *Goode's World Atlas,* 13th ed., rev., Rand McNally, Chicago, 1971, p. 189.

high density. Because few countries have their populations spread evenly over their territories, the population map shows countries in a new perspective. Huge territories such as the Soviet Union, China, Canada, and Brazil take on new shapes.

Except in Europe, the basic concentrations of population within countries are widely separated from those of their neighbors, and most political boundary zones are in sparsely occupied areas. What does such evidence indicate about the origins of these countries? And what might the implications be in terms of interaction and cultural similarity?

The Juxtaposition of Modern and Traditional Life

In terms of productivity, wealth, and power, the countries of the world can be considered from another approach using our examples of Washington and Ramkheri. It is obvious that all countries do not fall nicely into either the industrial group or the nonindustrial one. The world's countries ranging from the most affluent to the least productive can perhaps better be thought of as a continuum. The United States and Kuwait (Figure 6-15) occupy the high side of that continuum, Burundi and Rwanda the low. All other countries lie somewhere between. Some are nearer the United States, some nearer Burundi; others lie near the middle.

In the box on page 168 you were asked to list separately the industrial and the nonindustrial countries, but in reality there are no completely industrial or nonindustrial countries. Even the United States at the high end of the scale is not fully industrialized. It is just the most highly industrialized country according to the indices. And at the other extreme, Burundi is not totally undeveloped.

In a country such as the United States, modern life overwhelmingly dominates, both because of the large proportion of urban-metropolitan population and the inroads of modernity on rural people. Even farmers are specialists producing products for sale in other places and acting as consumers of the vast range of goods and services that are part of the modern system. Nevertheless, even in the United States, there are enclaves of locally based life on Indian reservations, on subsistence farms in the South, and in the ghettos of large cities where subsistence might come in the form of a welfare check.

On the other hand, in Burundi most of the people may live in agricultural villages even more primitive than those in India and coax the adjacent land to produce enough to keep their families alive. Yet Burundi has a capital city with a population of about 50,000 people and with modern buildings, automobiles and trucks, an airport, and telephones. Moreover, large numbers of the population work in government offices and in stores and shops in much the same way people do in Washington. In many ways the lives of these city workers, who look very different from Washingtonians, are more like those of people in the Washington metropolitan area than like the lives of the people out in the rural parts of Burundi. What is more, Burundi is represented by a delegation in the United Nations and in Washington. More and more of its farmers are raising coffee, cotton, and tea for shipment to markets in Europe and the United States. In Burundi, however, the traditional system still predominates.

And so in any country there are bound to be examples of both the modern and traditional ways of life. Just as the composition of a country cannot be determined by using national population-density figures, so the character of its life cannot be determined by examining only national indices of production and wealth. Some other way is needed to determine the distribution of the modern and traditional patterns of life.

SELECTED REFERENCES

Boyd, Andrew. *An Atlas of World Affairs,* 6th ed., Praeger, New York, 1971. An introduction to the world from a political point of view. Contrast this to the perspective presented in this chapter.

Espenshade, Edward, ed. *Goode's World Atlas,* 13th ed., revised printing, Rand McNally, Chicago, 1971. Available in paper and hard covers. A well-balanced world atlas containing world and regional reference maps, world and regional maps of selected measures of economic, environmental, and social characteristics, and an extensive place-name index.

Spencer, J. D., and Thomas, W. H., Jr. *Cultural Geography,* John Wiley, New York, 1969. This basic cultural geography contains an amazing array of thematic maps of the world. Although no sources are given, it has maps not available in the other atlases listed here.

Aldine University Atlas, Aldine, Chicago, 1969. An inexpensive college atlas that includes thematic maps such as those in this chapter.

Encyclopaedia Britannica Atlas, Encyclopaedia Britannica, Chicago, 1969. An elaborate atlas containing exceptionally good world and regional place maps. These maps are complemented by an extensive place index. There is also a very useful selection of world distribution maps on population, economic, cultural, and environmental characteristics.

Oxford Economic Atlas of the World, 4th ed., Oxford U. Press, New York, 1972. Probably the most complete set of thematic maps of the world of any atlas; covers the environment,

society and politics, demography, and disease as well as maps of primary and secondary production. Its maps of surface and air communications are especially helpful in indicating the reach of the modern interconnected system. Brief commentary is provided about the information displayed on some maps, and there is a valuable appendix of economic and population data covering all countries of the world for which there are United Nations statistics.

Statistical Sources:
Demographic Yearbook, United Nations.
Statistical Yearbook, United Nations.
Yearbook of Food and Agricultural Statistics, Food and Agriculture Organization of the United Nations.

These three yearbooks provide a comprehensive set of statistical information on population, food and other agricultural production, and economic characteristics for most countries of the world. As sources for recent comparable information for countries the United Nations references are among the best that are readily available.

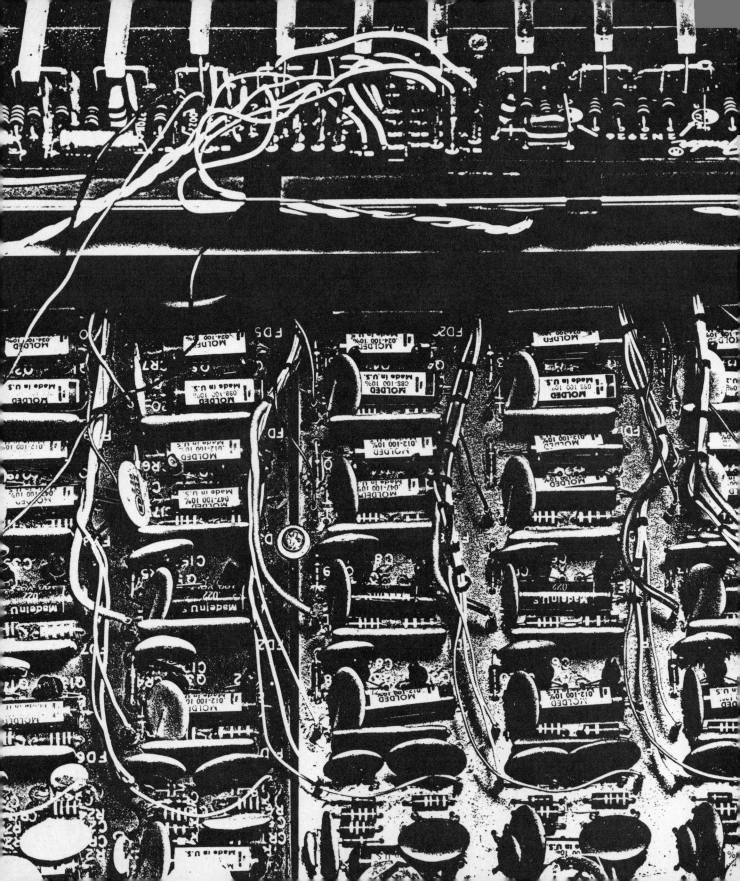

PART THREE   INDUSTRIALIZED AREAS:

WHERE THE MODERN,

INTERCONNECTED SYSTEM

DOMINATES

Chapter 7　The United States: Changing Evaluations of the Environment

T HE UNITED STATES is recognized as the most successful industrial country, at least in economic terms. By far the wealthiest large country in terms of gross national product (GNP) and per capita wealth, its GNP is almost as great as the total for the eleven other most populous countries in the world. Thus, the United States would seem to be the most likely place to examine the full workings of the modern, interconnected system.

The map of the world shows that the United States has the major asset of large physical size. Such a large area offers tremendous variations in environmental conditions, and regional variety is ideally suited to the workings of the modern system. With modern transportation and communication the resources of particular areas of the country can be readily drawn together. The huge environmental base of the United States offers sufficient national resources for industry without importing supplies. The United States is not, in fact, self-sufficient, but it is more self-sufficient than smaller industrial countries.

Today we see the United States as a modern industrial country more than 350 years after its first settlement by Europeans. In that time it has evolved into the modern, interconnected system we know today. Early settlers viewed only a small portion of the present area of the country; they also viewed it in very different terms. Life in the United States has evolved from this particular heritage, a heritage of more than just ideals and values. The present United States of cities, farms, transport and communications networks, and regional specialization has emerged from decisions and investments made in other times in response to very different economic and cultural conditions.

In this chapter we will examine the environmental base of the United States and will see how that base has been re-evaluated and utilized as times have changed.

It is 2,800 miles across the continental United States and 1,600 miles from the southernmost point in Texas to the 49th parallel at the Canadian border (see Figure 7-1). This, of course, does not include Alaska, which is twice the size of Texas, and Hawaii. Despite its large size, the continental United States lies entirely within the middle latitudes. Hawaii is the only part of the country in the tropics. In contrast, the Arctic Circle passes across northern Alaska, and the bulk of the state is north of 60°. Even the southern tip of the Alaskan panhandle reaches only as far south as 55°.

The Environmental Base: Natural Resources of Climate, Land, Sea, and Location

To examine the resource base of the United States, we start by constructing a model of environmental conditions. First, let us turn to agricultural potential. This means investigating the potential for plant growth in particular areas, ignoring for the moment the effect of technology. If the United States were a fertile plain with adequate soil moisture everywhere, remembering Chapter 5, what would be the plant growth potential of particular areas? What regions might we expect to be the most productive?

Plant Growth Potential

The most important factor in determining growth is evapo-transpiration, and it, in turn, depends largely on latitude. Potential evapo-transpiration is greatest along the southern border of the country, for the closer a piece of terrain is to the equator, the greater the sun energy for plant growth. The farther from the equator, the less energy is available. Thus, the tip of Florida, southern Texas, and southern Arizona appear to have the greatest potential

for evapo-transpiration of any part of the United States (refer to Figure 5-4). What would we expect the possibilities of Hawaii to be? Or the northern border of Alaska? In general, how much greater is the evapo-transpiration potential in the South as compared with that in the North? What is the relative potential of the present agricultural area of the Middle West?

The length of the frost-free season, also related to latitude (see Figure 6-9), is another dimension of growth potential. Compare the possibilities across the country. How much longer is the growing season in the South? How does it compare with the growing period in Iowa?

If we put the two measures of growth potential together, we can see that although the growing period through the year in the Middle West may be only half what it is in southern Florida, the Middle West has about as much potential, and plants can be expected to grow rapidly during the frost-free period. Still, farmers in the Middle West would have problems very different from those of farmers in the extreme South or from those farther north.

As we know, sun energy is just part of the growth story. There must also be adequate moisture in the soil for plants to grow. Remember that incoming moisture to the soil during the year does not regularly correlate highly with potential energy, and there are times of moisture deficiency and times of surplus. Commonly the surplus occurs during the cool season when there is little evapo-transpiration; the deficit occurs during the hot season when evapo-transpiration is maximum. Parts of the country near the Pacific Coast and in the Northeast have particularly large surpluses of moisture, while most of the western half of the country has a deficit. There is also less surplus in the southeastern part of the country than farther north.

This problem of moisture availability markedly changes the evaluation of different regions.

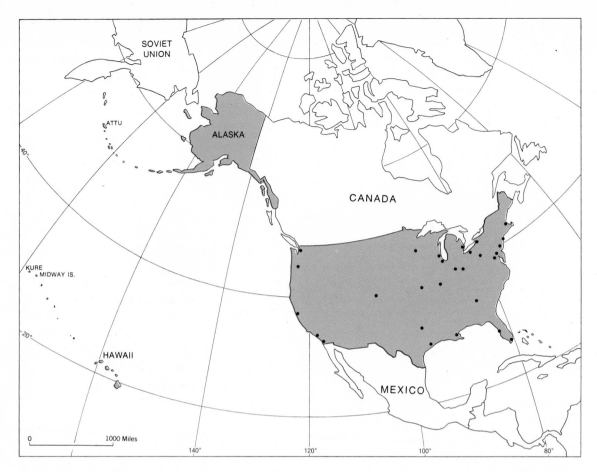

Figure 7-1 LOCATION AND GEOGRAPHIC EXTENT
OF THE UNITED STATES

• U.S. Metropolitan areas over 1 million population

In the Southwest, for example, artificial water supplies are necessary for agriculture. By contrast, New England and most of the rest of the East have enough moisture for the evapo-transpirational possibilities whatever they might be. Remember, however, that the moisture-holding capacity of soil may vary considerably. Any place in seemingly humid areas may suffer dry years or weeks of inadequate moisture during the growing season.

The world maps of evapo-transpiration and moisture deficiency have already indicated another feature of the environmental base that affects plant growth: the nature of the surface terrain. Temperature and moisture conditions vary markedly with differences in altitude. (Maps we have already seen establish the outlines of the major relief features of the country.) Except for the major mountain areas most of the rest of the country is level rather than rugged (Figure 7-2). There is some rough terrain, however, in the Ohio Valley, across southern Missouri and northern Arkansas, and in the Great Lakes area. But areas of lowland along the Atlantic Coast from Cape Cod southward and along the Gulf Coast lie so low that they are poorly drained and are swampy. Poorly drained land has also been a problem in the lower Mississippi Valley.

The natural vegetation that European settlers found in the United States as they moved across the country provided a reasonably good measure of environmental growth possibilities. Forest vegetation dominated the areas with moisture surpluses, but timber growth was much more rapid in the areas of high evapo-transpiration in the South. Fine forests of giant trees—some of the largest in the world—were found in the areas with surprisingly mild winters and high-moisture surpluses in the Pacific Northwest.

Along the western margins of the eastern forests was tall grass mixed with outlying woods. These grasses and woods, in turn, opened into almost treeless grasslands farther west, where moisture deficits were the rule. Trees appeared almost exclusively along stream valleys which contained an adequate supply of soil moisture. Westward toward the Rocky Mountains the grasses became shorter and shorter as moisture deficits grew more severe. Then, in areas of the Southwest with extreme moisture deficiencies, the grasslands gave way to the patches of xerophytic plants which form the incomplete cover of vegetation that characterizes the deserts of the world.

Forests are found on most mountain slopes of the country because higher altitudes result in lower temperatures and, thus, lower evapo-transpiration. At the same time there is usually greater precipitation in the cooler mountain temperatures. Only on the highest slopes of the West, where temperatures are so cold that growth is stunted, are mountains bare.

PRESENT EVALUATION OF GROWING AREAS

The location of agriculture, herding or ranching, and lumbering enterprises depends on variations in growth potential from place to place and their relation to markets. Decisions regarding location are based on far more than specific environmental tolerances and particular conditions for maximum growth of each commodity. Certain areas may be ideally suited to a range of crops and livestock and also to producing timber. Thus, there is potential competition for land uses in some areas, and particular choices have to be made. The question is not where will a crop grow best, but where can it be most profitably produced in terms of growing conditions, dietary preferences, market locations,

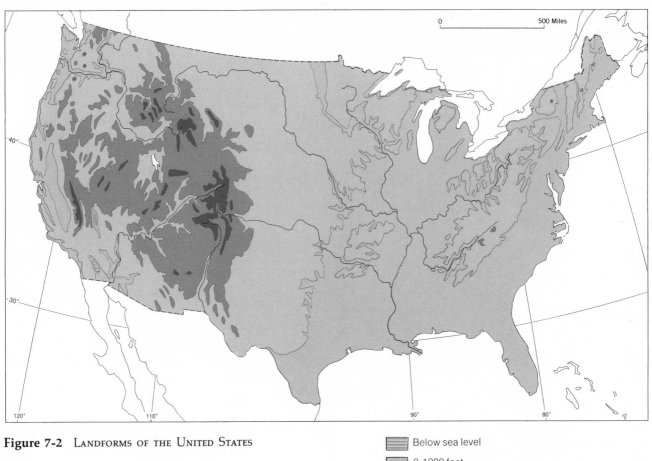

Figure 7-2 Landforms of the United States

0   500 Miles

Below sea level
0-1000 feet
1000-5000 feet
5000-9000 feet
Above 9000 feet

government farm policy, and other factors. Moreover, these evaluations must be made on a national scale rather than on a regional or local one.

Today the major wheat-producing areas of the country are not those where wheat grows best. Corn brings a better financial return than wheat on the best wheat lands; thus, wheat is grown near the outer limits of its range, beyond the range of corn and most other crops. On the Great Plains, wheat can be grown on large mechanized farms on land where few other crops within the dietary patterns of Americans will grow. Because of the necessity of raising wheat on marginal land, much effort has been made to breed strains that will grow in such areas and to develop farming practices and equipment especially for wheat in such an environment. In parts of the South, where both tobacco and corn grow well, the farmer's chief concern is with tobacco, a crop narrower in range than corn, one strongly supported by government subsidy, and one that brings a higher financial return per acre than corn—as long as tobacco smoking remains a part of the life style of most Americans. In general, the narrower the range of the crop and the more specialized its use, the higher its priority is as a crop to be planted near the location ideal for its growth.

Agricultural land use also brings a higher return per acre than does ranching or lumbering, and it generally has priority. As a result, ranching has largely been pushed to the drier margins, and forestry to the rougher, more poorly drained lands or those with soil ill-suited to agriculture.

Figure 7-3 shows the complexity of primary land use in the country. One can see the areas of grazing land beyond the western margins of middle western and southern agriculture, and beyond them the timber regions largely in the rougher areas of the country.

Notice how much of the country has been cleared and put into agricultural use. In the interior lowland almost all the land is in farm use, and virtually all of it is in crops. In contrast, the traveler through most of the agricultural regions of the Eastern Seaboard and the South is struck by the large amount of farmland kept as woods, pasture, and unused fields.

Actually, the amount of cropland in the country has declined slightly since 1930. In 1969 it was 6 per cent lower than in 1930, despite the fact that the population of the country had risen 55 per cent. This was a measure of the agricultural revolution that occurred in that period, thanks to the development of hybrid seeds, insecticides, fertilizers, and sophisticated machinery and to the greater capitalization of agriculture. The new technology greatly increased per acre yields, reducing the amount of land needed for production. The family farm was replaced by the corporate farm as the primary unit of agricultural production, and the new highly capitalized units invested money and labor in the best land, rather than in the marginal land. Here the evaluation was national, not local or regional.

In any case, there is little doubt that the important agricultural areas of the country lie in the South and the Middle West. Notice that the middle four regions in Table 7-1—Middle West, Great Plains, Southeast, and South Central—account of 69 per cent of the land in farm, 83 per cent of the farms, 82 per cent of the cropland, and 72 per cent of agricultural sales. Here is both the prime stronghold of agricultural America and the environmental base that produces most of the staple agricultural products upon which the country depends.

Table 7-1 also provides insight into the rural sector of United States culture and politics. One can outline the areas of the country where rural, small-town culture continues strong and where the "farm bloc" remains prominent in politics.

THE REGIONALIZATION OF AGRICULTURE

See what you can infer from Table 7-1.

1. What are the major agricultural areas of the country? How important are they?

2. The states of the Middle Atlantic, Middle West, and Southeast have a larger share of the nation's farms than of its farmlands. On the other hand, the Great Plains, South Central, Mountain, and Pacific areas show the opposite correlation. What conclusions do you draw?

3. The Middle West and Great Plains areas have a higher proportion of cropland than land in farm. What inferences do you draw from those figures?

4. The areas of Middle West, Southeast, Pacific, Middle Atlantic, and New England have a higher proportion of total agricultural sales than of either cropland or land in farm, but in the other areas the opposite is true. What characteristics of agriculture might these variations indicate?

5. Some areas show relatively high proportions of agricultural sales as compared with livestock sales; in others the reverse is the case. What is the significance of these variations? Keeping in mind that livestock sales nationally are about one-third larger than crop sales, indicate the relative importance of livestock and crops in each region.

6. See if you can develop from the table a general classification of agriculture in each region.

7. Compare regional agricultural sales with their share of population. What are the major surplus and deficit areas of agricultural production? What changes in crops and livestock would you predict?

Table 7-1 Regionalization of Agriculture in 1969 (per cent of totals)

Region	Number of Farms	All Land in Farm (acres)	Cropland* (acres)	Total Agricultural Sales, 1971 (dollars)	Crop Sales, 1971 (dollars)	Livestock Sales, 1971 (dollars)	Regional† Population
New England	1.1	0.6	0.6	1.6	1.2	1.8	5.8
Middle Atlantic	6.1	2.6	4.1	5.9	4.3	7.1	21.7
Middle West	32.9	17.8	35.6	31.6	27.7	33.8	25.3
Great Plains‡	9.5	21.5	23.0	11.3	10.4	13.0	2.5
Southeast	25.9	11.3	10.2	16.7	20.8	13.9	18.0
South Central†	14.6	19.2	13.5	11.9	11.2	12.4	9.5
Mountain	4.5	20.9	8.2	8.3	6.1	9.7	4.1
Pacific	5.4	7.1	4.8	12.6	18.3	8.3	13.1

*For cropland data, Alaska and Hawaii are not included in total data for the United States.
† Figures for "Regional Population" in Tables 7-1 through 7-3 were taken from: U.S. Bureau of Census, *Number of Inhabitants*, Vol. 1, Washington, D.C., 1971.
‡"Great Plains" includes North Dakota, South Dakota, Nebraska, and Kansas; "South Central" includes Arkansas, Louisiana, Oklahoma, and Texas.
Source: U.S. Census Bureau, *Statistical Abstract of the United States: 1972*, Washington, D.C., 1972, Tables 974, 993, and 1,011.

184 / INDUSTRIALIZED AREAS: WHERE THE MODERN, INTERCONNECTED SYSTEM DOMINATES

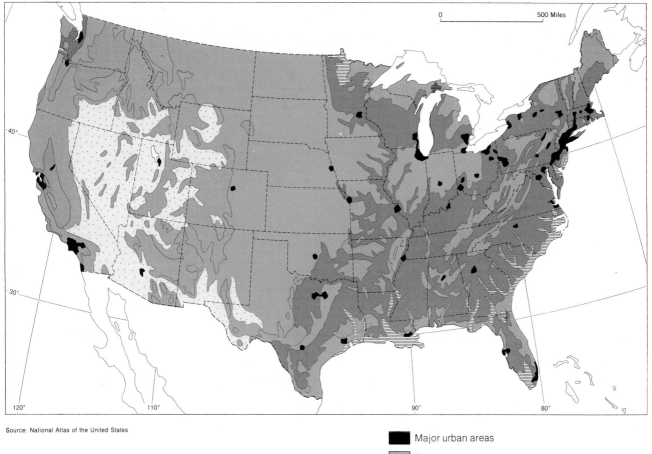

Source: National Atlas of the United States

Figure 7-3 PRIMARY LAND USES IN THE
UNITED STATES

Major urban areas
Crop and pasture land dominant
Crop, pasture and woodland intermixed
Forest and woodland
Desert and desert scrubland
Swamp and marsh

Table 7-2 REGIONALIZATION OF FORESTRY IN 1970 (per cent of totals)

REGION	LAND IN FOREST (acres)	COMMERCIAL FOREST LAND AS OWNED (acres)	LUMBER PRODUCTION (board feet)	REGIONAL POPULATION
New England	4.5	6.5	2.9	5.8
Middle Atlantic	7.1	9.9	3.5	21.7
Middle West	11.3	16.3	6.5	25.3
Great Plains	0.7	0.9	0.3	2.5
Southeast	20.8	30.6	25.5	18.0
South Central	8.2	10.3	12.1	9.5
Mountain	18.6	12.0	8.4	4.1
Pacific	28.8	13.5	40.8	13.1

SOURCE: U.S. Census Bureau, *Statistical Abstract of the United States: 1972*, Washington, D.C., 1972, Tables 1,041 and 1,042.

Turning to Table 7-2, we can see that the statistics substantiate what we already inferred from Figure 7-3, the map of land use: The distribution of forests and productive forest land is very different from that of agriculture. Surprisingly, almost half of the forests of the country are in the Mountain and Pacific regions, not in eastern United States, which once was densely forested. However, the Pacific figures also include Alaska with its vast forest tracts. The two western areas with all their forests account for less than 30 per cent of commercial timberland, indicating that much of their timber is inaccessible in today's economic terms. But these two areas still account for almost $5 out of every $10 of lumber sales. Obviously the forests that are commercially operated are very productive. In view of the giant Douglas fir and redwood trees in the Pacific Coast area, this is not surprising. Notice that the Pacific region produces almost half of all the lumber.

The Southeast is the other great forest and forest-products region. It has the largest share of commercial forest land of any region and accounts for one-fourth of all lumber production. The Southeast is also the most important region for pulp and paper production.

Like the agricultural producing areas, the major timber-producing areas are far from the major center of population in northeastern United States, and we can expect considerable long-haul movement of timber.

FISHING AND MINING RESOURCES

Growth potential for plants is essential for most primary production, but fishing and mining locate on the basis of different environmental factors. Fishing (limited in our discussion to saltwater fishing) depends, of course, on variations in the ocean environment: water temperature, salinity, depth, underwater terrain, and availability of plankton or other food. The waters off the United States have great variations, and fishing possibilities are strikingly different on either coast.

Not too distant from New England lie the Grand Banks, one of the most productive fishing grounds of the world. From New York south into the Gulf of Mexico, series of large shallow bays and estuaries are the habitats of seafood such as clams, oysters, and crabs. In the deeper water of the Atlantic and Gulf, fishing centers on the catching of menhaden, small fish that are a source of oil and meal which is used mainly as animal feed. In the Pacific the chief catch is large tuna and salmon.

Mineral deposits are produced by forces in the earth's interior and by energy from the sun. Stored in rocks from prehistoric times, they show little correlation with patterns of sun energy today. Moreover the conditions under which various minerals occur are usually very different. Areas with deposits of metallic minerals rarely have mineral fuels and vice versa.

In the United States, as elsewhere, the mineral fuels—coal, petroleum, and natural gas—are associated with sedimentary rocks. Although petroleum and natural gas are commonly found together, the major coalfields have a different pattern. Some petroleum is produced in important midcontinent coalfields of Illinois, Indiana, and western Kentucky, but little comes from the most important coalfields of the western Appalachians. The most important oil fields are along the Gulf Coast and in Texas, Oklahoma, and Kansas. Major production also comes from California. New fields of large proportions have recently been discovered along the remote "north slope" in Alaska. Questions of their development have included how to get the product to market from this distant, desolate place and, in addition, how to exploit the resource without severely damaging the fragile northern environment.

The major producing coalfields have been close to the large population centers of the country in the Northeast and Middle West, but oil and gas production has come from the South and Great Plains, far from the country's major markets. The result has been the development of an elaborate, long-range system built around oil and gas pipelines and special coastal tankers and river barges to move these fuels to the chief areas of consumption.

Metallic minerals—iron, copper, lead, gold, and silver—form in igneous and metamorphic rock, and appear in sedimentary rock only after they were eroded from the original deposit and redeposited as sediment. In the United States metamorphic and igneous rock are less common than sedimentary rock but are more widely scattered. They are generally in rougher, more remote areas, necessitating long-haul transportation. But, except for iron-ore, these minerals move in much smaller quantities than mineral fuels and carry higher per unit values. As a result, transportation is not as great an economic factor as it is with coal and oil.

More than three-fourths of the iron-ore, the most important metallic mineral, comes from mines around Lake Superior. Until recently this ore was shipped directly without processing, inasmuch as it had iron content of one-third or more. Now most of the ore of high iron content has been mined, and it is necessary to have large concentrating plants in the mine fields to reduce shipping costs.

In the same way, most of the specialty metallic minerals mined in the western United States are low-grade and require smelters and refineries to process the ores and reduce the bulk that must be shipped.

Limestone, clays, sand and gravel, and salt, among the nonmetallic minerals, are widespread, and major production occurs near large population centers; others are more localized. Sulfur comes from along the Gulf of Mexico

Table 7-3 REGIONALIZATION OF MINERAL PRODUCTION IN 1969 (PER CENT OF TOTALS)

REGION	VALUE OF MINERAL PRODUCTION (dollars)	REGIONAL POPULATION
New England	0.5	5.8
Middle Atlantic	8.8	21.7
Middle West	12.4	25.3
Great Plains	3.0	2.5
Southeast	8.5	18.0
South Central	43.7	9.5
Mountain	14.6	4.1
Pacific	8.5	13.1

SOURCE: U.S. Census Bureau, *Statistical Abstract of the United States: 1972,* Washington, D.C., 1972, Table 1,084.

in Louisiana and Texas, phosphate rock for fertilizer from Florida, and the specialty mineral salts from arid parts of California.

The regional locations of mineral production can be seen in Table 7-3. The populous area of the New England and Middle Atlantic states have less than 10 per cent of the national output of all minerals. The largest regional output, with almost 50 per cent of the total, is in the South Central area, showing the tremendous value of oil and gas as compared with any other mineral. The Mountain region of the West, with all its specialty metallic ores and such exotic minerals as uranium, is a distant second in mineral output. All other areas except the Great Plains show less than their share of mineral resources in terms of their population.

RESOURCE UTILIZATION AND MARKETS

The resource base for primary production, as the United States uses it today, shows that most primary production is outside the heavily populated Northeast. The populated area of the

Pacific generally seems to have an adequate and comprehensive base; we are dealing, however, with gross categories, and there are exceptions. This discrepancy between the places of production and the places of consumption necessitates a flow of food, timber, minerals, and other products toward the Northeast. It is generally recognized that there are one-way flows of raw materials and food from the underdeveloped countries into the United States in exchange for manufactured goods. But within the United States there is a similar trade flow between the highly industrialized, metropolitan areas of the Northeast and Midwest and the primary producing areas in other parts of the country.

ENVIRONMENTAL AMENITIES AS RESOURCES

There is also another aspect of the environmental base to consider. Man has long been conscious of the amenities of the environment and of their effect on his life style. The wild beauty of nature is valued by many people, and the recreational possibilities of the environment are important. Unlike other primary resources, however, the amenities cannot be harvested and carried somewhere else; the market must come to the environment.

Because of the limitations of transport and short vacations, the amenities that most people have been able to experience firsthand have been largely those of their own locality. But since the pattern of population is primarily the result of opportunities for employment, not recreation, most people have found themselves in crowded metropolitan centers or in vast areas of cleared agricultural land. Traditionally they have "made do" with whatever local amenities appeared in their area: nearby creeks and woods or man-made parks.

But modern transport, today's affluence, shortened work hours, longer vacations, and early retirement have opened up all but the most remote wilderness areas of the country. Now recreational resources of the country are being evaluated nationally in the same way that the environment has been judged in terms of agricultural and mining use (Figure 7-4).

Like many other American activities, recreation has become a tremendous business, developed and managed on a national scale. Even before 1900 the federal government set aside outstanding natural attractions as National Parks. In this century, conservationists have called for expanded policies to include forests and seashores. Recreational and other aesthetic use is still considered in the federal construction of dams for irrigation and flood control. States and local governments have also established parks and forest reserves.

The greatest development, however, has been undertaken by private interests, and the most successful of these projects have focused on pleasant climates rather than on scenic vistas. Areas in Florida, California, and Arizona, which had little value in terms of agricultural or other primary resources, have emerged as major population centers because of their subtropical climates. California's population almost doubled in the 30 years between 1940 and 1970, and the populations of Florida and Arizona increased two and one-half times during the same period. The Miami metropolitan area expanded more than three and one-half times.

The people who rush to these areas are not limited to the retired and vacationers and to the workers needed to provide services. Industry is following, too. Manufacturing firms not closely tied to suppliers and markets have been moving into areas of year-round "open" climate. Such climates offer savings on heating and building construction, but they are also major factors in attracting personnel, particularly executives and researchers. Manufacturing employment in

Phoenix increased over eight times between 1947 and 1967; in Miami the increase was over five times. These cities are not manufacturing centers on the order of major Northeastern cities, but the change has been significant in areas where there was almost no manufacturing before. Phoenix which had 7,000 manufacturing workers in 1947 had 59,000 twenty years later.

Besides industrial and retirement centers there are vacation centers where the population during "the season" may be more than ten times its size during the rest of the year. Others are centers for weekend and holiday visitors or even for single-day transients. Thus, there are national, regional, and local centers for recreation.

Evolution of the Modern System: Changing Views of a Given Environment

The socio-economic system presently functioning in the United States should be seen as something more than a single entity operating in today's world. It is the result of the evolutionary development which began in Virginia more than 360 years ago.

During this period, of course, there has been a tremendous revolution in the technology available for developing and managing the earth's environment. But even more important has been the shift in purpose. For more than 150 years the settled territory of the United States functioned as a series of colonies of first several and later one European country. Under those conditions, the primary purpose of the system was to support the "mother country" both economically and politically. Independence brought a shift in the goals of the system. The new nation sought to reorder its economy so that the primary benefits would be for itself rather than for an external power. In

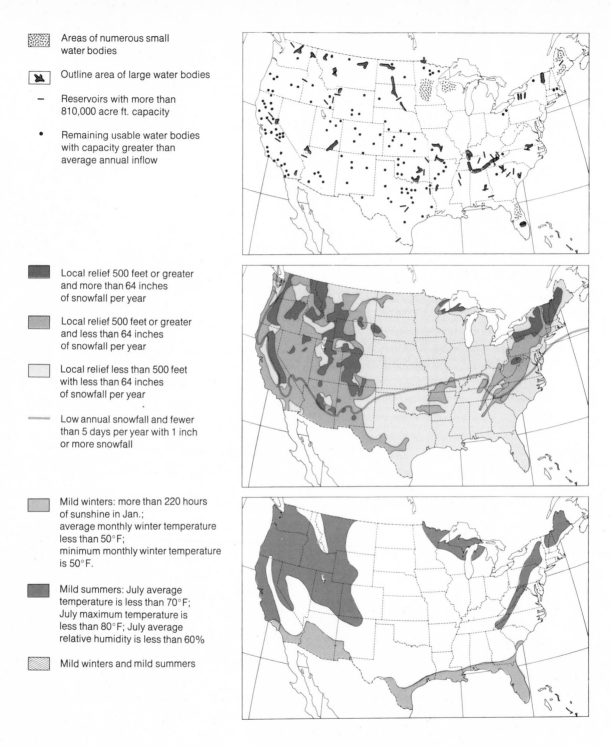

Areas of numerous small
water bodies

Outline area of large water bodies

Reservoirs with more than
810,000 acre ft. capacity

Remaining usable water bodies
with capacity greater than
average annual inflow

Local relief 500 feet or greater
and more than 64 inches
of snowfall per year

Local relief 500 feet or greater
and less than 64 inches
of snowfall per year

Local relief less than 500 feet
with less than 64 inches
of snowfall per year

Low annual snowfall and fewer
than 5 days per year with 1 inch
or more snowfall

Mild winters: more than 220 hours
of sunshine in Jan.;
average monthly winter temperature
less than 50°F;
minimum monthly winter temperature
is 50°F.

Mild summers: July average
temperature is less than 70°F;
July maximum temperature is
less than 80°F; July average
relative humidity is less than 60%

Mild winters and mild summers

Figure 7-4 Recreational Resources of the United States

addition, major efforts were made to absorb new regions into the nation.

The evolutionary changes that have occurred in the United States in the past 300 years reflect the same problems that the countries of the underdeveloped world are now faced with in terms of shifting from colonial rule to independence and control over their own socio-economic affairs.

TODAY'S ECONOMIC SYSTEM AND THE BURDEN OF THE PAST

The distribution of economic activities in the United States reflects many features that are remnants of the past when the objectives of the country were different. Yet these vestiges—places, people, ideas, and prejudices—are very much a part of economic interplay today. Perhaps the greatest problem in making the national economic system work is that no region can be wiped clean of these physical and psychological remnants of the past. Decisions made in a different cultural and technological environment have become established features of the landscape. Life today must take into account established railroad and road networks, cities, functioning industries, and farming areas.

This is a constant problem that we see in newspaper reports on the affairs of urban areas. Eastern cities, built originally in the era of the sailing ship, coach, and wagon, have had to adjust to the steamship, train, car, bus, truck, and airplane. Each advance produced a structural change, either removing or modifying the old. Today we hear of the need to build new freeways and to clear slums, but we know that these changes uproot families and affect the lives of hundreds of thousands of people. Moreover, usually those are the people who do not want the change and who will not benefit

very much from it. Whenever a new freeway or a new apartment complex is built, it creates new problems which require further adjustments.

Ideas inherited from the past often produce major deterrents to change, too. The racial views of middle-class America, whether seen from the perspective of a particular region of the country such as the South or of an urban ethnic neighborhood, stand as a case in point. But in discussions of governmental reorganization, one can cite presently accepted ideas of the separation of city and suburb, and the continuation of separate township, county, and state governments as present ideas that are not likely to change in the near future.

The Example of Appalachia

As the socio-economic system of the country has evolved, the whole basis on which a region or city has developed may change. Consider the case of the southern Appalachian Mountains (Figure 7-5). Before the coming of railroads, men had to walk, ride horses, or use Conestoga wagons to cross the mountains. Some of the most important routes from the Eastern Seaboard to the Middle West were through the valleys of western Virginia, eastern Kentucky, and Tennessee. Land along these routes harbored flourishing frontier settlements occupied by some of the most intelligent and adventuresome people of the time. Prospects for the future looked bright for these settlers with their access to important transport routes.

The opening of the Middle West and the coming of the railroads changed the pattern of movement between the East Coast and the interior of the country. The major railroads were built from New York and Philadelphia through upstate New York and Pennsylvania. Few crossed from Virginia or the Carolinas. The

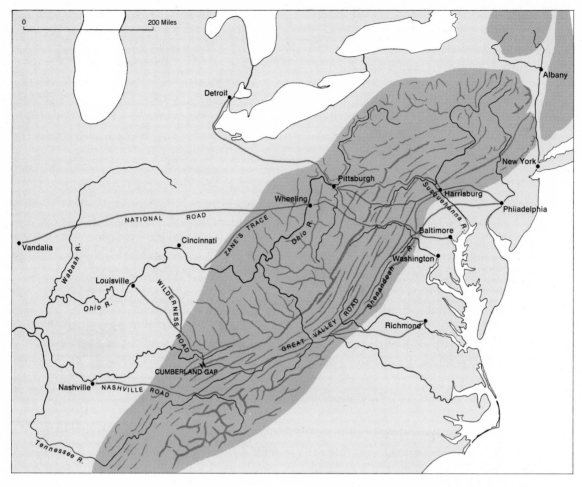

Figure 7-5 SOUTHERN APPALACHIA IN THE PRERAILROAD ERA

▦ Appalachian-Adirondack and New England highlands

〜 Main ridges and escarpments

⋝ Gaps

— Main trans-Appalachian routes

early routes had been abandoned, and the people in the valleys along these byways found themselves isolated from the rest of the country. As agriculture shifted more and more to commercial products, the farms of Appalachia proved too small and isolated to compete with new agricultural regions.

Later, a second wave of population entered parts of the region when coal was discovered: Railroads were built into the coal-producing districts, and for a time the new coal towns boomed. But in recent years coal mining has become highly mechanized; strip-mining techniques and automated equipment in underground mines have reduced mine employment as much as 90 per cent. Miners have found no job alternatives in these small communities built specifically as mining towns. The nationwide shifts to metropolitan life and to centers of modern transportation and communications have widened the gap between the people of Appalachia and the rest of the country, and today this is one of the nation's most seriously depressed areas.

The Example of Buffalo

We often speak of the city of Buffalo as if it were predestined by its environmental conditions. Located at the eastern end of Lake Erie, it has the advantages of being at the western end of the best natural route across the Appalachians in the Northeast, and of being just above Niagara Falls (Figure 7-6). Here is the ideal "break in bulk" point for goods and people changing between lake transportation to cross-country carriers, either land or canal.

With the opening of the Erie Canal from the Hudson River to Lake Erie, the rise of shipping on the Great Lakes, and the coming of the railroads, Buffalo rapidly grew to be a major center in the United States transport system. It became one of the most important grain storage terminals, the site of a large steel mill, and a major oil terminal. Low-cost electric power generated at Niagara Falls attracted industry. The industrial-transportation center of Buffalo has become the twenty-fourth largest metropolitan area in the United States with a population of 1.3 million people.

But suppose the transportation system of the United States were now being developed completely from scratch with no heritage of the past. There is a real question as to whether Buffalo would be needed. The St. Lawrence Seaway–Welland Ship Canal water route allows ocean-going ships and ore carriers to by-pass Buffalo and go directly on to Montreal or open water.

Ocean cargoes that once came to the port of New York for shipment by rail or barge to Buffalo can now go directly to Chicago. Moreover, the New York Central Railroad which developed the Hudson-Mohawk rail route has now merged with the Pennsylvania Railroad and offers more direct service between the Eastern Seaboard and the Middle West across Pennsylvania. New turnpikes across Pennsylvania also provide more direct connections east and west than the Hudson-Mohawk gateway, and even the electric power from Niagara Falls goes into a regional grid which can deliver electricity far from the power site. But Buffalo is too large to become a ghost town. Thanks to its origins in another era, Buffalo has remained an important American city.

THE COLONIAL SYSTEM: 1608–1776

Since the heritage of the United States has such influence on the present, we must examine its evolution. The Colonial Era lasted almost 200 years. Yet by the end of the Revolutionary War,

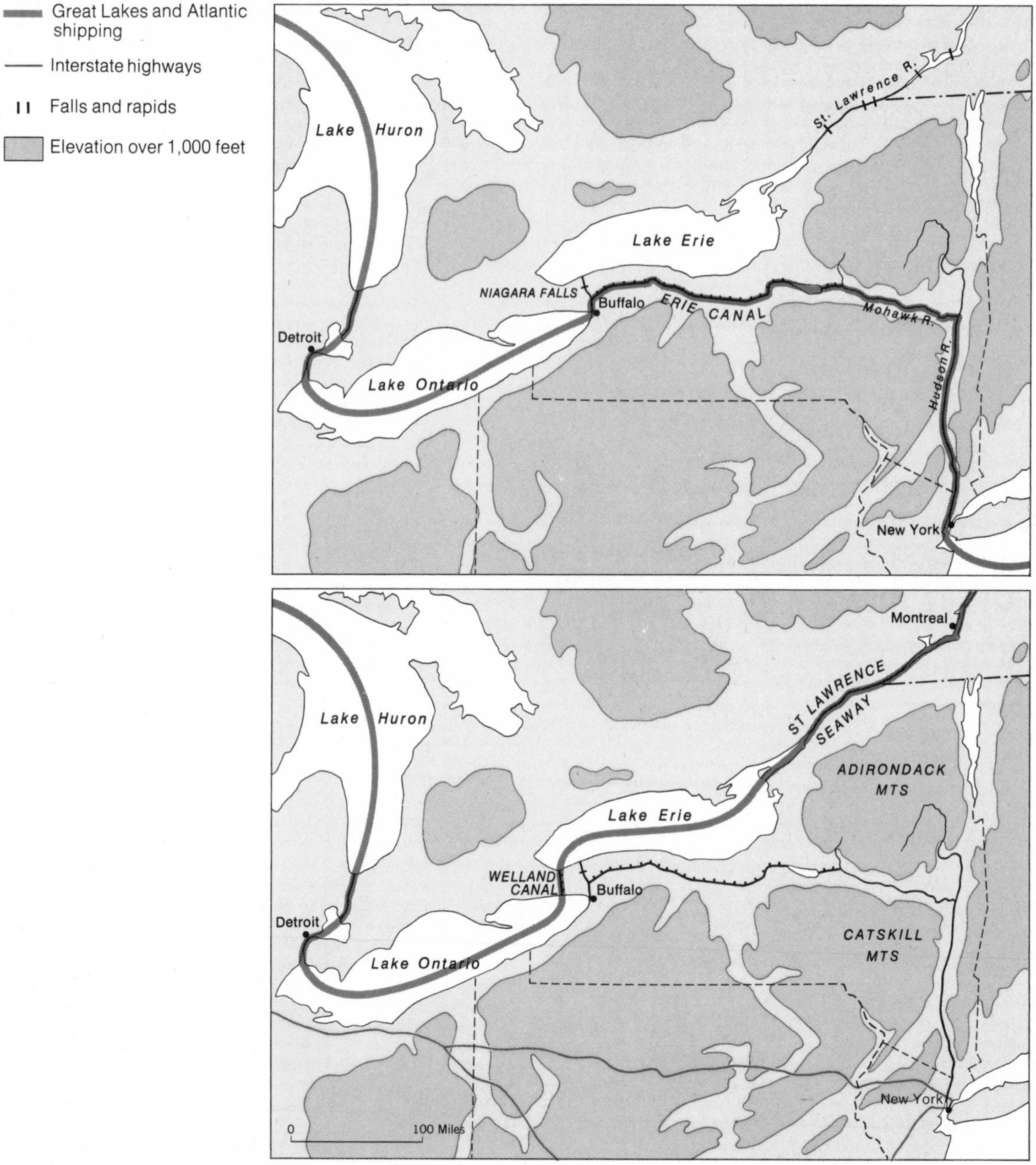

Figure 7-6 Buffalo before and after the St. Lawrence Seaway

there were only scattered settlements west of the Appalachian Mountains, and most European settlers lived within 200 miles of the Atlantic Coast (Figure 7-7).

The Colonial Era preceded the Industrial Revolution in America and, largely, in Europe itself. Manufacturing was a craft industry, and most people were farmers. Ocean travel occurred in sailing ships scarcely larger than many of today's cabin cruisers. Overland wagons and carriages were standard means of transport, but most people traveled by foot or on horseback. Wood and stone were used for construction, and the emerging factories used waterpower, not steam.

The economic and political focus of Colonial America was England. From the British point of view, the Colonies were an extension of England's resource base, providing useful primary products. America was not a single colony but a series of thirteen separate colonies, each with its own contributions to the English system.

For its part, each colony tried to develop a twofold economic base. It sought to produce goods that it could sell to the mother country. These items had to be high in value and low in bulk, such as tobacco, dried fish, and animal furs, to be carried in the small ships of the time. But the colony also needed its own primary support base. It required staple-food production, building materials, and basic crafts. There was little trade between colonies. As in underdeveloped areas today, little was produced in one colony that another colony needed in quantity, and transportation and communication were designed primarily for connection with England.

It is useful to recall that it was not the mid-latitude American colonies that were the prize possessions of the period. Rather it was the tropical islands of the Caribbean with their output of sugar and other products that Europe wanted. The English mid-latitude colonies had to search for products that could be sold to the mother country.

In those days the great environmental variations along the Atlantic Seaboard from Maine to Georgia were judged by each colony in terms of the products they offered for subsistence and for export. The value of the colony was measured in England by its contributions to her economy.

New England with its rough terrain and short growing season was in most respects more limited agriculturally than England. Thus, it could produce nothing of agricultural interest to the mother country. It was, however, adjacent to the rich fishing grounds of the Grand Banks, and its forests contained excellent trees for construction. With an interest in the sea and with the materials at hand for building, New Englanders soon built their own ships, carried their own cargoes to Europe, and engaged in trade with Africa and the West Indies (Figure 7-8). In the ports New England businessmen engaged in shipping and finance. They controlled a portion of the trans-Atlantic trade and gained some power over their own destinies. With the rise of populations in the port cities, the agricultural villages and towns back from the coast acted as suppliers of the basic food needs of the city dwellers.

The specialty products of the South were very different. So was the way of life. The growing season was longer, and an agricultural economy based on specialty crops developed. The primary crop was distinctly American: tobacco. Smoking had become popular in Europe, and tobacco was an almost ideal export because the leaves were light in weight and did not deteriorate badly in shipment. In South Carolina and Georgia long-season crops included rice and indigo.

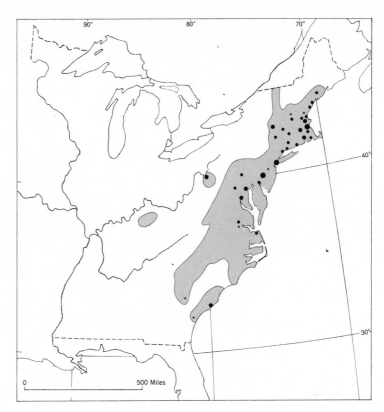

Size order of settlements

· 5th

• 4th

● 3rd

● 2nd

Area with at least 6 persons per sq. mi.

Figure 7-7 Settlement of the United States by 1790

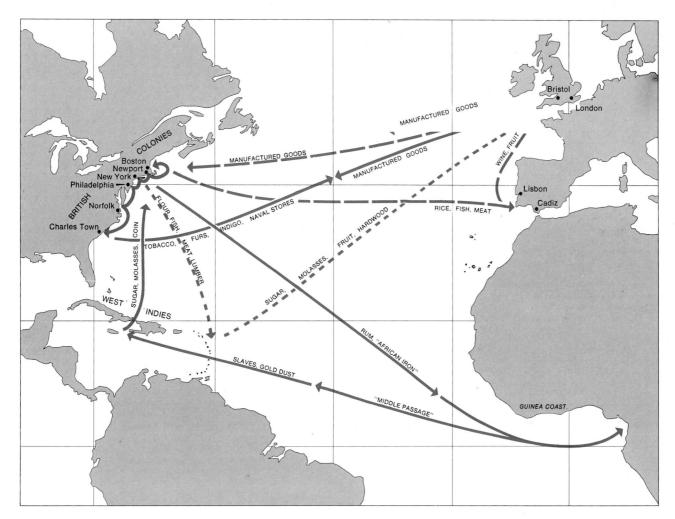

Figure 7-8 Overseas Trade of the Colonies

← Principal routes

◄─ ─ Secondary routes

Labels on the map:

MANUFACTURED GOODS

Bristol
London

COLONIES

Boston
Newport
New York
Philadelphia

BRITISH

Norfolk

Charles Town

MANUFACTURED GOODS

MANUFACTURED GOODS

WINE FRUIT

Lisbon
Cadiz

INDIGO, NAVAL STORES

RICE, FISH, MEAT

FLOUR, FISH

TOBACCO, FURS,

MEAT, LUMBER

FRUIT, HARDWOOD

SUGAR, MOLASSES, COIN

WEST

INDIES

SUGAR, MOLASSES,

RUM, "AFRICAN IRON"

SLAVES, GOLD DUST

"MIDDLE PASSAGE"

GUINEA COAST

The nature of the southern coastline with many small streams flowing into large estuaries and bays was an additional aid to shipping. Plantations on tidewater could be reached directly by the small sailing ships, eliminating the need for an urban intermediary. With direct shipment from dockside and the production of its basic crops and livestock, the tidewater plantation was in large part a self-contained unit connected only to Europe. In turn, Europe supplied the plantation with needed manufactured goods. But unlike New England, plantation owners did not own ships or handle the overseas trade. In fact, it was often New England vessels that carried their goods.

Colonies in the area of New York City and Philadelphia seemed to have few advantages. They did not grow exotic crops and had no particular nonagricultural resources. They could provide an adequate food base for themselves but had no crops in demand in Europe. Instead, flour and barreled pork from farms in the area were shipped to the West Indies to feed the slaves on sugar plantations. Local iron deposits were found, and iron smelting was developed in eastern Pennsylvania and nearby New Jersey, but this was for local use, not export.

In this Colonial Era, between 80 and 90 per cent of the population lived in rural areas or in villages supported by agriculture. City dwellers were a small minority, and there was no such thing as a metropolitan area. The largest cities each had less than 50,000 people. Since each colony functioned essentially as a separate entity, there was no national hierarchy nor prime city. Instead, each colony had its major center, usually the port connecting it with Europe. The size of the community depended largely on the importance of the trade moving through it. Each operated virtually independent of the others, and all were satellites of trade centered in London, the West Indies, and Africa.

Except for these port connections with Europe, each community was isolated by problems of transportation. Poor land routes meant dependence on the local area for the food base. They also meant that news traveled slowly. Moreover, each community had to provide the basic cultural life for its people. The church was the social as well as the spiritual center, and each town had its own days of celebration, fairs, and entertainments. Many towns developed their own weekly newspapers.

In the South towns and villages were less necessary than in other colonies. Each plantation functioned not only as a separate economic entity with direct ties to Europe but as a social and religious community as well. Like the haciendas that dominated rural settlements in Latin America, the plantation commonly had its own chapel and its own close social contacts. Workers seldom left the community. Thus, the large-scale plantation agriculture of the South established a society very different from small-town life in New England.

THE FIRST ATTEMPTS AT A NATIONAL
SYSTEM: 1776–1840

Independence from Britain brought the first attempts to reorient the system developed under colonial ties. Europe was still the cultural focus, the source of manufactured luxury items, and the chief market for goods. Breaking ties with England necessitated, however, a reorganization of the colonies into a functioning political and economic unit. To supplement the use of coastal waters, great effort was made to provide land routes from Georgia to Maine. And as settlers pushed out beyond the Appalachians, interregional trade took on a new dimension.

This period faced the beginnings of industrialization as well as national organization, but the effects of the technological change did not really take hold until later. Wagons, barges, and sailing ships continued to be the dominant forms of transportation, and it still took four days to get from New York to Washington. A journey to Cincinnati from the East took over two weeks. In rural America manpower and animal power without the aid of modern agricultural implements still did the work on farms.

There were new factories, but they represented only the first stage of industrialization, being powered by running water. Swiftly flowing rivers channeled into millraces turned the water wheels to provide power.

Regional specialties of colonial times continued to develop in the post-Revolutionary period, but they were conditioned by new technologies. Shipping and trading interests of New England flourished, and it became the country's first manufacturing area. Some manufacturing had been done in the area during the Revolutionary War, and by 1820 there was a boom of new mill towns at key waterpower sites, from Maine through Connecticut.

The hilly countryside with its humid climate had many small streams that could easily be adapted to waterpower. As in England, textiles developed as the leading new industry, first using wool and then turning to newly popular cotton. Sheep raised locally provided the raw material for wool, but cotton, which requires a long growing season, was not a crop that could be raised there. Local iron deposits and charcoal from the forests provided the base for a small metalworking industry.

In the South cotton became the base for an expanded plantation economy which moved back from the tidewater of southern Virginia into Alabama and Mississippi. Cotton, like tobacco, was suited to large plantation units, and a great deal of labor was required to weed and harvest it. The resulting importation of Negro slaves from Africa, often in ships operated by New Englanders, was the basis for the localization of black population in the rural South.

The plantation South did not organize itself to manufacture its chief crop into textiles, and the market for cotton even after the Revolution was mostly in England. Thus, the shift of plantations away from the tidewater resulted in the growth of new southern seaports. As New England mills emerged, important interregional trade began.

The Appalachians presented an almost unbroken barrier that had to be crossed to get to the interior of the county. From Atlanta as far north as Maryland, the Blue Ridge Mountains offered virtually no easy breaks, and no major routes directly west developed from the coast to the interior. In the Middle Atlantic states the Hudson-Mohawk gateway led to Buffalo, and there were also crossing points in the mountains farther south.

The Great Valley route behind the Blue Ridge and the Potomac route were popular, too. The Great Valley route crossed the famous Cumberland Gap into the Ohio River system. From there one could float west on a barge or move southwest to Knoxville, Nashville, and the Cumberland River leading to the Mississippi. The Potomac route followed river breaks through the series of parallel ridges that blocked westward movement farther north and then worked its way into the valleys leading to the Ohio River.

Using these routes the first settlers found not the rich plains but the hill lands that are the western extension of the Appalachians. They discovered only scattered valleys and lowlands rather than the best farmlands. Major settlements were in the rich basins of Kentucky and in the Nashville Basin. It was not until the opening of the Erie Canal and the development of the National Road to Illinois in the 1820s and 1830s that settlers had direct access to the rich agricultural lands farther west (Figure 7-9). In the far South settlers moved around the Appalachians at their end point and on into the lands

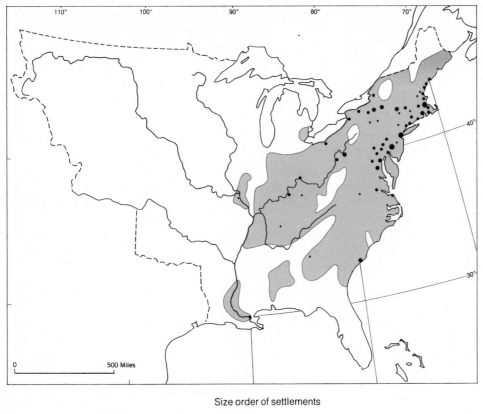

Size order of settlements

· 5th

• 4th

● 3rd

● 2nd

▨ Area with at least 6 persons per square mile

Figure 7-9 SETTLEMENT OF THE UNITED STATES BY 1830

of the Gulf Coastal Plain in Alabama, Mississippi, and Louisiana.

The problem of all settlers along the western frontier during this period was less in producing goods than in getting them to market. Without water transport and good roads it was almost impossible to ship bulky farm produce back over the mountains to eastern settlements, although herds of cattle and hogs could be driven over the trails and corn whiskey could be transported in jugs. Thus, the Ohio-Mississippi river system which led away from the settled areas toward New Orleans offered the best highway for goods moving out. Cargo could be floated downstream by raft or barge, and the raft as well as the cargo could be sold at New Orleans for shipment to Europe, the West Indies, or the coastal states. Not until 1840 was a network of canals developed to bridge the Appalachians and connect with the Great Lakes and the Ohio waterways.

The opening of the interior was a great boon to the Middle Atlantic states. The ports of Baltimore, Philadelphia, and particularly New York became the gateways to the interior, and their function as national trade centers was established. Settlers on their way from Europe to the frontier invariably came through these cities. Merchants and shippers prospered at these focal points of interregional trade, and from then on, the movement of immigrant settlers westward expanded the influence of these cities. Increasing trade, in turn, meant more jobs for more people in the seaboard cities. Moreover, iron manufacture in Pennsylvania and New Jersey and manufacturing in New England towns were stimulated by these new interregional contacts.

In the early nineteenth century transportation and communication were still very inefficient. Most regions remained in semi-isolation, and life continued to center around the local church, the local government, and local customs. Each community needed its own food base and its own craftsmen. New ideas filtered slowly westward from the port cities and Europe. Immigrant groups settling frontier areas tended to cluster together, with ethnic cohesion lending distinct character to settlements.

The lack of an easy flow of ideas on an interregional basis resulted in subcultures as well as regional economic specialization. This was most obvious in the contrast between the plantation South and the industrial and commercial North. Moreover, the agricultural Middle West evolved still other distinctive ways. Most decisions were in the hands of local entrepreneurs—shopkeepers, mill operators, and plantation owners. Only the ocean shippers in the large ports of the seaboard could really function efficiently on an interregional basis.

Towns and cities themselves were tightly packed. Since workers had to walk to work, residences were crammed close together. Only the wealthy who owned carriages could live any distance from places of business and shopping. The compactness of such cities of that day encouraged government officials in Washington to return a portion of the District of Columbia to Virginia. They did not believe the city of Washington would ever reach the limits of the District.

The Emergence of the Modern Industrial System

By 1870 the continental boundaries of the country as they exist today had been established. Except in the Southwest and northern Great Plains, Indian lands had been taken by the government, and the break between the North and South had been resolved. But the job of bringing this huge territory into an economic functioning unit remained. The tools of technology that could pull the regions of the United States together were at hand. The railroad and the telegraph for the first time made interregional mass movement possible. By 1860 railroads connected the three settled areas of the country (Figure 7-10). In addition, packet steamships sailed the rivers and coastal waters, and ocean-sized freighters carried traffic on the Great Lakes.

In the South the railroads were less a network and more a series of local routes connecting agricultural areas with port gateways. Notably absent were any trans-Appalachian lines between Atlanta and Washington, D.C. In the North an interstate system had developed that crossed the Appalachians along five different routes connecting Boston, New York, Philadelphia, and Baltimore with St. Louis, Chicago, and other midwestern cities. In the Middle West the railroad system crisscrossed the newly settled agricultural lands of what was quickly to develop into the Corn Belt. Almost from the start farmers found that they could easily market produce by rail to cities in the Northeast.

Interregional rail connections meant that regions did not have to depend on local food production any longer, and the production of staple foods in New England and the Middle Atlantic area began to decline. At the same time, specialized mining areas developed at coal, iron-ore, and copper deposits. For the first time, large-scale industry could draw on resources from areas thousands of miles apart. The iron and steel complex based on iron-ore from around Lake Superior and coal from the Appalachians is perhaps the best example.

Over the railroads from the Northeast came masses of immigrants from Europe. Large numbers of people from the East also settled farmlands or moved to there (Figure 7-11). This increased the close cultural as well as economic interaction between the two regions. At the same time the links between these two areas and the South were less close than even following the Civil War.

By 1870, before large-scale settlement of the frontier, a transcontinental railroad connection had been made with the Pacific Coast, and a rush of settlers to the Great Plains and the Far West began. Settlers were often recruited in Europe by railroad companies and then carried across country to their new homes by train. These people found the railroad ready to ship their products from the start, but the western areas had just begun to make a contribution to the economy of the country with some mineral production and limited livestock shipments. The important ties were still between the Midwest and the Northeast.

At this time, too, industry completed its changeover from wood and waterpower to coal. This required a major reassessment of the resource base of the country. New England, traditional center of manufacturing, was devoid of major deposits of coal and iron, the basic resources of the new industrialization. Although some deposits were found in the Middle Atlantic area along the east flank of the Appalachians, the major resource base emerged when the Great Lakes waterway linked the bituminous coal of the western Appalachians with iron near Lake Superior. Thus, the Middle West, developing as the great producer of agricultural staple foods, also became the most important center of metalworking.

Despite a lack of the raw materials needed for the new industrialization, New England's industry continued to expand. Coal was brought in from the Appalachians by rail and ship. More important, New England capitalized on its entrepreneurial experience, its skilled labor, and its access by rail to the markets of the Eastern

Figure 7-10 Extent of Railroad Connections by 1860

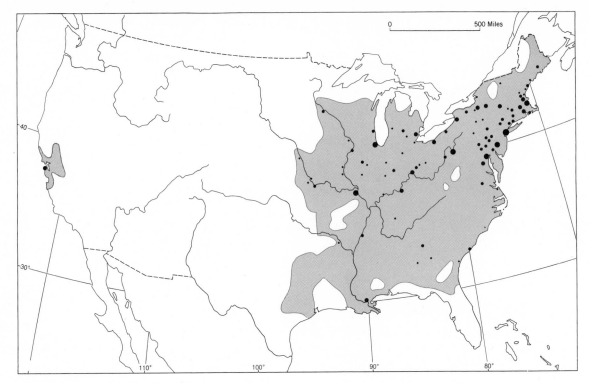

Figure 7-11 SETTLEMENT OF THE UNITED STATES BY 1870

Size order of settlements

. 5th

. 4th

● 3rd

● 2nd

● 1st

Area with at least 6 persons per square mile

Seaboard, of the Middle West, and overseas.

In the Middle Atlantic area the "hinge" function of the large gateway ports continued to develop rapidly. These centers were focal points in the growing interregional transportation system. The flow of people and goods meant an acceleration of business such as trade, finance, and supporting activities. The financial and commercial interests in these port gateways were in a position to manage interregional and international transactions, much as New England traders had been earlier. The businessmen in these cities provided the basic financial and management arrangements for economic developments in other parts of the country. Finally with new interregional communication and with the emphasis on political and military matters fostered during the Civil War, Washington, D.C., began to emerge as an important decision-making complement to the Middle Atlantic business centers.

The South remained dominantly rural despite attempts to build its own industrial base to support its war effort. Cotton and tobacco, the traditional cash crops, were the mainstays, and the European market was of first importance. Railroads connected agricultural regions with ports. Disruption and destruction caused by the Civil War handicapped the region, but the plantation system survived the freeing of slaves by shifting to a sharecropping system by which a former slave worked a tiny plot of ten or more acres of plantation land and paid rent in the form of a share of the crop. Such tenants, completely uneducated and inexperienced in business affairs, had virtually no hope of gaining an adequate living standard and were usually in debt to the local supply stores.

The parts of the South that did not produce cash crops on plantations—particularly the Appalachian uplands—were left out of the emerging interregional system. Their populations continued to live on a primarily subsistence basis, just as they had in frontier days.

For the first time, small towns could be in constant touch with the outside world by telegraph and rail. Local newspapers could print national news, and trains could bring big-city papers. Local stores could be sure of ready supplies of needed stock by train.

It was, however, the larger urban centers that benefited most from the new transportation and communication. These places were able to expand their economic and cultural influence at least as far as a train could travel in one night. Towns and villages within such a distance served as collecting points for agricultural products and other primary resources, and were the retail outlets for large urban-centered wholesalers. The big urban places not only were hubs connecting the small towns and rural areas around them but were also part of a network of interregional connections with large cities elsewhere. Because they were centers of transport and telegraph, big cities could develop regional businesses. Entrepreneurs and interregional management emerged in all the large cities of the Eastern Seaboard and the Middle West. Boston, New York, Philadelphia, Baltimore, Chicago, St. Louis, Cincinnati, and New Orleans each had over 100,000 people by 1860, and 15 per cent of the population of the country lived in cities of 10,000 or more.

In the largest cities horse-drawn streetcars provided the first municipal transport and allowed the city to sprawl slightly since workers could ride to work. But further expansion was not anticipated. When the present boundaries of St. Louis were established in the early 1870s, the city asked for county status, assuming that its borders were far beyond possible future growth.

Growth of the Modern Industrial System on a Transcontinental Base

From 1870 to 1920 the modern, interconnected system expanded rapidly. For the first time a national system operated from coast to coast. All major regions of the country were occupied and connected into the system, and the inter-regional connections themselves were greatly improved. Railways using steel rails, instead of iron, and large steam locomotives became the most efficient long-haul movers of both people and goods. With the invention of refrigerator cars, tank cars, and livestock cars, virtually all types of goods could be moved. Steam-powered, steel-hulled ships, larger and faster than earlier vessels, multiplied the volume that could be transported over water. The first oil pipelines were also completed.

Inanimate power was applied wholesale to secondary production and began moving into agriculture and mining. Machines for planting and harvesting crops were invented, and by the end of this period tractors had been developed to pull the equipment. Trucks slowly replaced horse-drawn wagons on the farms. In factories electricity provided easily controlled and flexible power for a great variety of operations, and large electrical generating plants were able to supply whole cities and their factories with electricity. Moreover, such organization enabled newly settled parts of the country to contribute specialties, while older producing areas became more specialized.

Regional Variations

By 1920 most of the present agricultural regions had been established (Figure 7-12). The Cotton Belt extended from North Carolina into western Texas. The Corn Belt of the Middle West was clearly defined as were the dairy regions in the northern Lake States, New York, and New England. Specialized tobacco-growing areas, each producing its own particular type of leaf, were found from Georgia to Connecticut. Citrus fruits came from both Florida and California. The newly settled lands of the Great Plains still produced cattle; however, with machines to work the land, the eastern margins had already become established as the country's most important wheat-growing areas. Orchard fruit came from the Pacific Northwest and upstate New York, and grapes from New York and California. In the mountain West newly irrigated areas using water from giant dams produced sugar beets, potatoes, specialty crops, and alfalfa.

The coalfields of southern Illinois and western Kentucky had also come into operation late in the nineteenth century, and the vast mid-continent and Texas oil fields were opened. The lead deposits of eastern Missouri proved the most productive in the world, and zinc was mined in a productive field along the border of Missouri, Kansas, and Oklahoma. The copper fields of western United States had been activated along with a variety of other new mineral sources in the western mountains.

Most of all, this was the period in which heavy industry developed: steel mills, metal foundries, railroad-equipment plants, oil refineries, chemical works, and meat-packing plants. These industries generally occupied sites in the largest metropolitan centers of the Northeastern Seaboard and the Middle West. A series of urban-centered districts of the new heavy industry sprang up near major cities from New York to St. Louis. These complexes sprawled over the landscape, belching forth noise, smoke, and dirt. Nearby were some of the most tightly packed workers' communities in the country, occupied mostly by new immigrant groups. Other features of the industrial districts were the miles of railroad tracks and gigantic switch-

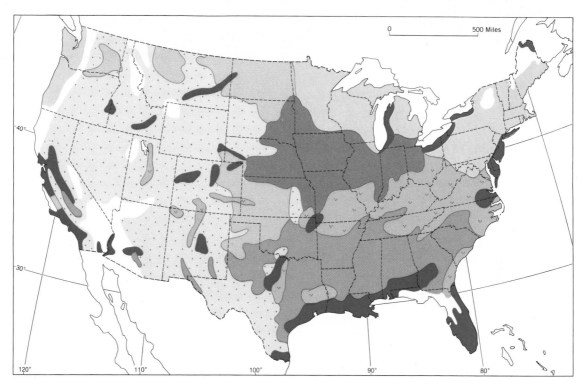

Figure 7-12 Regional Agricultural Specialty Products of the United States in 1920

Dairy, cattle, hay

Corn and livestock

Wheat and small grains

Cotton

General farming, tobacco

Fruit, truck, special crops

Grazing livestock

Little or no agriculture

yards needed for the delivery of raw materials and the shipment of their heavy finished goods.

The Rise of National Industries

Now national rather than regional control made its first appearance in industry. John D. Rockefeller gained virtual control of the United States oil industry until the government forced him to break up his empire. Andrew Carnegie laid the foundation for the United States Steel Corporation with plants in the Northeast, the Middle West, and the South. Multiplant corporations became the rule in the meat-packing, mining and smelting, metal-processing, and chemical industries. All of this was the result of improved interregional transport and communications and offered the advantage of industry-wide control of a firm. The heavy investment costs could be financed only at the highest banking level, and management control had to be sophisticated. Operations on such a scale were beyond the capabilities of the local or regional entrepreneur.

Cities became known not only for their specialized manufacturing operations but also as centers for the special financial and business services connected with those industries. Pittsburgh was the steel center; Chicago the center of meat packing, railroads, and railroad equipment; Detroit of automobiles; Akron of rubber; Cincinnati of soaps; and Toledo of glass. The headquarters of companies in these centers managed networks of plants. Banks and financial institutions specialized in loans and money management for the city's particular industry.

Manufacturing outside the Northeastern Seaboard and the Middle West was largely tied to particular resources, such as oil, lumber, and specialty metals. There was also a major migration of long-established northern industry into small Appalachian towns in the South— not everywhere but just along either side

of the mountains. Since the textile industry has no particular demands for massive power and special transportation, it moved south from New England in search of large quantities of unskilled labor. The usual textile mill produces different weights of standard "grey cloth" which is later dyed in other plants nearer to markets and so requires only simple machines and semiskilled workers. With the growth of other industry in New England, wages were rising. New England was one of the first areas to become unionized, and the unions called for better working conditions.

In the South there were no unions, few other industries to compete for labor, and communities eager to assist new factories with building subsidies and special tax relief. New regional electric-power networks provided all necessary electricity. This shift has continued until the South is now the overwhelming center of textile manufacturing in the country.

Other light industries dependent on quantities of unskilled labor moved to the South, to mining communities, and even to farm villages. It is common to find shoe and clothing factories in small rural towns. Such firms commonly employ largely female labor—wives who supplement the incomes of their farmer or miner husbands.

Notably, one major part of the clothing industry remained firmly centered in the heart of New York City. Here high-fashion clothing, particularly women's wear, is still made. This industry has really been a handicraft industry, rather than a factory one, with small shops of women operating cutting and sewing machines. It has always been located in New York City, center of the largest, most fashion-conscious market for clothing in the country. There it is in a position to adapt quickly to the latest style trends. If a new model sells well in New York shops and department stores, dozens of clothing makers can produce copies in quantity almost overnight.

The Development of Metropolitan Areas

By the beginning of this century the largest cities were becoming metropolitan areas. New developments in urban transportation—the electric street railroad, the elevated railroad, and commuter railroads—allowed cities to sprawl outward. Outlying communities were supplied by urban public utility networks. The first real suburbs grew up beyond city limits along commuter railroad routes. The wealthy and the upper middle class moved out along Philadelphia's Main Line or Chicago's North Shore.

In the cities streetcars could handle masses of people and allowed working-class people to move away from the neighborhoods of their work. By the end of this era, large apartment buildings were replacing the old row houses close to their routes. Outlying shopping districts appeared at major intersections of streetcar lines, and shops sprang up along important streetcar routes, forming commercial streets.

Slums, crowded with ethnic groups newly arrived from Europe, grew as the immigrants crowded into houses abandoned in the exodus of the wealthy to the suburbs.

Some aspects of city life improved. With water and sewer lines to individual homes, gas and electricity, and paved streets and sidewalks, cities were better places than ever. There were public schools and new civic centers with concert halls and museums. On the other hand crowds of low-income groups in tenements and row houses produced problems of health, congestion, and education on a scale never before experienced in any urban community.

Technological Changes and Mass Production

Since 1920 few new areas have been added to the system. Rather, the period leading up to the present has been one of continual adjustment to accelerating technological changes within the industrial system: highway and air transportation, regional electric-power nets, and the automation of agriculture, mining, and even many manufacturing processes. Most important, perhaps, has been the tremendous increase in the scale of operations. The population of the country has almost doubled from 105 million in 1920 to more than 200 million today. At the same time there has been a rapid rise in the living standard of most people. Unions have gained higher wages and better working conditions for laborers, and the demand for sophisticated services has called for better educated, better paid specialists.

The development of mass consumption has been facilitated by installment credit which allows people to pay for a relatively expensive item over a period of time. Mass consumption has caused a shift to mass production—not only in the factory but also on the farm and in the mine, and even in schools. Such a market, presently of more than 200 million people, has ensured the successful operation of large-scale, highly specialized enterprises turning out vast quantities of a single product. Thus, mass production has characterized not only manufacturing output but agriculture, mining, and even such cultural enterprises as schools, fashion, and the arts.

Mass production, then, has caused a further sorting of the productive units of the country. Large "factory farms" with corporate ownership have replaced family farms as the mainstay of agricultural production. Most of the steel of the country is produced by a dozen major corporations, most of which have only one or two basic manufacturing plants. All Cadillac automobiles are produced in a single factory, and so are all the Chevrolet Vegas.

Now the resources of particular parts of the

country are evaluated by national entrepreneurs to find the particular area best suited to the production of a particular item for consumption by the whole country. Industrial location specialists make studies of the cost of production and the distribution from different centers to find the location that offers "least cost" operations. Such studies do not "start from scratch." They build instead on existing patterns—markets, production sources, and transportation nets—and result in only a few localities being chosen. Thus, specialized areas of primary production have become sharply localized, and individual cities are steel centers, auto cities, and the like.

The change to large-scale units of primary production has brought a shift in the use of the resource base on which agriculture rests. Large mechanized operations place a premium on level land where machines can operate easily and where water is available for irrigation. This development has put a premium on middle western agricultural lands and on river lowlands such as the lower Mississippi Valley. In the West flood plains and river terraces have been supplied with irrigation water, and the great, almost level Central Valley of California has become the most important oasis of all. Even the high plains of Texas have been intensively farmed by using well water for irrigation, while hill lands have largely been abandoned.

The most highly prized level lands for large-scale farming are those with adequate moisture and long growing seasons. Thus, the irrigated lands of the Southwest, some of the most remote areas in terms of the major market for agricultural production, have been most successful in growing specialty crops. Florida, too, has been brought into the production of winter vegetables despite its lack of adequate soil. In contrast, much of the Great Plains remains a problem because rainfall does not provide adequate moisture. Vast stretches of the central and western margins of that region are without adequate irrigation sources and so remain pasture.

Under present technology soil is the least important ingredient in the environment. It is now possible to add chemical nutriments to soil "by prescription" to produce lush growth provided the land is flat and has adequate growth potential and moisture supply. South of Miami, Florida, land for winter vegetables has been created by moving rock-crushing machines across exposed limestone ledges.

In the American Midwest, a town that grew
up under the rules of a rural society, with
ties to river transport, must adjust under a
different set of rules to a world of TV, super-
markets, and interstate highways.

A person at the livestock show finds not only
prize cattle but advertisements for every product available in and out of town. Forty-year-
old cars, until recently in unremarkable common use, are now middle-class status symbols.
Ladies gather for bingo for many of the reasons they lately met to sew and pray. Small-
town traditions persevere set in suburban-style
bungalows.

   Giant towers marching across the landscape
bring power from many miles away and farming has adopted assemblyline techniques. The
river setting remains, but the large tows destined for cities upstream no longer stop here.

Cultural Changes and Contrasts

The national scale of the economic, political, and cultural life of the United States today is different from its more limited dimensions when the country first emerged from colonialism. The basic ideals of that age called for individual freedom in political, economic, and personal life. But in those days economic units were predominantly individual family units—the farmer, the shopkeeper, the lawyer. Everything was considered within a regional, or even local, framework.

With radio, television, and the telephone both the individual and the corporation are in constant touch with happenings over the entire interconnected system. Everybody is regularly pressured to buy through advertising and everybody is exposed to a steady stream of mass-produced culture. With the automobile the family can go across the city to see a sports event, to visit friends, to shop, or even to attend the old family church. Only children under driving age are restricted to neighborhood life, and it is only through them that many adults have contact with their neighbors.

Today, the scale of the national economic system, made possible by modern technology and communication, calls for a very different basic organizing unit. Efficiency in an economic sense has been achieved by developing units that can function within the nation as a whole. As a result, national corporations, financial institutions, and the federal government have evolved systems that enable them to produce, market, and manage in accordance with what the organization thinks best.

Virtually lost in this system is the individual who is still the basic working and consuming unit. Large corporate and governmental units make decisions that affect regions, communities, and individuals alike. The decision to place a new factory in a town may breathe new life into an area which has been declining. On the other hand, the decision to close a mine, or even to automate one, may mean the end of support to its community. Assembly lines provide the most efficient means of operation, but their impact on the lives of the workers who throughout their working lives must function as automatons on those lines is still unknown. Government decisions on agricultural subsidy payments, tax write-offs, irrigation dams, and power nets affect the lives of millions of people.

In a given area of the country, particular districts and communities have had healthy, often booming, economic growth while others lost their economic base. The South is a case in point. Traditionally it has been an area of agriculture and timber products, based, since the Civil War, on small farms and small, portable sawmills working a few acres. Today certain parts of the agricultural South are booming with lands devoted to specialty crops and winter vegetables. There one finds some of the most highly mechanized and most progressively managed farms in the country. These pockets of prosperity include processing plants, fertilizer dealers, banks, stores, and attractive residences sprawling out into the countryside. But over much of the South, the old small-scale agriculture struggles to support a decreased population. Congressional investigations in recent years have discovered many cases of malnutrition, generally characteristic of overcrowded agricultural areas in Latin America.

The same contrasts are found in the cities and towns of the South: Atlanta and Miami are among the most modern and rapidly growing metropolitan centers of the country, while such established centers as Savannah, Lake Charles, and Gadsden have suffered. The "solid South" of history books has become increasingly fragmented between these areas of increasing poverty and centers of great economic viability.

This one country, seemingly in a position to capitalize on the interconnected system, has found that the efficiently working system has created unforeseen problems. Pollution and congestion threaten the future of great metropolitan centers. Questions regarding the limitations of interconnections challenge modern industrial countries at present, but as we shall see in succeeding chapters, different countries are finding different answers.

SELECTED REFERENCES

Billington, Ray Allen. *Westward Expansion: A History of the American Frontier,* Macmillan, New York, 1949. The history of the dispersal of settlement across the continental United States is treated in a geographical manner. It offers insight into the factors associated with the sequence and spatial pattern of settlement of the nation.

Borchert, John R. "American Metropolitan Evolution," *Geographical Review,* Vol. 57, No. 3, July 1967, pp. 301–332. The development of the pattern of metropolitan places is analyzed from 1790 to the present. The relationships between transportation technology, industrial development, and physical resources are discussed in terms of their effects on the periods of growth peculiar to groups of metropolitan places.

Brown, Ralph. *Historical Geography of the United States,* Harcourt, Brace & World, New York, 1948. This is still the classic text on the historical geography of the United States.

Gottmann, Jean. *Megalopolis: The Urbanized Northeastern Seaboard of the United States,* Twentieth Century Fund, New York, 1961; M.I.T. Press, Cambridge, Mass., 1964. This analysis of the development of the Boston-to-Washington urbanized area focuses not only on the functioning character of this particular region but also on how its development has been part and parcel of the development of the national territory. The book provides evidence of the dominant role that this urbanized region has played in decision-making about the development of the rest of the nation.

Perloff, Harvey, and Wingo, Lowden, Jr. "Natural Resource Endowment and Regional Economic Growth," *Natural Resources and Economic Growth,* J. L. Spengler, ed., Resources for the Future, Inc., Washington, 1961, pp. 191–212. Also as Chapter 11 in *Regional Development and Planning,* John Friedmann and William Alonso, eds., M.I.T. Press, Cambridge, Mass., 1964. An overview of the changing role of resource endowments and their locations in the economic development of the nation is presented. A wealth of historical economic data has been integrated into this rather brief interpretative statement.

Pred, Allan R. *The Spatial Dynamics of U.S. Urban-Industrial Growth, 1800–1914: Interpretive and Theoretical Essays,* M.I.T. Press, Cambridge, Mass., 1966. Chapter 2 is an especially relevant analysis of the growth of major U.S. metropolitan centers from the impetus of industrial location and growth.

Ward, David. *Cities and Immigrants: A Geography of Change in Nineteenth Century America,* Oxford U. Press, New York, 1971. In contrast to Pred, this study looks at changes in American cities in the last century in terms of people and cultures.

Chapter 8   United States:
The Interconnected
System on a Grand Scale

THE MODERN, long-range, interconnected system depends on three fundamental components in addition to advanced technological skills: (1) the utilization of a large and varied environmental base from which to draw its needed resources, (2) a large, affluent consumer population, and (3) a modern transportation and communication system to tie all producing and consuming areas together.

Although certain of these characteristics were exhibited by such political-economic entities as the Roman Empire and the British Empire, the United States was really the first country in the world to exhibit all three. A single political entity of more than 3.6 million square miles, it is exceeded in size only by the Soviet Union, China, and Canada. Only China, India, and the Soviet Union have more people than the United States, and none of them has as many economically affluent persons able to buy consumer goods. Finally, all other large countries in the world have developed out of cultures that have been evolving for several thousand years. Their long-established patterns of settlement, routes of transport, producing areas, and cultural structure predate development of the modern, interconnected system. Change to new ways has had to come through remaking the man-made physical plant of past generations and through modifying cultural patterns to fit new conditions. Cities built for pre-industrial times have had to adapt to the airplane age. Methods of farming have had to take into account scientific developments. With it all has come pressure to change organizational, social, and personal patterns to fit the modern, interconnected system.

In contrast, only the Eastern Seaboard of the United States was settled by Europeans in pre-industrial times. The history of the United States has been one less of changeover from the traditional, closed system to the modern, interconnected system than of continued modification and improvement of the interconnected network in a rapidly changing technology. Basically the change in the United States has been one of scale of operation—from local, to regional, to national, and—finally to international interconnection.

With modern transportation allowing rapid movement of masses of goods and people from virtually any location in the country, and with modern communication providing instantaneous flow of information, it is possible for each place to make its particular contribution to the overall system. Yet, at the same time, it is possible for centralized political and cultural forces to supplant local cultural differences with a single, nationwide cultural structure.

Few other countries have the transportation network to draw together the overwhelming primary production from within their own borders. At the same time, the more than 200 million consumers, most of them unusually affluent by world standards, provide each specialized producer with a market of almost unprecedented size.

The modern, interconnected system, operating over the huge territory of the United States, has had the opportunity to develop separate distributions for production, consumption, and decision-making. Thus, primary production such as agriculture, mining, forestry, and fishing is closely tied to the variations in the environmental base, although it must take into account accessibility to customers, government regulations, and the location of competitors.

Secondary production which depends as much as primary production on availability of labor and market shows fewer ties to the environment, tending to concentrate in major urban nodes. At the same time, the customers are also workers so they locate where the jobs are, mainly in major metropolitan centers. The managers and decision-makers who now can

manage operations spread out over the whole country or even the world from a single location prefer the few very largest metropolitan centers for they have the best transportation and communication connections with the rest of the country; also in these centers there is the further advantage of the development of specialized management communities with unique supporting services.

A Distinctive Employment Structure

Table 8-1 shows the gross employment pattern of the United States in 1971, divided into four types, two production and two nonproduction. Production is either primary (those types that directly utilize the environmental resources) or secondary (those that make things out of environmental materials). Nonproduction activities are divided into supporting services (transportation, utilities, trade and other services) and management and decision-making activities.

Table 8-1 reveals the relative importance of the major economic specializations of the United States. Notice that only one-third of the working population is employed in the production of goods. Over 60 per cent of the workers are not employed in making anything! They neither extract earth resources in primary production nor manufacture goods. They are employed in servicing, helping and managing. A very large share of them "push paper" in offices.

The Map of Population as an Index of Economic Opportunity

The map of population (Figure 8-1) shows the contrast between two different distributions of people in the United States. On the one hand, the map shows the thin veneer of rural popula-

Table 8-1 EMPLOYMENT STRUCTURE OF THE UNITED STATES IN APRIL 1972

	EMPLOYMENT (millions)		PER CENT	
PRODUCTION				
Primary				
Agriculture	3.3		4.4	
Mining	0.6		1.0	
		3.9		5.4
Secondary				
Construction	3.2		4.2	
Manufacturing	18.7		24.9	
	21.9		29.1	
		25.8		34.5
CONSUMPTION SERVICES				
Transportation and utilities	4.5		6.0	
Trade	15.4		20.0	
Other services	12.2		16.2	
		32.1		42.2
MANAGEMENT AND DECISION-MAKING				
Finance, real estate	3.9		5.1	
Government	13.4		18.2	
	17.3		23.3	
Total		75.2		100.0

SOURCE: U.S. Bureau of Census, *Statistical Abstract of the United States: 1972*, Washington, D.C., 1972, Tables 340 and 361.

tion spread widely over the country with variations in density, presumably providing a measure of differing resource utilization. On the other hand, it presents the concentration of population not simply in urban areas but in the network of a relatively few large metropolitan centers.

From the map itself it is difficult to determine the relative importance of the metropolitan and non-metropolitan populations, but Table 8-1 leaves no question about which population is the greater. Less than 6 per cent of the working population of the United States is now engaged in primary production activities, which are essentially non-metropolitan.

Although we have noted that some types of manufacturing prefer non-metropolitan sites and the rural areas support villages, towns, and

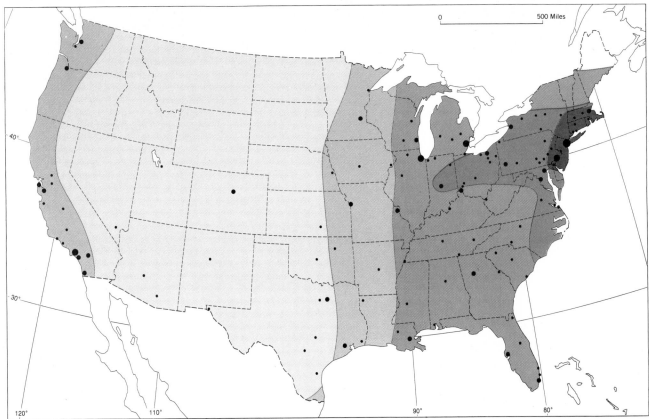

Figure 8-1 POPULATION DENSITY IN THE UNITED STATES

Rural population per square mile

- Less than 10
- 10-24
- 25-49
- 50-100
- More than 100

Population of metropolitan areas

- 250,000-1,000,000
- 1,000,000-4,000,000
- 4,000,000-16,000,000
- More than 16,000,000

Table 8-2 REGIONAL DISTRIBUTION OF NON-METROPOLITAN POPULATION
OF THE UNITED STATES, 1945 AND 1972

	1945		1972		CHANGE	
	POPULATION (000,000)	PER CENT OF NATIONAL TOTAL	POPULATION (000,000)	PER CENT OF NATIONAL TOTAL	NUMBER (000,000)	PER CENT
Northeastern Seaboard	9.5	15	11.8	19	2.3	24
Middle West	15.8	25	15.7	25	−0.1	. . . .
South	27.3	44	25.6	40	−1.7	−6
Great Plains– Mountain	6.8	11	6.9	11	0.1	2
Pacific	3.0	5	3.8	6	0.8	3
Total	62.4		63.8		1.4	2

SOURCE: U.S. Bureau of Census, *Statistical Abstract of the United States*, 1945 and 1972,
Washington, D.C., 1945 and 1972.

small cities, we can safely assume that the bulk of the population not engaged in primary production is employed in the predominantly metropolitan areas. In 1970 two-thirds of the population lived in large metropolitan areas.

RURAL POPULATION AND PRIMARY PRODUCTION

If we assume that the non-metropolitan population gives us the proportion of American people dependent on primary use of land resources, either directly using the resources or in supporting towns and cities, the distribution of this component should provide a reasonably good measure of agricultural, mining, forest, and fishing regions on which the interconnected system depends.

Notice in Table 8-2 that almost two-thirds of the non-metropolitan population is located in two areas of the country. None of the other three areas accounts for as much as 20 per cent of the total. Notice, too, that the non-metropolitan population of the country as a whole appears to have remained almost constant for over 30 years. This seems remarkable inasmuch as the total population of the country increased over 50 per cent in that period. The lack of growth surely does not fit the Malthusian idea of the geometric progression of population growth. Where are the population increments that should have occurred in the rural areas?

Table 8-2 also shows that there was almost no redistribution of the non-metropolitan population. The South and the Middle West, the most important areas of non-metropolitan population, were the only parts of the country with absolute losses. Gains in each of the other areas involved less than a million people, except in

Table 8-3 REGIONAL DISTRIBUTION OF PRIMARY PRODUCTION
(value of output in $ millions)

	LUMBERING* (1969)	FISHING (1969)	MINING (1969)	AGRI- CULTURE (1969)	TOTAL	PER CENT OF NATIONAL TOTAL	PER CENT NON- METROPOLITAN POPULATION (1970)
Northeast	48	124	2,533	3,890	6,595	8	19
Middle West	56	14	3,353	15,462	18,885	24	25
South	632	206	14,029	13,989	28,856	37	40
Great Plains– Mountain	314	0	4,726	9,290	14,330	18	11
Pacific	1,274	182	2,287	6,048	9,791	13	6
Total	2,324	526	26,928	48,679	78,457		

*Value estimated from board feet production.
SOURCE: U.S. Bureau of Census, *Statistical Abstract of the United States: 1971*,
Washington, D.C., 1971.

the Northeast where the growth was more than two million.

Notice in Table 8-3 the close correlation between the proportions of non-metropolitan population and primary production in the two leading regions, the South and the Middle West. Here is rural America. By contrast the Northeast has a higher proportion of non-metropolitan population than of production. We might infer that important numbers of this population are exurbanites who work in adjacent metropolitan areas.

Further examination of Table 8-3 shows that agriculture is by far the leading primary activity, with almost twice the importance of mining. Lumbering and fishing are much less important. The South is a major producer in all categories, while the Middle West is predominantly an agricultural area.

METROPOLITAN POPULATION AND THE BULK OF EMPLOYMENT OPPORTUNITIES

While non-metropolitan population has remained almost static during recent decades, metropolitan areas have shown tremendous growth, almost doubling between 1940 and 1970. This large growth shows a close correlation with the rise in employment in urban-centered industries. Figure 8-2 shows that total metropolitan employment more than doubled during the period 1940–1970. Growth was important in all categories, but the greatest growth occurred in consumption services and management. Manufacturing, which accounted for over one-third of metropolitan-centered employment in 1940, grew at a lower rate than other activities and totaled only slightly more than a quarter by 1970. It has been outstripped collectively by nonproductive metropolitan-centered

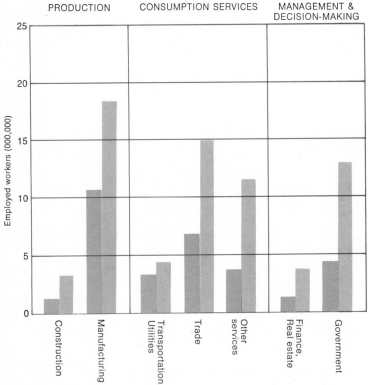

1940

1970

PRODUCTION   CONSUMPTION SERVICES   MANAGEMENT &
DECISION-MAKING

Employed workers (000,000)

Construction
Manufacturing
Transportation, Utilities
Trade
Other services
Finance, Real estate
Government

Figure 8-2 Changing Structure of
Metropolitan-centered Employment in the
United States, 1940 to 1970

activities, particularly government business, trade, and other services. Together these three activities have accounted for more than seven out of every ten new jobs in metropolitan areas since 1940.

Metropolitan growth has occurred in all parts of the country (Figure 8-3), but in contrast to the scarcely changed pattern of non-metropolitan population, there have been considerable shifts in the distribution of metropolitan population in recent decades. In 1940 almost three out of every four people in metropolitan areas were in the Northeast and the Middle West. By 1970 those two areas totaled less than three out of every five, despite the fact that the metropolitan populations of each had grown by more than 13 million. The largest absolute growth occurred in the South, especially in Texas and Florida. The Pacific surpassed the Northeast in total numbers of people added and showed a much greater percentage growth. The Great Plains-Mountain area, with the smallest total numbers, had the highest percentage increase. This metropolitan growth, coupled with the stable character of non-metropolitan population across the country, has caused sharp changes in the character of each region (Table 8-4).

In 1940, when the population of the country was almost balanced between metropolitan and non-metropolitan, only one section, the Middle West, was close to the national average. The Northeastern Seaboard and the Pacific areas were dominated by their metropolitan populations, and the South and the Great Plains–Mountain states remained predominantly non-metropolitan. By 1970, only one region of the country—the Great Plains and Mountain states—was mainly rural. On the other hand, the Northeast and the Pacific areas were over-whelmingly metropolitan. Four out of five of their residents lived in metropolitan areas.

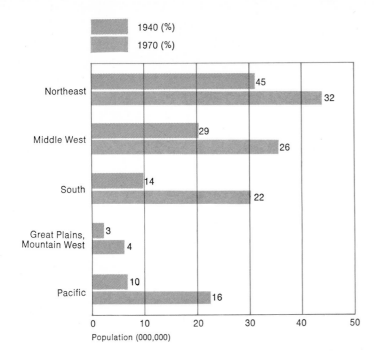

Legend:
- 1940 (%)
- 1970 (%)

Region		
Northeast	45	32
Middle West	29	26
South	14	22
Great Plains, Mountain West	3	4
Pacific	10	16

Population (000,000)

Figure 8-3 CHANGING DISTRIBUTION OF
METROPOLITAN POPULATION IN THE UNITED
STATES, 1940 TO 1970

Table 8-4 PER CENT OF POPULATION IN
METROPOLITAN AREAS BY REGION, 1945
AND 1971

	1940 (per cent)	1970 (per cent)
Northeastern Seaboard	77	79
Middle West	56	69
South	26	54
Great Plains–Mountain	21	48
Pacific	70	86
Whole of U.S.	53	69

SOURCE: U.S. Bureau of Census, *Statistical Abstract of the United States*, 1945 and 1971, Washington, D.C., 1945 and 1971.

As we look at the regions of the country, we see a contrast. Some areas, such as the Northeast, are largely metropolitan in population and, presumably, in functional base. The South and the Rocky Mountain region show a near balance between metropolitan and non-metropolitan populations. Only the Middle West shows the national ratio of two metropolitan residents to every one non-metropolitan.

Obviously these different parts of the country have different economic problems and aims, different political ideas, and even different cultural patterns. The large rural populations of the Middle West and South are greatly concerned with federal farm legislation such as acreage controls and subsidy payments. They follow development of hybrid seeds and feel threatened by attacks of environmentalists who object to the use of certain insecticides and fertilizers. They worry over the linking of cigarettes to lung cancer or the prospects of a large wheat crop in Australia. In turn, the political parties of the country recognize the importance of programs that will "carry the farm states" and the significance of the "farm bloc" in Congress.

On the other hand, the urban Northeast, Midwest, and Pacific Coast fight for revenue sharing between the federal government and cities, and struggle with welfare questions and labor strikes. These areas react to increasing imports of steel and automobiles from Western Europe and Japan. Politicians speak of winning the "industrial states" and the "big city vote."

The Nodal Nature of Metropolitan Centers

In contrast to rural population which is spread widely over the landscape, metropolitan populations are, by definition, nodal. Except in the most highly developed metropolitan areas (the megalopolis from Boston to Washington along the Northeastern Seaboard, or the Chicago–Milwaukee or Los Angeles–San Diego corridors), each metropolitan area is separated from others by the non-metropolitan world. These areas stand out as separate islands of concentrated population in a sea of rural population. The rural population extends almost without interruption over the eastern two-thirds of the country and breaks into separate isolated clusters only in the western Great Plains and most of the Mountain West. On the Pacific Coast it again forms a more continuous distribution. As Figure 8-1 shows, the metropolitan islands are sometimes large, sometimes small; sometimes closely clustered, sometimes not; they have their own pattern of distribution.

This separateness of the metropolitan nodes is important to keep in mind when considering their part in the functioning economy and culture of the United States. Such centers cannot operate in isolation. They depend on connec-tions with the primary producing regions for food and the basic resources used by manufacturers, but their most important connections are with one another. Modern business is a complex of interactions and transactions involving the producers, consumers, financiers, supporting business services, managers, and government support and regulatory agencies of the metropolitan nodes. But the nodes do not function as entities; rather they form a functioning network laced together by transportation and communication services.

TRANSPORT FLOWS AS INDICES OF THE
IMPORTANCE OF METROPOLITAN TIES

Examination of the transportation and communication system shows it is designed primarily to form links between the metropolitan nodes. Consider the interstate highway system shown in Figure 8-4. Although there are entry and exit points in a multitude of non-metropolitan centers, the pattern unmistakably focuses on major metropolitan areas. Table 8-5 lists the 100 leading centers in 1968 for the originating and terminating of airline passengers. Together these 100 points accounted for 90 per cent of all domestic airline passenger traffic in that year. Except for the three locations in Hawaii, all are metropolitan areas.

Figure 8-5, which shows the 100 leading airline routes linking pairs of cities in terms of passengers carried in 1968, puts the importance of intermetropolitan connections in sharper relief. Again, except on the islands of Hawaii, all origin-destination points shown on the map are major metropolitan centers. Only five air routes connect with smaller metropolitan centers, but it is obvious from the figures that the five are not a major part of the national air traffic flow.

The airline figures, of course, give only

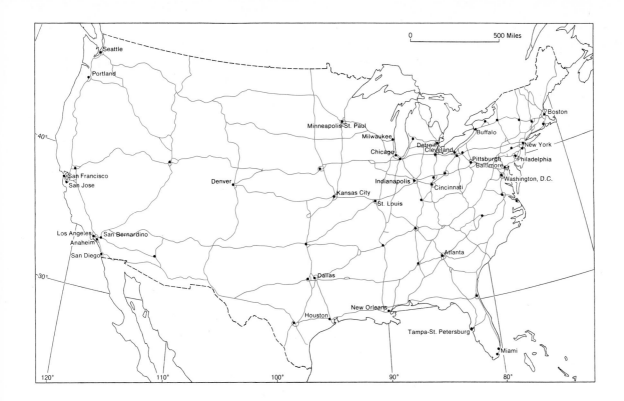

Figure 8-4 UNITED STATES INTERSTATE HIGHWAY SYSTEM

Table 8-5 100 Cities Originating or Terminating Domestic
Airline Trips, by Passenger Trips (in thousands), 1968

City	Trips	City	Trips	City	Trips
New York	22,696	Milwaukee	1,308	West Palm Beach	458
Chicago	13,340	Louisville	1,265	Kahului, Hawaii	457
Los Angeles	11,265	Dayton	1,222	Columbia, S.C.	444
San Francisco	8,031	Rochester	1,185	Charleston, S.C.	438
Washington, D.C.	7,674	Salt Lake City	1,152	Austin	417
Boston	7,023	Syracuse	1,045	Shreveport, La.	399
Miami	6,058	Omaha	1,039	Grand Rapids	397
Detroit	5,091	Norfolk	1,008	Toledo	378
Philadelphia-Camden	4,968	Jacksonville	980	Madison, Wis.	374
Dallas	4,283	Nashville	962	Huntsville, Ala.	373
Atlanta	3,924	Oklahoma City	943	Charleston, W.Va.	370
Cleveland	3,643	Charlotte	902	Fresno	369
St. Louis	3,611	Albany	784	San Bernadino	369
Seattle	3,553	Tulsa	777	Moline, Ill.	365
Denver	3,425	Albuquerque	763	Akron	354
Minneapolis-St. Paul	3,325	El Paso	763	Jackson, Miss.	350
Pittsburgh	3,269	Sacramento	736	Boise	344
Houston	2,939	Orlando	724	Harrisburg-York	336
Honolulu	2,651	Des Moines	696	Mobile	318
Kansas City	2,641	Providence	689	Colorado Springs	316
Las Vegas	2,439	Raleigh	681	Amarillo	315
New Orleans	2,164	Reno	678	Newport News, Va.	311
Baltimore	2,111	Tuscon	668	Augusta	310
Tampa	1,984	Birmingham	657	Chattanooga	307
Phoenix	1,909	Oakland	614	Salinas-Monterey	298
Buffalo-Niagara	1,787	Ft. Lauderdale	603	Lubbock, Tex.	298
Cincinnati	1,710	Wichita	566	Evansville, Ind.	296
Portland	1,705	Richmond	561	Corpus Christi	295
San Diego	1,660	Spokane	550	South Bend	288
Indianapolis	1,603	Greensboro	547	Cedar Rapids	284
Hartford-Springfield	1,460	Knoxville	521	Midland-Odessa, Tex.	284
Memphis	1,420	Lihue, Kanai, Hawaii	494	Fort Wayne	280
Columbus	1,412	Hilo, Hawaii	462		
San Antonio	1,349	Little Rock	458		

Source: *Handbook of Airline Statistics*, Civil Aviation Board, Bureau of Accounts and Statistics,
Washington, D.C., 1969.

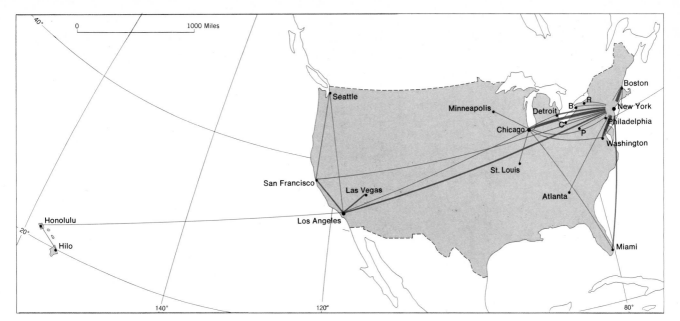

Figure 8-5 100 LEADING UNITED STATES
AIR ROUTES

Per cent of total domestic air passenger
travel in the United States

——— 0.4%-0.8%

——— 0.8%-1.6%

━━━ More than 1.6%

R-Rochester   B-Buffalo
C-Cleveland   P-Pittsburgh

movement of people. Some of these people are tourists; some are persons going to visit relatives or friends; some are military personnel; and others are students. But the predominance of male passengers and individual ticketholders suggests that by far the largest share of passengers traveling by air are moving from one metropolitan center to another on business. As a result, airline traffic, except when there is an obvious major flow of tourists to resorts, gives us a crude indication of intermetropolitan business connection (almost the only one readily available in public statistics), and it shows that the first-order movement of business personnel is between the metropolitan nodes rather than back and forth between the metropolitan centers and the multitude of non-metropolitan communities.

Although the primary movement of people is between metropolitan centers, we might expect the first-order movement of goods to be between metropolitan nodes and the non-metropolitan primary producing regions. What data we have, however, make this conclusion doubtful.

As with airline passengers, most of the goods from manufacturers remain within the network of metropolitan centers. This is really not surprising, since two out of every three people in the country live in these centers, and more than three out of every four dollars of manufactured output come from them. These individuals and firms are the major producers and consumers of goods in the country.

The major importance of metropolitan centers in the movement of goods and people in the functioning economy of the United States is apparent from Figure 8-6, the map of railroad freight traffic. The heaviest individual flows travel from coal-producing areas in the Appalachians and southern Middle West and from the iron fields in Minnesota, but, overall, the major traffic movements occur between the great metropolitan centers of the Northeast and Middle West. Elsewhere, more lightly traveled freight connections appear to form a network between these centers in the northeastern quarter of the country and outlying areas—regional metropolitan centers such as Dallas and Atlanta.

METROPOLITAN CENTERS AS THE KEY NODES IN THE FUNCTIONING SYSTEM

The flow pattern of traffic might be likened to the circulatory system of the body which is designed to deliver oxygen and nutrients to all parts of the body and to pick up and carry away waste. A main trunk system connects the major components of the system, and an outreaching system makes a minor tie to each tiny capillary throughout the body. At the same time there is little need to connect one outlying part with another; the system is designed to link the important centers with outlying areas.

Eighteen of the thirty-three metropolitan areas with one million or more people in 1970 were located in the Middle West–Northeastern Seaboard, including ten of the twelve largest. In these parts of the country, giant metropolitan agglomerations are closely spaced, virtually merging with each other.

Along the Eastern Seaboard from southern Maine to Washington, D.C. live almost two out of every five people in the country. Such a concentration of metropolitan areas so close that their population concentrations coalesce is called a "megalopolis." Five of the top twelve metropolitan centers and thirty-four smaller centers are located in this belt. This area is not merely important for its size—it is the functional interaction between these centers makes the region unique.

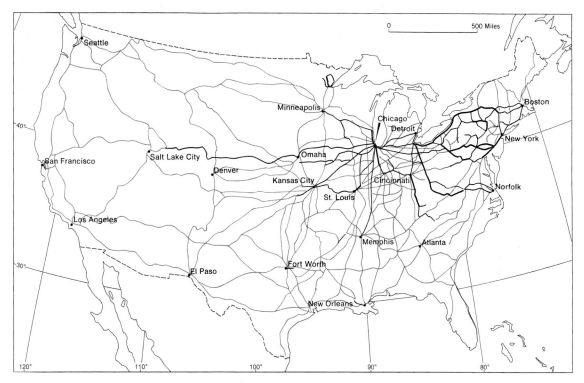

Figure 8-6 Railroad Freight Traffic in the United States

Traffic in tons per year

——— Less than 10 million

——— 10-25 million

——— More than 25 million

Outside of the Northeast-Midwest, the Pacific Coast area is the only other part of the country with a significant number of cities of over one million population. In the South, despite the recent growth of metropolitan population, there are only six metropolises of a million or more. In the Mountain West only Denver has slightly more than one million people. For the most part the individual large metropolitan centers stand well apart from each other as regional capitals for wide areas. They are the focus for regional transportation links and are nodal points which tie into the national system.

CENTERS OF ECONOMIC ACTIVITY

Metropolitan areas are more than places to live. Whereas the population living in metropolitan areas is 67 per cent of the total, figures for the different economic activities credit those areas with 70 per cent of retail sales and 87 per cent of wholesale sales. Personal services, business and repair services, entertainment, hotels, and manufacturing are overwhelmingly metropolitan activities in the functioning United States. Figures are not readily available, but we might suspect that metropolitan areas would show up equally well in the concentration of the nation's educational, medical, and other specialized services. (See Table 8-6.)

Notice from the table that the share of total national employment accounted for by the major metropolitan areas is greater than their share of population and that their share of national income is greater than their share of employment. Thus, we can see that they are more important as places to work than to live and more important as places to make money than to live. These areas are even more significant as centers of the economy of the country than as centers of population concentration.

They are also the dominant consumers of

Table 8-6 RELATIVE IMPORTANCE OF 223 METROPOLITAN AREAS TO THE UNITED STATES, 1968

	ALL METROPOLITAN AREAS (per cent)	30 LARGEST AREAS (per cent)
Population	67	38
Employment	68	. . .
Income	70	47
Retail sales	70	43
Wholesale sales	87	49
Selected services, receipts	83	59
Value added by manufacture	77	49
Bank deposits	81	58
Headquarters 500 largest industrial corporations	93	75
Headquarters 50 largest banks	100	88
Headquarters 50 largest insurance companies	96	56
Headquarters 50 largest retailing companies	100	92
Headquarters 50 largest transportation companies	100	84
Headquarters 50 largest public utilities	98	74
Scientists on national register of scientific and technical personnel 1966	77	47

SOURCE: U.S. Bureau of Census, *Statistical Abstract of the United States: 1971,* Washington, 1971; "Directory of 1,000 Leading Corporations," *Fortune,* May 1970; *National Register of Scientific and Technical Personnel,* National Science Foundation, Washington, D.C., 1969.

goods, services, and ideas. They are the marketplace where food and other products of primary production, resources, and manufactured goods, not only from the entire country but also from much of the world, are bought

Table 8-7 Top 15 out of 119 Cities with Headquarters of the 500 Leading Industrial Corporations

Cities	Rank	Assets of Corporations (dollars) (total for all 500: $401,593,482,000)	Per Cent Total Assets	Number of Corporation Headquarters
New York*	1	158,121,224,000	39.4	131
Detroit*	2	33,281,244,000	8.3	13
Chicago*	3	25,053,388,000	6.2	49
Pittsburgh*	4	18,589,526,000	4.6	14
Los Angeles*	5	17,066,990,000	4.2	20
San Francisco*	6	10,974,768,000	2.7	10
Cleveland	7	9,384,356,000	2.3	17
Philadelphia*	8	7,693,939,000	1.9	14
Akron	9	7,061,305,000	1.8	5
St. Louis*	10	6,636,532,000	1.7	11
St. Paul–Minneapolis*	11	5,627,752,000	1.4	11
Stanford–Greenwich, Conn.	12	5,046,261,000	1.3	7
Dallas*	13	4,843,540,000	1.2	7
Wilmington, Del.	14	4,512,965,000	1.1	3
Houston*	15	4,198,435,000	1.0	2

*Standard Metropolitan Statistical Areas of 1,000,000 or more population in 1970.
Source: "Directory of 1,000 Leading Corporations," *Fortune*, May 1970.

and sold. They are the major consumers of books, television and radio, fashions, political ideas, advertising, and other forms of culture. More and more the advertising agencies, innovators, politicians, and activists have been turning their attention to metropolitan audiences rather than to rural, small-town America. The result is an increasingly metropolitan-oriented culture and economy.

As Table 8-7 and Table A-2 in the Appendix show, metropolitan areas are the headquarters locations for the largest business firms in the country—industrial corporations, banks, insurance companies, retailing companies, trans-

portation companies, and public utilities. In all cases, over 90 per cent of the headquarters of the largest firms are located in metropolitan areas which are the management and decision-making centers for the country. When it is noted that the 500 largest industrial firms account for over two-thirds of all industrial sales, some measure of the importance of the concentration of their headquarters to metropolitan areas can be seen. The top fifty banks of the country—all headquartered in metropolitan areas—hold more than one third of the country's deposits.

As we have seen, the distinctive feature of this

modern, interconnected system is specialization. Each activity can be located where it seems to thrive best. Specialization results in Florida citrus growers concentrating on juice oranges, Texas citrus growers on grapefruit, and southern California orange growers on fruit for eating. One California district produces much of the country's lettuce during the winter months; another does so during the summer. In the same way manufacturers locate plants over the country in response to particular locational factors: availability of cheap power in one area, cheap labor in another, access to perishable food products in a third. The nearness of a supplier, a particular customer, and the availability of financial support are also factors. So it is with business management.

Producing and Marketing in Tandem

In the United States with its great range of environmental and consumer possibilities, many decisions about where to produce and where to sell must be made. But who makes those decisions?

Daily decisions are being made by each individual and each family—decisions that may be influenced by ideas from local schools, neighborhood discussions, or advertising on television or in magazines. But these decisions are made within the framework of the individual's local world, as we saw in the discussions of families in the Washington area. A life style, although influenced by outside forces, is formed within the framework of the daily world that rarely extends beyond one's own tiny space.

In the same way, the businessman in a small town only knows how to operate within his own local area. He knows his customers who all come from within a radius of twenty-five miles, and he deals with suppliers in a city less than

one hundred miles away. His other contacts are with local banks and businessmen. The banker in a small city may work within a larger scene. He knows a portion of his state or even several states. His loans may be to farmers and businessmen who live no more than a hundred miles away, and he has contact with banks in the large regional metropolis two hundred miles away.

But, as we have seen, the major decision-making of the country takes place in a few great metropolitan centers. These centers are the foci of transportation and communication networks that allow managers to be in constant touch with plants and markets throughout the country. In the industrial, financial, and trading world today, some of the giant corporations restrict themselves to operating within a narrow, specialized niche, such as clothing or motels. In so doing, they organize their operations on a national basis.

In recent years giant multi-faceted corporations have come to dominate a wide spectrum of businesses that may or may not show any relation to each other. The headquarters of such a corporation manage huge complexes of factories and production facilities scattered over the country or over the world. At the same time they control a marketing network that may have a very different distributional pattern. They produce where it seems best to produce and market where it is best to sell.

The Example of Potlatch Forests

Potlatch Forests, with headquarters in San Francisco, was the 284th largest company in the United States in sales in 1969. It deals in wood products and factory-built structures, in paperboard and packaging, in business and printing papers. Figure 8-7 shows how Potlatch has chosen to locate its subsidiaries across the

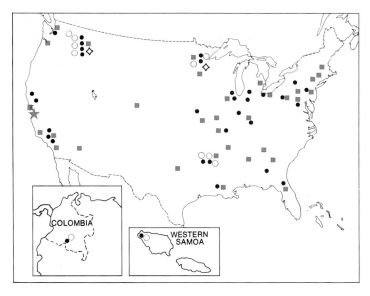

○ Timber reserves

◇ Research centers

● Manufacturing facilities

■ Sales offices

★ Executive offices

Figure 8-7 DISTRIBUTION OF FACILITIES OF
POTLATCH FORESTS

United States. Notice that they fall into several distinct categories, each with its own distribution pattern. Potlatch owns and manages timberlands, manufactures a variety of wood and paper products (each with particular locational needs), and then sells the products to customers throughout the country.

For the timber that forms its raw material, Potlatch has gone to three of the major timberlands of the country. Because logs are difficult to ship, its heavy manufacturing operations—lumber, plywood, veneer, and papermaking—are located in lumber towns adjacent to timberlands. But distinctions are made according to the particular character of the timber resource in a given area. Potlatch produces no lumber in Minnesota, where slow tree growth makes it difficult to carry on continuing lumber operations at one site. It makes veneer and plywood in Idaho where logs are large and of high quality. Wood is converted into pulp and the pulp into paper in Minnesota and Idaho, but paper is also made in northern California where there is no pulp mill. The latter operation specializes in stationery which does not depend entirely on wood pulp, and Potlatch purchases pulp from noncompany sources.

In contrast to the location of lumber and paper plants in the forests, the manufacture of paper specialties and the fabrication of lumber generally show locations close to markets. Here it is cheaper to ship the paper and lumber to the plant than to move the manufactured product to market. Paper specialties are made in California, milk cartons in California and Indiana, paper plates in Missouri and California, and facial tissues in Pennsylvania. Most of these plants are in metropolitan centers or strategically located towns. Cartons and boxes, the most difficult to ship, are manufactured in a number of plants in large metropolitan areas, located so that together they serve the major

General Motors Corporation—Largest industrial corporation in the United States in 1969
Headquarters—Detroit, Michigan.   Sales 1969—$24,295,000,000.   Employees—794,000.

1. *Automobile Assembly*
Buick Division—1 plant in Michigan
Cadillac Division—1 plant in Michigan
Chevrolet Division:

  a. Automobile assembly—plants in Georgia, Maryland, Michigan (2), Massachusetts, Wisconsin, Missouri (2), California, Ohio, New York

  b. Carburetors—1 plant in Michigan

  c. Boxes for automobile export—1 plant in New Jersey

  d. Axles—plants in New York, Michigan (2)

  e. Transmissions—plants in Ohio (2), Indiana, Michigan

  f. Sheet-metal stamping—plants in Ohio, Michigan, Indiana

  g. Forgings—plants in Michigan, New York

  h. Engines—plants in Michigan (3), New York

  i. Springs and bumpers—1 plant in Michigan

  j. Engine castings—1 plant in New York

  k. Grey iron foundries—plants in Michigan, New York

  l. Miscellaneous automobile parts—1 plant in Michigan

  m. Coke plant for foundries—1 plant in Illinois

Oldsmobile Division—1 plant in Michigan
Pontiac Division—1 plant in Michigan
Fisher Body Division:

  a. Automobile bodies—plants in Georgia, Maryland, Michigan (5), Massachusetts, Wisconsin, Missouri (2), California, Ohio, New York

  b. Truck bodies—plants in Michigan, Ohio

  c. Automobile stamping, trim—plants in Ohio (2), Michigan (5), Indiana, Pennsylvania, Illinois

GM Assembly—automobile assembly plants in Texas, Georgia, California (2), Missouri, New Jersey, Delaware (serves Buick, Oldsmobile, and Pontiac)
GMC Trucks and Coaches—1 plant in Michigan

2. *Automobile Distribution*
  a. More than 15,000 dealers

  b. 50 warehouses

3. *Automobile Components*
AC Spark Plug—1 plant in Michigan
Central Foundry—plants in Indiana, Illinois, Ohio, Arkansas, Michigan
Delco Moraine (brake components)—1 plant in Ohio
Delco Products (shock absorbers, etc.)—plants in Ohio, New York
Delco Radio—1 plant in Indiana
Delco-Remy (ignitions, batteries)—plants in California, Indiana, New Jersey, Kansas
Guide Lamp (lights)—1 plant in Indiana
Harrison Radiator—2 plants in New York
Hydramatic (transmissions)—1 plant in Michigan
Inland Manufacturing (steering wheels, brake linings)—1 plant in Ohio
New Departure–Hyatt Bearing (bearings)—plants in Connecticut (2), New Jersey (2), Ohio
Packard Electric (wiring, switches)—1 plant in Ohio
Rochester Products (carburetors)—1 plant in New York

4. *Defense*
A.C. Electronics—1 plant in Wisconsin
Allison—plants in Indiana, Ohio

5. *Engines*

Detroit Diesel Engine—1 plant in Michigan

Diesel Equipment (fuel injectors)—1 plant in Michigan

Electro-Motive (locomotives, large diesel engines)—2 plants in Illinois

Earth-moving Equipment—2 plants in Ohio

6. *Household Appliances*

Frigidaire—1 plant in Ohio

7. *Research and Training*

GM Defense Laboratories—California

GM Technical Center—Michigan

GM Training Centers—30 over the country

8. *Finance*

General Motors Acceptance, Motors Insurance, Motors Holding—offices over the country

General Motors is first of all an automobile manufacturing company. Making an automobile and then selling it call for a variety of operations: designing and engineering the car, making the components that go into it, assembling the car, and then marketing it. Each of these operations is distinct and can be done independently—either by a separate company or in a separate location.

Questions

1. Is there a distinctive geographic distribution to each of these separate operations as far as General Motors is concerned? Where are the design installations? Is the distribution of components plants similar to that of the assembly operations?

2. Buick, Cadillac, Oldsmobile, Pontiac, and GMC each has only one assigned assembly plant, but Chevrolet has 11. Why? How do the divisions other than Chevrolet get national distribution of cars?

3. Where are the parts suppliers for assembly plants in Georgia and California?

4. What other businesses is General Motors in besides the automobile industry? Does the distribution of the nonautomobile plants show any pattern different from the others?

5. Does the distribution of General Motors plants show any close relationship to the location of the company's headquarters in Detroit? General Motors has just completed in New York City a new office skyscraper in midtown Manhattan. Why do you suppose it has done this?

Sears, Roebuck and Company—Largest retailer in the
United States in 1969

Headquarters—Chicago, Illinois.   Sales,
1969—$8,863,000,000.   Employees—355,000.

1. The company operates 826 stores in the United
States and Puerto Rico.

a. 240 complete department stores in large metropolitan centers

b. 384 medium-sized department stores in
smaller markets

c. 202 appliance, sports equipment, and auto
supply stores in outlying shopping areas and
smaller communities

2. In addition, it has 2,131 catalogue offices in its
stores and in small-town locations.

3. It operates 11 catalogue-order plants to handle
distribution of catalogue merchandise.

4. It owns several subsidiaries manufacturing
products for its outlets.

a. A plumbing-fixture company with plants in
New Jersey, Texas, Pennsylvania (2), California,
Wisconsin, Georgia, Arizona; sales offices in
Boston, Atlanta, Sherman Oaks, California, Chi-
cago, Camden, New Jersey, Houston

b. A chemicals and furniture firm with:
(1) chemical coatings—plants in California
(2), Illinois (2), Ohio, North Carolina, Texas,
New York

(2) chemical products—Wisconsin, Illinois,
California

(3) engineering specialties—Illinois

(4) furniture—Florida, Oregon (3), Pennsyl-
vania, Illinois, Georgia, New Hampshire, New
Jersey, Texas (2), Washington, Mississippi (5),
Arkansas (4), Iowa

(5) other—New Jersey, California, Texas, New
York, Illinois

c. A home-furnishings firm with plants in Missouri (2), New York (2), Pennsylvania (3), Tennessee (2), Massachusetts, California (2), Illinois
(2)

d. A company that makes cooking ranges, lawn
mowers, and accessories with plants in Illinois
(2), Tennessee, Ohio, Maryland (3)

5. It has a research center in the Chicago metropolitan area.

6. Sears also has overseas operations (especially
in Latin America) with the following stores:

Mexico	27	Costa Rica	1
Puerto Rico	15	Honduras	1
Venezuela	14	Nicaragua	1
Brazil	12	Panama	1
Colombia	11	El Salvador	1
Peru	4	Spain	1

Questions

1. How do Sears stores adapt to selling to both
metropolitan and nonmetropolitan markets?

2. Why would Sears go into manufacturing?
And why into the particular types of manufacturing that it has?

3. Sears is obviously a nationwide retailer. How
does the distribution of its factories compare
with the distribution of population in the
country?

4. Does the distribution of factories show any
relation to the fact that Sears headquarters
are in Chicago?

5. Where would you place warehouses or distribution centers to serve Sears national
market?

regional markets. The one wood-fabricating operation, making relocatable buildings, produces structures that are also difficult to ship. To get coverage in major markets, three plants have been built, one near San Francisco, one in Indiana, and one near Pittsburgh.

The whole Potlatch operation with almost 12,000 employees is supervised from headquarters in San Francisco. There the final decisions on the purchase of new timberland, construction of new plants, closing of old ones, the development of new products, and strategies for selling are made. But notice that Potlatch operations show little orientation to San Francisco, or even California. It is a national firm producing for a national market. It has purchased its raw material sources and manufacturing plants to serve that market within present business conditions and the operations of a variety of competing firms. Its thirty-three major sales offices are spread across the country in the large metropolitan centers where the major retailers and industrial customers are found. Recently Potlatch has also gone overseas for additional sources of timber. It has timberlands and lumber mills in Western Samoa and Colombia, but has not yet opened up overseas sales offices.

GENERAL MOTORS AND SEARS AS THE
ULTIMATE EXAMPLES

Potlatch is far from being the largest industrial corporation, yet its operations are representative of how a corporation develops a network of resources, manufacturing plants, and sales units where management plans to fit them into the interconnected system. Some firms do not worry about immediate control of natural resources; some buy directly from other manufacturers; some sell to retailers and wholesalers; some only to other manufacturers. But all the large corporations that have been successful

have tended to gain control over their sources of supply.

The operations of two major corporations are outlined for study in the boxes on pages 238 to 240. Notice how General Motors, the largest industrial corporation in the country, has developed or purchased specialized divisions, each with its own part to play in the operations of the company.

Imagine the task of managing one of these operations. General Motors has 108 plants in the United States alone, each one of them a major operation in itself with the number of workers ranging from several hundred to several thousand. Although it produces a large share of its own component parts, General Motors buys products from hundreds of suppliers. In selling its products it must deal with more than 15,000 auto dealers, with appliance distributors, railroads, and industrial firms, and must enter into major contracts with the government.

Sears must supply 826 stores in the United States and 2,131 catalogue outlets. Even though it manufactures some of its own products, think of the number of suppliers it must have and all the customers it must bill!

The Game Headquarters Play

There is no doubt that the headquarters of these corporations must be in a center of communications. They must be able to communicate with each unit in the network, with suppliers and customers, with banks and financial institutions, and with supporting business services such as advertising agencies, shippers, and lawyers. Headquarters must be so placed that they know what their competitors are doing.

For them the metropolis, both as the center of a business community with all of its services and as a focal point for world-wide transportation and communications networks, is the key place from which to manage their complex operations. The larger the metropolitan center, the larger its supporting business community is likely to be and the better its transportation and communications service. Office buildings in the largest metropolitan areas have grown spectacularly in size since World War II, and the growth continues despite increasing congestion on city streets and pollution of city air. Indeed, it is likely to continue until there is a breakdown in transportation and communication. Workers and executives in New York City complain about the increasing problems of commuting, the rise of crime in the streets, the greater number of smoggy days each year, and rising taxes. Yet skyscrapers continue to increase in number and size.

In 1969, however, the first real threats to business in New York began to appear. The telephone system in Manhattan became overloaded, there were delays in getting installation on new phones, and the message-switching system broke down. Then the airports serving the city became overloaded, and it was necessary to ration the number of commercial flights during certain hours. Even so, at rush hours twenty or more planes were waiting to take off. And finally there was an electric-power failure which virtually stopped the whole business community and its supporting services for several hours.

All of this has caused some companies to move their headquarters. Up to now most have moved from the skyscrapers of downtown or midtown Manhattan to suburban locations in Westchester County, Long Island, or southwestern Connecticut. A few have completely abandoned the New York metropolitan area for smaller centers such as Houston and Phoenix that have attractive climates. It is notable that the move in such cases has been out of the East Coast megalopolis altogether.

Indeed, the key management question for the future is whether the increasing dominance of New York City, encouraged in recent years by the mergers of industrial corporations into ever larger management units, will continue. Theoretically, improvements in transportation and communications that link metropolitan centers together make it increasingly possible to manage a financial-industrial complex from any major metropolitan center in the network. With modern telephone connections and jet aircraft, the firm based in Phoenix is within instantaneous reach of the entire New York business community. The only thing missing is the direct personal contact which businessmen claim is so important. (It is said that many of the country's most important business transactions have been carried out over lunch or on the golf course.)

New York City has also generally been recognized as a special place to live as well as to work. As the theatrical, musical, and artistic center of the United States, it has had a special attraction for the wealthy and ambitious of the country. As the largest city and primary connecting link with the rest of the world, it has a varied and cosmopolitan atmosphere. But now these urban attractions are competing with the attractions of outdoor living with its recreational possibilities. Moreover, with increased congestion, pollution, crime, and other problems, the quality of life in New York has deteriorated.

The decision of a corporation to locate where it chooses is considered an unchallenged right despite the fact that the addition of thousands of new workers and hundreds of new telephones may overload a city's infrastructure. But these urban problems have been seen in the United States as part of a different "sector" of the economy—the public as contrasted to the private sector. Often the private sector makes a decision in terms of its own economic self-interest, leaving the burden of providing back-up facilities to the public sector. It is said that many of the real costs of production, both supporting facilities and costs in terms of the quality of life, are not being met by the private sector in the United States. Rising taxes and deterioration of public services stand as evidence to support this position.

Up to now there has been no attempt to control the ever-growing centralization of economic activities in New York City despite the problems of overextension of supporting services, the growing congestion on city streets and in airports, and pollution. In contrast, the government of Britain has been conscious of the problems of overcentralization in London. It has placed limitations on the amount of new industrial and management expansion that can take place in the capital. New relocation in London can take place only after the government has studied its impact and issued a certificate of necessity.

The Management Network

Table A-2 shows the concentration of corporate management in the megalopolis of the Northeastern Seaboard. Newark, Philadelphia, Hartford, and Boston all stand out along with New York. (Surprisingly Baltimore and Washington do not appear. The big corporations managed in Washington total just less than $3.2 billion assets and in Baltimore less than $1.3 billion.) Large corporations headquartered in the megalopolis probably account for about 30 per cent of total assets of all businesses. Chicago and the large metropolitan areas of the Middle West from Pittsburgh to Minneapolis–St. Paul and St. Louis are less than half as important.

San Francisco and Los Angeles are less than half again as important as the centers in the Middle West. Regional centers elsewhere total less than the figure for Los Angeles alone.

It is apparent, too, that most of the metropolitan headquarters centers are highly specialized. Only a few other centers besides New York have important management control in the full range of corporation types shown. Chicago, Los Angeles, Boston, Dallas, Minneapolis–St. Paul, and St. Louis, and perhaps Philadelphia and Atlanta, show such diversification.

Perhaps we can get a feel for the functioning United States system from Table 8-7. New York stands alone as a management center; thus, it is the prime business-financial center of the country. Elsewhere, Chicago and Los Angeles stand as secondary centers for major producing and population regions of the country—the Middle West and California. Backing up New York are the major diversified centers of the megalopolis of the Northeast—Philadelphia and Boston. In the same way Minneapolis–St. Paul and St. Louis serve the Middle West. They are less important as regional centers than as gateways to the flow of goods, money, people, and ideas. In the less significant business areas of the country, Atlanta, Dallas, and Portland play a similar role.

Other major centers contribute their own specialties to the business system. Detroit is the capital of the automobile. Pittsburgh is the traditional center of steel. On a smaller scale Akron is the same for the rubber industry. Houston contributes its specialty of oil, petrochemicals, and natural gas. Cleveland, Cincinnati, and Milwaukee are manufacturing-business centers whose roles in management control are less easy to define. The firms headquartered there are smaller and dominate less important specialty industries such as machine tools, metalworking, brewing, soap, and cosmetics.

Newark and Hartford, with their large life insurance companies, function as specialized financial satellites within the megalopolis dominated by New York.

Although New York is the primate city in a hierarchy that stretches over the entire country, there are other major control centers of important subsystems. Linkages and transfers exist between one subsystem and another. The whole is the functioning unit, not the subsystems. There is also an overlap of function rather than the establishment of particular spheres of influence. Firms in any one of the centers may operate national organizations as do General Motors headquartered in Detroit, Sears in Chicago, and Potlatch in San Francisco.

"Nesting" of Decision-making Centers

The big centers constitute only the highest level of management in the United States. Decisions are made and management functions at all levels—the metropolitan center, the city, the small town, the village, the farm unit, and the individual family. In the same way there is not only a hierarchy but also links, transfer, and overlap. The classic model in regional economics shows a nesting of villages within the trade area of a town, of towns within the larger trade area of a small city, of cities within the larger territory of a metropolitan area. But in the model the movement of goods, people, and ideas takes place simultaneously at all levels. The farm family deals with the village for some things, the town for others, the city for others, and with the metropolis for some. City dwellers know that the same kind of multiplicity exists between the neighborhood, the large shopping center, and "downtown." The housewife with an automobile operates within all three markets. She may frequent the local supermarket, go to the regional shopping mall, or shop downtown, depending on her needs and mood. Either way, the system works through a series of nodes which serve both as a point of transaction and a point of decision-making and management.

The whole operation is, of course, dependent upon transportation and communication. The better the transport and communication, the larger the scale on which the system can work. With better transportation the village can be by-passed for the city or metropolis; with communication contact can be made directly to and from the largest metropolitan center and each home.

Perhaps the flexibility and overlapping of the system can be seen in the banking system. There are more than 13,500 federally insured commercial banks in the country. They range in size from huge international banks, such as the Bank of America in San Francisco with more than $22 billion in deposits, to small-town banks. At the top, the fourteen largest banks in the country control about 25 per cent of total deposits. The bottom quarter, the smallest banks, has less than 2 per cent of deposits.

These banks provide service to all financial customers in the country ranging from the largest corporation to the smallest child. Customers include corporations, the federal government, local governments, and pension accounts as well as individuals. Money flows from one bank to another as needed. For example, a farmer in the Corn Belt may need $55,000 of operating capital a year, and in the surrounding community there may be dozens of farmers with similar requirements. This demand for loans exceeds the resources of the small-town local bank. But if the loans are approved, the money can usually be obtained by the bank through larger correspondent banks in regional centers.

At the other end of the scale, a giant corporation may need a multimillion-dollar loan for expansion purposes. This may be beyond the capacity of any single bank in terms of its lending capacity for any single loan. If the bank joins with other banks, a "package" of money can be put together to provide the necessary funds.

Banks are classified as international, regional, or local in terms of the basic range of their operations. Only twelve to fourteen United States banks are recognized as operating regularly on an international scale. That is only one-tenth of one per cent of all commercial banks. Over half of these international banks are in New York City, and all but one other are in Chicago, Los Angeles, and San Francisco.

About 500 banks—less than 5 per cent of all banks—operate on a national scale (not necessarily those termed "national bank," meaning that they have a national rather than a state charter). They deal with other banks and national corporations. On the next level are 1,500 regional banks, mostly in metropolitan areas and the larger regional cities. Their range is regional rather than national. Finally there are the local banks—totaling over 90 per cent of all banks—that operate only in a small town or a neighborhood of a city. A small town or village may have only one small local bank. A large metropolitan area such as New York has local banks in neighborhoods, regional banks, national banks, and international banks, each serving a particular function.

Money must flow throughout the system and even enable international transactions. United States international banks have offices in foreign capitals—such as London, Paris, Rome, Tokyo—and foreign banks have offices in New York, San Francisco, and Chicago.

ORGANIZATIONS AND COMMUNITIES AS "INTERMEDIARIES"

The bank serves as a marketplace or an intermediary between persons and organizations wishing to save money and earn interest on deposits and those wanting to borrow money. The bank makes a profit by charging for its services as a catalyst or conduit. This is also true of other organizations in the system. All depend on a two-way interchange: an inflow of goods and ideas on the one side and an outflow of finished products on the other. This is the way a manufacturer operates. A retailer or wholesaler is likewise an intermediary. So is the life insurance company which collects on policies and invests money. So, too, is the publisher who takes manuscripts fed to him by authors and markets the manufactured book. Government takes in taxes and distributes services; or perhaps it provides services and must take in taxes to pay for them; it establishes rules of conduct that society follows and then metes out punishment to those who do not conform.

Communities also serve as intermediaries, providing a service to some and receiving services from others. Each organization and individual, then, is dealing on two levels. He collects on one and distributes on another, operating both on a local level and on a broader one. The grain elevator in the local community collects produce grown by many farmers in the area and ships it to a warehouse or user; it receives fertilizer or feed from a company and distributes that product to local farmers. But the largest center, like the international bank, simultaneously operates all levels—international, national, regional, and local.

THE NETWORK OF CULTURAL AND POLITICAL IDEAS AND DECISIONS

Government activity and cultural influences have impact beyond the effects of the mere movement of goods, services, and money. Ideas move through the system from the individual rural family to the largest metropolitan area. At the center are people in New York whose social and economic contacts are largely international. Others function mostly at a national level. The sphere of most people is restricted to their own neighborhoods.

Individual values are shaped by local schools, churches, and social groups, by regional influences such as state universities and church dioceses and conferences, and by national forces such as television commercials, church denominations, book publishers, and fashion houses. In the communications field a local newspaper may be read along with a regional big-city newspaper or the *New York Times.* Local radio and television stations intersperse local news and advertising with programs from the national broadcasting networks. Before the era of instantaneous communication, national influence came only after days or weeks of delay; local influences were primary; communities were distinctive. Now, with network programs conceived in New York, Washington, and Los Angeles, national forces are reflected in any community. Popular topics spread from center to center overnight. However, each community, each family, each individual has his own filter for accepting or rejecting to one degree or another these stimuli. Moreover, each community presents its own feedback, its own tiny input, to the system. In turn, each community is judged not in terms of its own life style or the happiness of its people but by how it fits economically, politically, and culturally into the national system on the basis of values established at the national level.

The freedom of individuals within the hierarchial system varies sharply, too. The wealthy have maximum choice regarding sources of supply. The middle class has more limited choice, and the poor have least of all. Consider the purchase of furniture or food. The wealthy individual furnishing a house calls in an interior decorator with access to the full range of furnishings, from local shops, department stores, and specialty shops throughout the system. Goods may be purchased in the community where the buyer resides, from a New York importer, or from an antique shop in the part of the country where a "period piece" might most easily be found. The middle-class individual is likely to make the selection of goods himself, so his choices are limited to goods in the stores in his community. The family of modest means without a car has to choose from the neighborhood furniture dealer or to buy from a catalogue outlet.

The range in food buying is much the same. The wealthy party-giver may telephone to have fresh Maine lobster flown in; the middle-class shopper may drive to the supermarket in a city twenty miles away; the low-income buyer walks to the neighborhood grocery. The choice of recreation follows a similar pattern: the wealthy can fly to any part of the world; the middle-class family buys a camper and goes off to see the United States; the low-income family goes to the local movie.

With the impact of national cultural and economic influences has come the rise of national government. As we saw in our earlier study, Washington, D.C., has grown from a rather minor metropolitan area and become second only to Los Angeles in population growth. Now it ranks seventh among the largest metropolitan areas of the country. It has reached this position despite its unimportance in manufacturing, wholesaling, and banking. But it is *the* decision-making and management center of the federal government.

Combating the problems of industrialization is an expensive and complex task. In many respects people in the modern, industrialized countries are as set in their ways as people in traditional societies, and the changes which have been undertaken in recent years are tentative but visible.

ANTI-POLLUTION TEST VEHICLE
operating on natural gas

U.S.MAIL

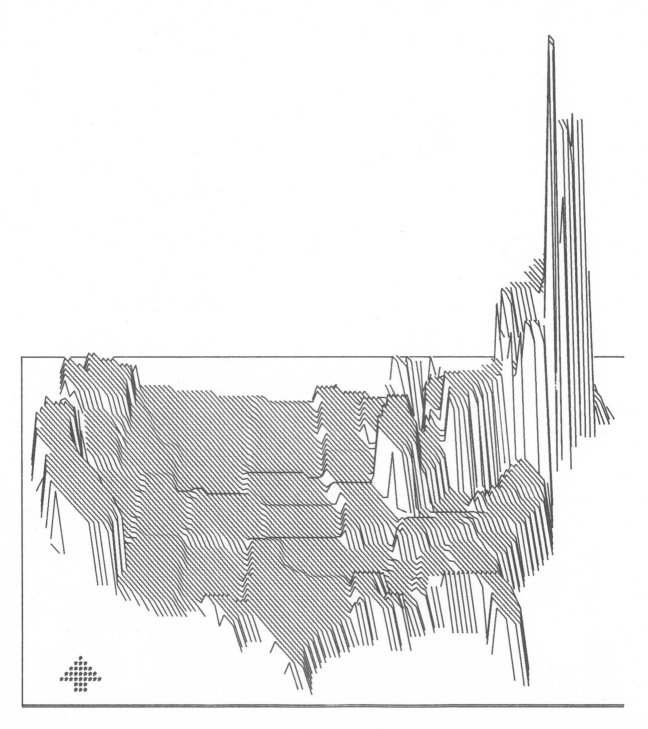

POPULATION DENSITY IN THE CONTERMINUS UNITED STATES BY STATE, 1970

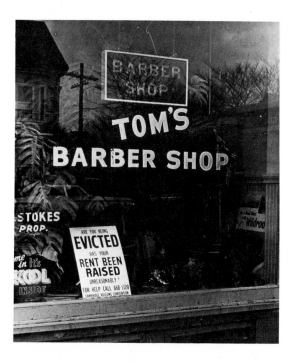

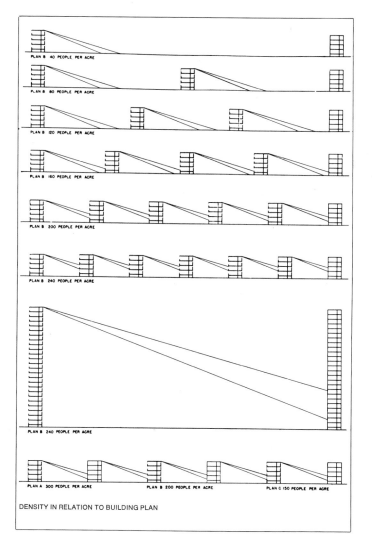

PLAN B 40 PEOPLE PER ACRE

PLAN B 80 PEOPLE PER ACRE

PLAN B 120 PEOPLE PER ACRE

PLAN B 160 PEOPLE PER ACRE

PLAN B 200 PEOPLE PER ACRE

PLAN B 240 PEOPLE PER ACRE

PLAN B 240 PEOPLE PER ACRE

PLAN A 300 PEOPLE PER ACRE   PLAN B 200 PEOPLE PER ACRE   PLAN C 150 PEOPLE PER ACRE

DENSITY IN RELATION TO BUILDING PLAN

The challenge of preserving the environment often conflicts with our need to provide inexpensive housing, efficient transportation, and rewarding jobs for millions of people. But the preservation of personal dignity for mankind remains one of the difficult challenges facing the industrialized world.

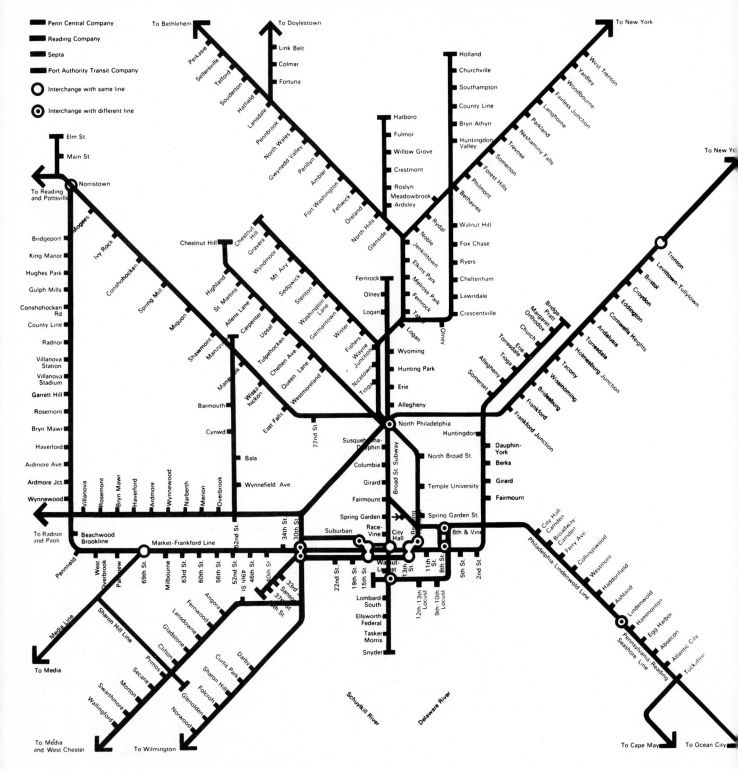

Legend

- Penn Central Company
- Reading Company
- Septa
- Port Authority Transit Company
- ○ Interchange with same line
- ◉ Interchange with different line

To Bethlehem
To Doylestown
To New York

Elm St.
Main St.
To Reading and Pottsville
Norristown

Perkasie
Sellersville
Telford
Souderton
Hatfield
Lansdale
Pennbrook
North Wales
Gwynedd Valley
Penllyn
Ambler
Fort Washington
Oreland
North Hills
Glenside

Link Belt
Colmar
Fortuna

Hatboro
Fulmor
Willow Grove
Crestmont
Roslyn
Meadowbrook
Ardsley

Holland
Churchville
Southampton
County Line
Bryn Athyn
Huntingdon Valley
Philmont
Bethayes
Forest Hills
Somerton
Trevose
Neshaminy Falls
Langhorne
Parkland
Woodbourne
Fairless Junction
Yardley
West Trenton

To New York

Trenton
Levittown-Tullytown
Bristol
Croydon
Eddington
Cornwells Heights
Andalusia
Torresdale
Holmesburg Junction
Tacony
Wissinoming
Bridesburg
Frankford

Bridgeport
King Manor
Hughes Park
Gulph Mills
Conshohocken Rd.
County Line
Radnor
Villanova Station
Villanova Stadium
Garrett Hill
Rosemont
Bryn Mawr
Haverford
Ardmore Ave.
Ardmore Jct.
Wynnewood

Mogees
Ivy Rock
Conshohocken
Spring Mill
Miquon
Shawmont
Manayunk
Wissahickon
Barmouth
Cynwd
Bala
Wynnefield Ave.

Chestnut Hill
Chestnut Hill
Gravers
Wyndmoor
Highland
St. Martins
Allens Lane
Carpenter
Upsal
Tulpehocken
Chelten Ave.
Queen Lane
Westmoreland
East Falls
22nd St.

Mt. Airy
Sedgwick
Stenton
Washington Lane
Germantown
Wister
Fishers
Wayne Junction
Nicetown
Tioga

Fernrock
Olney
Logan
Logan
Wyoming
Hunting Park
Erie
Allegheny

Rydal
Noble
Jenkintown
Elkins Park
Melrose Park
Fernrock
Tabor
Olney

Walnut Hill
Fox Chase
Ryers
Cheltenham
Lawndale
Crescentville

Allegheny
Tioga
Erie
Torresdale
Church
Margaret Orthodox
Bridge Pratt

Somerset
Frankford Junction

North Philadelphia
Huntingdon
North Broad St.
Temple University
Girard
Fairmount
Spring Garden St.
8th & Vine

Susquehanna-Dauphin
Columbia
Girard
Fairmount
Spring Garden
Suburban
Race-Vine
City Hall
Walnut-Locust

Dauphin-York
Berks
Girard
Fairmount

Broad St. Subway

City Hall, Camden
Broadway, Camden
Ferry Ave.
Collingswood
Westmont
Haddonfield
Ashland
Lindenwold
Hammonton
Egg Harbor
Absecon
Atlantic City

Philadelphia Lindenwold Line

Pennsylvania Reading Seashore Line

To Radnor and Paoli
Beachwood Brookline
Market-Frankford Line

Penfield
West Overbrook
Parkview
69th St.
Milbourne
63rd St.
60th St.
56th St.
52nd St.
46th St.
40th St.
34th St.
30th St.
33rd St.
37th St.
Spruce St.
22nd St.
19th St.
15th St.
13th St.
11th St.
8th St.
5th St.
2nd St.

Villanova
Rosemont
Bryn Mawr
Haverford
Ardmore
Wynnewood
Narberth
Merion
Overbrook
52nd St.

Media Line
Sharon Hill Line
Angora
Fernwood
Lansdowne
Gladstone
Clifton
Primos
Secane
Morton
Swarthmore
Wallingford

Curtis Park
Sharon Hill
Darby
Folcroft
Glenolden
Norwood

Lombard South
Ellsworth Federal
Tasker Morris
Snyder
12th-13th Locust
9th-10th Locust

Schuylkill River
Delaware River

To Media
To Wilmington
To Media and West Chester
To Cape May
To Ocean City

MAP OF PHILADELPHIA SUBWAY & RAIL SYSTEM

© Group for Environmental Education, Inc

Parts of the Country "Outside" the Modern System

So far we have presented the modern metropolitan, interconnected system of the United States as reaching out over the entire resource base to its many diverse regions. If we examine the system more closely, however, we find that this is not really true. As local, regional, and national industries have made decisions about the best places to locate, they have centralized activities in a few key locations and have discarded the marginal places. We saw this in the early nineteenth century when New England discovered that agricultural products from the fine farmlands of the Middle West could be delivered to its cities more efficiently than by local efforts and began abandoning its marginal farming.

The problem is that marginal areas, including whole regions such as Appalachia, do not fit the system any longer. Nor does it look as if their situation will improve in the future. They are no longer being considered by local or regional decision-makers. As places they are being measured against national—and even international—alternatives and found wanting. Efforts at regional economic development have been going on for years. Farm assistance programs and special government projects try to provide incentives for development. Governmental organizations such as the Tennessee Valley Authority and more recently the Appalachian Regional Commission have also made efforts. But the most depressed areas of the country are in the least promising places. Agricultural possibilities are poor, most manufacturing and service activities will not locate in small-town, rural environments, and mining and manufacturing are getting more automated all the time. Moreover, the people of these areas are among the most poorly educated and therefore are least prepared to take their places in the modern system.

Figure 8-8 shows the regions of the country that the federal government has designated depressed areas, that is, areas eligible for financial assistance from the government. They are areas of persistent high unemployment or continuing low farm incomes. These measures of "depression" provide an index of the degree to which local counties are not keeping pace economically with the nation.

Notice there are three types of depressed areas. Some are rural resource-based, primary producing areas; some are one-industry mill towns of labor intensive industries such as textiles, shoes, and steel; and others are Indian reservations. American Indians have never been integrated into the system. Instead, they have been shunted off onto the poorest lands and largely ignored.

The depressed rural areas are predominantly agricultural since agriculture is the most important primary activity. However, one can also find lumbering and mining areas where resource depletion or automation has left populations without an economic base. The farm areas include those of Appalachia which were settled early and then bypassed as the modern, interconnected system developed. Areas of the South, where machines and large-scale farming have displaced intensive sharecropping of cotton also fit the pattern.

It may be surprising that such large portions of the United States are depressed, and Americans were shocked a few years ago by disclosures that large numbers of their rural families were not only poor but malnourished.

Noting that the depressed areas cover a sizable portion of the map of the United States, we recall from Chapter 7 that these areas are

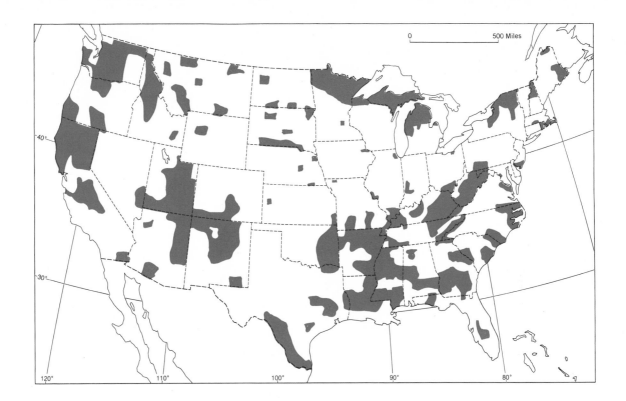

Figure 8-8 Areas Eligible for
Financial Assistance from the
United States Government in 1970

largely inhabited by the minority that is still rural in a predominantly metropolitan world. Still the presence of depressed areas in a modern, interconnected system which has the technical capacity to service all areas shows how selective the modern, interconnected system has become. It uses only those resources and those regions that it finds useful. In the process it has abandoned other areas which once supported viable economies when the system operated at local or regional levels. And, in the process of abandoning regions, the system abandoned the people who lived there.

One of the major problems of the country today is what to do with the people in these largely abandoned areas. Government programs such as those for Indian reservations and Appalachia call for investment in new roads, new water and sewage facilities, loans for building factories and subsidies to manufacturers who could locate in such areas. But small towns and villages in remote areas populated by unskilled workers have little to offer the metropolitan-oriented business community.

Some people have advocated increased encouragement of migration from depressed areas. They suggest abandoning the regions. But this does not seem to be a feasible answer either. Certainly it is not a politically attractive answer to local politicians and to representatives to Congress from such regions. Some people presently living within the depressed regions of the country love their home territory. They do not want to move, and the challenge of finding them a meaningful life style remains.

Those people who do emigrate from rural America have almost nothing to offer the system operating in metropolitan areas. In their rural schools they have not been educated to handle the specialized office jobs needed in business or taught the technical skills required in modern manufacturing.

Like peasant European refugees in the past, they are qualified only for unskilled jobs. But unlike the situation of fifty to seventy-five years ago, expanding industries no longer need masses of unskilled workers in packing plants, steel mills, and clothing shops.

Since these emigrant people do not have the skills necessary to fit into the modern, interconnected system of the city, they have developed a sort of subsistence life style, living—or existing—on far below what is necessary for an adequate life in a metropolitan economy. They are supported by a combination of welfare payments, part-time or menial jobs, and a counter-economic system involving panhandling, prostitution, and robbery.

Through little fault of their own, they have not become part of the modern, interconnected system any more than have Indians on reservations or Appalachian farmers. In cities they are concentrated in a small area instead of being spread over the rural areas of many states. The New York metropolitan area has more families on public assistance than have all the southern states east of the Mississippi.

Generally this "inner city" is in old residential structures which have been allowed to deteriorate through the years. Houses and apartment buildings have been divided into multiple units with several families per floor—often per apartment—and population densities are extremely high.

With so many of these ghetto dwellers crammed together, the problems produced by a background of rural depression are intensified. Buildings have inadequate plumbing and waste disposal. Rural people who are used to being able to dispose of their waste in the countryside find it piling up around them—and swarming with vermin. In such unsanitary conditions disease of epidemic proportions is a constant threat.

In rural, small-town America people of various social strata came into contact with one another if only in rather impersonal ways. But in big-city ghettos masses of poverty-stricken people occupy block after block and mile after mile completely separated from the rest of the city except for their television sets. There is little opportunity to break away from the mold, and life in the counterculture becomes a norm by which the ghetto lives. There is no place to go, no way to get away from the world in which they are confined.

The Modern System: The Problems of Success

The modern, interconnected system operates on its most massive scale within the United States. The result has been affluence and the replacement of physical labor by machines. But the results vary in different parts of the country and on different levels of the whole society. The largest metropolitan centers have grown spectacularly both in numbers and importance, while most rural areas have declined. Metropolitan centers themselves have reached such a peak that, as in the case of New York City, it has become difficult to continue to supply essential transportation and communication services on the scale needed. Moreover, the mass consumption of energy and goods in great metropolitan areas has produced such a disposal problem that the system must re-evaluate the whole question of pollution control and waste disposal.

SELECTED REFERENCES

Berry, Brian J. L. *A Geography of Market Centers and Retail Distribution,* Prentice-Hall, Englewood Cliffs, N.J., 1967. A basis survey of retail trade and cities as central places with case studies and results of methodological studies.

Brezezinsku, Zbigniew. *Between Two Ages: America's Role in the Technotronic Era,* Viking Press, New York, 1970. A popular book on the order of *Future Shock* provides another dimension of the question of America and the way it functions in the world.

Galbraith, Kenneth. *The New Industrial State,* Signet, New York, 1967. The well-known economist looks at the industrial system that dominates the United States.

Gottmann, Jean. *Megalopolis: The Urbanized Northeastern Seaboard of the United States,* Twentieth Century Fund, New York, 1961. A thoughtful study of the large metropolitan centers of the Eastern Seaboard that dominate the country's decision-making process. See the chapters on the main street of the nation, the continent's economic hinge, the white-collar revolution, and the conclusion.

Higbee, Edward C. *Farms and Farmers in an Urban Age,* Twentieth Century Fund, New York, 1963. An agricultural geographer looks at the revolution in agriculture created by the rise of urban centers.

Patterson, J. H. *North America,* 4th ed., Oxford U. Press, New York, 1970. This volume by an English geographer provides both a systematic and a regional treatment of the U.S. and Canada in very traditional form. Contrast his approach with the one used in this text.

Starkey, O. P., and Robinson, J. L. *The Anglo-American Realm: A Geographical Analysis of the Economies of the United States and Canada,* McGraw-Hill, New York, 1969. An attempt to give a different perspective to the geography of the U.S. than that given by Patterson. Again, compare to the approach here.

Chapter 9 Europe:

The First Example of the

Modern, Interconnected

System

EUROPE IS NOT just another example of the modern, interconnected system. It is the part of the world first dependent on trade with other continents and where industrialization began.* Sea exploration, begun in the Atlantic Ocean by Spaniards and Portuguese in the fifteenth century, brought trade connections with all other continents over the next two hundred years. With these early explorations began European political and cultural domination over the Americas, Africa, most of Asia, and Australia, which has only begun to fade in this century. Industrialization—the harnessing of nonliving energy—was a European development which provided the base of even greater economic and political dominance over most of the world in the nineteenth and early twentieth centuries.

Greater Prosperity but No Longer a World Power

Today the people of Europe are more affluent than ever. Only three of the twenty-six countries in Europe—Yugoslavia, Portugal, and Albania—have a per capita gross national product below the world average, and Europe in general is the most affluent area of the world after Anglo-America and Australia–New Zealand. Sixteen countries have per capita GNP more than twice the world average, and three of those—Sweden, Switzerland, and Luxembourg—average more than four times the world average. American travelers feel at home in most European cities, with their fashionable

---

*In this chapter when we speak generally of Europe, it is to be understood that the Soviet Union is excluded. Our purposes are better served by one discussion of the Soviet Union as an entity in Chapter 10.

shops and department stores, streets congested with automobiles, and residents dressed according to the latest trends.

Industrially, European countries remain among the most important in the world. European-made automobiles, cameras, optical goods, fashions, and fabrics are highly prized by people in the United States. European financial organizations, among the most successful in the world, have world-wide investments. European capitalists even control such giant industrial organizations in the United States as the Shell Oil Company and Lever Brothers.

Yet the largest and most economically advanced countries of Europe—Great Britain, France, West Germany, and Italy—are no longer among the great powers of the world, either economically or politically. They have lost their empires and have been far surpassed by both the United States and the Soviet Union since World War II.

A Different Geographic Base

Although the modern, interconnected system dominates most of Europe, the basic geography of the continent is very different from that of the United States. In the first place, Europe is very small and crowded by United States standards. Its population of 462 million people in 1970 was more than twice that of the United States and 15 million above the combined populations of the United States and the Soviet Union. Yet this large population is crowded into a land area only half the size of the United States. As a result, European population density is more than four times that of the United States. In 1970 this density was more than 240 persons per square mile, compared to less than 60 in the United States and less than 30 in the Soviet Union. Pressure on the environmental

base of Europe to produce primary products would seem to be much greater than that of the United States or the Soviet Union, unless, of course, Europe's interconnected system draws more resources from other continental areas.

MODEST ENVIRONMENTAL ENDOWMENTS

Although Europe, like the United States, is a mid-latitude area, its environmental base, in terms of both climate and terrain, is very different. It also has a dissimilar land configuration resulting in growth potential different from that of the United States.

High Latitude Location

Several important facets of Europe's environmental base can be seen in Figure 9-1. Notice that Europe is located in higher latitudes than the United States. The southernmost part of Europe, the Greek island of Crete in the Mediterranean Sea, is 35° north of the equator, the same latitude as North Carolina. Rome has the same latitude as Chicago. Moreover, a large share of Europe lies north of the 49th parallel, the northern boundary of most of the United States. What would we expect this to mean in terms of energy potential for crop growth?

A Peninsula of Peninsulas and Islands

As Figure 9-1 shows, Europe appears as a large peninsula extending out from the Eurasian landmass with seas on three sides. Only the east is landlocked. The European "peninsula" comprises a series of secondary peninsulas—Norway and Sweden, Denmark, northwestern France, Spain and Portugal, Italy, and Greece—and numerous islands, including the British Isles. On a smaller scale these and other portions of Europe encompass many more local peninsulas, promontories, and islands. As a re-

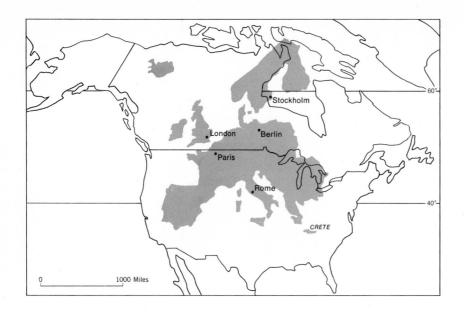

Figure 9-1 Europe and the United States Compared in Area and Latitude

sult the sea seems to lap around and into Europe, and most places are no more than 400 miles from the sea, whereas parts of interior United States are more than 1,000 miles away.

MARITIME CLIMATE

Not only is there comparatively easy access to the sea from most places in Europe, but the winds from the seas have a major influence on the climate and the plant growth potential of the continent. In these latitudes, as in the United States, air flows predominantly from west to east, exposing Europe to full maritime influences from the Atlantic. Since the major mountains and highlands of Europe follow an east-west axis except in the north, there is no terrain obstructing the air from the sea as it crosses the continent. This penetration of maritime air across the relatively small European landmass results in the general availability of moisture throughout the continent. Europe is the only continent without great tracts of desert, although southern Europe, like California, receives moisture only during the winter half of the year.

Moreover, the maritime influence "air-conditions" Europe. Since water heats and cools more slowly than land, air coming off the water is comparatively warm in winter and cool in summer. This fact is particularly important in winter because of Europe's high latitude. Instead of winter lows of −20° F as in Canada, temperatures in most of Europe rarely fall below 0° F. Average January temperatures in London are higher than those in New York City and almost as high as those in Washington and cities farther south. Farther inland from the Atlantic coast, winters are more severe, but they do not approach those of the northern Great Plains, except in the highlands of northern

Scandinavia. Cities in Poland have January averages roughly like those of Chicago and Detroit. In summer the cooling effect of the sea means both lower temperatures and more cloudiness than in most of the United States. Over virtually all of Europe in July, average temperatures are less than 70° F, at least 10° lower than those in most of the United States.

Under these climatic conditions growing potential is different from that of the United States. The length of the growing season is, however, much the same as that in the United States. Southern Spain and Greece are essentially frost-free. Small parts of Spain, Portugal, Italy, and Greece are bothered with only one or two cold spells a winter, and the rest of Europe, except for the Alps and areas far to the north, has more than ninety days of summer growth. But the cooler summers and high latitudes result in considerably less potential energy for growth (see Figure 5-4). The potential evapo-transpiration of the lands in the south around the Mediterranean Sea is much the same as that of the southern United States north of Florida and southern California, but the gradient drops off sharply inland.

Most of British Isles, northern France, the Low Countries, northern and central Germany, Denmark, Poland, and southern Sweden have annual potential evapo-transpiration figures similar to those of Canada north of the agricultural frontier (Figure 5-4). Even in the more interior areas of Hungary, Rumania, and Bulgaria, potential evapo-transpiration is comparable only to that in New England and the northern Great Lakes area of the United States.

The westerly winds off the Atlantic supply continual moisture over most of the continent. Water deficits during the growing season are not present in eastern Europe (see Figure 5-5); they are very small in western and northern

Europe and generally less than deficits in the Eastern United States. Thus, actual crop growth in most of Europe can reach the full potential of evapo-transpiration for each district. Unlike the situation in the United States, it is the relatively low total annual evapo-transpiration, not moisture, that presents the major limitation to crop growth.

Only southern Europe is faced with serious moisture deficiencies. But since this is the region of Europe with the best growing season, man has worked for thousands of years to overcome the problem. There is little water surplus during the winter wet season except in the mountains (see Figure 5-6). During the hot summer, the time of greatest growth potential, there is almost no precipitation, and the deficit becomes great. Southern Europeans since the time of the ancient Greeks and Romans have developed reservoirs to hold winter rainfall in the highlands for use in summer irrigation, have chosen crops and livestock that can adjust to the moisture-poor environment, and have carried out dry farming.

In summary, we might speak of the climate of Europe as "manageable," but it is not outstanding in its potential for agricultural and timber production. While water management is a problem only in the Mediterranean area, it is there that one finds energy potential for growth most like that of the southern United States. Over most of Europe there is no moisture shortage to hamper plant growth, but sufficient radiant energy is simply not available for high evapo-transpiration in summer. The relatively high crop yields in much of Europe reflect both the adequate moisture for full plant growth and the extremely intensive, enlightened care given the land by a hard-working, knowledgeable labor force. For fertilizer farmers rely on large amounts of both animal manure and plant compost.

Usable Land Surface and Soil

Although the European terrain is largely usable, there are no vast lowlands comparable to those in the interior of the United States. The European Plain (Figure 9-2), which extends from western France along the southern margin of the North and Baltic seas, is generally less than 200 miles wide; in eastern Europe it is even narrower because there it is broken by a maze of glacial deposits that have blocked drainage and made the terrain like the rolling landscape of southern and central Michigan.

The major upland extends east-west almost without interruption from the Pyrenees between Spain and France through the massive, spectacular Alps to the Carpathian Mountains of eastern Europe, extending through Czechoslovakia, Hungary, some of the Soviet Union, and Rumania. The Apennines, which form the backbone of the Italian Peninsula, are an offshoot of that system. The Alps, Europe's highest mountain system, separate southern Europe from the European Plain. Much of the British Isles and the Scandinavian Peninsula consist of old Appalachian-type mountains. Europe's landscape is thus characterized by great variety within surprisingly short distances. In almost any part of the continent there are valleys and hills, usually within sight and always within a few hours' drive.

The soil is generally adequate but is not the best. Most soils have modest amounts of plant nutrients. In the north, particularly in Finland, much of the original soil cover was removed by the action of continental glaciers long before human settlement. The plain that lies south of the North and Baltic seas is covered with glacial deposits providing a varied soil cover: sandy in some areas, rocky in others, and poorly drained in other places. Moist regions in northern and western Europe were covered with forests be-

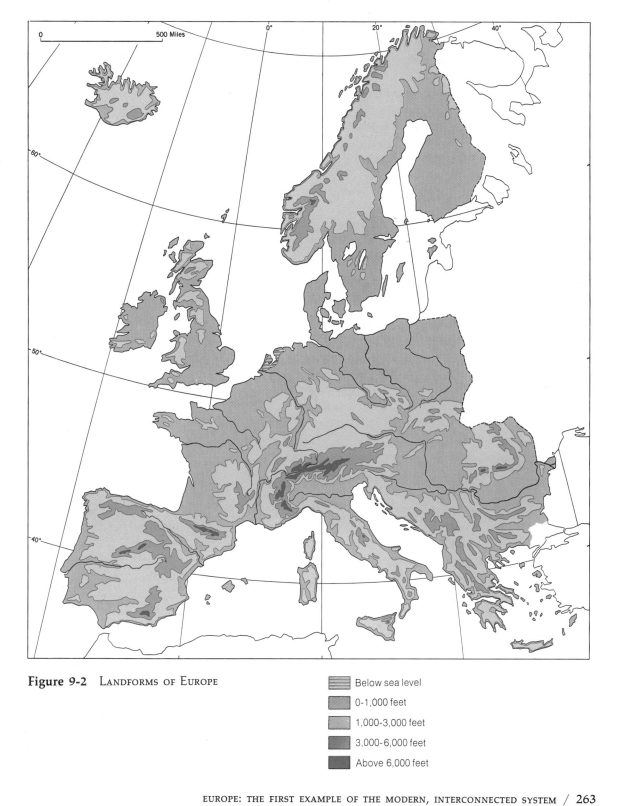

Figure 9-2 LANDFORMS OF EUROPE

Below sea level

0-1,000 feet

1,000-3,000 feet

3,000-6,000 feet

Above 6,000 feet

fore man began clearing the land. In the mid-latitudes forest soils are inferior to those found in grasslands like those of the United States. Around the Mediterranean Sea the combination of rugged terrain and semiarid conditions along with a sparse vegetative cover has resulted in soil of varying quality. However, in those areas where soils have been worked for thousands of years on more rugged slopes, there has been severe erosion.

A Manageable Environment

In Europe man has found a usable environment rather than a bountiful one. Its most important characteristic is the relative ease with which its land resource base can be managed. There is adequate moisture during the growing season and a low probability of drought, flood, tornado, and hail.

Furthermore, the European environment has generally been improved by human management. Through thousands of years of land use, the population increase has imposed ever-greater pressure on it to produce. Along with this has come a growing technology for agriculture and forestry. As a result, Europeans have worked harder to make their land productive than have people in any other part of the world except Asia. Since the Industrial Revolution Europeans have applied to production a greater scientific knowledge of land use, crops, and livestock than anyone except the Japanese. To a high degree the agricultural land of Europe today is man made, the result not only of clearing but also modifying both the surface and the soil.

Europeans have also led the world in forest management. Most of the forests of central and northern Europe are the result of replanting after the original timber had been cleared. Forests have been restocked with the most valuable species, and much of the timberland is expected to yield an annual harvest, as any other crop does.

The combination of workable environment and tremendous human effort has resulted in a fuller utilization of the land in Europe than on any other continent. Almost one-third of Europe is cultivated, while on no other continent is the proportion more than 18 per cent. Only slightly more than 20 per cent of Europe is considered wasteland. Elsewhere the proportion is at least 26 per cent, and frequently it is over one-third. However, this is as much a result of intensive farming as it is an indication of the quality of the environment.

Europe's land surface has presented few real barriers to human movement. The presence of a rather complex system of rivers, particularly in humid central and northern Europe, has provided man with both waterways to the sea and land routes along the valleys through rough country. Yet it has been the desire of Europeans to cross barriers as much as having the easy routes that has provided interregional connections. Even the Alps have been crossed by established routes since the days of the Roman Empire.

Moreover, the environment has provided essential resources for building and industry from early times. Timber from European forests has been a basic building material since prehistoric days, and rock has been used in construction for just as long. Waterpower from small streams has been employed since the beginnings of industrialization, and windpower has been used on land and sea even longer. Although it is no longer considered a major center of world mineral production—particularly not of oil and nonferrous minerals—Europe has adequate and accessible deposits of coal and iron. These provided the basis for the spread of the Industrial Revolution in the nineteenth century. Coal and iron are produced in quantity in some countries, and for the first time significant deposits of oil and natural gas are now being developed.

The Human Component

To a large extent the great contrast between the geographic base of Europe and the United States is human, not environmental. Europe, half the size of the United States but with more than twice the population, is divided into twenty-six sovereign states and a few tiny city-states (Figure 9-3).

POLITICAL AND CULTURAL FRAGMENTATION

Each of these countries represents a separate functioning unit not only in political terms but also, to a high degree, in economic and cultural terms. The citizens of each country represent a different group that has been conditioned through training at home and in school to think of itself as loyal to the country, its government, and its heritage, rather than to Europe as a whole. Instead of a single version of the modern, interconnected system to be contrasted with the United States, there are twenty-six different versions, each with its own solutions to its own geographic problems.

Unlike the United States, Europe has never developed as a single functioning entity. Its twenty-six separate countries stand today as evidence of that fact. Indeed, it was in Europe that the concept of the national political state, functioning in terms of its own self-interests, emerged. Each country in Europe has established itself as a political, economic, and cultural entity distinct from its neighbors. As such, it has spent extra effort managing its particular resources. Throughout much of European history certain all-European economic and cultural

forces, such as the Hanseatic League and the Roman Catholic Church, have been important. Some of these forces are still present, but they are in conflict with deeper-rooted nationalist sentiment.

POLITICAL DIVERSITY AND DIFFERENT ECONOMIC SYSTEMS

The importance of the political sovereignty of individual European countries must not be underestimated. Each government has authority over the people and institutions within its territory. Each establishes its currency and sets down the rules and laws under which institutions of all sorts operate. Different European governments have very different ways of looking at economic and cultural goals and also have different political aims. Thus, the east European countries under the dominance of the Soviet Union since World War II have established one-party political regimes and have followed the Soviet economic system under which essentially all land and structures of the business and manufacturing community are controlled by the central government. In other parts of Europe there are a variety of political and economic forms. In some countries kings and queens still rule as vestiges of ancient traditions, but most have only ceremonial power. In others, dictators are in control, granting varying degrees of freedom to both individuals and business enterprises.

In Britain there are only two major political parties, but in Italy and France party structure is highly fragmented. Socialistic economic systems have developed in some European countries since 1900. In Britain such a system allows individual ownership of rural properties but calls for strict control of land around urban centers and for government ownership of economic services deemed to be essential to the

Figure 9-3 POLITICAL MAP OF EUROPE

public good. Thus, the British government not only operates the postal service and radio but also coal mining, electricity production, the railroads, and the airlines. At the same time shops, large businesses, most industrial enterprises, and farms as well are privately owned but operate under strict government regulations.

In Europe the management question has been how to develop each of the particular parts of the continent within the limits imposed by twenty-six political entities. The resources and people of the continent are not being organized into a single functioning system despite the recent changes that have pulled eastern European countries into a Soviet-dominated bloc and have brought some aspects of economic unity to the countries of the European Common Market.

In size European countries are comparable to individual states within the United States. France, the largest European country in area, is four-fifths the size of Texas, and tiny Luxembourg is about the size of our smallest state, Rhode Island (Table 9-1). Imagine Rhode Island or Maryland or even California and Texas functioning as separate sovereign entities!

In densely populated Europe most countries have larger populations than states of comparable size. Even so, each of the four most populous countries has only slightly more than one-fourth the population of the United States. Over half the countries of Europe have populations smaller than Ohio.

DIVERSITY: A HANDICAP TO THE MODERN SYSTEM

Whether measured by size or population, European countries are diminutive when compared to the United States or the Soviet Union. As a result, Europe operates on a much smaller scale than either of the great powers. And from our consideration of the concept of mass pro-

duction and mass consumption, European countries would seem to be at an important disadvantage in economic terms. The modern, interconnected system in the United States is based on mass production and mass consumption for a market of over 200 million Americans. Europe would offer a market over twice as large. But since it is divided into twenty-six separate entities, each with the power to control the goods moving in and out of its territory, the real Europe comprises twenty-six different markets, many of which have just a few million people. The economy of large-scale operation, so important in American mass production, is presently unattainable in Europe.

Even with the Common Market, which in recent years has broken down important national barriers in the economic scene, the lack of European economic and cultural integration can be seen by the importance attached to traditional local and national interests. Europe still has fifty or more different languages. Twenty different types of currency are used. Its people are governed by twenty-six different sets of laws and business codes. We can get some idea of the situation by imagining each of the states of the United States east of the Mississippi River functioning as a sovereign power with its own money, language, laws, businesses, and military forces. It is obvious that none of these states could operate entirely in isolation from the rest. Goods, ideas, and people would move across political borders, particularly between adjacent states. But each state would have as its chief concern developing its own economy and raising the standards of its own people. Thus, in competing with strong neighbors, it might well encourage trade with areas overseas rather than expand interstate trade to its fullest. In any case, the problem of developing the area east of the Mississippi would be seen in terms very different from those of today.

European countries do have economic, social,

Table 9-1 SIZE AND POPULATION COMPARISON OF EUROPEAN COUNTRIES WITH STATES IN THE UNITED STATES

SIZE: square miles

Larger than any U.S. state but Alaska and Texas

		U.S. state	
		Alaska	586,000
		Texas	267,000
France	210,000		
Spain	195,000		
Sweden	174,000		

The size of states in the western U.S.

		U.S. state	
		California	159,000
Finland	130,000		
Norway	125,000		
Poland	121,000		
Italy	116,000		
Yugoslavia	99,000		
West Germany	96,000		
United Kingdom	94,000		
Rumania	92,000		
		Washington	68,000

The size of states in the Middle West and South

		U.S. state	
		Illinois	56,000
Greece	51,000		
Czechoslovakia	49,000		
Bulgaria	43,000		
East Germany	42,000		
Iceland	40,000		
Hungary	36,000		
Portugal	35,000		
Austria	32,000		
		South Carolina	31,000

The size of states in the northeastern U.S.

		U.S. state	
Ireland	27,000		
		West Virginia	24,000
Denmark	17,000		
Switzerland	16,000		
Netherlands	14,000		
Belgium	12,000		
Albania	11,000		
Luxembourg	1,000	Rhode Island	1,000

POPULATION 1970: millions of people

Larger than any U.S. state

		U.S. state	
West Germany	58.6		
Britain	56.0		
Italy	53.7		
France	51.1		
Spain	33.2		
Poland	33.0		
		California	19.9

The size of the more populous U.S. states

		U.S. state	
Yugoslavia	20.6		
Rumania	20.3		
		New York	18.1
East Germany	16.2		
Czechoslovakia	14.7		
Netherlands	13.0		
		Illinois	11.1
Hungary	10.3		

The size of medium-population states

		U.S. state	
		Ohio	10.6
Belgium	9.7		
Portugal	9.6		
Greece	8.9		
Bulgaria	8.5		
Sweden	8.0		
Austria	7.4		
Switzerland	6.3		
		Massachusetts	5.7

The size of small states

		U.S. state	
		North Carolina	5.0
Denmark	4.9		
Finland	4.7		
Norway	3.9		
Ireland	3.0		
Albania	2.2		
Luxembourg	0.4		
		Alaska	0.3
Iceland	0.2		

SOURCES: Size: *Goode's World Atlas*, 13th ed., revised printing, Rand McNally, Chicago, 1971; Population of European states: Population Reference Bureau, Inc., *1970 World Population Data Sheet*, Washington, D.C., 1970; Population of U.S. states: U.S. Bureau of Census, *Number of Inhabitants*, Vol. 1, Washington, D.C., 1970, Table 8, p. 48.

and political dealings with one another, but they do so as separate entities with freedom to encourage or cut off such contacts, to ingratiate or antagonize. Their interrelations are similar in character to those each holds with the separate political units in other parts of the world. In some cases ties with more distant places may be even stronger than the ties with European neighbors.

Some European countries, particularly those that have been politically and economically powerful, have found their basic ties outside the continent. In so doing they have been encouraged by negative factors of economic and sometimes military rivalry within Europe and the positive factor of ready access to the seas of the world—the cheapest and most flexible trade routes. When closer economic ties were being planned between western European nations after World War II, Britain stood aloof, feeling that its ties to Commonwealth countries such as Canada, Australia, New Zealand, and India were more important. The Commonwealth still maintains closer language, currency, trade, and legal ties among its members than can be found between European countries. France also continues trade preferences and cultural ties with former colonies in Africa and other parts of the world.

This does not mean that there are not important movements of people, ideas, and goods between parts of Europe, but when these occur, it is because they appear to be of advantage to the particular political units involved. The smaller countries, less viable as individual entities, often feed products into the economies of larger countries. Denmark, for instance, provides pork products and eggs for Britain and other western European countries, and Ireland raises cattle for British markets.

The resources of the continent, either environmental or human, have never been assessed in terms of a unified Europe. Europe simply has not been developed as a single, modern, interconnected system in the way that the United States and the Soviet Union have. Instead, the continental resources have been measured and developed in a compartmentalized fashion that reflects the political separation.

Traditional Society as a Basis for Diversity

Why hasn't Europe, faced with competition from the much larger interconnected systems of the United States and the Soviet Union, organized itself into a single political and economic unit? With more population than the two rival powers combined, a skilled scientific community, and a highly productive plant, a united Europe would indeed be a formidable rival to the great powers of today. The answer may lie in the cultural diversity of the continent. This in turn raises the question: Why is there such cultural diversity in Europe which is so much smaller than the United States?

Through the thousands of years of Europe's development, limited technology for transportation and communication determined that most activities followed the traditional, locally based pattern for managing the environment. Even though Europe was crisscrossed with trade routes dating back to antiquity, each group developed its own ties to the resource base, its own values, and its own language. These groups were aware of their neighbors and of groups in distant corners of the continent, but contacts were very infrequent. Everyone else was a "foreigner."

These cohesive group forces still are very significant in the European way of life. Unique to each political state are its interrelations, memories of traditional friendships and hatreds, and carefully developed traditions and skills. No unifying force—political, military, economic, or spiritual—has been able to overcome the deep-rooted localism. Many have tried—the Roman Empire, Holy Roman Empire, Catholicism, Napoleon, and Hitler—but each has failed. To an important degree the present political states are too large, not too small. Scots and Welshmen call for separation from Britain, and the Bretons from France. Tyrolese and Serbians also think of themselves as separate peoples. It remains to be seen if the strong economic pressures for international European cooperation and even for a "United States" of Europe can succeed where other forces in the past have failed.

Physical variations in the topography of Europe, geographic barriers, and sheer distance have tended to separate the continent into different regions. The east-west direction of the Alps and the seas that separate northern Europe from the rest of the continent have made north-south contacts more difficult than east-west ones. Early migration into Europe apparently came from the east and soon diverged into distinctive streams—one into southern Europe south of the Alps, another across eastern and western Europe north of the Alps, and a third, less well defined, into northern Europe beyond the Baltic and North seas. Westward movement within southern Europe was made difficult by the rugged, broken terrain, and migrations here were commonly by sea. North of the Alps, however, overland movement was comparatively easy on the east-west plain along the northern seas and on the uplands along the northern margin of the Alps.

Along each of these east-west paths, different peoples occupied particular territories. In southern Europe the Greeks and Romans could be distinguished from east to west, and the Franks, Germanic peoples, and Slavs each occupied certain lands north of the Alps. Farther north the Scandinavians formed another dis-

tinctive cluster of cultures. These subcultures can be seen in the physical appearance of the people of Europe, in the distribution of languages, and in the distribution of religions. Roman Catholicism is centered in the Mediterranean lands from Italy westward. Protestantism has its stronghold in the northern areas, and the domain of the Eastern Orthodox church extends from Greece northward.

Even as late as five hundred years ago Europe functioned largely in local terms. Under the feudal system kings ruled through arrangements with local landlords, and for most people life was centered within the local fiefdom. The area provided all the necessities of life. Only the nobles and their entourage traveled much beyond the fief or had access to foreign goods. The nation-state, a European invention, did not really emerge until the fifteenth century and the beginning of industrialization. Italy and Germany, two of the largest countries of Europe, were not consolidated until the nineteenth century.

Thus, the cultural and political diversity of present-day Europe, so very different from the situation in the United States, is largely a product of roots extending far back in time. In this regard it is important to remember that Europe's pattern is typical of world diversity. Asia and Africa, the other parts of the world where culture groups trace their beginnings back to earliest time, show a similar cultural diversity. In India more than one hundred and fifty different major languages are spoken, and Nigeria, a large African country, has more than forty recognized languages. In view of these conditions, it is not surprising that Europe is culturally fragmented. Large-scale cultural homogeneity, such as one finds in the United States, is the exception in today's world. Moreover, this unity stems from the fact that the United States was settled in modern times.

In Europe, waterways are a valued part of the modern transportation system that ties all the interconnected systems together.

Europe is a continent of peninsulas and islands, and its success as the first interconnected system was due largely to its water system. The combination of navigable rivers, protected harbors, and deep channels enabled this small region to dominate world trade for centuries. But within this single system there are economic, political, and cultural distinctions. The tension between a single "community" on the one hand and its separate components on the other helps to define **Europe** in the twentieth century.

Table 9-2 DISTRIBUTION OF POPULATION, 1970

DIVISION	PER CENT OF LAND AREA*	POPULATION (in millions)	PER CENT OF POPULATION	POPULATION PER SQUARE MILE
Western Europe (west of Iron Curtain)	28	207.4	45	398
Southern Europe (including Yugoslavia)	27	128.2	28	253
Eastern Europe (east of Iron Curtain)	20	103.0	22	277
Northern Europe (not including Denmark)	25	19.8	5	42

*From *Goode's World Atlas,* 13th ed., revised printing, Rand McNally, Chicago, 1971.
SOURCE: Population Reference Bureau, Inc., *1970 World Population Data Sheet,* Washington, D.C., 1970.

Distribution of Population

The population of Europe, like the population of the United States, is unevenly distributed. The continent can be divided into southern, eastern, western, and northern areas which in large measure reflect different earth environments and distinct cultural variations. Each of these four divisions constitutes roughly one-quarter of the total land area of the continent (Table 9-2). But almost half of the population is in western Europe. Notice that the population there is just slightly larger than that of the United States. Two other divisions—southern and eastern Europe—each have close to their expected quarter of the total population. It is northern Europe that is comparatively sparsely populated with only 5 per cent of the total.

Figure 9-4 shows that there is considerable variation in population density within any of the four large divisions of Europe. Moreover, one can see bands of dense population that extend almost all the way across Europe. The most densely populated areas have densities similar to those in the eastern megalopolis and in other metropolitan areas of the United States. These areas cross Europe in both east-west and northwest-southeast axes which merge along the English Channel between England and France and Belgium. One axis extends from the Scottish Lowlands, across the Channel, up the Rhine, and then south of the Alps into northern Italy. The other axis reaches from southern England eastward through Germany and on into the Soviet Union. The most sparsely populated area is northern Scandinavia, but much of the rest of Europe has densities comparable to most non-metropolitan counties in the eastern United States.

In Europe, as in the United States, the distribution of population is the sum of two different components. First, there is the population that has spread over the landscape and is engaged in agriculture and herding in rural, small-town settings. Second, there is the population that is involved in manufacturing, services, and management, and has settled overwhelmingly in large metropolitan centers.

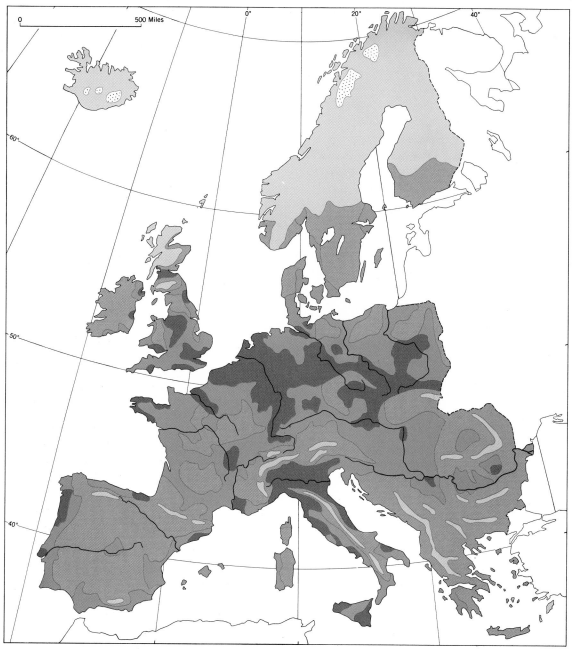

Figure 9-4 POPULATION DENSITY OF EUROPE

Persons per square mile

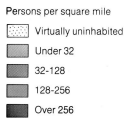

Virtually uninhabited

Under 32

32-128

128-256

Over 256

The Rural, Agricultural Component

In Europe, where settlement has developed in-place for centuries, the traditional, locally based society was deeply rooted long before the rise of trade. Although hundreds of years ago there were trade routes and trading centers such as Venice in the twelfth century, most of the people in Europe have by tradition been part of an agriculturally based, peasant society. Without modern energy sources and machinery, great human effort was needed to turn the mediocre European agricultural base into production. As in Asia, a tremendous amount of human energy was expended in intensive agriculture. The need for so much manual labor resulted in small working units, large families, and high rural and village densities. One needs only think of medieval, feudal Europe, built around an agricultural society, to recognize that the village population of Europe rests today on the same base. There, in the Middle Ages, the population clustered together in almost completely separate, self-contained communities surrounded by farmland within walking distance of the castle or town. What wealth and power there was remained in the hands of the landowning class, and the great mass of the population was made up of landless peasants working the fields with hand tools and simple plows pulled by animal power.

The locally oriented, traditionally European culture we see today was based on the feudal system. The peasant farmer, first serving as a serf under a local feudal landlord, then working as a tenant or freeholder on his own small farm, was concerned primarily with providing basic food and other family needs from his land and then selling the surplus in the local market center. Sometimes his goods moved on to other markets or cities, but he and his family, like the Indian farmer today, lived their lives within the local environment. It was their community, their local officials, and their priests or ministers that set the cultural temper of their lives.

This peasant tradition of agriculture and small-town living established the basic pattern of population over Europe. Higher densities meant better, more intensely used agricultural areas, and lower densities implied areas where rough terrain, a short growing season, dryness, or some combination of unfavorable circumstances reduced the possibilities for agriculture.

Agricultural Lands and Farm Population

The most important agricultural area through European history has been southern Europe; western and eastern Europe emerged later. Table 9-3 shows that southern Europe is still the most agricultural area of Europe in terms of land use. It has more total productive land than any other European area. In productive land western Europe is close behind, and eastern Europe ranks up with the other two areas in total cultivated acreage. Northern Europe is far less important than the others. It has less than half the productive farmland of the two ranking areas and only 4 per cent of cultivated land. It is significant only for its pasture and woodland, as might be expected in a cool region with a short growing season, rough terrain, and considerable marginal soil.

Farm population does not show a perfect positive correlation with productive farmland. Notice in Table 9-3 and Figure 9-5 that both eastern and southern Europe have larger proportions of farm workers than of total productive agricultural land and arable land. Combined, these two areas have almost three out of four of all farm workers in Europe, yet they total only about half of Europe's population. Western Europe with 48 per cent of total population has only 23 per cent of farm workers. Thus, the distribution of farm population of Europe is quite different from that of total population, just as it is in the United States. How-

Table 9-3 THE DISTRIBUTION OF AGRICULTURE IN EUROPE, EARLY 1960S

	SOUTHERN EUROPE		WESTERN EUROPE		EASTERN EUROPE		NORTHERN EUROPE	
LAND USE	Million Acres	Per Cent	Million Acres	Per Cent	Million Acres	Per Cent	Million Acres	Per Cent
Productive land (cropland, pasture, and woodland)	290.9	32	279.4	31	217.1	24	125.8	13
Arable land (cropland and cultivated land)	131.0	35	113.1	30	117.1	31	17.4	4
Pasture and woodland	160.0	30	166.2	31	100.0	19	108.4	25
AGRICULTURAL OUTPUT, CROPS	Million Tons	Per Cent	Million Tons	Per Cent	Million Tons	Per Cent	Million Tons	Per Cent
Cereals	38.5	26	62.0	41	45.0	30	5.1	3
Other major crops (sugar beets, tubers, pulses, grapes, cotton)	36.3	20	71.8	38	75.2	40	4.2	2
Total major crops	74.8	22	133.8	40	120.2	35	9.3	3
LIVESTOCK	Million Head	Per Cent	Million Head	Per Cent	Million Head	Per Cent	Million Head	Per Cent
Cattle	22.1	18	65.0	54	26.9	23	5.8	5
Horses	6.6	39	3.6	21	6.0	36	0.6	4
Swine	18.0	16	50.5	45	41.1	36	3.1	3
Goats	16.7	79	2.1	10	2.2	11	0.1	....
Sheep	47.0	38	45.3	36	31.0	24	3.0	2
FARM WORKERS	Millions	Per Cent	Millions	Per Cent	Millions	Per Cent	Millions	Per Cent
	20.9	36	13.3	23	22.4	38	1.7	3
TOTAL POPULATION	102.3	24	200.9	48	99.1	24	16.1	4

SOURCE: *Encyclopaedia Britannica World Atlas International,* Encyclopaedia Britannica, Chicago, 1967, pp. 191–193.

ever, in most European countries the rural component is much more important than in the United States.

Further study of Table 9-3 reveals that neither southern nor eastern Europe is the most productive farmland of the continent. Western Europe is the leading agricultural area in productivity of total major crops including cereals. Some inferences about agricultural methods can also be drawn from the table. The higher agricultural productivity of western Europe is accomplished with far fewer farm workers and fewer horses. The low farm population indicates that western Europeans have moved to city jobs, leaving a modern, productive agricultural system. The small population of horses supports this view, as we can assume that horses have been replaced by machines.

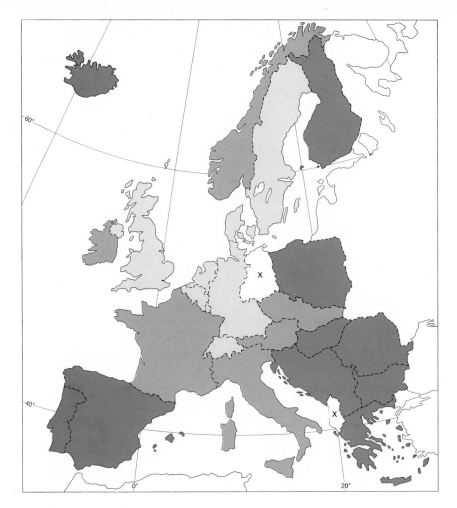

5 to 25%

26 to 45%

More than 45%

X   No data available

Figure 9-5   Percentages of European Population, by Country, Employed in Agriculture, Forestry, Fishing, and Hunting

In southern and eastern Europe, even now, a high dependence on traditional farm methods continues with considerable reliance on the labor of man and animal. This is true even in eastern Europe where after World War II the properties of large landowners have been confiscated to form collective farms after the Soviet model and where former peasants are now part of work teams on large-scale farms. In Greece, Spain, Portugal, and Ireland, traditional peasant agriculture still dominates.

The Urban, Metropolitan Population Component

In agricultural Europe as early as the eras of the ancient Greeks and Romans, there were already long-range interconnections across Europe and by sea. Cities, though small by present standards, emerged as the major nodes in this network. Nobles wanted luxuries from other lands, the Church needed to communicate its decrees, and rulers wished to keep up to date on political happenings. These cities served as the intermediaries between the long-distance, interregional connections, on the one hand, and the rural towns and villages of their own region, on the other.

Cities in Agrarian Europe of the Past

Since the people, goods, and ideas moving in interregional trade were largely limited to those involving the wealthy nobility and religious orders and since the local area served by any center was small, few large cities developed. Even as late as 1800, only seventeen cities in all of Europe had 100,000 people, and they accounted for only 3 per cent of Europe's population (Table 9-4). Of these, ten were political capitals, five were ports, and the other two, Milan and Lyon, were ancient trade centers. The locations of these cities of less than two hundred years ago show their ties to the traditional

centers of cultural leadership. Only four countries had more than one large city: what is now Italy had four, France had three, and Spain and what is now Germany had two each.

Today, after the change in European life that resulted from the Industrial Revolution, metropolitan centers have risen to dominate much of the continent. Now there are more than one hundred metropolitan centers in Europe with 250,000 people, and they total more than 35 per cent of Europe's population. Thirty-seven metropolitan areas in Europe have at least a million people, and greater London has more than 11 million.

Rural Europe has not disappeared, but metropolitan Europe has risen with a considerably different population distribution. As Figure 9-6 shows, nearly two-thirds of the population in large metropolitan areas is found in western Europe. We can see that southern and northern Europe each have about equal metropolitan and non-metropolitan populations.

Just as the Northeast and Midwest contain the large metropolitan populations of the United States, so western Europe dominates Europe. This is not surprising since here industrialization had its beginning and since the countries located here are still the industrial leaders.

Industrialization and the Concentration of Metropolitan Areas in Western Europe

The Industrial Revolution completely changed man's capacity for creating energy, hence increasing productivity. First he developed water-powered factories, then he learned to utilize coal for factories and transportation. There followed the railroad, the steam-powered ship, electricity, mass-produced steel, the radio, the internal-combustion engine, the use of petroleum, and a myriad of other engineering developments. At the same time a system of banking,

Table 9-4 Large Metropolitan Centers of Europe
(population in thousands)

	1800*	1900	1964	1968*		1800	1900	1964	1968
WESTERN EUROPE					**SOUTHERN EUROPE**				
London	959	4,537	11,025	11,025	Milan	170	539	2,775	3,365
Paris	547	2,774	8,450	8,850	Madrid	160	540	2,575	2,980
Essen–Dortmund–					Rome	153	423	2,500	2,810
Duisburg		252†	5,200	5,150	Barcelona	115	533	2,175	2,475
West Berlin	172	1,889	2,955	2,893	Athens			1,975	2,100
Manchester		544	2,850	2,890	Naples	350	621	1,765	1,875
Birmingham		522	2,640	2,665	Lisbon	180	356	1,375	1,500
Hamburg	130	706	2,300	2,335	Turin		330	1,350	1,480
Brussels		599	1,975	2,070	Genoa	100	378		(no
Vienna	247	1,675	2,025	2,020					data)
Glasgow		762	1,885	1,860					
Amsterdam	201	511	1,730	1,805					
Liverpool		685	1,685	1,750					
Cologne		373	1,550	1,655	**EASTERN EUROPE**				
Munich		500	1,500	1,625					
Leeds–Bradford			1,360	1,530	Budapest		732	2,265	2,350
Frankfurt am Main		289	1,450	1,520	Katowice–Zabrze–				
Stuttgart			1,415	1,470	Bytom			1,960	2,025
Copenhagen	101	401	1,380	1,385	Warsaw	100	638	1,575	1,700
Newcastle-upon-					Bucharest		276	1,400	1,550
Tyne			1,155	1,400	Prague		202	1,110	1,150
Mannheim–					East Berlin		See	1,070	1,082
Ludwigshafen–							West		
Heidelberg			1,170	1,210			Berlin		
Lyon	110		1,000	1,125					
Rotterdam		319	1,010	1,095					
Antwerp			1,015	1,040	**NORTHERN EUROPE**				
Düsseldorf	n.a.	n.a.	n.a.	1,075					
Marseille	111	459		1,025	Stockholm		301	1,180	1,275

*Only those cities over 100,000 are listed in 1800 and 1900, and those over 1,000,000 in 1964 and 1968.
†Essen and Dortmund only.
SOURCES: 1800 and 1900: Woytinsky and Woytinsky, *World Population and Production: Trends and Outlooks*, Twentieth Century Fund, New York, 1953; 1964 and 1968: *Encyclopaedia Britannica International Atlas*, Encylopaedia Britannica, Chicago, 1970, table p. 280.

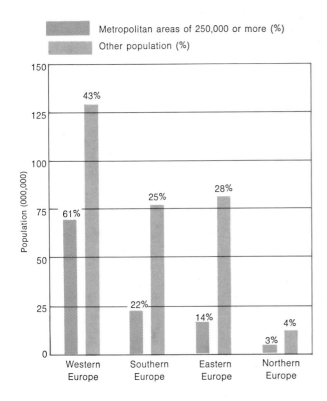

Metropolitan areas of 250,000 or more (%)

Other population (%)

Figure 9-6 METROPOLITAN AND NON-METROPOLITAN POPULATIONS OF PARTS OF EUROPE, 1963

business management, and trading functions was developing to take advantage of technological change. The result was a revolution in the life of western Europe, and the first modern models of the long-range, interconnected system emerged in Britain, France, Germany, Belgium, the Netherlands, and Switzerland.

Cities grew, based on employment in manufacturing, trade, government, and services. They were centers of cultural as well as economic life. They were publishing centers and the generators of ideas in art, music, and politics.

Western Europe has emerged as a largely metropolitan-centered society, while the other regions of the continent are still rural (Figure 9-7). Britain is by far the most metropolitan area with over half its population in such centers, but most other western European countries have over one-fourth of their populations in big cities. Greece is the only European country outside of western Europe with such a large metropolitan component. In southeastern Europe Rumania, Bulgaria, Yugoslavia, and Albania still have less than one-tenth of their populations in major metropolitan centers.

The concentration of employment in western Europe in urban-type activities such as manufacturing, construction, trade, transportation, and services can be seen in Figure 9-8. Most western European countries have over three-fourths of their employment in urban-type activities. Only Sweden and Norway in northern Europe and Italy in southern Europe show similar concentrations. Eastern and southern European countries in most cases have less than half of their working populations in urban-type employment.

POPULATION VARIATIONS AND POLITICAL STATES

Variations in the importance of agriculture and urban activities are not unusual for any large

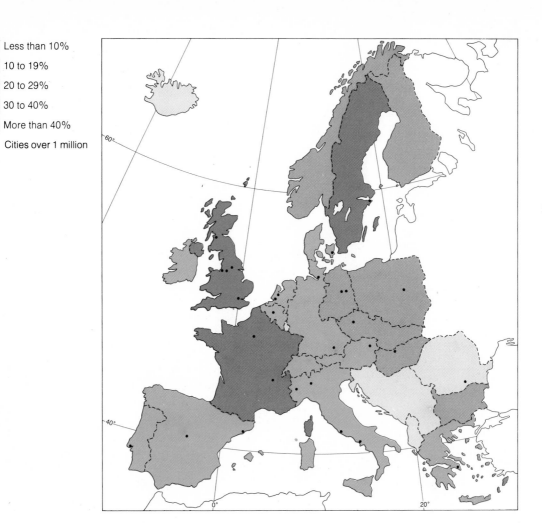

Legend:
- Less than 10%
- 10 to 19%
- 20 to 29%
- 30 to 40%
- More than 40%
- • Cities over 1 million

Figure 9-7 PERCENTAGES OF EUROPEAN
POPULATION, BY COUNTRY, LIVING IN
METROPOLITAN AREAS OVER 250,000

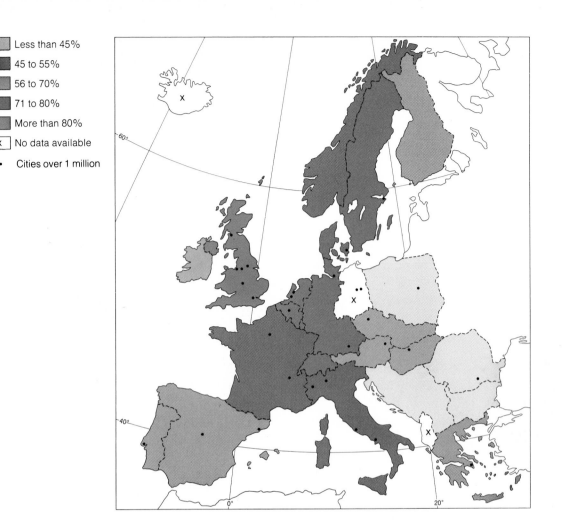

Figure 9-8 Percentages of European
Population, by Country, Employed in
Urban-centered Activities

territory such as Europe. We certainly saw this to be the case in the United States where the more rural South stands in contrast with the metropolitan Northeastern Seaboard. But in Europe the contrasts are more significant. Each area standing distinct on a map or in a table is a separate country and also an economic entity. Variations on the maps and in the tables indicate countries with high living standards and those with low.

Figures 9-9 and 9-10 show per capita GNP and consumption of industrial energy sources. The contrast between affluent, energy-consuming western and northern Europe and eastern and southern Europe is striking. Perhaps Italy, East Germany, and Czechoslovakia fit the affluent, energy-consuming mold as well, but most of the countries of the south and east show much less intensive industrialization. Ireland is the only country in western Europe that is not highly industrialized. We are not thinking in world terms now: all countries in Europe except, perhaps, Albania would rank above the world average. The point is that most of those in south and east Europe are much less developed than their European neighbors to the west and north.

The Importance of Political Capitals

A look at a map showing major metropolitan centers (Figure 9-11) also points up the difference between the more industrialized countries of Europe and others. Notice that in the still largely agricultural countries, virtually the only major metropolitan centers are the political capitals. Excepting Italy, in eastern and southern Europe, only three industrial cities in Poland and Barcelona in Spain are noncapitals that have reached the level of one million people. In the urban, industrial countries, such as Britain and West Germany, however, there are a half dozen or more such cities besides the national capitals. Even there, except in West Germany, the capital is also the largest metropolitan area in the country.

One would expect capital cities in even largely agricultural countries to be major metropolitan centers. Government, after all, is a metropolitan function. A capital must be at the focal point of transportation and communication if it is to carry out its role of decision-maker for the country. It is the country's point of contact with the rest of the world and the collection and dissemination point for dealing with all places within its boundaries. It is also likely to be the place in the country that will have international airline connections and domestic air and overland transport links to all possible parts of its territory. It will be the heart of most communications with the rest of the world and the central node of its own mail and other national communications links. A country may have only a few road and railroad connections across its boundaries, but its internal network of roads and railroads will focus on the capital. As a result it is often necessary to travel via the capital even when going from one smaller city to another.

The capital of an agriculture-oriented country is more than a political center; it is the cultural axis that tends to dominate smaller regional fragments of the country; and, as the largest center, it is the most important marketplace for goods produced within the country or outside. Further, it is the intellectual hub and focus of the most advanced technical thought in the country. As the largest labor source and marketplace, it is the logical point at which to carry on whatever manufacturing might occur other than special-purpose manufacturing tied to particular resources.

Such an exceedingly important role for the capital has been traditional in European history

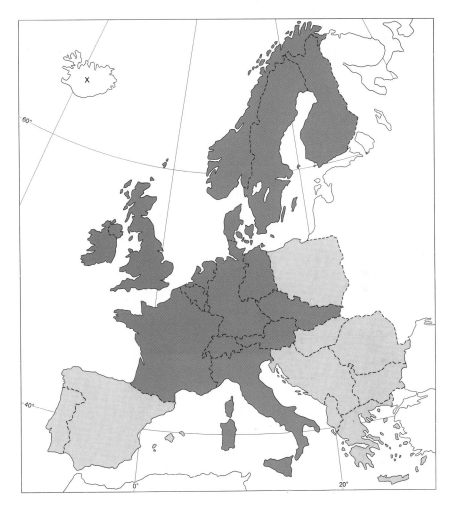

Figure 9-9 Per Capita GNP, by Country, for Europe in 1967

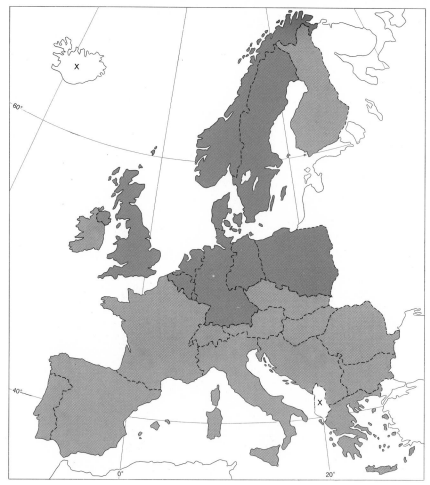

Figure 9-10 PER CAPITA INDUSTRIAL ENERGY
CONSUMPTION FOR EUROPE IN 1968

Figure 9-11 EUROPEAN CITIES WITH 250,000 OR MORE PEOPLE IN 1960

- • 250,000 to 1,000,000 population
- ● 1,000,000 to 4,000,000 population
- ⬤ More than 4,000,000 population

since the rise of the national state, and we shall see it emerging now in most of the developing areas of the world. In only a few countries such as West Germany and Switzerland is the capital not the major city.

INDUSTRIAL METROPOLITAN CENTERS

The pattern of industrialization in western Europe grew out of the established commercial system; the political capitals of the pre-industrial era were the logical bases for the new industrial-business world—the centers of decision-making and management, both governmental and private. Moreover, they were major markets in themselves and focal points for transportation and communication. The same was true of the ports of the earlier period which had not only physical facilities for shipping but also management and financial expertise.

Then the development of modern industry, commerce, and management changed these old cities into very different places. They have grown many times larger in size and have also become centers of a much bigger and more complex system. If cities in the eastern United States have had problems fitting in railroads, highways, and new public utilities, imagine the difficulties of European cities dating back to medieval times. Here buildings built five hundred or more years ago now house industrial and commercial activities or serve as homes for city families. Streets built for use by pedestrians and horsecarts must carry modern traffic.

However, the old network of capitals, trade centers, and ports did not prove sufficient to meet the needs of the modern large-scale industrial system. New industry used great quantities of power and minerals; industrial cities developed at waterpower sites, in the coal and iron fields, or at transport sites nearby.

Today Britain has seven metropolitan centers of more than one million people; West Germany has nine; Italy has four; and even tiny Belgium and the Netherlands each have two. (See Figure 9-11.)

Traditional trading centers and ports are among the large western European metropolitan areas today. Cologne, Frankfurt am Main, and Mannheim were traditional trading centers in the older system. Hamburg, Liverpool, Glasgow, Newcastle-upon-Tyne, Rotterdam, and Antwerp were pre-industrial ports.

But the importance of new industrial sites can be seen by the fact that the Ruhr metropolitan complex is the third largest metropolis in Europe and that Manchester and Birmingham now rank among the seven largest metropolitan areas as well. Like Detroit or Pittsburgh, they are highly specialized industrial nodes where outgrowth industry linked to initial heavy industries has contributed to expansion. More important, they have developed as management and financial centers serving the interests of their particular manufacturing operations.

The importance of early resource sites can be seen in the location of these centers. Manchester and Leeds-Bradford, at the sites of waterpower, were centers of the early textile industry that opened the industrial era. The Ruhr complex and Manchester grew up in coalfields, and Birmingham was adjacent to iron deposits as well as to coal.

The World-wide Resource Base of European Countries

Britain more than any other country built its modern, interconnected system on world-wide connections. In the center of the vast network of sea lanes that reached to Canada, Australia and New Zealand, India, and to colonies in Africa, Latin America, the Middle East, and the Pacific was London. Ports such as Liverpool, Glasgow, and Manchester (by means of a canal) were busy links in the system. Yet London was without doubt the focal point because the Empire was a vast political system, and basic management decisions were made there. In actuality the Empire was also the core of a great economic system through which Britain tapped the resources of regions throughout the world.

THE BRITISH EXAMPLE

After its industrialization Britain became increasingly an area of urban-centered activities. Like the United States, less than 10 per cent of Britain's population is engaged in primary production. But this is less the result of mechanization of primary production than a shift away from dependence on the resources of its own country to dependence on the rest of the world. In a sense, Britain functions in much the same way as the megalopolis of the Northeastern Seaboard of the United States. It concentrates on business, trade, and industry and draws its basic food, raw materials, and specialty manufactures from other areas. In Britain's case, the basic producing areas lie outside the country and to a large degree outside Europe. The British have long depended on the resources of their Empire and on those of other areas with which they developed trade ties. Although the Empire is gone and the British Commonwealth, which replaced it, is a weak political organization, the economic system developed during Empire days still remains in many respects.

Just as the eastern megalopolis depends on the Middle West and Great Plains for grain and meat, so Britain utilizes the mid-latitude farming regions of Canada, Australia and New Zea-land, and Argentina and Uruguay. As megalopolis gets cotton, citrus fruits, wines, and tobacco from the South and California, Britain draws on production from subtropical areas throughout the world, including those of the United States. Megalopolis is tied to oil fields in Texas and the Gulf Coast; Britain is tied to those of the Middle East. Just as the industrial megalopolis ships goods throughout the United States, British motor vehicles, machinery, chemicals, and textiles are shipped overseas to the specialized primary producing areas that do not have adequate manufacturing to meet their own needs.

Since Empire days the whole system of world-wide economic outreach has been managed from Britain, essentially from London. Britain not only built the ships but owned and operated them, financed the operation through its banks and insurance companies, controlled trade through its great commercial organizations, and provided supporting legal, accounting, and other services. London, as the management and decision-making center, was the business capital of the international, interconnected system until the rise of the large international corporations headquartered in New York. It is still second only to New York.

THE IMPORTANCE OF EXTRA-EUROPEAN TRADE

The model of world-wide dependence on resources and markets overseas also fits the other industrialized countries of Europe, although on a smaller scale. France, Germany, Italy, Belgium, and the Netherlands had overseas empires and trading ties. Today, despite the loss of political empire, these world-wide relations continue.

Some measure of the importance to European countries of extra-continental trade can be seen

Table 9-5 Extra-European Trade on a Per Capita Basis, 1963

Country	Per Capita Value (dollars)
United Kingdom	282
Belgium	244
Switzerland	244
Netherlands	236
Sweden	236
Denmark	184
West Germany	165
Finland	160
France	151
Norway	135
Ireland	91
Italy	76
Austria	70
Portugal	51
Greece	41
Spain	35

Source: *World Trade Annual: 1964,* Statistical Office of the U.N., New York, 1964.

in Table 9-5 in per capita data for some European countries. As a point of reference, overseas trade of the United States (total foreign trade except with Canada and Mexico) totaled $59 per person in 1963. From this perspective, almost all European countries depend more on overseas resources and markets than does the United States. However, within Europe the range is very great. Western and northern European countries except Ireland and Austria are two to three times as dependent on overseas trade as southern European countries. Britain remains the most dependent, but the small countries of Belgium, Switzerland, the Netherlands, and Sweden find overseas trade scarcely less important. Notice that extra-European trade

is much less important to West Germany and France.

Table 9-6 shows the gross pattern of overseas trade for European countries outside the Eastern European Bloc. The great range in size of total overseas trade is striking. As might be expected, Britain has by far the largest trade volume. Notice how much greater it is than the other large industrial countries. But regardless of the size of the trade there is a remarkable similarity in the pattern. The leading trading partner for all countries except France, Finland, and Portugal is not the underdeveloped areas supplying raw material; rather it is industrialized Anglo-America. What must this mean in terms of the kinds of goods moving in trade? (Remember that Table 9-6 is based on dollar value, not tonnage.)

Notice also that trade with Southwest Asia and North Africa is usually the greatest of any connection with nonindustrialized areas. This indicates the importance of petroleum shipments from that part of the world to Europe, which has few productive oil and natural gas deposits within its boundaries.

The isolation of the Communist countries is evident. China, with the largest population of any country, is insignificant in the total trade of all non-Communist European countries. The Soviet Union, the largest industrialized country in the world and a close neighbor to European countries, accounts for less than 10 per cent of the trade of any of the countries in Table 9-6 except for three of the smaller non-Communist ones. Trade with the United States and Canada is many times greater than that with the Soviet Union.

One can also see in Table 9-6 that virtually all European countries have a negative trade balance in overseas trade. Only Switzerland, Finland, and Norway show a larger value of exports than imports, and all the largest trading countries buy significantly more than they sell

Table 9-6 OVERSEAS TRADE CONNECTIONS OF SELECTED EUROPEAN COUNTRIES, 1963

Country	Subsaharan Africa (per cent)	South and East Asia (per cent)	Southwest Asia–North Africa (per cent)	Oceania (per cent)	Latin America (per cent)	China (per cent)	Anglo-America (per cent)	USSR (per cent)	Imports (in $ billions)	Exports (in $ billions)	Total Trade (in $ billions)
United Kingdom	10	16	13	20	12	1	25	3	8.8	6.6	15.7
West Germany	7	14	16	6	18	1	35	3	5.5	4.0	9.6
France	19	6	34	6	11	1	20	3	4.1	3.1	7.3
Italy	7	8	23	7	16	1	32	6	3.3	1.7	5.1
Netherlands	10	13	20	4	17	1	33	2	1.8	1.0	2.8
Belgium	15	11	17	8	14	1	33	3	1.5	0.8	2.3
Sweden	6	11	10	6	21	1	36	8	0.9	0.7	1.6
Switzerland	4	18	11	6	18	1	40	2	0.6	0.8	1.4
Spain	4	8	20	3	25	....	39	1	0.9	0.2	1.1
Denmark	7	14	13	3	18	1	38	6	0.5	0.4	0.9
Finland	1	4	7	4	11	1	20	52	0.3	0.4	0.7
Austria	5	12	10	7	13	1	30	22	0.3	0.2	0.5
Norway	12	11	10	6	26	1	26	8	0.2	0.3	0.5
Portugal	46	6	11	4	7	....	26	....	0.3	0.2	0.5
Greece	3	7	21	3	9	....	42	15	0.2	0.1	0.3
Ireland	4	16	16	7	9	....	46	2	0.2	0.1	0.3

SOURCE: *World Trade Annual: 1964*, Statistical Office of the U.N., Walker & Co., New York, 1964.

overseas. This seems to contradict the usual view that advanced countries such as the industrialized countries of Europe have a trading advantage in buying low-value raw materials and foods, and in selling high-value manufactures. What reasons might there be for the unfavorable balance? How is this imbalance resolved?

Intra-European Trade

We have made the point that the twenty-six countries of Europe function politically, culturally, and economically as separate entities in contrast to the fifty states of the United States which operate as a single functioning whole. This is obvious in the political and also, to a high degree, cultural realms. But perhaps we have overdrawn the economic situation by stressing overseas ties rather than intra-European economic links. As we have just seen, the overseas ties of at least the highly industrialized countries of western and northern Europe are very great. Despite political barriers there is a busy intra-continental trade, too.

In fact, as Table 9-7 shows, intra-European trade is of greater value for most European countries than overseas trade. Britain is the single exception. But notice that, whether in

Table 9-7 Intra-European Trade Compared with Extra-European Trade on per Capita Basis, 1963

Country	Intra-European (dollars)	Extra-European (dollars)
Belgium–Netherlands–Denmark	543	229
Switzerland–Austria	446	140
Norway–Sweden–Finland	446	179
Ireland–Iceland	417	104
West Germany	298	165
France	196	151
United Kingdom	168	282
Italy	129	76
Spain–Portugal	54	38
Yugoslavia–Greece–Albania	45	*
Poland–Czechoslovakia–East Germany	29*	*
Hungary–Rumania–Bulgaria	27*	*

*Does not include trade within Iron Curtain.
Source: *World Trade Annual: 1964*, Statistical Office of the U.N., New York, 1964.

absolute terms or terms of relative importance, it is the small, advanced countries of western and northern Europe that depend most on intra-European trade. What are the ratios of intra-European to extra-European trade for small industrial countries?

While all countries of Europe have extra-European connections, the large European industrial countries that had nineteenth-century empires are most concerned with intercontinental ties today. The smaller countries feed goods into the interregional trade of Europe and draw other products in return. This is particularly true of Austria, Switzerland, Norway, Sweden, Finland, and Ireland. Like the outlying regions of the United States, they play a role complementary to major metropolitan centers.

INTRA-EUROPEAN TRADE FLOW

Figure 9-12 presents in diagrammatic form the movement of intra-European trade except for the trade between the countries of the Eastern European Bloc. The importance of the major industrialized countries of western Europe is immediately obvious. The greatest interchange occurs between France, Belgium-Netherlands-Denmark, and West Germany with a major secondary extension of that triangle of trade to Italy. These major industrial countries are now trading partners within the European Economic Community. Notice that Britain is much less significant than even Italy in the pattern of flow. The other blocs of countries are "outsiders." Like the outlying regions of the United States, their significant trade flows are not between each other. The important ties are really between each outlying area and the countries in the inner trade flow. So the French-German-Benelux interactions with Denmark and Italy form a trade hub, and the links with the outer areas are like the spokes of a wheel. There is virtually no rim of interconnections around the European trade wheel.

Germany, the most populous and industrial country on the continent, appears most important within the inner trade hub. But perhaps the most surprising fact is the importance of the Belgium-Netherlands-Denmark cluster. These three small countries appear more important on the pattern of trade flows than France and compare favorably with West Germany. In view of the locations of the three, why might this be so?

As Figures 9-13, 9-14, 9-15, and 9-16 indicate, it is the flow of manufactured goods—machinery, transportation equipment, and basic manufacturing, including textiles and chemicals—

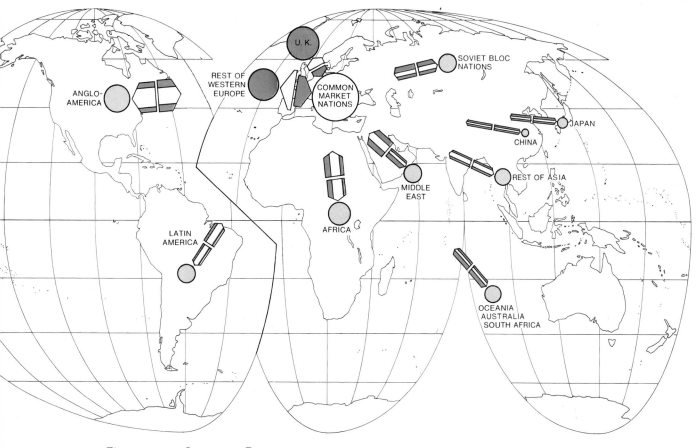

Figure 9-12 Schematic Representation of
Trade to and from European Countries, 1963

Width of arrow based on volume of trade

Import and export

Rest of Western Europe
European Economic Community
United Kingdom

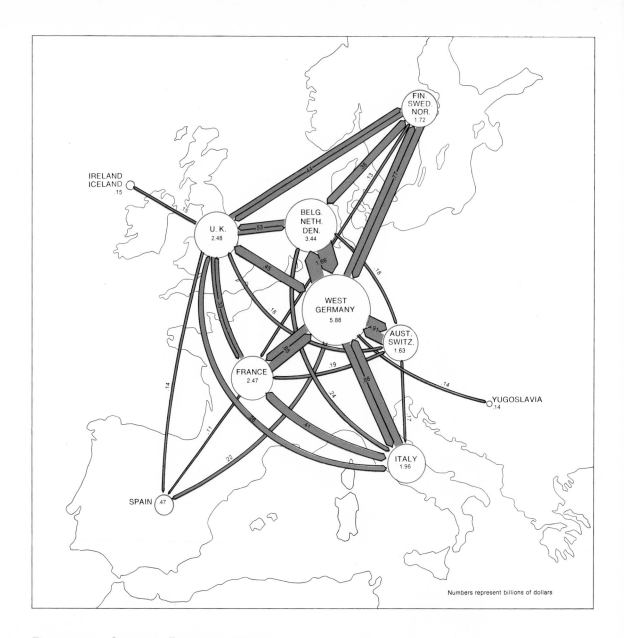

Figure 9-13 SCHEMATIC REPRESENTATION OF
INTRA-EUROPEAN TRADE IN MACHINERY AND
TRANSPORTATION EQUIPMENT, 1963

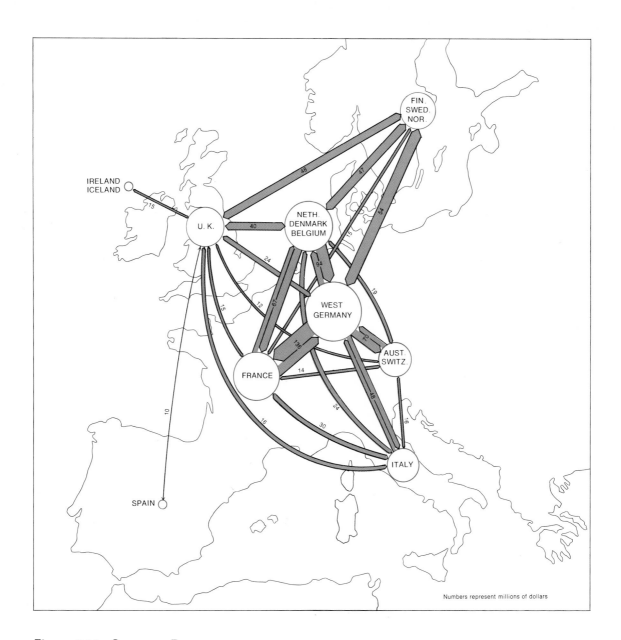

Figure 9-14 Schematic Representation of
Intra-European Trade in Basic Manufactures, 1963

Numbers represent millions of dollars

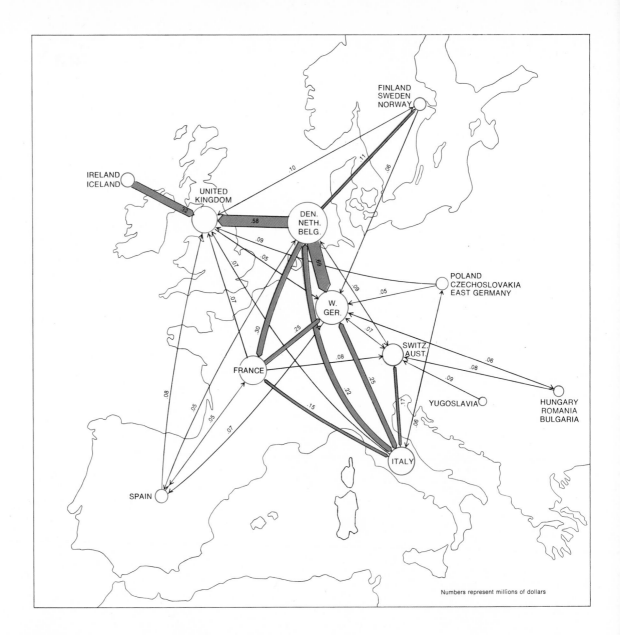

Figure 9-15 Schematic Representation of
Intra-European Trade in Food Products, 1963

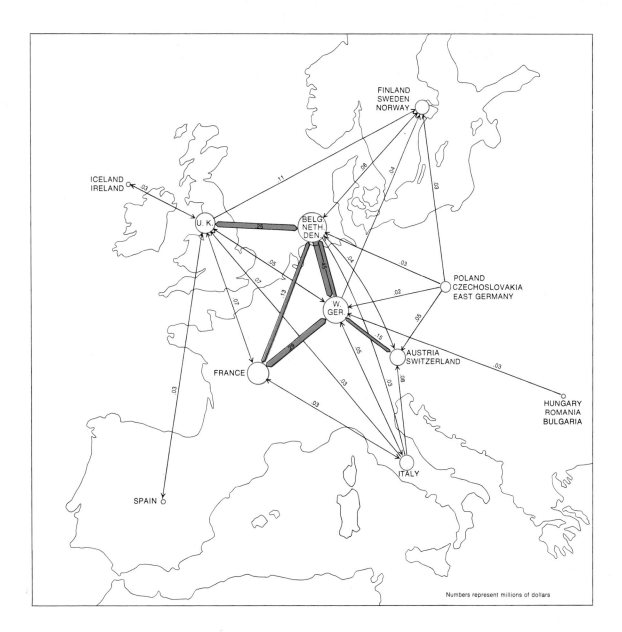

Numbers represent millions of dollars

Figure 9-16 Schematic Representation of
Intra-European Trade in Fuel, 1963

that dominates intra-European trade. These goods move mostly within the inner hub of European trade, although the peripheral industrial areas appear significant, too. Except for Italy, the nonindustrial nature of the Mediterranean countries is obvious.

Food also flows largely within the inner hub or to Britain. Notice the importance of Belgium-Netherlands-Denmark as a supplier of food products to the large, highly developed countries. Some of this activity undoubtedly involves imports from overseas moving through the ports of Rotterdam and Antwerp, but it also must reflect the intensely developed and highly efficient specialized agriculture of the three North Sea countries. With great effort and the application of scientific technology to the land, farmers of these areas supply a large share of industrial Europe's deficit in dairy products, pork products, poultry, and eggs. In contrast, the agricultural areas of the Mediterranean other than Italy fare rather poorly even though their distinctive agricultural environment is basically the same as that of California.

We can now begin to see that both the economic and political facets of the modern, interconnected system of Europe are quite different from those of the United States. Three distinct types of interaction are all taking place at once.

1. The basic order of interaction is within each of the individual countries of the continent. As political entities they make great effort to educate their populations on the importance of being Frenchmen, Germans, Swedes. The national government reinforces the local cultural heritage and determines the basic pattern of daily life and the degree of interaction with other countries. Most European countries have also worked to integrate their countries economically. The capital, usually the largest city, is both the major market for production from outlying parts of the country and the center of the industrial-business economy. The first-order importance of national integrity can be seen in the transportation and communication networks of each nation that provide a channel linking the capital with other parts of the country.

2. The large western European countries that formerly had empires still depend more heavily on overseas connections than the United States does. Their most important overseas ties are not, however, with underdeveloped nations as sources of raw materials and tropical products. Trade with them continues, but now the United States is the center of economic action both as supplier of advanced manufactures and as a market for European products. Furthermore, the migration of Europeans to the United States has resulted in social and cultural ties, and affluence in both areas has created interest in similar activities.

3. With improved transportation and communication and with greater affluence, total intra-European interaction is more significant than overseas ties. As we have seen, European trade centers around the major industrial countries of western Europe, particularly West Germany, France, Belgium, and the Netherlands. As in the United States there is first-order interaction, both of products and business communication, between the major industrial metropolitan areas of western Europe. Second there is movement between the centers at the core and the largely less developed producers of specialized food and raw materials in eastern and southern Europe. In recent years there has also been greater movement of workers shifting jobs; large numbers of Italians, for instance, have moved to France and Germany. And more affluent Europeans are traveling to other European countries for recreation. Eastern Italian and Yugoslav resorts cater to Germans; the

coast of Spain and the island of Majorca attract the British; and Switzerland and the Riviera are the mecca for a full range of Europeans.

Figure 9-17 is an attempt to diagram the network of European interaction as it focuses on the large metropolitan centers of the continent. The basic national networks can be seen, particularly in France, Britain, Spain, and West Germany. The interstate links are less obvious. But one must remember the actual flows of intra-European trade which we saw in Figure 9-12. Finally, the large number of different overseas links resulting from the political divisions of Europe is obvious. Most European countries have their own ports connecting them to the rest of the world, in contrast to the United States where the eastern ports of New York, Philadelphia, and Baltimore serve most continental hinterlands. The major international trade link within Europe is the Rhine River system; the Dutch port of Rotterdam and the Belgian port of Antwerp serve this system that reaches into Germany, Luxembourg, eastern France, and Switzerland.

Even though Europe has not been as consciously developed as an economic entity as the United States has been, regional trade has been built in the interchange of local specialty production, as Figures 9-13, 9-14, 9-15, and 9-16 clearly indicate. Northern Europe, particularly Sweden and Finland, provides timber products, especially paper and pulp. The Mediterranean with its warm, sunny climate provides agricultural specialties, citrus fruits, wine, and tobacco, mostly for the metropolitan core in western Europe.

Minerals, too, move from one part of Europe to another. Coal and iron-ore have long moved across national boundaries between France, West Germany, Belgium, and the Netherlands. Iron-ore moves into the metropolitan core from Sweden by way of a Norwegian port

and from northern Spain; coal comes from Poland.

Moves toward Economic Integration within Europe

Eastern European countries were once important food suppliers to western Europe, particularly to Germany, but Soviet policy since their take-over near the end of World War II in reordering the economic ties of those countries toward the Soviet Bloc appears to have been effective. Under pressure from the United States there also has been an effort to isolate eastern Europe, and thus the Soviet Union, from the receipt of industrial goods that might have military value. Consider the part of Europe that the Soviets have pulled into their sphere of influence. What could it best contribute to the Soviet economy? Also, consider what the effects might have been if the Soviets had been able to extend their sphere of influence farther westward.

The trade map (Figure 9-12) therefore points up the political and economic changes that need to be added to the story of functioning Europe today. The tiny trickle of trade between the countries of eastern Europe and the rest of the continent shows the effectiveness of the division between eastern and western Europe. This division acts as an economic as well as a political barrier, preventing all but a small amount of food and fuel from moving westward and allowing almost no manufactured goods to move eastward.

European Common Market

Followers of happenings in Europe since World War II can offer another dimension to this explanation of the important flow of trade be-

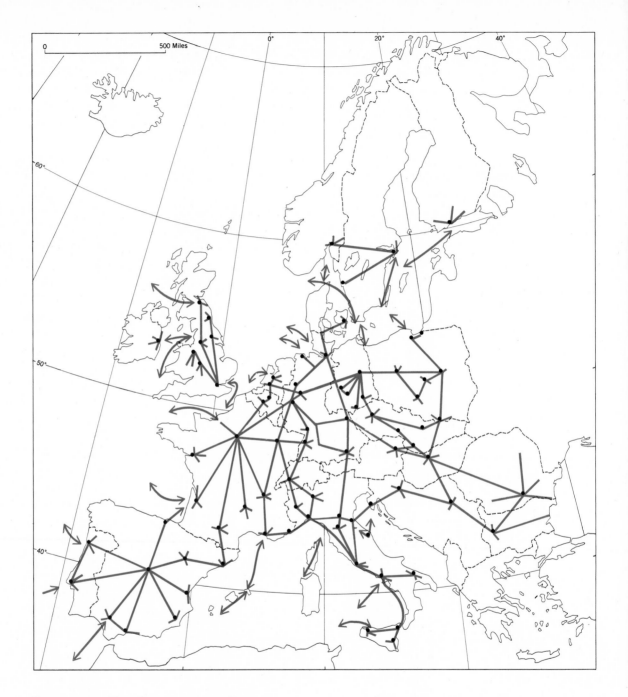

Figure 9-17 The Connectivity of Europe

tween West Germany, France, Belgium, the Netherlands, and Italy. For these countries, together with tiny Luxembourg, were the original members of a remarkable economic enterprise: the European Economic Community, or Common Market.

Initiated in 1958 the E.E.C. has gradually resulted in the elimination of tariff barriers between these countries and the establishment of a single set of duties around them for goods entering any of them. Essentially, while the countries remain separate political units, each with its own rules governing the conduct of business and its own currency, there are no longer any real barriers to the movement of goods from one of the member countries to the others. Thus, instead of six separate national markets for goods, products can flow freely throughout the six countries.

What is significant about the founding of the Common Market is the countries that were its charter members. All were recognized as among the leading examples of industrialization and the modern, interconnected system, and they included three of the largest countries in Europe: West Germany, France, and Italy. Common Market countries have an affluent population comparable to that of the United States or the Soviet Union. The Common Market overcomes what has been a major handicap to the development of modern mass production: the small size of European national markets.

It should be added that the Market developed from an earlier agreement of the original six nations to eliminate barriers to trade in coal and steel, the basic ingredients of modern industrialization. This called for the rationalization of resource production: the closing of inefficient mines in one country and the expansion of efficient ones in another, and the development of steel capacity wherever it seemed best to do so regardless of country. It also called for the free movement of workers to wherever they were needed. This freedom of movement has now been generally expanded to include all industries and is resulting in international movements of workers in addition to rural-urban movement which is common during industrialization. Large numbers of Italians, in particular, have moved to industrial districts in other member countries.

NEW MARKET AND LARGE CORPORATIONS

Surprisingly, at first thought, the organizations that have taken greatest advantage of the Common Market have not been companies of the countries themselves but large international corporations, most of them American. By 1967, $14 billion of United States investment had been made in plants and equipment in Common Market countries. This total is one-third of all such overseas investment since 1958. All types of major corporations have been involved: automobile manufacturers, food processors, chemical companies, oil companies, and even hotel and motel chains. Particularly important have been industries involving the most advanced technology such as electronic equipment, semiconductors, computers, and integrated circuits.

The reason it has been American firms rather than European companies that have been taking advantage of the opportunities offered by the Common Market is that few European corporations are of the large scale of the biggest American firms. Only 40 European companies ranked with the 100 largest in 1969 in terms of total sales, and the largest share of them were British. These largest corporations tend to dominate world markets. They have the research staffs which turn up new products, market analysts, and the resources for obtaining necessary financing.

Table 9-8 COMPARISONS WITH THE EUROPEAN ECONOMIC COMMUNITY (millions of units)

		WEST GERMANY	FRANCE	ITALY	NETHER-LANDS	BELGIUM AND LUXEM-BOURG	E.E.C.	UNITED KINGDOM	UNITED STATES	USSR	JAPAN
1970	POPULATION	61.6	50.7	54.5	13.0	10.0	189.8	55.7	203.2	241.7	103.7
	AGRICULTURE										
	Farmland (acres)	34.0	82.8	48.5	5.5	4.5	175.3	30.8	1,089.0	1,495.0	16.5
	Arable land (acres)	20.4	48.0	35.9	2.3	2.2	108.8	18.5	599.0	568.1	14.2
1969	Wheat (metric tons)	5.6	12.9	9.6	0.6	0.8	29.6	4.2	37.5	99.5	0.5
	Rye	2.7	0.3	0.1	0.2	0.1	3.4	0.0	1.0	15.0	0.0
	Barley	5.1	9.5	0.3	0.4	0.6	15.9	8.7	9.2	32.6	1.0
	Oats	3.0	2.3	0.5	0.3	0.3	6.4	1.3	13.8	13.1	0.1
	Maize	0.4	5.7	4.5	0.0	0.0	10.6	0.0	116.4	12.0	0.1
	Potatoes	16.0	9.0	4.0	4.7	1.6	35.3	6.2	14.2	92.0	3.6
	Sugar beets	13.4	17.9	10.6	5.0	4.4	51.3	6.0	25.2	71.2	2.1
1968–69	Cattle	14.1	21.7	10.0	4.3	2.9	53.0	12.4	110.0	95.7	3.5
	Pigs	18.7	10.5	7.3	4.8	2.9	44.2	7.8	60.6	49.0	5.4
	Sheep	0.8	10.0	8.2	0.6	0.1	19.7	26.6	21.2	130.7	0.1
1969	ENERGY CONSUMPTION										
	(metric tons coal equivalent)										
	Coal and lignite	135.2	59.3	12.8	9.5	24.8	241.6	159.2	446.5	437.2	86.3
	Crude petroleum	140.8	98.6	94.5	31.3	25.4	390.6	114.3	887.4	318.7	189.6
	Natural gas	15.9	12.2	15.9	19.2	3.8	67.0	8.4	801.8	240.4	3.4
	Hydroelectric, nuclear electric†	3.2	7.1	6.2	0.0	0.3	16.8	4.2	33.6	14.3	10.1
	Total	295.1	177.2	129.4	60.0	54.3	716.0	286.1	2,189.3	1,010.6	289.4
1969	MANUFACTURING										
	Steel (metric tons)	45.3	22.5	16.4	4.7	18.3	107.2	26.8	128.2	110.3	82.2
	Aluminum (primary)	0.3	0.4	0.1	0.1	0.0	0.9	0.0	3.4	1.1	0.6
	Sulfuric acid	4.5	3.5	3.4	1.5	1.8	14.7	3.3	26.8	10.7	6.8
	Cement	35.1	27.7	31.3	3.3	6.5	103.9	17.4	69.6	89.7	51.4
	Cotton fabric	0.3	0.3	0.2	0.1	0.1	1.0	0.2	1.6	1.6	0.5
	Synthetic staple	0.4	0.2	0.2	0.0	0.1	0.9	0.3	1.1	0.3	0.8
	Autos (units)	3.4	2.2	1.5	0.1	0.1	7.3	1.8	8.2	0.3	2.6
	Merchant vessels (gross tons)	1.6	0.8	0.5	0.6	0.1	3.6	1.0	4.0	n.a.	9.3
1970	GNP ($ billions)	186.4	147.6	93.2	31.3	26.6	485.1	170.0	974.8	370.0	230.0

† Excluding thermal electricity.
SOURCES: Statistical Office of the European Communities, *Monthly Bulletin*, No. 12, 1971, Tables 0 and 1; Statistical Office of the U.N.: *U.N. Statistical Yearbook: 1971, U.N. Demographic Yearbook: 1970*, and *World Energy Supplies: 1966–1969*, Series J, No. 14; and *U.N. Production Yearbook: 1970*, Vol. 24, Food and Agricultural Organization Office, Rome, 1971.

BRITAIN AND THE COMMON MARKET

Although its leaders were among those pushing for international economic cooperation in Europe after World War II, Britain turned down the opportunity to join the Common Market in early days. Among other things it felt it was a world power whose Commonwealth ties were of first importance.

Recall that Britain is the only European country that depends more on overseas trade than on intra-European transactions. However, Britain found it increasingly difficult to compete with the new economic scale of the Common Market. To strengthen its position Britain joined with six other European countries—Norway, Sweden, Denmark, Austria, Switzerland, and Portugal—in a more limited trade organization aimed primarily at increasing trade among its members by reducing tariffs between them. All but Portugal were advanced industrial-trading countries, and the total market within the seven countries was more than $96 million in 1970.

The limited success of the British-sponsored Free Trade Association led Britain to push for membership in the Common Market, and in 1972 Britain, Norway, Denmark, and Ireland were invited to join. Britain, Denmark, and Ireland accepted the invitation, but Norway turned it down. Britain's joining was significant. It was the most important extra-European trading and business country of the continent. The population of the Common Market surpasses that of the United States and is comparable to that of the Soviet Union. Table 9-8 shows that economically the Common Market compares favorably with the two great powers.

Thus, most of Europe outside of the Communist Bloc has become a single economic market—at least the more industrialized countries have. All this movement toward a single functioning economy would seem to portend some form of political unity in Europe. Framers of the original Coal and Steel Community had this in mind, and the Common Market has a single institutional framework which includes a Council of Ministers, a Parliament, and a Court of Justice. But the political and cultural divisions of Europe are so deeply rooted in history and in the psyche of its people that little evidence of progress along these lines has accompanied the development of economic integration. One of the great questions for the remainder of the twentieth century is whether or not Europe will integrate politically.

SELECTED REFERENCES

East, W. Gordon. *An Historical Geography of Europe,* 5th ed., E. P. Dutton, New York, 1966. An English historical geographer tries to place in perspective the changing structure of European life.

Gottmann, Jean. *A Geography of Europe,* 4th ed., Holt, Rinehart and Winston, New York, 1969. A basis regional geography with insights into ideas as well as information on Europe and its individual countries.

Parker, Geoffrey. *An Economic Geography of the Common Market,* Praeger, New York, 1969. An in-depth probe into this new dimension to Europe's economy from a geographic perspective.

Reynolds, Robert L. *Europe Emerges,* U. of Wisconsin Press, Madison, Wisc., 1967. An historian presents the antecedants to the Industrial Revolution.

Servan-Schreiber, J. J. *The American Challenge,* Atheneum, New York, 1967. A French journalist-politician sees the invasion of large American corporations into the Common Market countries as a challenge to the future of Europe.

Chapter 10 The Soviet Union: A Highly Centralized Interconnected System

I N TERMS of the essentials for success, the Soviet Union, the third example of the modern, long-range, interconnected system, would seem to have all the ingredients. A single, highly centralized government has complete control over the economic, cultural, and political development of the world's largest country—a country about the size of all North America and the islands of the West Indies. Such a huge area is likely to have the natural resources necessary for the functioning of a modern, interconnected system.

Nevertheless, the Soviet Union has not yet achieved the material prosperity of the United States and most European countries. With more people than the United States, its gross national product is less than one-third as large.

Per capita GNP is only about half that of West Germany's or Britain's and is even lower than Italy's. As a result, the Soviet Union's position as the second great political and economic power in the world rests more on the size of the country's economy than on the intensiveness of its industrial development. If the per capita GNP increases to the level of West Germany's, the Soviet Union's total GNP will be almost twice its present level.

Self-contained Variation of the United States Model

The Soviet Union more closely resembles the United States model of the interconnected system than it does the models of European countries. Its environmental management problem is much the same as that of the United States and Canada: how to develop the resources of a continental-sized territory for the benefit of a population relatively large by world standards but relatively sparse in terms of density. In fact, the population densities of the Soviet Union are

Table 10-1 INTERNATIONAL TRADE, 1970

	TOTAL TRADE ($ millions)	PER CAPITA* ($)	TRADE WITH USSR ($ millions)	PER CAPITA ($)
European countries outside the Iron Curtain	271,450	1,116	2,170	8.9
United States	83,180	407	105	0.5
Japan	38,200	371	270	2.6
USSR	23,463	97		
Trade with other Communist Bloc countries	16,530	68		

*Population data from *Goode's World Atlas,* 13th ed., revised printing, Rand McNally, Chicago, 1971, p. 189.

SOURCE: *Encyclopaedia Britannica, Book of the Year: 1971,* Encyclopaedia Britannica, Chicago, 1971, pp. 728–729; *Book of the Year: 1972,* p. 688.

almost the same as that of the area of the United States and Canada combined. Each has about 33 persons per square mile.

More than any other major industrial country of the world, the Soviet Union is dependent on the resources of its own territories. The major capitalist countries have not looked with favor on the Soviet communist political and economic experiment. The experiment has been avowedly anticapitalistic and threatened an onset of world revolution against private capitalism. Moreover, the Soviets have tried to carry out their method of governing and managing the country's economy in isolation from the contamination they saw in capitalism. For these reasons Soviet international trade totals less than one-third that of the United States. As Table 10-1 shows, in 1968 over half of Soviet trade was with other countries within the Communist Bloc in Eastern Europe and with China, North Korea, Mongolia, and North Vietnam.

A Vast Inheritance of Past Empire

The huge size of the Soviet Union is difficult to comprehend. It covers 15 out of every 100 units of land area on earth and is the only large country that is an important part of two continents: Europe and Asia. It stretches almost halfway around the earth from 20° east longitude to 170° west. A person can take a train from Leningrad on the Baltic coast in the west to Vladivostok on the Pacific Ocean in the east. That trip, across the longest dimension of the country, is similar in distance to traveling on a train from San Francisco straight to London, England. The north-south dimension is not as great as that of North America but at its maximum would be like a trip from Memphis northward to the northernmost island off the coast of mainland Canada (Figure 10-1).

Although the Soviets have reoccupied land that had been lost in World War II, the vast

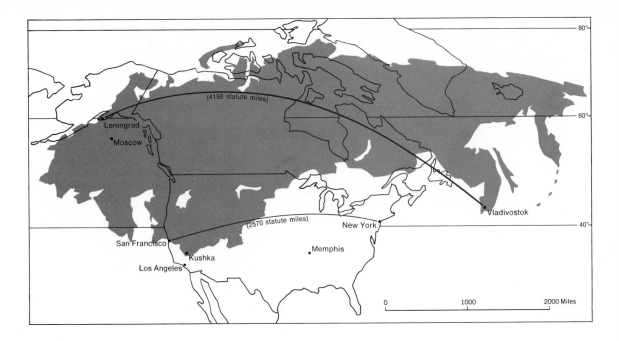

Figure 10-1 THE SOVIET UNION AND THE UNITED STATES COMPARED IN AREA AND LATITUDE

expanse of national territory is not the creation of an aggressive outreach by the Soviet Union. Rather it is the inheritance of the empire of the Russian czars, who, at the same time that the Western European countries were developing their sea empires, pushed a land frontier and European culture eastward from the area of Moscow-Leningrad to the Pacific Ocean and beyond to Alaska and the coast of northern California.

CULTURE: A VARIETY OF PEOPLES

As the sea empires did, the czars brought peoples of very different cultural backgrounds under their control. The czars, with their home base near the Baltic in the northwest, were Great Russians, a Slavic people. Slavic people are found over much of eastern Europe from

Poland, Czechoslovakia, and Yugoslavia eastward and are distinguished by a particular language family. All Russian Slavs are descendants of an early group that occupied land near the present city of Kiev in the Ukraine, but, through time, they diffused to the north, west, and south.

The Great Russians, the northern group later, pushed south and southwestward to occupy lands of two other Slavic peoples, the Byelorussians who lived east of Poland and the Ukrainians who lived north of the Black Sea. In the same way they brought in non-Slavic Europeans along the Baltic Sea and on the Rumanian and Finnish borders.

Movement to the southeast and east brought in non-Europeans: the peoples of the Caucasus Mountains and central Asia, who in language, religion, and economic practices were Middle

Table 10-2 ETHNIC COMPOSITION OF THE
SOVIET UNION, 1970

ETHNIC GROUP	PER CENT	
Slavs	74	
Russians		53
Ukrainians		17
Byelorussians		4
Other Europeans	7	
Jews		1
Baltic peoples		2
Moldavians		1
Armenians		2
Iranians	1	
Turkic peoples	13	
Caucasian peoples	2	
Uralians	2	
Mongols and other primitives	less than 0.5	

SOURCE: For Tables 10-2, 10-3, and 10-5: *Translations on the
USSR: Political and Sociological Affairs*, No. 143, Joint Publications Research Service #53061, July 1971; nondepository
entry 11311–29.

Eastern. Mostly Moslem rather than Christian,
they were nomadic herders and farmers experienced in irrigation agriculture. To the east other
Turkic peoples had spread over the flatlands of
southwestern Siberia; then various scattered
Mongol tribes, some very like the American
Indian in their primitive ways; and even Chinese.

This ethnic variety can still be found in the
population of the Soviet Union (Table 10-2).
The dominance of the Slavic people, and particularly the Great Russians, persists even today.
Moreover, when other European groups are
added, the total of Europeans in the population
is over four out of five.

Soviet Reorientation of the Management of Empire

While holding to the boundaries of the Czarist
Empire, the Soviets markedly changed the concept of management of the territory. The Empire had been run in the fashion of a traditional
early European empire, strictly for the benefit
of the ruling power—the Slavs of the Moscow
area and the European northwest. The resources
of outlying regions were the property of the
Crown and thus were exploited for shipment
to the center of power in the area of Moscow
and Leningrad (then St. Petersburg). Moreover,
the non-Slavic peoples were considered subjugated, and their ways were at best tolerated.
The Russian language and Russian customs
were the instruments of political and economic
power in all parts of the Empire, with Russian
officials in charge and Russian culture superimposed on native groups. As was the case in
most European empires, the native peoples
were expected to learn the official language—
Russian—if they hoped to function in the economic system on a manpower level above the
brute.

POLITICAL ORGANIZATION

When they ascended to power, the Soviets
were faced with much the same problem that
confronts many underdeveloped countries
throughout the world today: how to build a
functioning modern state with a population
divided among a great many minority cultures.
Empires traditionally subjugate minorities, particularly those geographically isolated in different parts of the country, but the doctrine of the
Communist Party called for the equality of peoples and an international communist movement.

The Soviets replaced the Empire with a political organization whose name portended

Figure 10-2 POLITICAL MAP OF THE SOVIET UNION

<u>Moscow</u> National capital

<u>Kiev</u> Capitals of Soviet Socialist Republics

• Metropolitan areas over 1 million

1: Lithuanian S.S.R.   2: Latvian S.S.R.   3: Estonian S.S.R.
4: Moldavian S.S.R.   5: Georgian S.S.R.   6: Armenian S.S.R.
7: Azerbaidzhanian S.S.R.

Table 10-3 MEMBER REPUBLICS OF THE SOVIET UNION: THEIR SIZE, POPULATION, AND LEADING ETHNIC GROUPS, 1970

| | LAND AREA* (per cent of national total) | POPULATION | PER CENT OF POPULATION THAT IS | | |
			NAMESAKE ETHNIC GROUP	RUSSIAN	SLAVIC
Russian SFSR	76.7†	54	83	83	86
European republics					
Ukraine SSR	2.7	20	75	19	96
Byelorussia SSR	0.9	4	81	10	94
Estonia SSR	0.2	1	68	25	28
Latvia SSR	0.3	1	57	30	39
Lithuania SSR	0.3	1	80	9	19
Moldavia SSR	0.2	1	65	12	28
Caucasian republics					
Georgia SSR	0.4	2	67	9	10
Armenia SSR	0.1	1	89	3	3
Azerbaijan SSR	0.4	2	74	10	10
Central Asian republics					
Kazakhstan SSR	12.4	5	32	43	52
Turkmen SSR	2.1	1	66	15	16
Tadzhik SSR	0.6	1	56	12	13
Uzbek SSR	1.8	5	65	13	14
Kirghiz SSR	0.9	1	44	29	34

*Data from tables in each regional chapter of Cole and German, *A Geography of the U.S.S.R.*, Butterworth & Company, Ltd., London, 1961.   †Includes Western and Eastern Siberia and the Far East.

change. In 1923 the Russian Empire became the Union of Soviet Socialist Republics with the "union republics" designated along ethnic lines. Since ethnic groups had been sharply localized in particular parts of the country, it was comparatively easy to draw the boundaries of the republics with regard to populations of the major groups (Figure 10-2). There are now fifteen Soviet Socialist Republics, six named after European peoples, three after people living in the Caucasus Mountains between the Black and Caspian seas, and five after peoples of central Asia. Note that the Great Russian Republic occupies 76 per cent of the land area and contains 54 per cent of the population. It stretches from the Baltic to the Pacific and from the eastern shore of the Black Sea to the Arctic Ocean. It therefore encompasses many of the smaller ethnic groups. Although it is named for the Russians who dominate its population, the presence of other ethnic groups is recognized by calling it the Russian Soviet *Federated* Socialist Republic (RSFSR).

Notice in Table 10-3 how minor most of the

member republics are in relation to the total population of the country. Eight of the fifteen republics have only 1 per cent of the population each. Only the RSFSR and the Ukraine have over 5 per cent. From the table we can see how local ethnic groups dominate each member republic. Only in the Kazakh and Kirghiz republics of central Asia, where new agricultural settlements have been introduced in formerly sparsely settled areas by largely Slavic settlers, does the ethnic group for which the territory is named constitute a minority of the population. But in almost all republics, Slavs, and particularly Russians, constitute a significant minority group within the population. Although some of these people settled in the territories under the czars, most of them have migrated since the Revolution. This, too, reflects the problem of the development of a land with minorities having different traditions and a different technology. The modern state of the Soviet Union has been built on a complete life style involving the economics and social structure that evolved within one particular ethnic group, the Slavs.

Each ethnic group has had to be taught the Soviet life style, and its economy has had to be developed to fit the modern, interconnected system controlled from Moscow. This has meant that there had to be cadres of trained leaders. Since trained leadership was scarce among ethnic groups which had been subjugated by the Czarist Empire, Slavs were sent to the member republics to assume positions of leadership. Today young non-Slavic people are trained and educated in schools in the republics and in Moscow, but Slavs are still in key political and economic positions in member republics. Moreover, the small populations of the ethnic minorities mean that there are not masses of people to draw on for new development schemes such as the agricultural settle-

ment of the "virgin lands" at Karaganda or new mining-industrial complexes in non-Slavic areas. In these programs the Slavs dominate, thus adding to the nonnative minority of the republic.

Soviet policy has encouraged the ethnic groups to develop their local cultures. Groups have been urged to continue national costumes at festivals and holidays. They have developed written languages where there were none and written ethnic histories. But all this has been done within the strict limitations of the doctrines of national government and the Communist Party. Thus, a local language has become a device for communicating communist ideas to minority peoples, and their local history is written to emphasize ethnic bonds to communist doctrine. Ethnic identity is confined within the narrow constraints of Soviet perspective of the proper values of life. At the same time, however, they have maintained strong centralized political and economic power in Moscow and retained Russian as the economic language in all parts of the country.

The ostensibly democratic constitution of the Soviet Union allows member republics to secede and maintain independent control over agriculture, law, education, and public health. Although the Ukrainian and the Byelorussian republics are in the United Nations, the law of the central government prevails over the law of a union republic, and most republican governments have little autonomy. It is not surprising that Moscow, the national capital, is identified as both the medieval center of the Russian state and the capital of the huge RSFSR.

Smaller ethnic groups are recognized by "Autonomous Republics," most of them subdivisions within the outlying areas of the RSFSR. They are represented in one of the two national governing bodies, "the Soviet of the Nationalities." Moreover, much is made of the local elected soviets—as the popularly elected councils are called—at regional, city, town, and other community levels. Representatives from these soviets attend the Congress of Soviets held every two years; they also maintain law and order at the local level, and carry out policies of the national government.

Centralized Economic Control

Among modern industrial countries the distinctive feature of the Soviet Union is the virtually complete control maintained over the national economy. As soon as they were able to consolidate power after the Revolution, the Soviets rejected the systems of private enterprise and capitalism and the idea that supply and demand should control the production and sale of goods and services. Instead, the government took charge of economic production and concentrated almost all decision-making within the central government. From that point on, it set the economic course for the country and established growth priorities. In the Soviet Union the government operates as landlord, banker, traffic manager, and chairman of the board for the whole country.

The national interest, as determined by the government, has basically replaced individual and corporate interests as a basis for decisions. The economy functions on a national plan which determines what is to be produced and in what quantities. Then the state allocates resources to enable the components of the plan to be carried out. Central planners working under priorities approved by the central government establish production quotas for all types of goods and services in the country. These national totals are then broken down into subquotas for each region and each producing unit—factory, mine, and farm. Each unit calculates what supplies and materials will be needed to fulfill its obligations and sends the report to the central planning agency which allocates the necessary material as approved.

Prices, too, are fixed centrally in terms of what is considered best for the state, rather than in terms of the cost of production. Thus, consumer items in short supply, such as automobiles, may be deliberately priced out of the reach of most people, while buses, considered more important to the country's economy, may carry a low price when sold to governmental transport agencies.

One might think that in such a controlled economy the goal of the plan would be to have each producing unit meet its quota exactly. In that way, all the economic parts would mesh. But, surprisingly, producing units and individual workers are given incentives not only to meet quotas but also to exceed them. Scorn and even punishment go to units and individuals who do not meet their quotas, and special awards and recognition go to those who surpass them. Presumably, if all units surpassed their quotas, the plan would operate on a level higher than can reasonably be expected, which would benefit the state. In some industries, however, products are dependent on the manufacture of a variety of different parts. If, for example, those who make engines for buses surpass their quotas, but those who make the tires do not, the total number of buses will remain limited. This is not true of food quotas or quotas on small items, such as shirts, that are made in one factory.

Emphasis has usually been on quantity rather than quality of output. Annual production quotas are part of larger, long-range plans, usually covering five-year intervals. These plans establish priorities and determine what to produce and even where to build new factories and other units needed to produce the desired output.

In recent years there has been some loosening

FREIGHT TRAFFIC: A REGIONAL ANALYSIS

We have talked about regions of the Soviet Union in terms of core areas, sources of supply, and regional self-sufficiency. Table 10-4 presents basic railroad freight traffic figures with which to test the character of particular parts of the country.

Examine the individual economic regions of the country in terms of their proportions of the total trade of the Soviet Union. Do the areas of high freight traffic correspond to our core areas of the country?

In what regions is intraregional trade greater than interregional? What is the significance of such a trade pattern?

What are the major centers of interregional trade? Are they net shippers or receivers?

Are there areas of "balanced traffic" with interregional traffic, shipments, and receipts about equal in total?

What are the problems of using figures for rail freight traffic reported in tonnage? What evidences of these problems do you find in the regional pattern of freight in the Soviet Union?

Do you find major differences in the Soviet pattern as compared with what you might expect to find in the United States?

of the extreme centralization of control. Agencies in the union republics are now given more say about capital investments and work forces in their areas. Instead of placing emphasis on sheer physical output of goods, factories are being asked to operate in terms of "profits," or the value of production measured against the capital invested and the amount of sales. In this way the output of goods is somewhat responsive to demand.

By no means, however, do the slight decentralization and shift from physical quotas represent a turning away from the centralized, state-controlled economy. In examining the environmental base of the country and its development by the Soviets, we must consider them in terms of centralized planning by the state in contrast to the multifaceted economies of the United States and Europe with their many corporate and individual entrepreneurs. The central government is the landlord, entrepreneur, and decision-maker for virtually the entire economy of this huge country.

Soviet production priorities are somewhat different from those in the United States. The first order of production is to produce those things that are needed to enhance the strength of the Soviet Union as a modern industrial power. The Soviets have been pushing hard to catch up with the United States and European countries. Because the Soviets have been industrializing for less than fifty years, their new economic programs give highest priority to the military and economic needs of the state. Consumer demands are not the first concern.

On the other hand, the Soviets have made a conscious effort to eliminate as much as possible the sharp economic classes that characterized Czarist Russia when a small upper class lived in wealth while the masses of the people, either as peasants or city workers, had a very low living standard. Now goods and services in the consumer sector are broadly distributed,

Table 10-4 RAILROAD FREIGHT CIRCULATION IN THE SOVIET UNION, 1961 (millions of metric tons)

Area	Total Circulation		Interregional Circulation				Intraregional Circulation	
	Tonnage	Per Cent	Total Tonnage	Per Cent	Tonnage Originated	Tonnage Received	Total Tonnage	Per Cent
EUROPEAN SETTLEMENT								
Traditional Slavic homelands								
Northwest	176.5	6.4	96.1	6.3	50.7	45.4	80.4	6.5
Center	423.8	15.4	n.a.	n.a.	n.a.	n.a.	n.a.	n.a.
Center	258.4	9.4	139.4	9.2	48.0	81.3	119.0	9.7
Upper Volga	74.5	2.7	56.8	3.7	27.3	29.5	17.7	1.5
Black earth	90.9	3.3	63.6	4.2	23.0	40.5	27.3	2.2
Byelorussia	52.99	1.9	34.9	2.3	6.2	28.7	18.0	1.5
Ukraine–Moldavia	711.4	25.9	n.a.	n.a.	n.a.	n.a.	n.a.	n.a.
Donetsk–Dnieper	459.1	16.7	194.6	12.8	138.2	60.4	264.5	21.7
Southwest	167.3	6.1	105.7	6.9	34.3	71.4	61.6	5.1
South	69.4	2.5	58.1	3.8	22.1	36.0	11.3	0.9
Moldavia	15.6	0.6	11.0	0.7	3.4	7.7	4.6	0.3
Other European homelands								
Baltic	61.4	2.2	35.5	2.3	7.8	27.6	25.9	2.1
More recent European settlement								
Volga	154.6	5.6	114.3	7.5	70.1	44.2	40.3	3.3
Urals	382.8	14.3	193.1	12.7	101.3	91.8	189.7	15.6
Western Siberia	199.5	7.3	112.8	7.4	77.0	35.8	86.7	7.1
North Caucasus	152.0	5.5	92.8	6.1	51.2	41.6	59.2	4.9
TRADITIONAL NON-EUROPEAN SETTLEMENT								
Caucasus	63.9	2.3	23.8	1.6	8.6	15.1	40.1	3.3
Central Asia	206.0	7.6	n.a.	n.a.	n.a.	n.a.	n.a.	n.a.
Central Asia	61.8	2.3	30.8	2.0	8.6	22.2	31.0	2.5
Kazakhstan	144.2	5.3	91.7	6.0	50.1	41.6	52.5	4.4
Eastern Siberia	99.8	3.7	52.0	4.2	31.8	20.2	47.8	3.9
Far East	59.4	2.2	17.0	1.1	5.9	11.1	42.4	3.4
Total	2,743.1		1,523.1		761.6	762.2	1,220.9	

SOURCE: M. I. Galitskiy, S. K. Danilov, and A. I. Korneyev, *Ekonomicheskaya Geografiya Transporta*, S.S.S.R., Moscow, 1965.

and no one is to live in poverty. The Soviet housing program, for example, aims at mass production of living quarters, not only for middle class workers but for everyone. When measured against middle-income housing in the United States, Soviet apartments seem Spartan indeed. But the Soviets seem to have virtually eliminated slums. Medical and educational services are provided to all citizens.

Because per capita productivity is still below that of the United States and Western European countries, the amount of consumer goods also remains well below Western standards. But the workers of the Soviet Union have a much higher living standard than did their forebears under the czars. From their point of view this may be the most meaningful measure.

Government decisions concerning the development of the Soviet Union have political as well as economic bases. The Far East has been developed as a largely self-sufficient base despite the fact that it is half a world away from Moscow. Western Siberia was first developed in the late 1930s as a hedge against a feared invasion by Germany—a decision that appeared very wise when most of the European part of the country was lost to the German armies in 1941 and 1942. Lost were the country's prime agricultural base west of the Volga River, the iron and coal resources of the Ukraine, and major industrial centers. But grainfields of Siberia, its new iron and coal supply, and the recently built factories east of the Volga helped provide the base for an offensive that drove out the Germans. This area is also strategic because it is on the Chinese border not far from Korea and Japan.

At first glance, one would expect that the central planners of the Soviet Union who are in control of the whole economy could organize production very differently from the technique used in the United States. All the planners have to do is analyze the potentials of different parts of the country and then allocate capital and other resources to complete the development. In fact, the constraints in Soviet decision-making are almost the same as those of any large corporation making locational decisions. First, they must take into account the sum of investments from the past that are still present in the economy. Functional or not, they form the base upon which new investment decisions must be made. Second, although political and other noneconomic factors can be considered in setting planning priorities, only so much capital is available to the state, and it must be invested wisely. New developments out in the wilderness must be built from scratch, and this calls for investment not only in the vital plant needed but also in transportation links with established sources of supply and markets. Whole towns for workers and their families must be created because workers must come from other areas. As in the United States, it is more expensive to develop remote resources, rich as they may be, than to use those in or near already established centers.

The vast size of the Soviet Union presents particular problems to developmental planning. With the Arctic Ocean and north-flowing rivers frozen most of the year and outlets to the Black and Baltic seas opening only into roundabout international waterways that depend on passage through narrows controlled by other countries, the major burden of tying the huge land area of the country together rests with overland transportation or air service.

There is, then, greater pressure on the Soviet land transport system than there is on the systems of other countries. In order to lessen this burden, the Soviets have had to modify the traditional modern, interconnected system that calls for each region to specialize in the things that it can produce best, while, in return, depending on other regions for its basic needs. In the Soviet Union each region is encouraged to produce the specialties the country needs but, at the same time, is asked to supply as many of its own needs from its own local resources as possible. Each region is to have its own agricultural food base and its own essential industrial economy so as to reduce the need for transportation hauls from other parts of the country.

The situation is somewhat like the economic development in California, the area of the United States most remote from the productive Northeast and Middle West. Faced with the high prices of goods shipped from the East, California has developed a remarkably well rounded economy but still ships its specialties to the rest of the country. But all of central Asia has fewer people than California, and the Far East has less than a fifth as many. Thus, it is questionable whether they have sufficient markets to support such major investments as steel mills, oil refineries, and chemical plants.

The Soviet experiment has not been carved from a wilderness; rather it has been superimposed on a long established base that developed in place over thousands of years. New agricultural, mining, and urban areas have been developed, but the most important of these regions today are, in large part, those inherited from the czars. The major agricultural areas remain the Ukraine and the lands south of Moscow in the RSFSR. The most productive iron and steel complex is still that established at the turn of the century by tying together coal of the Don River basin and iron-ore near the Dnieper River in southern European Russia.

THE EUROPEAN MAJORITY

The distribution of population in the Soviet Union largely reflects the European heritage. Over three-fourths of the people live in European Russia, and, as Tables 10-5 and 10-6 show, almost half live in the historic lands of the Great Russians, Byelorussians, and Ukrainians. In fact, as Figure 10-3 indicates, most of the population lies within a great triangular zone with its base along the boundary of eastern Europe from the Baltic Sea to the Black Sea and stretching eastward into western Siberia to the city of Irkutsk on the eastern shores of Lake Baikal. This is the land of the European peoples of Russia, their homelands in the west and their zone of eastward migration after the empire had been established. If the area between the Black and Caspian seas north of the Caucasus Mountains and the "virgin lands" area of north Kazakh are included, this zone of primarily European settlement includes 76 per cent of the population of the Soviet Union.

In contrast, the areas of the traditional non-European peoples who were encompassed by the Czarist Empire are of minor importance. As Table 10-6 shows, south of the triangle the Caucasus area accounts for only 5 per cent of total population; central Asia, for 14 per cent; and the vast area of eastern Siberia and the Far East, for about 5 per cent.

The contrast between the land area of primarily European settlement and the traditionally non-European areas is striking. European areas account for only slightly more than one-third of the land area of the country, but four out of every five people live there.

VARIATIONS IN DENSITY AND THE
LOCALIZATION OF POPULATION

Population densities are greatest in the traditional Slavic homelands (see Table A-5). Not only do these areas contain almost half the population of the country, but each, except the extreme Northwest, has a density of more than 100 persons per square mile. The only other area in the country with such a density is the Caucasus, a non-European area. It is important to keep in mind that these densities are much

Table 10-5 POPULATION DISTRIBUTION OF THE SOVIET UNION, 1970

AREA	TOTAL POPULATION (000)	NUMBER OF METROS OF 100,000+	POPULATION IN METROS OF 100,000+ (000)	NON-METRO POPULATION	PER CENT OF POPULATION OF AREA IN METROS
EUROPEAN SETTLEMENT					
Traditional Slavic homelands					
Northwest	12,160	10	5,679	6,481	47
Center	43,998	36	16,198	27,800	37
Byelorussian SSR	9,003	9	2,216	6,787	25
Ukrainian-Moldavian SSRs	50,708	44	13,768	36,940	27
Subtotal	115,869	99	37,861	78,008	33
Other European homelands					
Baltic	7,583	7	2,302	5,281	30
More recent European settlement					
Volga	18,377	16	6,604	11,773	36
Urals	15,184	16	5,728	9,456	37
Western Siberia	12,110	14	4,991	7,119	41
North Caucasus	14,285	15	3,632	10,653	15
Subtotal	59,956	61	20,955	39,001	35
Total	183,408	167	61,118	122,290	33
TRADITIONAL NON-EUROPEAN SETTLEMENT					
Caucasus	12,292	10	3,843	8,449	31
Central Asia	32,804	27	7,096	25,708	22
Eastern Siberia	7,464	8	2,402	5,062	32
Far East	5,760	9	1,718	4,062	30
Total	58,340	54	15,059	43,281	26
Grand total	241,748	221	76,177	165,571	32

less than half those of Western Europe. They are more like those in the Middle West of the United States.

Most of the areas of more recent European settlement in the Soviet Union have less than half the population densities of the traditional Slavic lands. Only the North Caucasus region

with eighty-five people per square mile approaches the crowding of the older, established centers (Table A-5).

Except for the Caucasus, the traditionally non-European areas are the most sparsely settled of all. This is particularly true of the vast regions of the north and east which include over

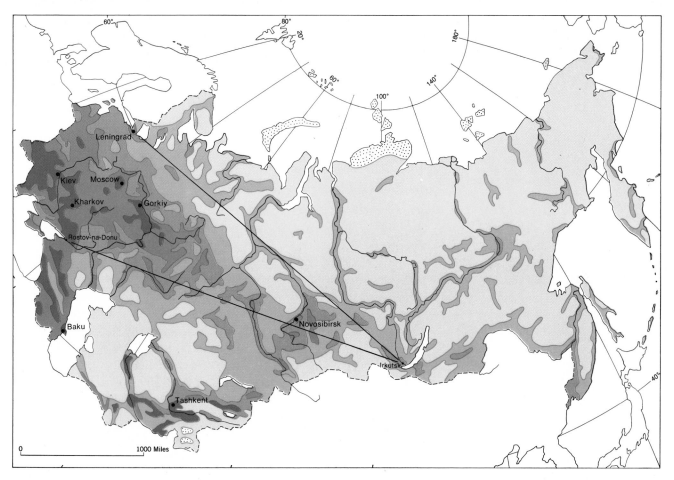

SOURCE: Atlas Naradov Mira 1964

Figure 10-3 POPULATION DENSITY OF THE SO-
VIET UNION. *Note the "Population Triangle" in
Eastern Europe.*

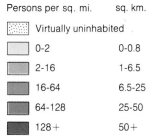

Persons per sq. mi.	sq. km.
Virtually uninhabited	
0-2	0-0.8
2-16	1-6.5
16-64	6.5-25
64-128	25-50
128+	50+

• Cities over 1,000,000 population

half of the total land area of the country. There densities are around five persons per square mile or less, like those in Montana (see Table A-5).

In area the traditional centers of European Russia occupy a territory about equal to that part of the United States east of the Mississippi River and have a population slightly less. The full land area encompassed by the population triangle is almost nine-tenths that of the United States. Thus, the Soviet Union's major functioning areas are, in fact, quite comparable to the United States.

In the United States and Europe we found that the distribution of total population was the sum of two distinctive population patterns: the non-metropolitan population largely tied to the country's primary resource production and the population of large metropolitan centers mainly engaged in urban-centered activities. In both the United States and Europe the populations had very separate patterns; some areas were dominantly metropolitan, others largely non-metropolitan.

Although the same situation exists to a limited extent in the Soviet Union, the variations in the two kinds of populations are much less marked (Tables 10-5 and 10-6), perhaps because agricultural employment is much higher there. The last Soviet census reported that 35 per cent of the labor force was employed in agriculture. Western European countries have up to 25 per cent of their labor force in agriculture; in the United States the figure is less than 10 per cent. At the same time, the populations of cities of 100,000 or more in the Soviet Union account for only 32 per cent of the total. Even considering differences in definition of *metropolitan* and *city*, there is no doubt that the Soviet Union is far less metropolitan in its population distribution than is the United States or Europe.

Metropolitan Population

REGIONAL VARIATIONS

With the national percentage of people in metropolitan centers totaling 32 per cent, only a few parts of the country vary far from that average. Northwest and Western Siberia are more highly metropolitan, and Byelorussia and the North Caucasus are less metropolitan. Still, a pattern of metropolitan importance can be seen by grouping the regions by the relative importance of their population that is metropolitan.

It can be noted in Table 10-7 that the more metropolitan areas form an east-west extension across the zone of heavy population from the old Russian centers of the northwest and center through the Volga to the Urals and western Siberia. These five regions of higher than average metropolitan population account for a higher share of metropolitan population than of total population. At the other extreme, the more rural areas lie to the south of the metropolitan zone except for lightly populated areas of the north and, again form a zone from Byelorussia and the Ukraine to the southern boundaries of the country. The areas close to the national average are generally both peripheral and low in population.

It should be remembered that while the percentage of population in metropolitan areas in the Ukraine-Moldavian region is below the national average, this region still has the second largest metropolitan population in the country. In fact, as we saw from Table 10-6, the Center and Ukraine-Moldavia together have 39 per cent of the metropolitan population.

The Soviet Union has far fewer large metropolitan centers than the United States, even though the former has a larger total population. In the United States there are thirty metropolitan centers with a million or more people;

Table 10-6 Distribution of Metropolitan and Non-metropolitan Population in the Soviet Union, 1970

Area	Non-metropolitan (per cent)	Metropolitan (Metros of 100,000+) (per cent)	Total Population (per cent)
EUROPEAN SETTLEMENT			
Traditional Slavic homelands			
Northwest	4	7	5
Center	17	21	18
Byelorussian SSR	4	3	4
Ukrainian-Moldavian SSRs	22	18	21
Subtotal	47	49	48
Other European homelands			
Baltic	3	3	3
More recent European settlement			
Volga	7	9	8
Urals	6	8	6
Western Siberia	4	7	5
North Caucasus	6	5	6
Subtotal	23	29	25
Total	73	81	76
TRADITIONAL NON-EUROPEAN SETTLEMENT			
Caucasus	5	5	5
Central Asia	16	9	14
Eastern Siberia	3	3	3
Far East	2	2	2
Total	26	19	24

SOURCE: Derived from Table 10-5.

in the Soviet Union there are only 12. Moscow, with 8 million people, is the largest metropolitan center and the major center of political, economic, and cultural decision-making, but it has only slightly more than half the population of the greater New York metropolitan area. Moreover, the United States has 126 metropolitan areas with 250,000 people or more, while the Soviet Union has only 72.

CITIES TODAY AND CZARIST URBANISM

To outsiders unfamiliar with the Soviet Union, it would seem that the planners could simply

Table 10-7 Variations in Metropolitanism by Areas in the Soviet Union, 1970

More Highly Metropolitan: 35 Per Cent and Above		Average Metropolitan: 27–35 Per Cent		Low Metropolitan: Below 27 Per Cent	
Northwest	47	Eastern Siberia	32	Byelorussia	25
Western Siberia	41	Caucasus	31	Central Asia	22
Center	37	Baltic	30	North Caucasus	15
Urals	37	Far East	30		
Volga	36	Ukraine–Moldavia	27		

Source: Derived from Table 10-6.

create an urban development wherever it is needed. But, like any major investment, the establishment of a new development requires a tremendous amount of capital. Not only must the structure of an entire local community and its industrial-commercial base be built, but also essential transportation and communication links to the outside world must be provided. Despite the fact that the population of the Soviet Union has doubled since the Revolution and the country has become primarily an urban nation, there have been few large new cities established from scratch. In fact, only six of the seventy-two cities with populations of 250,000 or more were not settled in 1920. Four of them were the result of the further expansion of new mineral and heavy industrial developments in the Ukraine and the Urals. Only two were new enterprises in relatively remote areas which have emerged since the Revolution—Karaganda, site of a new coal complex in northern Kazakh, and Dushanbe, the capital of a new republic.

Communities have been created either at new resource locations or at particularly strategic sites. However, only eight new cities having populations of over 100,000 have appeared out-side of the population triangle since 1920 (see Figure 10-4). Of these, only three represent developments in areas that are remote from existing cities. These cities in Siberia show that if development is important enough, cities in the wilderness can be established. Yet, new cities in out-of-the-way places are the exception, not the rule. They are not well tied into the Soviet transport system; they have no rail connections with the rest of the country and must be served by airplane and road.

The Hierarchy of Large Metropolitan Places

In contrast to the United States, the hierarchy of major cities in the Soviet Union is simple. Since political, economic, and cultural decision-making are all combined under state control, each of the most important cities functions as both an economic and a political administrative center. Only one of the thirty-one cities in the Soviet Union in 1967 with 500,000 or more people (Figure 10-5) was not the center of its own administrative unit. The only large cities that are not administrative centers are those in mining and industrial districts where urban developments are close to one another.

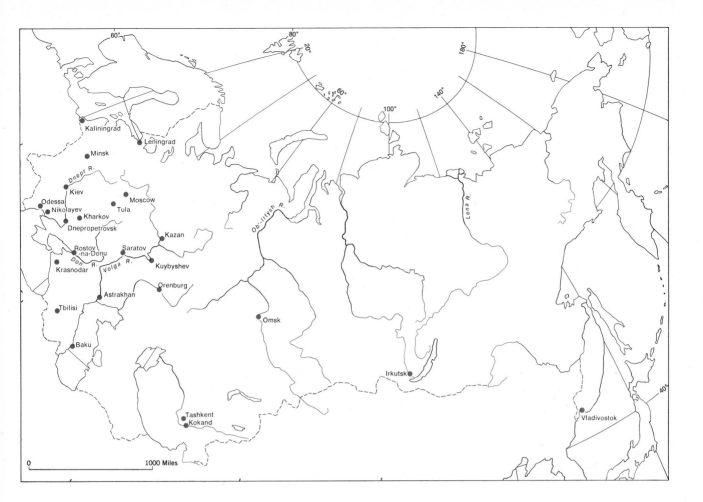

Figure 10-4 CITIES OVER 100,000 IN 1920

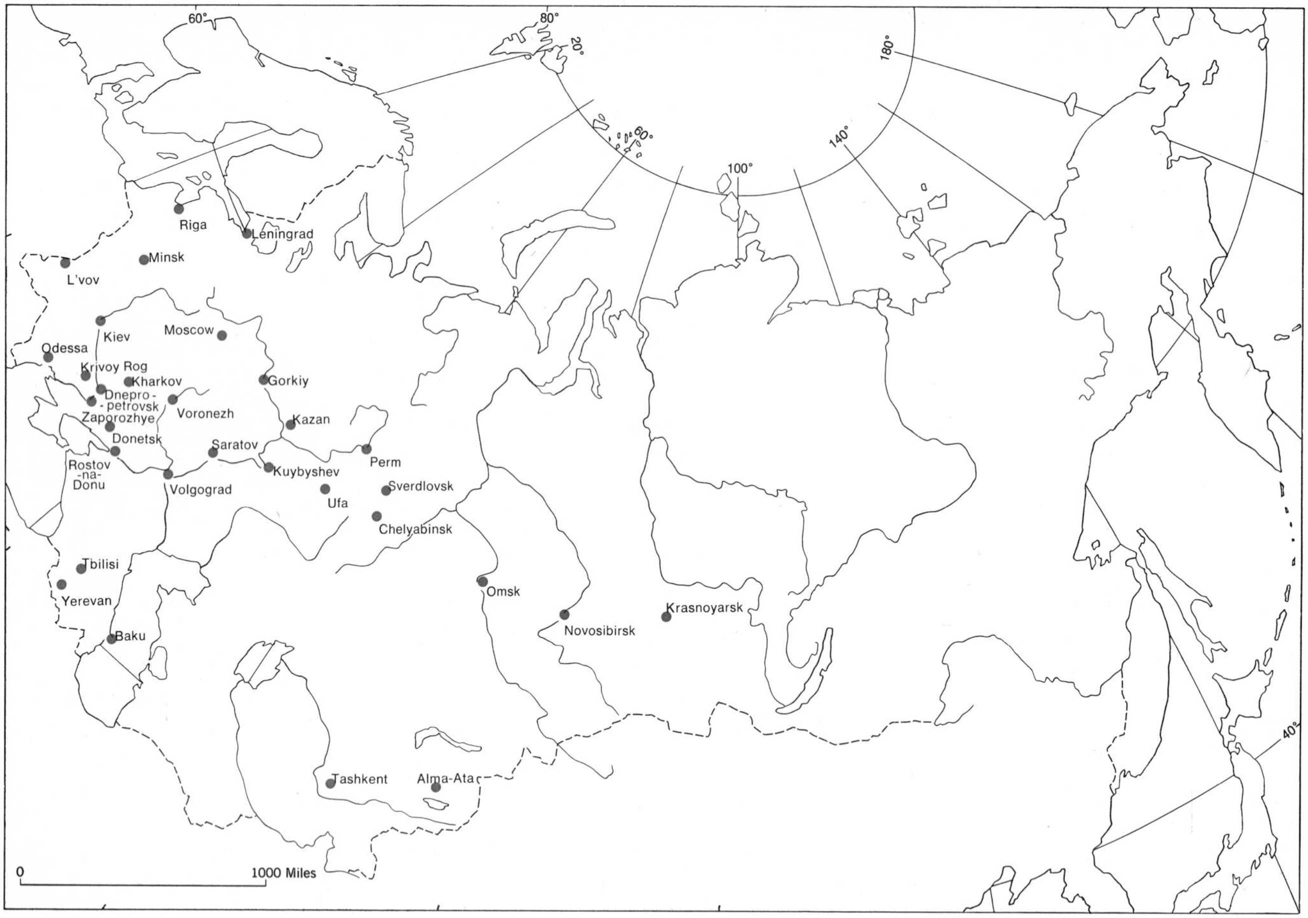

60°
80°
20°
180°
60°
140°
100°
40°
Riga
Leningrad
Minsk
L'vov
Kiev
Moscow
Odessa
Krivoy Rog
Kharkov
Gorkiy
Dnepro-
-petrovsk
Voronezh
Zaporozhye
Kazan
Donetsk
Saratov
Perm
Rostov
-na-
Donu
Kuybyshev
Volgograd
Sverdlovsk
Ufa
Chelyabinsk
Tbilisi
Yerevan
Omsk
Krasnoyarsk
Novosibirsk
Baku
Tashkent
Alma-Ata
0
1000 Miles

Figure 10-5 Cities over 500,000 in 1967

The most important of these are the areas of coal mining and metalworking in the eastern Ukraine and western Siberia and the iron-mining and industrial areas of the Urals.

The hierarchy follows economic lines more closely than political ones. Six of the twelve cities of over one million population are old established centers in the RSFSR, and four are in the Ukraine. The two remaining are capitals of non-European republics, Baku and Tashkent. But Baku, like Houston, is the center of a major oil field, and Tashkent is the leading industrial-commercial center of central Asia. Only one of the fifteen capitals has less than 250,000 people.

Non-metropolitan Population

If the non-metropolitan population can be taken as a measure of the people most closely tied to primary production and hence the resource base of the country, the resources must be either unevenly distributed or unevenly developed. Almost half of the non-metropolitan population resides in the traditional homelands of the Slavs which is only 9 per cent of the country's land area. On the other hand, the north, eastern Siberia, and the Far East with 52 per cent of the land area have only 7 per cent of the population. Almost four out of five of the non-metropolitan people of the Soviet Union are concentrated in the population triangle.

Primary Production and the Environmental Base

Since the emptiness of an area can usually be taken as a prime indicator of its lack of agricultural possibilities, it would seem that most of the Soviet Union has poor agricultural re-

Table 10-8 Distribution of Sown Land in the Soviet Union, 1967

	Per Cent of Total	
	Sown Area	Land Area
The North and East		
North	2	13
Eastern Siberia	1	25
Far East	1	14
	4	52
The South		
Caucasus	1	1
Central Asia (including Kazakhstan)	9	18
	10	19
Settled European areas		
Northwest	1	1
Center	18	4
Byelorussian SSR	3	1
Ukraine–Moldavia	20	3
Baltic	3	1
Volga	10	2
North Caucasus	8	2
	63	14
Urals–Western Siberia		
Urals	9	3
Western Siberia	11	5
Krasnoyarsk–Irkutsk	3	7
	23	15

Source: *Narodnoe Schziaistvo SSSR*, 1967.

sources. Table 10-8 shows the lands sown to crops and confirms that these areas are little used for crop agriculture. Only 4 per cent of the country's cropland is in the north and east of the country, which total 52 per cent of the land area. Only 10 per cent of the sown area is in the south, which accounts for 19 per cent of the country's area.

These uncultivated areas are hardly the result of a lack of interest in agricultural potential. The Soviet Union has been struggling to increase agricultural production since the Revolution and has experimented with agricultural colonies in all areas where scientists and agricultural specialists see possibilities for farming.

FAVORABLE TERRAIN; DEAD-END DRAINAGE

The problem of the largely nonagricultural regions is not primarily one of terrain (Figure 10-6). Notice that the major mountain systems in the Soviet Union form a rim around its southern and eastern peripheries. Only in the Far East do they actually extend inland any great distance from the border. The Ural Mountains, forming the traditional border between European and Asiatic Russia, extend north and south across the great European–west Siberian plain, but they are low like the Appalachians and have presented less of a barrier to movement than have the American mountains.

Because of the mountain rim to the south, most rivers of the Soviet Union drain northward into the Arctic Ocean. The Volga, the most famous of Russian rivers, flows southward into the Caspian Sea, with no outlet to the oceans; the north-flowing rivers of Siberia are all larger, about the size of the Mississippi system, but the north-flowing streams indicate one of the problems of Russia's northland: poor drainage. There rivers flow from areas with longer growing seasons into those with less. In the spring, melting snows occur in the upper reaches while the mouths and lower channels are still icebound. Particularly in western Siberia is tremendous spring flooding with very poor drainage. The situation is typical of the northern lands of Canada too.

THE SEVERELY LIMITED NORTH AND EAST

More serious than the drainage problem are the relatively small quantity of sun energy for evapo-transpiration and the severe temperature in winter. Along the northern coast and in most of eastern Siberia and the Far East, the growing season is less than ninety days, which precludes mid-latitude crops. In addition, the land generally north of 60° latitude has an annual potential evapo-transpiration of less than half that of the American Middle West (Figure 5-4). Even though most of this potential is seasonally concentrated in the summer months when daylight is almost continuous, it is still generally inadequate for growing anything but grasses and root crops such as beets.

The combination of the short, cool summer and the long, severely cold winter results in permanently frozen subsoil called "permafrost." The Soviet Union has more area in permafrost than any other country. It is almost ever-present in Siberia north of 65° latitude and is found in patches as far south as the Chinese border. Permafrost impedes soil drainage in summer and stunts tree roots. It makes mining and any type of excavating, even plowing, difficult and expensive. The Soviets have experimented more on how to carry out agriculture and engineering in permafrost areas than any other country has, yet the problem remains a major one.

In the north, even where there is no permafrost, soils are low in organic matter. Thus, all but the most southern areas of eastern Siberia have severely limited agricultural potential under today's technology. There are forests, but this tremendous resource is difficult to develop because of the expense required in providing transportation and communication deep into almost empty wilderness. The woodlands of the north form the largest existing forest area of any country in the world, but they are exploited in

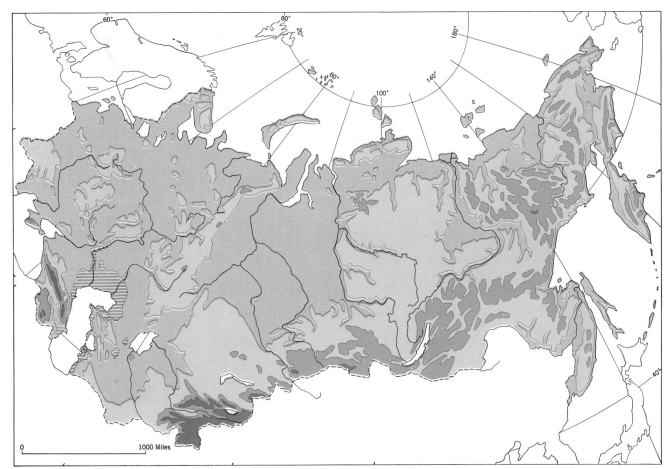

Figure 10-6 LANDFORMS OF THE SOVIET UNION

Below sea level
0-650 feet
650-3,000 feet
3,000-10,000 feet
Above 10,000 feet

only a relatively few places. It is only the northern forests of Europe relatively close to population centers that are significantly developed.

REMOTE INDUSTRIAL RESOURCES

The same story holds for mineral resources (Figure 10-7). Since the empty lands of the north account for half of the country's land area, they can be expected to include much of the potential mineral wealth. But, once more, these resources have been tapped in only a relatively few places. The north has vast resources of coal, petroleum, and iron; however, most remain virtually untouched, while other, often less rich, deposits close to population centers are utilized. Most of the mineral development of the north has taken place either close to the main populous regions of the country or at remote sites where particularly strategic minerals such as gold are found. These mining sites can usually be served by air, and the tremendous expense of developing roads and railroads through the wilderness can be avoided. There are virtually no railroads in all of eastern Siberia and the Far East other than the Trans-Siberian which follows close to the Chinese border.

The north also has the major waterpower resources of the Soviet Union, but these are tapped only along the southern margin close to the population triangle. The largest hydroelectric stations are at the eastern apex of the population triangle, where the Soviets have put tremendous effort into developing what was largely a forest wilderness. This area is the site of major power-using industries, particularly aluminum, atomic power, and chemical plants.

THE DRYNESS OF CENTRAL ASIA

South of the population triangle the Soviets face another climatic problem: the dryness associated with the region's location in the heart of the world's largest landmass. Walled off from the south by mountains, the major source of moisture is the westerly winds bringing air from the remote Atlantic. Winters in central Asia are cold, but summers are sunny and hot. Moisture supply, rather than temperature or frost-free season, is the critical factor, for this area has the greatest evapo-transpiration potential in the Soviet Union, comparable to that in the northern Great Plains of the United States.

In central Asia, as in the western United States, there is a continuing deficiency of moisture which in most places makes agriculture impossible without irrigation. Where streams or underground water sources can be tapped, this is the most productive area of the Soviet Union, but only 5 per cent of the land area has been irrigated despite heavy Soviet investment in facilities. As in the north, settlements form tiny islands in a vast, empty area. Most of the irrigated sectors lie along the southern margin where streams coming from the moist mountains are tapped in individual oases. Such irrigation dates back hundreds of years, but modern dams and distribution facilities have increased productivity.

Cotton is the most important of many specialized irrigated crops. The Soviet Union is now the world's leading cotton-producing country, and nine-tenths of its production comes from central Asia. Like similar parts of California, these irrigated areas, far from the country's major markets, also specialize in other high-value crops such as orchard products. This is one of the most important livestock regions in the country. This interest rests on the early development of nomadic herding along the desert fringe and in the mountains by native tribesmen. Cereals and fodder crops on irrigated lands are raised to supplement natural pastures.

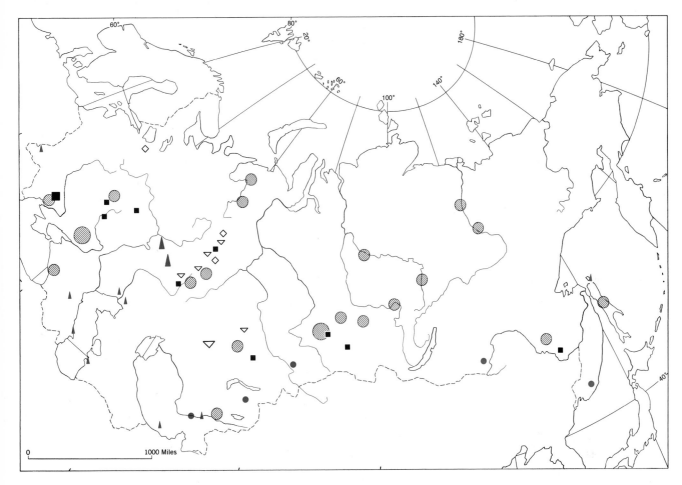

Figure 10-7 Some of the Important Mineral Deposits in the Soviet Union

Coal
Lignite
Petroleum
Iron ore
Copper
Bauxite
Lead and zinc

Important mineral deposits have also been exploited in dry central Asia. The southern margin of deserts is remote from the major population centers, so emphasis has been on high-value, strategic minerals—copper, gold, tin, and mineral salts—which are not readily available in the more accessible areas of the population triangle. Furthermore, in keeping with the Soviet program of regional self-sufficiency, minerals needed in the local economy of the region, such as coal and oil, are also produced.

The Caucasus has been more closely tied to the core of the country than other non-European areas because of its proximity and accessibility via the Caspian Sea–Volga River water route. The Baku and north Caucasus oil fields were the first important fields to be discovered in the country (in Czarist days); for many years they remained the basic oil-producing regions of the country. Moreover, a tiny area on the east coast of the Black Sea has a subtropical and humid climate. It stands as the Florida of the Soviet Union, producing citrus fruits and tea.

SUPPLEMENTAL, COLONIAL REGIONS

Over half of the country functions essentially in a supplemental, or "colonial," position in terms of the Soviet economy. The more important outlying areas—ports in the European north, coal and oil fields there, oases and mining areas of central Asia and the Caucasus, and strategic settlements along the southern Pacific Coast near China and Japan—are linked by railroad to the population triangle. But as Figure 10-8 shows, the flows are generally much smaller than those within the populated regions. The smaller, even more isolated developments in the Asian north are linked largely by air transportation. Efforts are also made to use the Arctic and Pacific oceans and the Siberian rivers, despite the fact that they are usable only for a few weeks during summer. One of the great problems of the Soviet Union has been the development of a transport network to make the vast overland connections needed to tie the huge expanse together. Even central planning has found the task overwhelming.

THE PRODUCTIVE POPULATION TRIANGLE

The essential Soviet population and economic development lie within the population triangle of European settlement, an area smaller than the United States. Its economic importance can be seen from the flow of traffic on the railroads (Figure 10-8), the basic means of freight transportation in the Soviet Union, carrying about 80 per cent of all movement of goods. From this area comes the primary agricultural and industrial output of the country.

As Tables 10-9 and 10-10 show, the population triangle produces the basic agricultural food needs of the country. It accounts for 86 per cent of the sown cropland, and its share of total output is even greater. Emphasis is on grain, sugar beets, potatoes, hay, silage, and sunflowers. In the drive toward national self-sufficiency, sugar beets have replaced dependence on tropical cane sugar, and sunflowers replace soybeans and peanuts in providing essential vegetable oil. Except for sheep, which are prominent in central Asia and the Caucasus, the basic livestock production of the country comes from this area, too.

THE EASTERN UNITED STATES WITH LIMITATIONS

Even within the population triangle, which is the agricultural base of the Soviet Union, the environment for plant growth suffers compared to that of the United States. The problem is that in the northwest where there is adequate moisture for full crop production, summers are cool

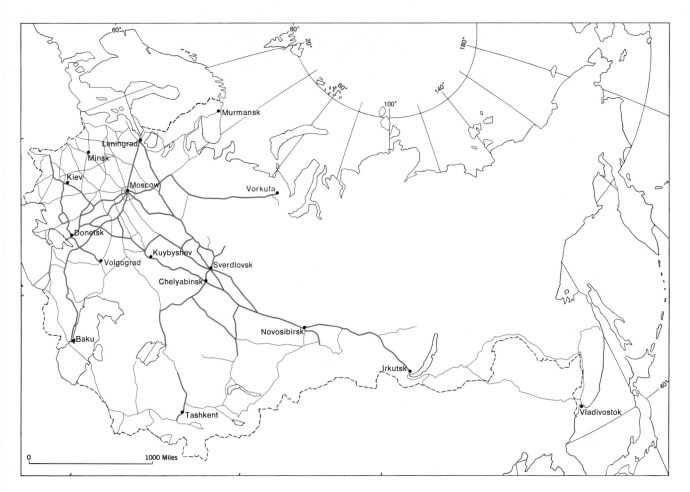

Figure 10-8 RAILROAD CONNECTIONS IN THE
SOVIET UNION

—— Railroads
—— Lines carrying heaviest freight traffic

as in Europe. Thus, there is not enough sun energy for wheat to mature properly during the growing season (Figure 5-4). On the other hand, along the southeastern margin where there is a growth potential, in terms of sun energy, of more than 50 per cent, moisture deficiences of as much as 30 per cent exist (Figure 5-5), and despite the summer maximum rainfall, great shortages of soil moisture normally occur during the growing season. Thus, in the population triangle where it is wet enough to grow crops, there is not enough sun energy for many mid-latitude crops; where it is warm enough for good mid-latitude crop growth, conditions are like those of the western margins of the Great Plains in the United States: there is a critical moisture shortage in most years, and in some, drought kills the plants.

Thus, the environment of the northwest near the Baltic Sea resembles that of central Michigan in the United States, while that along the southern border of the population triangle east of the Volga is similar to the western margin of agriculture in Montana or the western prairie provinces of Canada.

The Soviet "Middle West" lies south of Moscow and west of the Volga, extending southward to include the Ukraine. This is the "black earth" country where rich prairie soil covers largely flat areas. There over three-quarters of the land is in agricultural use, and 70 to 80 per cent of that land is in crops (Table 10-9). Yields here are the highest per acre in the country outside of irrigated regions. Under the czars this was the traditional wheat-producing region of the country. It still produces almost a third of the country's wheat (Table 10-10) and most of its winter wheat, but wheat land has been shifted to corn, sugar beets, sunflowers, potatoes, and other higher-return crops. An attempt has been made to establish a "corn belt" type of grain-livestock

economy, and the area includes over one-third of all the cattle and almost half of all the hogs in the country.

As it has in the United States, wheat in the Soviet Union has been shifted to the drier fringes in the lands east of the Volga and into the "virgin lands" on the dry margin of the population triangle, largely in northern Kazakh. These lands which were uncultivated in Czarist times now account for almost half of the wheat crop, most of it spring wheat. Production of wheat east of the Volga in 1964 was more than three times that in 1940. As in the United States, production is possible only by using tractor-driven plows and mechanized combines that can cover hundreds of acres of land.

The areas north and west of Moscow are much less important because of the more limited physical environment. But like New England, they are close to market and have increasingly specialized in market oriented products such as dairy products and vegetables. Here, however, much of the land remains in forest, and large parts of the farmland are planted for hay crops.

During the past fifty years the Soviet Union has been transformed from a poor agricultural society to a formidable industrial state.

The Soviet Union has organized its resources and carried forward the process of industrialization by using centralized planning on a massive scale. One can find remnants of an earlier, traditional way of life in Russia, but the life style of the modern, interconnected system now predominates.

338

Table 10-9 Sown Area in the Population Triangle
and Kazakhstan, 1958

	Total Area (000 sq. mi.)	Sown Area	Per Cent of Total of Sown Area in Triangle	Sown Area of Total Area (per cent)
Old established centers				
Northwest	82.0	7.7	1	9
Baltic	67.0	17.0	3	25
Byelorussia	80.0	21.1	3	26
Center	370.0	121.5	18	33
Ukraine-				
Moldavia	245.1	134.8	19	55
North Caucasus	137.1	68.1	10	50
Total	981.2	370.2	54	
Recent areas of settlement				
Volga	191.0	71.4	10	37
Urals	291.0	63.9	9	22
Western Siberia	409.0	76.1	11	19
Kazakhstan	1,064.0	110.6	16	10
Total	2,936.2	692.2	46	

Source: Cole and German, *A Geography of the U.S.S.R.*, Butterworth & Company, Ltd., London, 1961, pp. 88–89.

Variations in Agricultural Use and Productivity

Thus, there is considerable variation in the use of agricultural land within the population triangle. Table 10-9 shows the area of sown crops to be almost equally balanced between the established areas west of the Volga and the areas to the east in which the Soviets have encouraged settlement. But notice that the areas along the western margin of the triangle—the Northwest, Baltic, and Byelorussia—together account for only 7 per cent of sown cropland. Each of the other areas contributes at least 9 per cent, and the Ukraine and Moldavia, the Center, and

Kazakh each approaches 20 per cent.

Table 10-9 shows that the proportion of each region used for sown crops varies greatly. It is as low as 9 per cent in the Northwest and 10 per cent in Kazakh in the southeast, around 20 per cent in the Urals and western Siberia, and about 25 per cent in the Baltic and Byelorussian areas. In the Center area it reaches one-third, and in the North Caucasus and Ukraine-Moldavia it peaks at 50 per cent or better. This description gives some measure of the evaluation by the Soviets of the agricultural possibilities of the population triangle and some idea of the greater difficulties that are apparent in the northwestern, western, and southeastern margins.

The importance of the central areas in the triangle—the Center and the Ukraine—can be seen by looking at output of major agricultural products rather than at area sown to crops. On the margins where it is too cool or dry for maximum crop growth, one would expect to find smaller crop areas and lower yields—thus, less production. Table 10-10 shows regional proportions of production of the three major types of crops plus meat and whole milk in the population triangle. First, however, it is important to note that the triangle accounts for virtually all production of these agricultural products. Notice how much more important the areas west of the Volga are in total output of these major food products than areas to the east. The western portion accounts for 70 per cent of the triangle's meat, 72 per cent of its milk, 80 per cent of its potatoes, and 92 per cent of its sugar beets. It is only in grain production that the more recently settled eastern areas hold an equal share.

Only when the combined output of these five major agricultural products from the Center and Ukraine-Moldavian regions alone is considered, can the importance of the central portion of the triangle in the agriculture of the Soviet Union be appreciated. These two regions account for 80 per cent of the sugar beets, 53 per cent of the potatoes, 47 per cent of the whole milk, 44 per cent of the meat, and 35 per cent of the grain of the triangle. This is the Soviet "Middle West," by far its major food-producing area.

On the other hand, the importance of the Soviet drive to increase agricultural production east of the Volga, particularly in western Siberia and northern Kazakh, can be seen. These areas obviously show that the program to turn them into grain "factories" has resulted in the production there of almost half of the country's grain needs. In fact, in 1964 they produced 60 per cent of the total wheat crop of the country. The crop choice on these drier lands is limited, but they contributed an amount equal to almost 20 per cent of the potato crop of the triangle and around 30 per cent of its whole milk and meat output—certainly significant figures in a country striving for increased agricultural output.

The areas along the western margin of the triangle, while generally less important, are still significant producers of some agricultural products. Notice that over 20 per cent of the potatoes and almost 15 per cent of milk and meat products come from those regions. Thus, cool, damp areas also contribute to agricultural output.

Soviet Agricultural Reorganization

Agricultural production under the czars was just emerging from a feudal state before the Revolution. Most of the productive land had been held in large estates of individuals in favor at the Emperor's court. The peasants who worked the land lived in farm villages and went out daily to till small pieces of soil, using manpower and animal power and such ancient equipment as scythes and sickles. In the period before and after the Revolution, the peasants gained title to the land, but like sharecroppers in the southern United States after the Civil War, they continued to work tiny plots in traditional ways.

Reorganization according to the United States Model, Except . . .

The Soviets saw the need to modernize the agricultural system that existed after the Revolution. Manpower from the farms was needed in the factories that were the first priority of the Soviet crash program to join the ranks of the modern industrial powers. Moreover, with foreign loans from capitalist countries eliminated as a result of their communist experi-

Table 10-10 PRODUCTION OF MAJOR CROPS AND LIVESTOCK PRODUCTS BY
AREA WITHIN THE POPULATION TRIANGLE AND KAZAKHSTAN, 1964

AREA	PER CENT OF PRODUCTION IN POPULATION TRIANGLE AND KAZAKHSTAN				
	GRAIN	POTATOES	SUGAR BEETS	WHOLE MILK	MEAT
Old established centers					
Northwest	1	4	. . . .	5	3
Baltic	2	7	1	8	7
Byelorussia	1	14	1	6	6
Center	13	29	20	22	17
Ukraine-Moldavia	22	24	60	25	27
North Caucasus	11	2	10	6	10
Total	50	80	92	72	70
Recent areas of settlement					
Volga	16	8	4	9	10
Urals	9	5	. . . .	8	7
Western Siberia	9	4	1	7	5
Kazakhstan	16	2	2	4	8
Total	50	19	7	28	30
Per cent of total production of the USSR within triangle	94	96	97	91	88
Relative importance of each type of crop compared to other two, based on tonnage output	.46	.29	.25	n.a.	n.a.

SOURCE: Gregory, *Russian Land, Soviet People*, Pegasus Publishing, Indianapolis, 1968, Appendix table 3, pp. 887–890.

ment, they needed to draw as much capital for industrialization from agriculture and other primary production as possible. Finally, they had to maintain agricultural output in order to feed their own population.

As they looked at their environmental base and its possibilities, they turned for their new agricultural system to the United States. They took as their example the large mechanized farms that had been developed in the Corn Belt and Great Plains. The Soviets wanted to have mechanization similar to that used on midwestern farms.

COLLECTIVE FARMS

In organizing they rejected the private ownership of land, whether by farm family or corporate organization, and instituted collective farms. These farms were not as revolutionary as they seemed to people in the United States. Unlike American farmers, Russia's rural population had traditionally settled in villages where fifty to several hundred farmers lived close together, each with a barn, a garden, and a few animals on a lot. Like farmers in Europe and Asia, they went out every day to work farmlands around the village. Under the Czarist landlords the village was considered the working unit. Villagers shared a common pasture, and families did not work the same allotted land each year. Even when the large estates were broken up, there was little attachment to particular family fields. The focus of ownership was on the home plot in the village.

Not surprisingly, Soviet planners chose the farm village as the basic unit in their new agricultural plan. Through collectivization the Soviet government became the new landowner, and the village, with its farmlands, was turned into a single functioning agricultural unit which worked the land around the village as a team. The work force of the village became the work team of the collective farm. The villagers were to elect their own leaders who, in turn, would manage the farm, according to quotas assigned by the central planners, and assign individuals to the particular tasks needed to produce crops.

Like the landlord of the past, the government made the decisions about what to grow and received a share of the crop as "rent" or taxes. Each farm was given production targets for particular crops, and, in addition, basic quotas of farm products had to be turned over to the government at very low fixed prices, whether the harvest was large or small. The government, like the old landlord, was assured of its share, and the risk of farming was placed in the hands of the farmers on the collective. If they worked efficiently and growing conditions were favorable, the net output to be shared by the workers in proportion to the hours worked was large. If for any reason the crop was bad, so was the return to collective workers.

Collectivization meant the loss to farm families of all private agricultural property. As might be expected, the collectivization drive carried out in the early 1930s was met by considerable peasant opposition. There were riots, and a large proportion of the livestock, farm wagons, and other equipment was destroyed.

Faced with such opposition, the Soviet government compromised by allowing individual families to keep title to their individual small plots in the village that had represented the norm of private ownership under the czars. Garden crops, livestock, and poultry raised on the home plot were not part of collective farm production; they could be used to feed the family or sold through open markets.

Private plots remain an important part of the Soviet agricultural system. Although they total less than 4 per cent of all sown area, they account for almost half of the country's production of potatoes and vegetables, over half the dairy cows, and nearly a third of the hogs. This productivity is in part the nature of small-plot production: gardens carefully tended are among the most intensive forms of agriculture in the world, and many types of livestock production fit well on small units. However, the incentive to produce on "your own land" is much greater than the incentive to work as part of a team managed by someone else, particularly since that work has no fixed reward. The individual farmer can feel that he is in control of production on his plot.

Collectivization was designed not only to give the government direct control over the distribution of a large share of agricultural production through its quotas but also to release farm workers for work in newly rising factories and other jobs in the modern system. Thus, mechanization of farming was essential.

The big collective farm units were well suited to the layout of large fields adapted to the use of tractors and mechanized planting and harvesting equipment, but the machinery itself had to become available. This involved more than the tremendous task of purchasing equipment abroad and then building production lines to produce the machines in the Soviet Union; it required a system of allocation among the collectives of the country.

In the early days, mechanization was brought to the farms. It did not come by turning equipment over to the individual collectives. Rather, the government established machine tractor stations, each designed to serve a large number of collectives in a particular area. Equipment was outfitted with lights to be used in shift work, and the stations had trained operators and mechanics to keep operations moving as efficiently as possible. Collectives contracted with the stations for tractors for plowing and combines for harvesting. In this way the scarce equipment could be kept in almost continuous use. In the 1950s, when agricultural equipment finally was in ready supply, the stations were liquidated, and the equipment was distributed among the collectives.

Since the mid-1950s the number of collective farms has been reduced drastically. While there were more than 254,000 collectives in 1950, the number had been reduced to less than 37,000 by 1965. This was done by combining collectives centered in individual villages into larger units consisting of the lands of several villages. The average size of the collective has risen from about 1,000 acres to over 15,000 acres. It originally had been hoped that the villages could be combined into "agricultural towns" of 5,000 or more people with more urban services, but the latter plan has not been fully successful.

STATE FARMS

Collectives were the solution to the problem of mechanizing the traditional agriculture of the villages, but a different unit was developed for

Table 10-11 REGIONAL DISTRIBUTION OF ARABLE LAND IN COLLECTIVE AND STATE FARMS OF POPULATION TRIANGLE, 1964 (000 hectares)

AREA	COLLECTIVE FARMS	STATE FARMS
Old established population centers		
Northwest	1,811	991
Center	21,587	9,537
Baltic	3,376	1,416
Byelorussia	3,690	1,865
Ukraine-Moldavia	27,309	6,524
North Caucasus	9,341	6,694
Recent areas of settlement		
Volga	15,891	12,133
Urals	8,538	9,722
Western Siberia	6,132	11,173
Kazakhstan	4,153	26,491
Total	101,828	86,546

SOURCE: Gregory, *Russian Land, Soviet People*, Pegasus Publishing, Indianapolis, 1968, pp. 900–903.

lands that had never been settled before or that were those the government fully took over from estates. This was the "state farm," a unit completely controlled and managed by the government; workers were paid wages just as factory workers were. Not only was this more in keeping with the Soviet philosophy, but it was particularly well suited to new agricultural units on the dry margins where the risk from year to year is high, making collective farm operation almost impossible. State farms are designed also for the operation of highly specialized undertakings where the maximum amount of government management and technical skills

Table 10-12 Manufacturing and Fuel Consumption in USSR, 1960 (per cent)

Area	Manufacturing Workers	Primary Fuel Consumption
Population Triangle		
Old established centers		
Northwest	6	7
Baltic	3	2
Byelorussia	2	2
Center	27	16
Ukraine–Moldavia	19	22
North Caucasus	4	5
	61	54
Recent areas of settlement		
Volga	6	6
Urals	11	16
Western Siberia	5	6
	22	28
Total	83	82
Outlying areas		
North	3	*
Eastern Siberia	3	4
Far East	2	3
Transcaucasia	3	3
Kazakhstan	3	5
Central Asia	3	3
Total	17	18

*Included in totals for Northwest.
Source: E. J. Stanley, *Regional Distribution of Soviet Industrial Manpower: 1940–1960*, Praeger, New York, 1968, p. 95 and Tables 26 to 39; P. E. Lydolph, *Geography of the U.S.S.R.*, 2nd ed., John Wiley, New York, 1970, Table 1,511, p. 480.

can be applied. Since a large number of these farms are located in the drier areas where per acre yields are low, they are larger than collectives. The concentration of land on state farms in the newer developments and on collectives in the better established agricultural areas can be seen in Table 10-11. Notice the importance of collectives in the two most important agricultural areas, the Center and the Ukraine-Moldavia, and the concentration of state farms in Kazakh, site of the virgin lands scheme.

Manufacturing, the First Priority

It is industrial, not agricultural, output that is of first importance to the Soviet Union. Agricultural output contributes less than one-fourth of the total domestic production of the Soviet Union, while manufacturing and mining total twice as much, or slightly more than half of all production.

Again, Concentration in the Population Triangle and Czarist Origins

Manufacturing, too, is highly concentrated within the population triangle. The latest available figures (Table 10-12) report the localization of 82 per cent of primary fuel consumption and 83 per cent of workers involved in manufacturing in the population triangle, not including Kazakh. In contrast to agriculture, manufacturing is a much more localized activity. Not only does it tend to concentrate in focal points, usually urban and metropolitan places, but certain regions develop as major centers of manufacturing. Recall the localization of manufacturing in the United States.

The same condition holds even in a state-controlled economy such as that of the Soviet

Union despite the avowed purpose of providing a minimal industrial base in all outlying regions (Table 10-12). The table shows over half of the capital and workers of the entire country concentrated in just three regions of the population triangle, the Center, the Ukraine-Moldavia and the Urals. No other part of the country contributes more than 6 per cent of either measure of industrial development.

Big Three and Their Origins

All three of the leading industrial districts date back to the Czarist attempts at industrialization. Manufacturing was highly centralized in the Empire. It was in the Center and neighboring northwest, where St. Petersburg, the capital, was located, that artisans producing the manufactured needs of both the court and the government were concentrated. Following the western European model, the first factories were based on waterpower, and the production of textiles was most important. At first production was based on local supplies of wool and flax, but as in western Europe, cotton manufacture soon followed. As the most advanced part of the country and its most important market, this area pioneered new manufacturing and engineering undertakings. Moreover, it was the hub of transport links from all major parts of the Empire and so had access to necessary resources. The development of railroads in the nineteenth century tended to centralize this manufacturing even more.

However, the Central area, like the Northeastern Seaboard of the United States, did not contain much of the basic resources of modern industrialization—coal, iron, and petroleum. The first real iron industry developed in the Urals on the basis of local deposits of iron-ore and charcoal, but that area, too, lacked the necessary large-scale coal deposits required to shift

over to the mass production of steel in the late nineteenth century.

It was the Ukraine rather than the Central area or the Urals that developed as the first locale of large-scale, modern steel production. Like the Middle West of the United States, this fine agricultural area also had the massive deposits of coking coal and iron first to be used as a base for a modern iron and steel complex. Coal was mined in the southwest corner of the Ukraine known as the Donets Basin. Iron was found some two hundred miles to the west. By the time of the Revolution, the Ukraine was the dominant steel center of the country, producing two-thirds of the country's output. Most of the development was undertaken with western European capital and technical knowledge.

Industrial Regions under the Soviets

The Soviets have been on a crash program of industrial expansion ever since the earliest days. They have seen the development of manufacturing as providing the base for the political and economic power of their state. They have pushed expansion in the established centers inherited from the czars and, at the same time, have established new areas largely related to accessible resources with good market orientation (Figure 10-9).

It is not surprising in this centralized state that the most important industrial complex exists in the area known as the Center. This is where the market is concentrated, and it is the focal point of the entire transport system of the country. Here is the intellectual, scientific, management, and engineering center and the largest labor pool. The only major obstacle, as noted earlier, is the relative sparsity of the resources needed for modern industry, and this can be readily overcome by importing the needs from other areas. The Center meets some of its en-

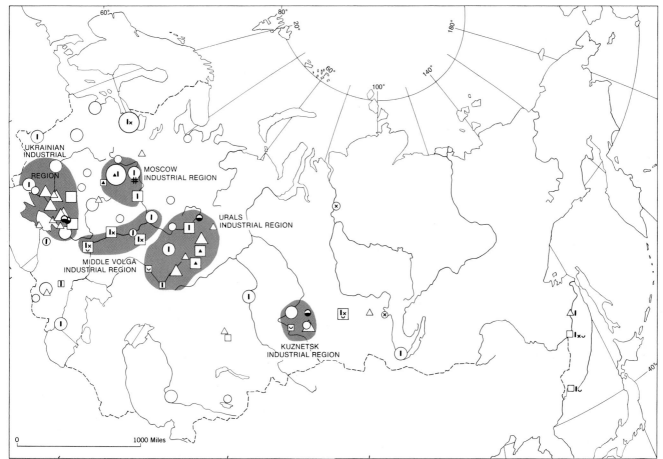

Figure 10-9 MAJOR INDUSTRIAL REGIONS OF THE SOVIET UNION

▨ Major industrial region

Major types of industrial centers
Classified by most significant type of manufacturing

△ Iron and Steel

□ Machine building

○ Diversified industries

Other important industries

❙ Petroleum refining

◒ Chemicals

✕ Wood industries

Textiles

◡ Food processing

ergy requirements by utilizing local deposits of peat and lignite but draws most of its energy from coal, oil, and gas brought in by rail and pipeline. Like industry on the Northeastern Seaboard of the United States, industrial production in the Center emphasizes sophisticated engineering equipment and consumer goods rather than steel and chemical production which requires large-scale resources.

Today the Ukraine remains the Soviet Union's leading producer of crude and rolled steel. Supporting this complex are resources of manganese, salt, and natural gas and large-scale hydroelectric power on the Dnieper River. Also, the region's productive agricultural base serves both as a supplier of raw materials and as a market for farm machinery, fertilizer, and other farm chemicals. The result is a middle-western type of industry emphasizing metalworking activities, closely linked to the Center by the country's busiest transport routes.

Despite its Czarist industrial origins, the Ural center is much more a Soviet creation. Ural development was seen first of all in strategic terms. Here the Soviets could build an industrial complex far from their European borders and one well located with regard to developments farther east in Siberia. Here they have had to establish a whole new modern industrial base, despite the presence of a basic infrastructure of rail lines, cities, and mines. But the Urals' industry is largely based on the utilization of the region's rich mineral resources, particularly iron and nonferrous metals. This is the country's second primary steel base, and it produces more than a third of the crude and rolled steel. The Urals also have the advantage of being near the industrial-government Center and along the major transport links.

It was on the site of a major iron deposit in the southern Urals that the Soviets built a steel industry and a new city, Magnitogorsk, in the 1930s. Connection was made with coking coal deposits near the eastern apex of the population triangle. This became the Soviet's proud "second iron and steel base." But it was a most expensive operation to move coal westward such a distance and then return iron eastward. Today, the Urals industry has found a closer coking coal source in northern Kazakh, and iron mines have been developed at the steel end of the line in Siberia. With its resource orientation the Urals' industry is predominantly metalworking.

The other smaller industrial districts of the population triangle also consist of largely resource-oriented heavy industry. They include the steel center and other industrial communities in western Siberia, cities along the middle Volga River, and cities near hydroelectric power at the eastern apex of the triangle. The Siberian regions suffer from their remoteness from the major centers of population. Despite rich resources, the steel industry in the east is less than a third as important as the Urals', and the Siberian hydroelectric power development resembles the specialized industrial district of the Pacific Northwest of the United States.

The lower Volga cities could be called the Soviet "Gulf Coast." They have always had ready access to Caucasus oil, and now the most important oil field in the country has been developed east of the Volga along the western flank of the Urals. Thus, these cities specialize in oil refining and petrochemicals as well as in producing the equipment needed in oil production.

The Functioning, Interconnected System Today

We have seen the area of the population triangle first of all as largely covered with a veneer of non-metropolitan people living in collective farm villages and on state farms, and working the land for crop and livestock production. In the better agricultural areas, rural population densities are high, and in the northwest and on the fringes, they are lower. Other important agricultural outposts lie in separate oases within Transcaucasia and central Asia.

Superimposed on the widespread agricultural population is a series of metropolitan and urban nodes of two sorts. There are administrative centers like Moscow and the capitals of the other fourteen member republics, and industrial centers, largely tied to the development of particular nonagricultural resources. Some of these are in traditional, early industrial locations which had their beginnings before the development of coal and steel manufacture. Some were developed in Czarist times, and others are the result of recent expansion under the Soviets. Some are part of a cluster of cities in interconnected industrial regions such as the Center around Moscow. Others of lesser importance stand alone.

INTERNAL OPERATION

The metropolitan centers stand as more than industrial producing centers. They are the marketplaces, distributing centers, and administrative and cultural centers for rural areas and smaller communities around them. The national transportation system provides first-order connecting links between them and second-order ties to the communities and rural areas around each of them.

There is also a hierarchy of places. In the Soviet Union, however, with its highly centralized control, the hierarchy is both simpler and more structured. There are not separate entrepreneurs and corporations competing for national and regional markets or in the development of particular resources. The government determines the resource development plan, the production centers, and the distribution network. Through its system of production goals, allocation quotas, and consumer controls, it pushes for conformity to its master plan for the working of the system. In recent years there has been pressure to build incentives into the system by allowing plant managers more freedom in terms of procurement and production. Still, the whole Soviet producing system can be equated to the operations of a gigantic corporation with the goal of achieving the most successful and efficient production.

Moscow, the national capital of the highly centralized Soviet system, is the country's first-order center; it is by far the largest metropolitan center, the dominant national city, and the hub of the national transportation and communication systems. Moscow is also the administrative center of the largest republic, the RSFSR.

On a level below Moscow, but more than twice the size of any other metropolitan center, is Leningrad, only 300 miles away. As Moscow's closest port, the former Czarist capital, Leningrad, continues to be an extension of the political, cultural, and economic control center of Moscow. It is primarily a national center, although it is the regional center of the northwest.

At the lowest level of the system, two different components determine the importance of a city in the hierarchy: administrative connections to the member republics and economic links to the various producing centers. Although all major metropolitan areas carry out both of those functions in the administrative centers, manufacturing employment is generally less than 50 per cent of the work force, while in most industrial centers at least 50 per cent of employment is in manufacturing.

Remember that in the state-controlled Soviet

economy, there are far fewer retail and service outlets. The size of a metropolitan center in either line of the hierarchy is a measure of its relative importance.

There are fewer administrative centers than industrial ones. Among cities of over 500,000 population, eight are administrative centers and twenty-one industrial. Of course, the administrative centers as capitals of member republics are widely spread out, while the industrial areas are more concentrated. All industrial centers are within the population triangle, eight of them in the Ukraine alone.

Usually, the larger the member republic or the economic region is, the larger the number of places of smaller order. In the Ukraine are twenty-nine other metropolitan centers of over 100,000 people, four of them in the coal mining area and two adjacent to a major industrial center on the Dnieper. In the Center are thirty-one centers over 100,000, eight of them within the immediate Moscow district.

Isolated outlying places, administrative centers in eastern Siberia, and mining centers outside of the populated zones may have less than 100,000 people and be tied into the national system only by airline and radio communication; it all depends on their importance.

As you might expect, parts of the country are virtually outside of the present-day functioning system. This is true of the vast expanses of the empty north and east in Siberia, of the central Asian desert beyond economic islands such as the new coal mining center, and of the high mountains of the south and east. Such areas have been the home of primitive peoples similar in level of development to the American Indians and the Middle Eastern Bedouins.

It must be said, however, that the Soviet system sees all people as tied to the state and its goals, and the state tries to bring them into the political and economic system as much as pos-sible. Efforts have been made to indoctrinate even the most isolated peoples in the principles of the Soviet political-economic system and to provide them with the basic social and health services of the country. Mountain peoples and Arctic tribes now have their own collective farms. In the same way, efforts are made to encompass all city populations in the modern, interconnected system.

People are expected to work in the system, and, in turn, the system provides for their basic needs. It must be added, however, that the government-controlled system allows less freedom of movement from rural areas to cities; workers need working permits. As a result, there is less movement of unskilled rural people than in the United States or Europe. At the same time, schools and training institutes in rural areas provide schooling which is designed to fit young people not only into rural work situations but also into urban ones. Moreover, major efforts have been made to raise standards of life and employment in rural areas. Doctors, engineers, and other technicians are commonly assigned to work in outlying cities and on collective farms.

Limited External Contacts

Because of traditional ideological differences with the capitalist world, the Soviet Union is a much more self-contained functioning unit than even the United States. As we noted in Table 10-1, the value of Soviet international trade is less than half that of the United States. Moreover, two-thirds of that trade takes place within the Soviet Bloc, virtually all with Eastern Europe. These areas function in much the same way that Canada does for the United States. They provide needed resources such as coal and food, but particularly manufactured goods, especially machinery and consumer goods. The modern industrialized areas of East Germany, Poland, and Czechoslovakia contribute importantly to the Soviet economy in much the same way that industrial districts within the Soviet Union do.

The United States has viewed trade by the Soviets with industrial countries as having strategic importance, since they have been mainly interested in obtaining machinery and other items of advanced technology. For many years after World War II, the United States had an embargo on trade with the Soviet Union and extended pressure on Western European countries to hold down their trade with the Soviets.

The Soviets, like other industrial countries, have used programs of foreign aid to gain the support of underdeveloped countries. Soviet assistance has gone to India, the United Arab Republic, Indonesia, North Vietnam, and Cuba. Since Cuba's break with the United States and with most Latin American countries, the Soviet Union has provided basic military and technical aid to that country and has become the primary source of foreign trade for the Castro regime. In the same way, the Soviets sent both capital goods and technicians to assist China when the Chinese Communist regime gained power. This aid stopped only when the Chinese broke with the Soviets and asked that assistance be discontinued. The Soviet Union has also provided India with grain and built a steel mill there. All of these moves have had political as well as economic motives. In this respect Soviet moves are similar to foreign-aid programs of the United States and other industrial countries.

Since World War II, the Soviet government has also undertaken programs of technical and military assistance to underdeveloped countries in Africa, Asia, and Latin America. It built the new Aswan Dam in Egypt after the United States refused to contribute, and in addition it provided that country with military equipment.

Because there is no private sector in the Soviet Union, all foreign dealings are government controlled. This is the aspect that other industrial countries and their private trading enterprises fear. The Soviet government can enter any trading market in the world and undersell corporations to gain entry in a market or dump underpriced goods on international exchanges, thereby disrupting existing patterns. Thus, Soviet trade with underdeveloped countries is looked upon with great suspicion not only by capitalist governments, but by corporate management as well. This trade is not terribly significant, however. It is true that the Soviet Union needs tropical products, but the underdeveloped world accounts for only a small fraction of Soviet trade.

The Soviet Union has also been pushing to open diplomatic and trade channels with Western Europe. There are still gaps in Soviet production that can be filled by trade with those countries, particularly with West Germany and France. European and United States corporations have obtained licenses to trade with the Soviet Union and to produce their products there. Automobiles are being built there, and the United States truck manufacturers have been discussing production also. Plans are being made to produce and market Pepsi Cola there.

Thus, the Soviet Union seems to be moving toward greater participation in the modern, interconnected system on the international level. In so doing, it again seems to be following the model of the United States—trading overseas where there is an advantage in obtaining resources or selling products, and where political pressures dictate. And, like the United States, there seems to be little doubt that the Soviet Union will continue overwhelmingly to depend on its own resource base and market in expanding its modern, interconnected system.

SELECTED REFERENCES

Cole, John P., and German, F. C. *A Geography of the U.S.S.R.,* 2nd ed., Rowman & Littlefield, Inc., Totowa, N.J., 1971. An impressive compendium of systematic and regional information on the USSR from a geographic point of view.

Hooson, David J. *A New Soviet Heartland,* Princeton U. Press, Princeton, N.J., 1964. A geographer looks at the geopolitical question of the interior of Eurasia as a place of great strategic significance.

Laird, Roy, ed. *Soviet Agriculture: The Permanent Crisis,* Praeger, New York, 1965. A number of different experts look at the problems of Soviet agriculture from both environmental and economic views.

Lydolph, Paul E. *Geography of the USSR,* 2nd ed., John Wiley, New York, 1971. A systematic and regional treatment with a wide range of thematic maps and tables.

Salisbury, Harrison, ed. *The Soviet Union: The Fifty Years,* Harcourt, Brace & World, New York, 1968. Experts from a variety of disciplines view the progress of Russia under the Soviet regime.

Chapter 11 Japan:
An Oriental Culture
Adapted to the
Interconnected System

I F THIS book had been written a hundred years ago, Japan would be described in much the same terms used for India today. At that time Japan, long a traditional Oriental culture isolated by royal decree from the outside world, was in the initial stages of industrial development. While there was great wealth and pomp in the Imperial Court, 80 per cent of the population were subsistence farmers who lived much as Indian farmers do now. The Japanese exerted great effort in growing crops on tiny farm plots which formed the resource base for each family. The living standards of most of the people were little better than those of present-day Indian villagers, and any disruption of crop production (flood, drought, fire) brought famine. With 34 million people Japan was overpopulated.

The Changing Population Equation

Only forty years ago Japan was one of the leading economic and military powers in the world. Japanese ships carried textiles, silk, tea, and toys and novelties throughout the world, even to the industrialized countries. Japanese industry built airplanes, tanks, warships, and other military supplies for the armies that had overrun much of China and held Korea and Manchuria. With 72 million people Japan was not overpopulated.

TWENTY YEARS AGO:
OVERPOPULATION THREAT AGAIN

If this book had been written in the early 1950s, a few years after the Japanese defeat in World War II and during the United States' occupation, it might have followed the popular view of the time that Japan was one of the problem areas of the world where population had outrun the means of production. In those days Japan's population had badly outdistanced the ability of the land to produce. Japan's industry, destroyed during the war, was unable to compete with modern technology, and Japanese goods were not welcome in many areas. At that time their economy was supported in part by economic grants from the United States. With 80 million people it was again overpopulated.

Since 1950: A New-found Affluence

Now Japan is again one of the most successful of the industrialized countries of the world, the one that showed the greatest increase in output during the 1960s. Its gross national product in 1970 was greater than any individual European country and stood third in the world after the United States and the Soviet Union. With 103 million people Japan is not overpopulated. Table 11-1 tells the story.

There is no question that Japan is one of the great examples of the modern, interconnected system. It is the third largest steel-producing nation and a leader in textiles and clothing; its electronic equipment, optical goods, and complex machinery—radios, television sets, cameras, sewing machines, calculators, and automobiles—are marketed in the United States, Europe, and elsewhere. In fact, Japan has a trade surplus of exports over imports, particularly when compared with other industrial countries. It has twelve of the world's largest commercial banks and twelve of the largest corporations in terms of assets.

As Table 11-1 shows, the results of Japan's rapid growth in production during the 1960s are reflected in the increase in per capita GNP. In 1960 it was about one-fourth that of the major industrial countries. But by 1968 Japan's per capita GNP had increased almost five times and

Table 11-1 National Income and Income per Capita of Leading Industrial Countries

	Total GNP ($ millions)	Per Capita GNP ($)		
		1958	1963	1968
United States	720.0	2,115	2,560	3,580
USSR		1966: 860		
Japan	112.2	285	560	1,110
West Germany	100.6	840	1,255	1,670
France	96.2	1,005	1,270	1,925
United Kingdom	79.9	1,015	1,300	1,445
Italy	60.6	480	765	1,150
Canada	46.7	1,505	1,600	2,250

Source: *Encyclopaedia Britannica, Book of the Year: 1970,* Encyclopaedia Britannica, Inc., Chicago, 1970, "National Income," Table 3, p. 407.

was not far from that of any of the leading European countries. We can assume that this has meant the shift of a majority of the Japanese people from a subsistence economy to a monetary economy in just one short decade.

A Possible Prototype for Underdeveloped Countries Today

Japan is the one non-Western country that has accomplished an industrial revolution. It can now compete as an equal with the United States, the Soviet Union, and Europe. Japan is an example of what other non-European underdeveloped countries might do with similar hard work and skillful organization. But to evaluate Japan's success as a model for others, we must first examine the country's resource base and notice the circumstances that made the changes possible.

Japan's Geographic Base

Although Japan ranked third in GNP behind the vast countries of the United States and the Soviet Union, it is more comparable to the small European countries. Japan is two-thirds the size of France and 10 per cent smaller than California. At the same time, Japan is one of the most populous countries in the world. Whereas California has 20 million people, Japan has 103 million. It has almost twice the population of any of the European countries and ranks sixth in the world in population. Still, Japanese population densities are less than those of the smaller, most densely populated countries like the Netherlands and Belgium.

If, however, the amount of usable land is considered, Japan is probably the most densely populated country in the world. Only one acre in more than five is in crops, which means there are seven persons for every acre of cropland. Even in the Netherlands there are only five persons per arable acre.

If the population were dependent on the resources of its own country, as it was a hundred years ago, Japan would be greatly overpopulated. But today Japan has one of the most successful and rapidly growing economies in the world. It has achieved this by completely shifting its relations to the earth environment. Today the long-range, interconnected system is utilized to tap raw material and food resources in many parts of the world, and, in exchange, the products of Japanese industry and business flow into world-wide markets.

Rugged Mountaintop Islands

Japan is essentially a series of mountains that stand above the Pacific Ocean off the coast of Asia, forming a series of rugged islands. As Figure 11-1 shows, there are four large islands, each separated from the next by no more than ten miles: Hokkaido, in the north; Honshu, the largest and main island of settlement; Shikoku, south of Honshu; and Kyushu, in the southeast. Honshu occupies over half the land area, Hokkaido over one-fifth, Kyushu over one-tenth, and Shikoku only one-twentieth. In the northwest the tip of Hokkaido is about 150 miles from the Russian coast, and in the southwest Japan is only 125 miles from the Korean Peninsula.

A Mid-latitude, Air-conditioned Climate

Like other highly industrialized countries, Japan lies in the middle latitudes. Southern Kyushu is in the latitude of the coast of Georgia, and northern Hokkaido is about as far north as southern New Brunswick, Canada, just across the coastal boundary from Maine (Figure 11-2). With such a great latitudinal range, one would expect important climatic variations within Japan, and so there are. But the islands are very narrow so the full air-conditioning effect of the sea is felt. No place in Japan is more than 75 miles from the sea.

The result of the sea effect is that although the growing season is long in the lowlands—as much as 300 days in the extreme south and 150 days in northern Hokkaido—the relatively cool summers reduce the potential evapo-transpiration. The evapo-transpiration potential is no higher than in southern France or in the Great Lakes region of the United States. But with the long growing season it is possible to grow rice in Hokkaido, oranges in the southern half of Honshu, and tobacco and tea over most of Honshu. In small portions of the extreme south, it is even possible to grow two crops of rice.

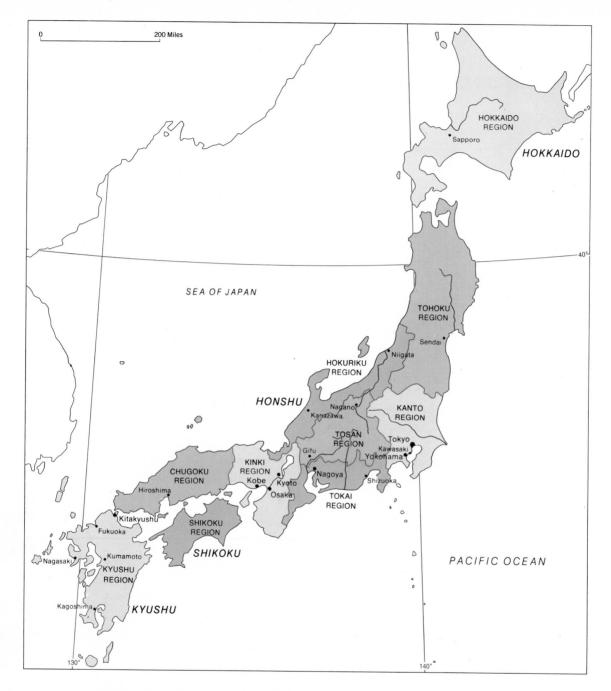

Figure 11-1 POLITICAL MAP OF JAPAN

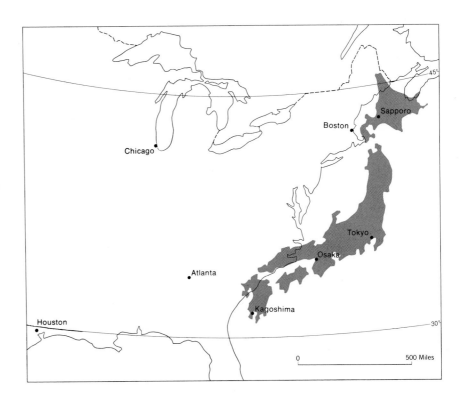

Figure 11-2 JAPAN COMPARED TO THE EASTERN
UNITED STATES IN LATITUDE AND AREA

Winter grain crops are grown in rice fields as far north as central Honshu (Figure 11-3).

With its maritime location there is no shortage of moisture even in summer. Rainfall is over 40 inches a year in all areas except eastern Hokkaido and over 100 inches along the eastern side of the mountains in central Honshu (Figure 11-4). Much of that amount comes as snow. Japan is in the belt of typhoons—tropical storms similar to hurricanes—in the spring and fall. Like hurricanes, typhoons often do considerable damage with their high winds and torrential downpours.

LIMITATIONS OF TERRAIN

It is the terrain rather than the climate that limits Japanese agriculture and settlement (Figure 11-5). Mountains in central Honshu reach heights over 10,000 feet, and only one-fifth of the total land area of Japan is lowland—small coastal areas and interior basins. The highlands are generally too rugged for agriculture. More than 60 per cent of the island is considered productive forest land, but large sections are scrubby and useful only for fuel wood. About 10 per cent of Japanese timberland is in such rugged and inaccessible mountain terrain that it is unused (Figure 11-6).

TRADITIONALLY A SELF-CONTAINED POPULATION

Since the fourteenth century Japan has functioned with an essentially centralized government. Moreover, since that time Japanese culture has been virtually homogeneous. Until late in the nineteenth century Japanese were prohibited from leaving, and foreigners were not allowed to move freely through the country or settle. Since then there has been almost no in-migration from other areas. Except for the

period of military occupation after World War II, Japan has functioned as an independent political state for hundreds of years. As a result, the story of Japanese development is one of a single people under a given central government developing its own future.

The Changing System during the Last 125 Years

If one were looking for the Asian country least likely to industrialize in the middle of the nineteenth century, Japan would have been a leading candidate. It fitted the traditional Oriental mold of a largely subsistence, agricultural people. Moreover, until the latter half of the nineteenth century, the Japanese government tried to keep it sealed off from the outside world. Trade was allowed only through the single port of Nagasaki in the southwest and was limited to a few Dutch and Chinese merchants who were restricted to the limits of the port.

THE ISOLATED FEUDAL KINGDOM OF THE EARLY NINETEENTH CENTURY

Essentially the cultural-economic system was a self-contained one, and the people of Japan lived off the resources of their own small country. The Japanese government was basically an absolute monarchy after the model of Czarist Russia. The shogun, or military minister of the emperor was the real head of state, living in a palace in Tokyo. The titular emperor, whose role was limited to religious ceremonies, lived in splendid isolation in Kyoto. Although the shogun controlled vast lands, most of the country was divided up into almost three hundred feudal realms each under the control of hereditary families. Like European feudal lords, these local rulers taxed their people, had their own

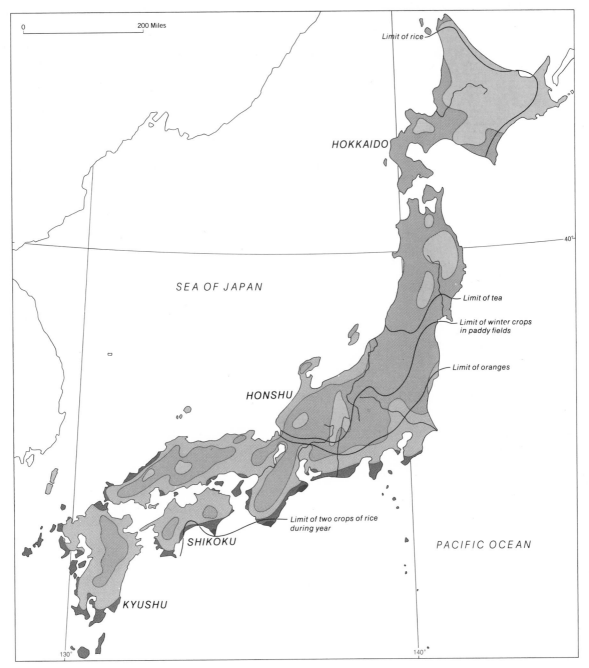

Figure 11-3 Frost-free Season in Japan

Number of days between average dates of last spring frost and first autumn frost

- Under 150 days
- 150–200 days
- 200–250 days
- 250–300 days
- Over 300 days

359

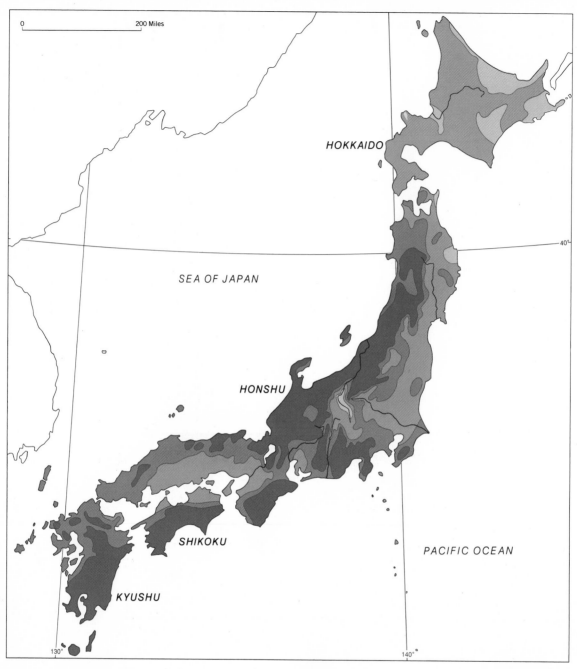

Figure 11-4 AVERAGE ANNUAL RAINFALL IN JAPAN

Under 40 in.

40-60 in.

60-80 in.

Over 80 in.

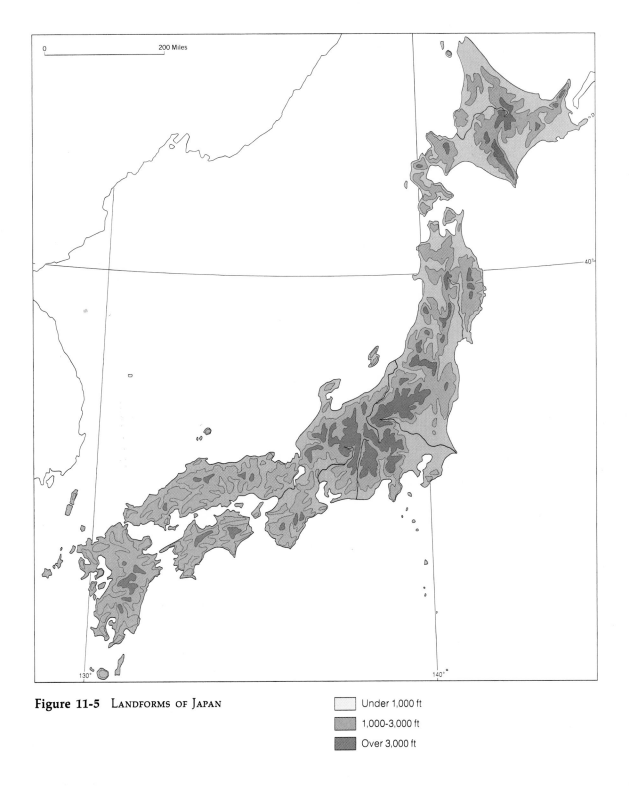

Figure 11-5 LANDFORMS OF JAPAN

Under 1,000 ft

1,000-3,000 ft

Over 3,000 ft

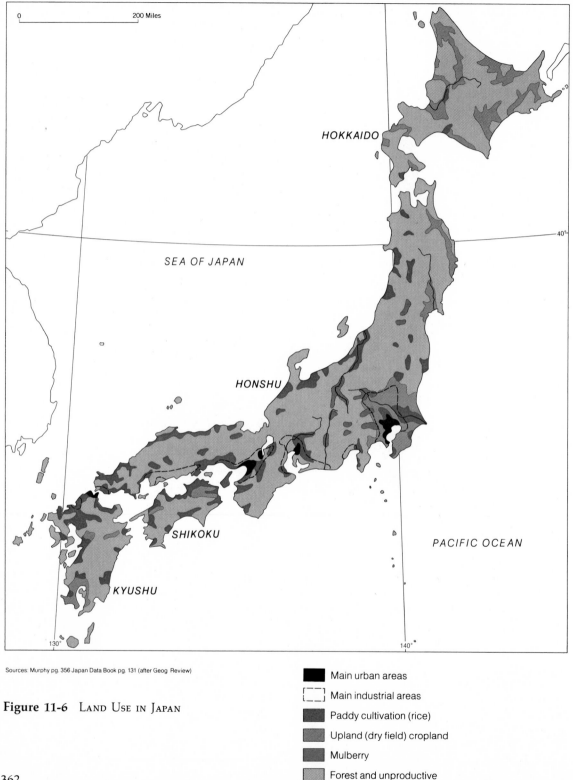

HOKKAIDO

SEA OF JAPAN

40°

HONSHU

SHIKOKU

PACIFIC OCEAN

KYUSHU

130°   140°

Sources: Murphy pg. 356 Japan Data Book pg. 131 (after Geog Review)

Figure 11-6 LAND USE IN JAPAN

■ Main urban areas

┆ ┆ Main industrial areas

Paddy cultivation (rice)

Upland (dry field) cropland

Mulberry

Forest and unproductive

military force, and were responsible for the maintenance of roads and public works in their domains. Their rule depended on the authority of the shogun. To guarantee this authority, the local rulers were expected to live half of the time in Tokyo, and even when they returned to their own localities, their families remained in Tokyo as hostages.

An Agricultural Base and Local Orientation

In the nineteenth century the economic base of the system was agricultural. About 85 per cent of the population were farmers who, except for the payment of a portion of their crop as taxes, lived on a subsistence basis. Traditional agriculture followed the Oriental model of farming tiny plots to produce the basic needs of the family. There was almost no livestock except for beasts of burden and the horses used by the nobility for travel. The tiny plots were more gardens than farms. They were carefully tended by all working members of the family, and although animals were used in plowing, most of the cultivation and harvesting was done by hand with simple implements.

The chief crop was rice, grown in virtually all parts of the country. Vegetables were usually the supporting crops, and grains other than rice were planted in the paddies during the winter season. Little meat was eaten, but near the coast fish provided important protein for the diets of most of the population.

Population Distribution and the Land

Because the population was predominantly agricultural, a map of population density (Figure 11-7) provides a measure of the varying quality of farmland. Where growing conditions were good, population was dense. Where conditions were poor, population density was low. In a sense the population map was also a map of agricultural production. Figure 11-7 which shows the local population densities for 1903, gives some idea of distribution, for even at that time most of the Japanese population was engaged in agriculture. Thus, except for the densities in Tokyo and Osaka, the variation in population can be attributed to agriculture.

What conclusions about variations in distribution can be drawn from the map? The differences in densities between the north and south and between coastal plains and more rugged areas show up. The greater agricultural productivity of the south is obvious. This shows the importance of winter crops which increased production capacity.

To feel the pressure of people on the land, it is helpful to compare it with California. At the end of the nineteenth century there were 34 million people in Japan—70 per cent more than in California today, despite the fact that Japan is slightly smaller in land area. But most Californians live in cities and depend on the long-range, interconnected system to provide needed resources. Only 5 per cent of the California population, in contrast to 85 per cent of pre-industrial Japan, depend on agriculture for a living. What is more, it has the benefit of the latest scientific know-how. With its predominantly rural population, it is not surprising that Japanese farms were only a few acres in size.

A METROPOLITAN MINORITY AND A NATIONWIDE SYSTEM

The small urban population presented a different pattern of the interconnected metropolitan system of the time. Most lived in one of three large metropolitan centers: Tokyo, the residence of the shogun; Kyoto, the residence of the emperor; and Osaka, the port for inland Kyoto. In 1800 Tokyo was probably one of the three largest cities in the world.

Here was a population functioning on a na-

Inhabitants per sq. mi.	sq. km.
Under 128	Under 50
128-256	50-100
256-512	100-200
512-1,280	200-500
Over 1,280	Over 500

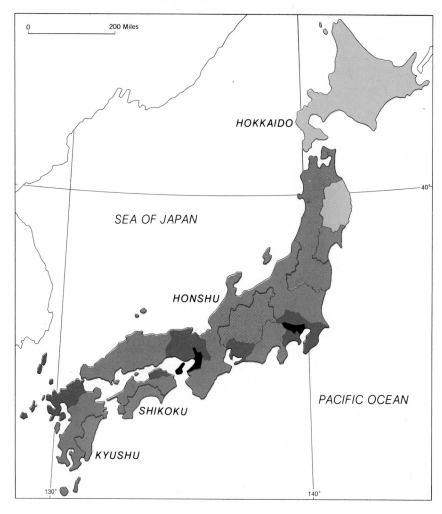

Source: Japan Date Book pg. 120

Figure 11-7 POPULATION DENSITY OF JAPAN

tional rather than a local base, although an important part of the food came from local areas. The big cities were connected by roads to all the feudal centers of the country, and over these roads moved not only the feudal lords on their annual processions to Tokyo but also food, porcelain, and other luxury products of Japan's local artisans. One traveler described the road between Tokyo and Kyoto as crowded with as many travelers as the streets of any European capital. In addition, coastal ships carried passengers and goods to Tokyo and Osaka.

The interconnections and metropolitan centers were primarily political and cultural, rather than economic. Tokyo and Kyoto were important places because they were the seats of national components of governmental and cultural authority for the whole country. Movements between these centers and the local feudal centers largely involved political control and management.

There was economic movement as well. The urban population had to be fed and otherwise supplied. Moreover, these centers housed the wealthiest people in the country who demanded the best work of artisans, the most exotic foods, and the best services. As the centers grew, they fostered increasing trade despite the fact that under the Japanese system of social classes (based on the value of one's services), merchants were the lowest of the four leading groups.

The importance of political ties can be seen by the fact that even as late as 1890 only four of the urban centers designated as cities were not centers of local government (Figure 11-8). Under the traditional system Tokyo, Kyoto, and Osaka, the national centers, had their links to the local feudal centers throughout the country. In turn, each of these was the center of a locally based agricultural system. The trade flows were

those needed to support the wealthy, their supporting warriors, their servants, specialized artisans, and the growing merchant class.

Cultural Classes and Rules of Behavior

The basic socio-cultural patterns established in the nineteenth century are important because of their implications for the later rapid modernization and the functioning of Japan today. Socially the population was divided into sharply distinguished classes based on economic function, similar to the Indian caste system. Highest were the warriors who provided military support for the feudal system. They were landless urban dwellers. Next was the mass of the peasant farmers who provided the food, followed by the artisans who provided necessary goods to the two higher groups. Lowest of the important groups was the merchant class which, in a society with little trade, seemed least needed. Below all of these, as in India, were those dishonored by the disagreeableness of their jobs: the ragpickers, tanners, and the like.

Rules of behavior were based on a hierarchy of loyalty bonds between these classes and also within the family. Wife was inferior to husband, son to father, vassal to lord. Not only did the inferior offer loyalty to his superior, but the superior had paternalistic obligations to subordinates. The whole system ascended to the power of the shogun and his loyalty to the emperor as head of the family state.

Under such a system little individuality was permitted, and personal freedom was further circumscribed by a dictatorial legal system. Groups rather than individuals were held responsible for crimes; whole families would be punished for the activities of an individual. But, despite the fact that the personal freedom usually associated with artistic societies was lacking, considerable attention was paid to educa-

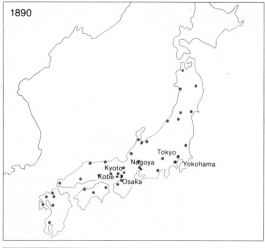

1890

Tokyo
Nagoya
Kyoto
Kobe
Osaka
Yokohama

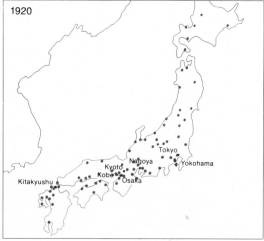

1920

Tokyo
Nagoya
Kyoto
Kobe
Osaka
Yokohama
Kitakyushu

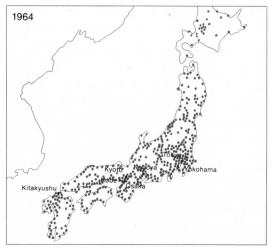

1964

Tokyo
Nagoya
Kyoto
Kobe
Osaka
Yokohama
Kitakyushu

Figure 11-8 CITIES IN JAPAN: 1890, 1920, 1964

tion and culture. By the end of this period, probably 40 per cent of the males were literate, and there were centers of higher education, art, and theater in the large urban places.

THE NEW JAPAN: 1868–1945

In 1868 the whole political structure of Japan, which for more than two hundred years had been attempting to develop a stable world of its own, was overthrown. Japanese leaders had awakened to the power of the industrial world of the Europeans. They were trying to maintain Japanese autonomy at a time when India and most of Southeast Asia had become colonies and China was locked into trading treaties which gave Europeans virtual economic control over resource development.

The Thrust toward Modernization

The result was a change almost as great as what Russia experienced under the Soviets. Centralized political and economic power was combined in the hands of a national leadership determined both to compete economically with the industrialized countries and to maintain Japan's political and cultural individuality. To an extent even greater than in the Soviet Union, the policy of the new Japan was to leave agriculture alone and concentrate on building modern industry. But the strategy of tiny, resource-poor Japan was very different from that of the giant Russia. Instead of economic isolationism, Japan followed the path of colonial England. Emphasis was on trade with the outside world, even with the industrialized countries themselves.

In the early days of the new policy, Japan had virtually no steam engines or machines and no knowledge of modern industrial organization. Thus, its early exports were traditional products of farmers and native artisans: silk, tea, porce-lain, curios, copper products, and pearls. Japanese businessmen traveled all over, studying modern systems of law, government administration, banking, and manufacturing.

Agricultural Specialties Pay Early Bills

From the beginning tea and silk formed the basis for trade with the industrialized world. Both were traditional agricultural products of China as well as Japan, but the new Japanese regime, by concentrating on the improvement of grading and processing, soon became the world's leading exporter. Silk accounted for between a quarter and a half of the value of all Japanese exports until the depression of the 1930s. (Remember that before the coming of nylon in the late 1930s, silk was the luxury fabric of the world, and Japan was the world's leading producer of raw silk.) This high-value product paid many of the bills for early Japanese investment in industry.

Textiles and the Beginnings of Industrialization

As in other developing countries, it was the textile industry that first surpassed the shipment of raw silk as the leading export. The Japanese not only manufactured silk into finished products but also shifted to cotton textiles and clothing. Farm girls were employed in the mills of Osaka for periods of two or three years, receiving food, lodging, and enough wages to provide them with a dowry when they left to marry. Cheaper Japanese textiles and clothing drove Europeans from the Asian market. Osaka and its outport, Kobe, became the leading ports for textiles. At the same time Japan began manufacturing cheap pottery and glassware, matches, low-cost novelties, and paper products for export, but until the 1930s textiles accounted for over half the employment in Japanese manufacturing.

The comparatively low wages of Japanese

workers, the tradition of hard work under un-questioned authority, and the education of trained technicians were the keys to Japanese success. Equally important was the Japanese effort to maintain organizational control over external trade. Like New Englanders in Colonial times, they not only produced needed products but gained increasing control over the financing, selling, and shipping of their products. New profit also came from overseas shipment and sales.

Decision-making in Japan was more decentralized than in the state-controlled Soviet Union. But the group making decisions was the small hereditary group which had been the nobility of the old closed system. Its members were both the political and economic leaders of the country, so whether decisions were made by the government or by private corporations, and whether production was state-owned or family-owned, the same people were involved in the decisions and reaped the benefits. These new industries gradually developed into giant industrial-commercial-financial organizations largely controlled by the leading families of the new regime.

But this leadership was not content with specialization in light, high-labor-input industries when its advantages over other industrial countries seemed clear. By 1930 Japan was producing steel, cement, and chemicals, and before long it had productive capacity in the full range of the heavy equipment needed by any modern military power.

The Japanese farmer benefited somewhat from all this. His children found jobs in cities, he sold more of his crops for cash, and he could buy some fertilizer and essential machine-made tools. But he was an accessory to the system, not the essential component. His living standard remained low, and he still was first of all a subsistence peasant farmer and was not an important market for Japanese industrial goods. According to Japanese traditions, however, he remained, accepting his place both in the society and in the economy.

Changes in the Wake of Industrialization

The links to other parts of the country were soon improved by building modern roads and railroads, but the essential ties shifted to overseas markets. As industry evolved, the fledgling interconnected system centered on Tokyo, Kyoto, and Osaka expanded, although Kyoto became less important when the emperor moved to Tokyo.

One of the most densely populated and one of the most recently modernized countries, Japan has experienced the burgeoning of the modern urban-centered world in the midst of a traditional agricultural society.

Ongoing conflicts result in a struggle for new
cultural identity that blends the charm of
traditional paternalistic life with the mass
culture of the world-scale modern interconnected
system.

374

Table 11-2 POPULATION CHANGE IN JAPAN BY REGION, 1950–1960

	1950		1960		CHANGE	
	POPULATION (000)	PER CENT	POPULATION (000)	PER CENT	TOTAL	PER CENT
Hokkaido	4,296	5.2	5,039	5.4	743	17
Honshu						
Tohoku	9,022	10.8	9,326	10.0	304	3
Kokuriku	5,179	6.3	5,201	5.5	22	0.4
Chugoku	6,797	8.2	6,945	7.4	148	2
Tosan	4,417	5.4	4,402	4.7	−15	−0.4
Kanto	18,242	22.0	23,003	24.7	4,761	26
Tokai	7,323	8.9	8,448	9.1	1,125	15
Kinki	11,607	13.9	14,031	15.0	2,424	21
Shikoku	4,220	5.1	4,121	4.4	− 99	−2
Kyushu	12,097	14.5	12,904	13.8	807	7

SOURCE: *Japan Data Book,* High School Geography Project, miscellaneous publication, National Science Foundation, 1966, pp. 74–77.

As these new urban-centered activities developed, other cities became part of the interconnected system. But overseas connections were overwhelmingly through the ports on Tokyo Bay and with the Osaka and Kobe combination. The new industrial centers, by and large, were the established urban places of the past, but the new industrial areas, almost without exception, arose along the southern coast from Tokyo on the west to the island of Kyushu in the east. Thus, the traditional population centers of the country became more populous under industrialization. New agricultural, forest, and mining developments took place in Honshu's sparsely populated north and on Hokkaido, but there was little of the new industry. An "industrial belt" similar to those of the United States and the Soviet Union developed on southern Honshu and northern Kyushu, while the rest of the country continued to focus on primary production.

The resource base of Japan, like that of European industrial countries, expanded to include all the resources of the world. Sixty per cent of Japanese imports in the 1930s were raw materials and another 20 per cent semifinished raw materials. Japan imported iron-ore, petroleum, nickel, salt, and many other essentials. But, despite the doubling of the Japanese population since 1868, food accounted for less than 10 per cent of imports during the 1930s.

MODERN JAPAN SINCE 1945

Although Japan made the initial transition from a traditional, closed society to a modern industrial country in the seventy years or so prior to World War II, the emergence of the country as an affluent industrial power has been a post-World War II phenomenon. From a country whose cities and industry were badly destroyed, whose people were in a state of shock, whose markets had been closed off by the war, and whose territory was occupied by a foreign power, Japan has emerged as the most rapidly expanding of all industrial powers.

Population Shifts from the Land to the City

Even by 1960 the dramatic change in Japanese life could be seen in population shifts. There was a dramatic shift of people from rural areas to the largest metropolitan centers. And, as in the United States, the change resulted in regional shifts as well. Although the Japanese population increased by more than 10 million between 1950 and 1960, most regions of the country experienced relatively little growth, and some even declined. As Table 11-2 shows, four of the ten regions had growth rates above the national average of 12 per cent, and three of those were the regions of the country's largest metropolitan centers. Together they form the country's most populous area with almost half of the total population.

The only other region of rapid growth is the island of Hokkaido in the north, Japan's frontier area and the least densely populated part of the country. In most of the rest of the country, the rural areas and small cities showed little net population growth. Obviously people were migrating in large numbers from such areas to the largest metropolitan areas (Table 11-3).

Metropolitan Japan

Even in small Japan, the metropolitan population is sharply localized in one section of the country. Japan has essentially developed its own megalopolis along the southern coast of Honshu from Tokyo to Kobe. The four largest metropolitan areas of the country, with about four out of every ten Japanese, are located in this

366-mile stretch. These areas accounted for 80 per cent of the population increase in Japan between 1950 and 1960.

The only other metropolitan area with over one million people is part of an urban extension of the megalopolis along the southern shore of Honshu, northern Shikoku, and western Kyushu, which also includes seven other cities of over 250,000. In fact, there are only four cities in Japan of over 250,000 population that are not in the great metropolitan-industrial belt from Tokyo to Kyushu Island.

The Two Different Population Components

The contrast between the metropolitan and non-metropolitan regions of Japan shows up in employment patterns (Table 11-4, Figure 11-9). The non-metropolitan areas, which account for only 39 per cent of total employment, have 45 per cent of their production in primary activities, overwhelmingly in agriculture. The metropolitan regions have an even larger share in traditional urban-centered occupations of secondary production and tertiary activities. Moreover, primary production is the leading employment category in all non-metropolitan areas, while the two urban-centered activities dominate the metropolitan areas. It is perhaps useful to note that now Japan (like the United States, but not the Soviet Union) employs more people in tertiary services than in secondary production.

Like that in other major industrial countries, Japan's non-metropolitan population is spread widely over the country (Table 11-4), and the metropolitan population is concentrated in a few large nodes. As a single functioning municipal entity Toyko is the largest city in the world and one of the most densely populated. Whereas New York City has about 128 square feet per inhabitant, Toyko has only seven! Moreover, the Osaka-Kobe metropolitan area is larger than Los Angeles, Chicago, Moscow, or Paris.

Table 11-3 POPULATION OF THE LARGEST METROPOLITAN AREAS OF JAPAN BY PREFECTURE, 1950–1960

| | 1950 | | 1960 | | GROWTH | |
AREA	POPULATION (000)	PER CENT OF NATIONAL TOTAL	POPULATION (000)	PER CENT OF TOTAL	(000)	PER CENT
Tokyo						
Tokyo	6,228	7.5	9,684	10.4	3,456	55
Kanazawa	2,488	3.0	3,443	3.7	955	38
Saitama	2,146	2.6	2,431	2.6	285	13
Chiba	2,139	2.6	2,306	2.5	167	8
Subtotal	13,001	15.7	17,864	19.2	4,863	37
Osaka						
Osaka	3,857	4.6	5,504	5.9	1,647	43
Hyogo	3,310	4.0	3,906	4.2	596	18
Subtotal	7,167	8.6	9,410	10.1	2,243	31
Kyoto	1,833	2.2	1,993	2.1	160	9
Nagoya						
Aichi	3,391	4.1	4,206	4.5	815	24
Kita Kyushu						
Fukuoka	3,530	4.2	4,006	4.3	476	13
Yamaguchi	1,541	1.9	1,602	1.7	61	4
Subtotal	5,071	6.1	5,608	6.0	537	11
Total	30,463	36.7	39,081	41.9	8,618	

SOURCE: *Japan Data Book,* High School Geography Project, miscellaneous publication, National Science Foundation, 1966, pp. 74–76.

"Japan, Inc.": Traditional Control of a Modern Industrialized System

Observers note that Japan's economic decision-making is almost as centralized as that of the Soviet Union, with the government functioning as a close partner and decision-maker with in-dustry. Even after the democratization of political elections, the traditional close liaison between government and business in Japan seems to have re-emerged with the few prominent Japanese families again in leadership positions.

Business analysts in the United States speak of "Japan, Incorporated" in explaining Japanese

Table 11-4 EMPLOYMENT PATTERNS OF METROPOLITAN AND NON-METRO-POLITAN REGIONS OF JAPAN, 1960

	WORKERS EMPLOYED (000)							
REGIONS	PRIMARY PRODUCTION		SECONDARY PRODUCTION		TERTIARY PRODUCTION		TOTAL	
Metropolitan								
Kanto	2,420		3,680		4,670		10,770	
Tokai	1,118		1,595		1,493		4,206	
Kinki	1,098		2,568		2,830		6,496	
Kyushu	2,345		1,176		2,046		5,467	
	6,981	26%	9,019	33%	11,039	41%	26,939	100%
Per cent of national totals		47%		71%		66%		61%
Non-metropolitan								
Hokkaido	782		523		882		2,187	
Tohoku	2,560		689		1,335		4,584	
Hokuriku	1,145		664		818		2,627	
Tosan	946		614		674		2,234	
Chugoku	1,355		829		1,206		3,390	
Shikoku	878		393		653		1,924	
	7,666	45%	3,712	22%	5,568	33%	16,946	100%
Per cent of national totals		53%		29%		34%		39%

SOURCE: *Japanese Statistical Yearbook: 1961*, Bureau of Statistics, Office of the Prime Minister, Tokyo, 1961, pp. 53–55.

success in capturing world markets. They make the analogy of a giant United States corporation such as General Motors to describe both the close alliance between the Japanese government and Japan's large industrial corporations, and the intimate cooperation that often exists between firms in business practices. The Japanese Supreme Trade Council is presented as similar to the corporate board and headquarters staff that does the planning and decision-making for GM. Various commercial banks, industrial firms, and corporations operate like the semi-autonomous divisions of GM. Just as GM's Chevrolet, Pontiac, Buick, Oldsmobile, and Cadillac divisions essentially compete with one another but share know-how, facilities, and

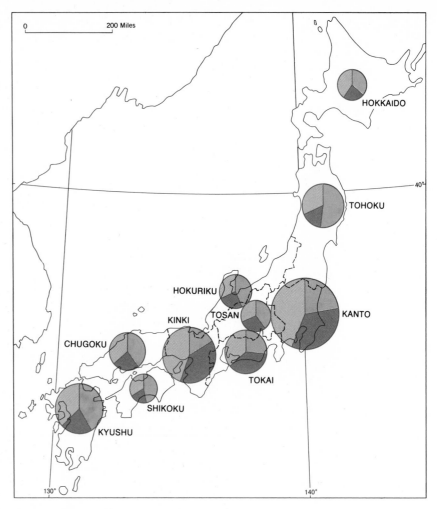

Figure 11-9 REGIONAL DISTRIBUTION OF TOTAL
LABOR FORCE IN JAPAN, 1960

overall management of the corporation, so do the giant Japanese companies.

The key to economic operations is the Supreme Trade Council chaired by the Japanese head of state, the premier, and made up of top business and government leaders who decide how firms will divide up world markets and set annual goals for every major product and country. Then to boost exports the government backs corporations with whatever financial assistance may be needed. Groups of exporters meet regularly to fix prices and lay plans for operating in overseas markets against foreign competitors. A government-owned company operates on a global basis to promote products and provide companies with export intelligence. The government foreign-aid program is another wedge into the economies of developing countries. This program provides long-term credit for development using Japanese goods and money for direct Japanese investment in foreign countries.

The unique international trading firms are also important. These three companies operate in countries with which Japan trades and serve as "the eyes and ears of industry." Each has about a thousand employees and a hundred offices over the world, and serves a thousand different Japanese companies. These firms can carry out almost any kind of overseas trading arrangement, financing, or investment. They report back market trends, industrial developments, customer habits, and investment opportunities from all over the world.

The great family-owned combines of the past which function across the whole range of industrial, financial, and trading operations also play an important role. Table 11-5 shows the major companies and business firms in Japan, each bearing one of three famous family names. Notice the range of activities in which they operate. In terms of holdings all are primarily involved in financing rather than in manufac-

turing, but notice the importance of the trading firms within each family grouping. The holdings given for the Mitsubishi name are greater than those of General Motors, although less than half those of American Telephone and Telegraph. They are reminiscent of the holdings of famous United States families—the Carnegies, Mellons, Du Ponts, and Rockefellers—but represent a larger share of their country's total business holdings.

Although the family holdings in pre-industrial Japan were in the rural areas, today they are in the largest cities. Land reform carried out under military occupation by the United States after World War II eliminated the control of largest landlords over farmland, but the re-emergence of Japanese business and industry after the occupation once more found the traditional families in charge. Just as the management function of large United States corporations is centered in New York City and Chicago, so the Japanese family enterprises are headquartered in Tokyo and Osaka. The Mitsui and Mitsubishi holdings are based in Tokyo, but the Sumitomo empire is headquartered in Osaka. These family firms control over 80 per cent of the holdings of the 100 leading industrial corporations in Japan. Tokyo alone holds almost two-thirds. Twelve of the fifteen largest banks in Japan are based in the two areas as are most of the largest trading companies. Family enterprises in Japan are as clustered together as the big business corporations of any modern industrial country.

Manufacturing for a World Market

This close interaction of government and business is primarily designed for the world market rather than for Japan's domestic market. It assures that goods manufactured in Japanese factories travel overseas in Japanese ships and that

Table 11-5 HOLDINGS OF MAJOR FAMILY-NAME FIRMS IN JAPAN, 1969
($ millions)

MITSUI FAMILY		MITSUBISHI FAMILY		SUMITOMO FAMILY	
Mining/manufacturing		Mining/manufacturing		Mining/manufacturing	
Tuatsu chemicals (68)*	265	Heavy industries (1)*	2,612	Metal industries (11)*	984
Shipbuilding and		Electric industries (12)	857	Chemicals (30)	576
engineering (46)	478	Oil (32)	325	Electric industries (38)	269
Mining and smelting (58)	253	Chemical industries (33)	510	Metal mining (69)	200
Mining (90)	212	Rayon (43)	282	Cement (109)	210
Petrochemical		Metal mining (48)	271	Machinery (129)	141
industries (118)	220	Petrochemical (88)	325	Light metal (138)	113
Total	1,428	Mining (144)	143	Coal mining (196)	121
		Paper mills (151)	129	Bakelite (258)	58
		Cement	110	Total	2,672
		Steel manufacturing (190)	106		
		Plastics (247)	60		
		Edogawa chemicals (357)	63		
		Kakoki (393)	32		
		Total	5,825		
Banking		Banking		Banking	
Bank (7)	3,969†	Bank (3)	5,920†	Bank (2)	6,041†
Trust and banking (3)	727	Trust and banking (1)	844		
Total	4,696	Total	6,764		
Insurance		Insurance	0	Insurance	
Mutual life insurance (7)	528			Marine and fire (8)	149
				Mutual life (3)	1,210
				Total	1,359
Warehouse (2)	32	Warehouse (1)	36	Warehouse (3)	25
Shipping (2)	338				
Real estate (1)	264	Real estate (2)	376		
Trading firms (2)	2,588	Trading firms (1)	2,357	Trading firms (6)	889
Construction (11)	115			Construction (29)	61
Grand total	9,989		15,358		11,047

* Figures in parentheses indicate ranking among leading Japanese companies in that category.
Most are rated in terms of sales or revenues rather than holdings.
† Deposits, not holdings. Holdings would be greater since they would include deposits plus
physical holdings.
SOURCE: *The President Directory*, The President Magazine, division of Time, Inc., New York,
1969.

the financial and trading arrangements are made by Japanese firms.

Japanese manufacturing has been geared to foreign markets. Textile exports to the United States have been so great that textile mills in the American South have been lobbying for quota limitations or protective tariffs. Just as textile manufacturing arose in the South on the basis of low-cost production compared to high-wage New England, so cheap Japanese labor threatens the South today. In contrast to American workers, Japanese textile laborers make about 45 cents an hour.

The Japanese factories are also new and very efficient. New plants without windows offer precise lighting and temperature controls. Machinery is highly automated and includes a water-jet loom developed in Japan that is three times as fast as conventional looms and much quieter. Moreover, the textile industry has been organized by the industrial combines on a much larger scale than have most locally owned mills in the United States.

Japanese firms have invested heavily in research on synthetic fibers and processing. They are now moving production to areas where labor is even cheaper: Taiwan, South Korea, and Thailand. There wages are only 10 to 15 cents an hour. A Korean-made sweater may sell for $2 retail in the United States when a comparable American-made sweater costs $4.95. And, of course, the Japanese commercial intelligence system knows the style preferences of the United States, and Japanese manufacturers make clothing to such specifications.

Today textiles account for only 11 per cent of Japan's exports. Industry concentrates on the most advanced products of modern technology. Motor vehicles, ships, electronics equipment, electrical appliances, machinery, component parts, and steel products. In 1952 Japanese decision-makers decided to shift the economic base to heavy industry. A Machinery Industry Promotion Law was passed; banks made low-interest loans to companies in these industries.

Japan is now third in the world in steel production, with more than 93 million tons in 1970—more than nine times the production of 1955. As with textiles, most production comes from the most modern and efficient mills using some of the world's largest blast furnaces, modern oxygen converters, and the latest finishing facilities. These enable Japan to compete despite the fact that it must import more than 80 per cent of its iron-ore and 60 per cent of its coking coal. Japanese steel production is the base for the country's new diverse metalworking-machinery industry, but basic steel products are among the country's leading exports. Again, agreements have been necessary to control Japanese steel imports to the United States. But Japanese combines and trading companies have been creating markets for Japanese steel by helping to develop steel-using industries in newly developing countries such as Ethiopia and Venezuela.

Japan has also become the world's leading shipbuilding country, specializing particularly in the new, giant oil tankers of 300,000 tons or more that have revolutionized the shipment of oil since the closing of the Suez Canal during the Arab-Israeli war. But it has been in advanced electronic equipment and automobiles that Japanese manufacturing has gained its greatest reputation. Japan specialized early in electronic components—transistors, printed circuits, and semiconductors—for American manufacturers. Then it moved into making radios and television sets, first marketed under American labels and now under Japanese brands such as Sony and Panasonic. Again, Japan capitalized both on the technology that could be borrowed from other countries and on innovations by its own researchers. In motor vehicles there was

RAIL TRAFFIC AND THE JAPANESE INTERCONNECTED SYSTEM

As do other modern industrial countries, Japan has a core region and outlying peripheral regions. From our examination of the workings of the Japanese economy today, you should be able to identify the Japanese core. But examine the maps of railway traffic (Figures 11-10a and 11-10b) to test your generalizations. Remember that Japan is small in size, so air service within the country is not important, and automobile ownership, though rapidly increasing, is still low. Thus, trains form the major means of intercity passenger and freight movement. The map of railroad traffic, then, is a much more accurate measure of movement within the system than it would be in the United States.

How would you describe the pattern of railroad traffic? Does it fit in with other evidence of the distribution of the modern system? From it can you identify producing and consuming areas?

a progression from bicycles to motorcycles and motorbikes to automobiles. Automobile exports totaled more than one million in 1970, more than a quarter of them destined for the United States.

THE RISING DOMESTIC CUSTOMER

Until recent years Japanese businesses concentrated on overseas markets, but growing industrial and business activity in the country has begun to create an important consumer market at home. While prior to World War II the urban population had remained secondary to the rural family, the Japanese population is now predominantly metropolitan. Wages have also risen. In 1950 the basic food diet of rice and other essentials took 57 per cent of family income. Under these circumstances few Japanese had any hope of owning a refrigerator, vacuum cleaner, or washing machine, much less an automobile. But with higher incomes the share of income spent on food has dropped to about one-third, and the new city workers, though poorly paid by American standards, now have a new affluence that sparked domestic demand not only for clothing and food but also for television sets, radios, household appliances, motorcycles, and even automobiles. More than three-fourths of Japanese automobiles are now sold in Japan, and the leading export industries actually produce from 60 to 90 per cent of their output for the home market.

Such a market would seem ready for imports of consumer goods from the other industrial countries of the world, and the United States and European countries have been trying to enter it. But the Japanese decision-makers who have so efficiently worked their way into the markets of other countries have built a protective barrier against foreign competition with Japan. Automobile tariffs have been very high. Until recently there was a quota of one thou-

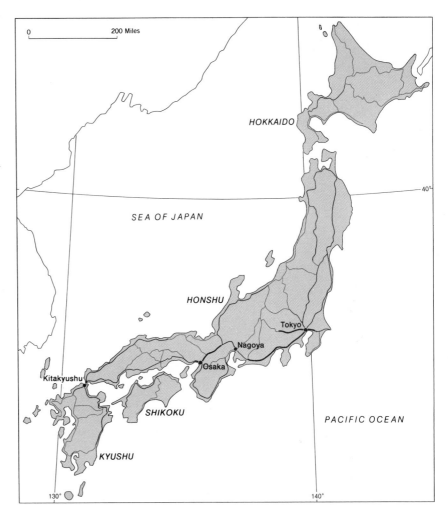

Millions of passenger

——— Less than 1,000

——— 1,000–5,000

——— Over 5,000

0   200 Miles

HOKKAIDO

40°

SEA OF JAPAN

HONSHU

Tokyo

Nagoya

Osaka

Kitakyushu

SHIKOKU

PACIFIC OCEAN

KYUSHU

130°

140°

Figure 11-10a Railway Traffic in Japan, Passenger

sand on automobile engines that could be imported. Moreover, there have been sharp limitations on foreign investments in Japan.

Manufacturing Operations

To an important degree the whole industrial-business system of Japan operates on the traditional basis of loyalty arrangements dating back through the centuries. It is said that the Japanese have been brought up with the concept that their own needs are secondary to serving others. The worker feels that his job is to help make more and better products. This will lead to the prosperity of the company and the greater prestige for the country. Workers hold great loyalty to their employer and operate with machinelike discipline. Employment with one company is most likely to last the worker's full career. In turn, the company provides the traditional paternalistic posture of the feudal lord. Workers are often given family benefits so that

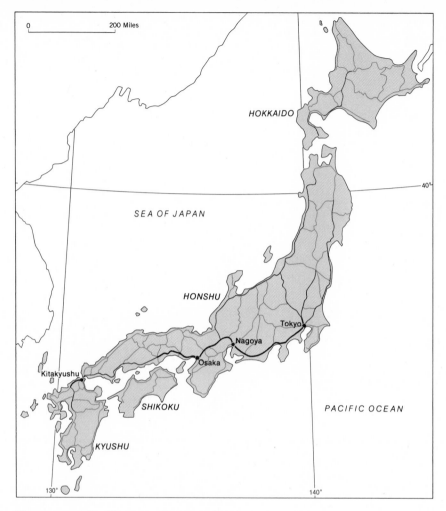

Average tons per day
— Less than 10,000
— 10,000-30,000
━ Over 30,000

Figure 11-10b Railway Traffic in Japan, Freight

one with children is paid more than an unmarried worker. There are medical services and hospitalization, day care nurseries, and sometimes even sports facilities and employees' clubs. It is also traditional to pay bonuses twice yearly, usually equivalent of two to three months' pay. The side benefits are generally greater than those received by workers in the United States, a factor that somewhat narrows the wage gap between the two countries.

Although industry is dominated by the huge industrial-business combines, Japan retains a large share of traditional "cottage-type" industries. These are the Japanese sweatshops where work is from dawn to dusk with perhaps two days off a week. Wages are extremely low, and workers are mostly ill-educated people from rural areas. A firm may employ only a few workers outside of the owner's family, or it may have up to 50 employees. These are the traditional "subcontractors" making component parts—bicycle wheels or electronic plugs—used

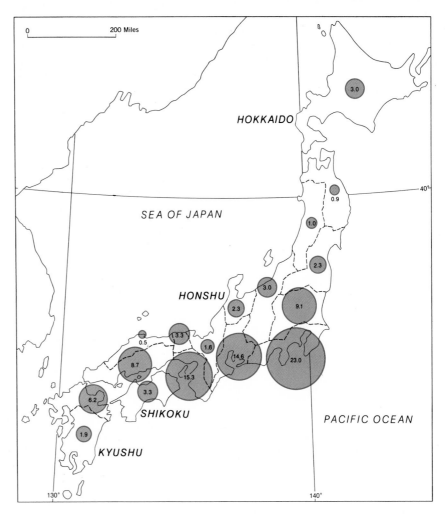

Proportional circles represent
percentage of national total

Figure 11-11 Manufacturing Distribution in
Japan by Percentage of Value of National Total, 1970 Estimate

on the assembly line of some large combine.

Figure 11-11 shows the tremendous amount of manufacturing output concentrated in the large metropolitan centers. Roughly two-thirds of the country's manufacturing production comes from the five largest metropolitan centers and less than a quarter of its output from regions beyond the southern coast. This high concentration is not surprising in a country that has been oriented to external trade. Even now, two-thirds of all Japan's foreign trade moves

through the ports on Tokyo Bay, Osaka-Kobe, and Nagoya. When one recalls that Japanese industry depends on both foreign sources of raw materials and foreign markets, a position in and around the major ports would seem a prime requisite for most industry.

The Agricultural Revolution

In the midst of the great postwar industrial expansion there has also been a quiet revolution in agriculture. Basically the farms remain small, family-owned enterprises, and the garden plots are still handled with utmost care. But now the benefits of modern scientific know-how are being applied to agriculture with the same intensity that has been applied to industry. The result is that the productivity of Japanese agriculture increased about two and one-half times between 1890 and 1965, and it has increased further since then. Although some additional land was put into production during this period, the increase in output has come essentially from improved irrigation systems, better fertilizer, better insecticides, and plant breeding, which increase productivity per acre. Japan probably has the highest per acre agricultural production of any country in the world.

There has been a major shift in the products of agriculture as well as in productivity. Rice production has increased sharply as the result of a government program of price supports and the introduction of new "miracle" strains. Productivity increased by 50 per cent in the late 1960s. By 1968 rice production was half again the total for the early 1950s. And, in fact, for the first time in Japanese history, rice was in surplus. Despite the fact that more than one million tons were made available as livestock feed, there still was an eight-month supply stored in government granaries. Faced with the cost of its rice program, the Japanese government is beginning to pay farmers for not raising or harvesting rice. But since rice is the traditional crop on 90 per cent of Japanese farms and the government price subsidy makes it one of the highest-value crops in Japan, the farmers continue production.

The new affluence of the Japanese worker has brought an important change in other crops of Japan's farming system. In recent years other grain crops have decreased rapidly, even though it is necessary for Japan to increase imports of feed grains. With city customers changing their diets toward more meat, dairy products, fruits, and vegetables, the output of these products has risen sharply. The production of oranges increased more than six times, grapes four times, and apples have more than doubled. Milk production is up six times, and the number of cattle slaughtered has doubled, swine tripled, and chickens quadrupled. Egg production has increased more than nine times.

Modern roads and trucks enable farmers to organize in much the same way that the commercial farmers have organized in the United States and Europe. They are planting the crops that do well in their particular area and can be marketed in the cities. There has also been a rapid move toward mechanization and electrification. Almost all Japanese farms use powered garden cultivators and power sprayers, and there are electrical connections for both farms and homes. On larger farms full-sized tractors are in operation.

Time has brought a change in the traditional surplus of farm labor; migration to the cities has left a shortage of farm workers. By tradition the eldest son became the manager of the farm, and the younger boys moved to other jobs. But now most young people prefer city life, and, as in the United States, the farming population gets older and older each year. Rural areas are still major producers of children but not for their own lands.

Fish and other seafood provided the basic protein in the Japanese diet for centuries. As they have done with other things, the Japanese have expanded and modernized fish production. The country has become the leading fishing country in the world, and it is also the owner of one of the largest, most modern fishing fleets. Since the oceans are international waters, Japanese fishermen work almost all the major fishing areas of the world from the coast of California and South America to the whaling grounds off Antarctica.

A Cultural Revolution

The rapid development of the modern, interconnected system in Japan has produced a cultural revolution as well. Like people in other industrial nations, the Japanese have increased materialism and have greater material expectations. But there has been a more fundamental change than that.

Before modernization Japan was a homogeneous culture largely isolated from outside influences. With world trade has come the influx of outside ideas and values, particularly those from the United States and Europe. Many western ways have taken root in Japan: films, modern music, dress, and even baseball. Present, too, are the very different philosophical and moral values of the West.

The result has been cultural conflict in Japan probably on a greater scale than in any other industrialized country, even the Soviet Union. This conflict has both sociologic and geographic dimensions. As might be expected, the new ways take hold particularly among the young. Thus, the generation gap within Japan is one between the traditional Oriental values of parents and the strongly westernized young people. At the same time, Japanese industrialists whose families set the standards in traditional Japan find that they must accommodate to the modern industrial world. They, more than almost any other Japanese, live in two worlds, two worlds in which they take leadership roles. The contrasting values of old and new show up in any modern country between the people of the city and those who remain on the land. Thus, modern Japan has risen in the cities, while traditional Oriental practices have remained entrenched on the farmsteads and in the rural villages. But even there, electricity and radios and television sets are bringing the new world into rural homes.

There seems little doubt that the outward appearance of Japanese culture will shift more and more toward what is becoming a world-wide metropolitan-centered culture. Just as the Japanese have developed their own particular version of the modern, interconnected system, so we might expect that the culture of the "new Japan" will continue to bear the particular stamp of the Japanese heritage which has been so distinctive up to now.

Japan is the last of the industrial regions of the world that we will examine. As we begin to turn our attention to the underdeveloped areas of the world, it will be important for us to keep the Japanese situation sharply in mind. For Japan, of all the industrial countries, is the one country that has emerged from non-European roots. Its circumstances may be unique, but it is seen as offering promise to non-Western peoples throughout the world. It may also foreshadow an increasing threat to the traditional Western monopoly on modern affluence.

SELECTED REFERENCES

Dimock, Marshall E. *The Japanese Technocracy,* John Weatherhill, New York, 1968. An in-depth treatment of the Japanese system of industrial development and control.

Downs, Ray. *Japan: Yesterday and Today,* Bantam Books, New York, 1970. A basic introduction to Japan from an historian's viewpoint.

Isenberg, Irwin. *Japan: Asian Power,* Wilson, New York, 1971. History of modern Japan since 1868.

Stone, Peter. *Japan Surges Ahead,* Praeger, New York, 1969. An examination of industrial growth, industry by industry.

Trewartha, Glenn. *Japan: A Geography,* U. of Wisconsin Press, Madison, Wisc., 1965. An exhaustive traditional geographic study dealing with man and environment.

Chapter 12   The Outreach
of the Modern,
Interconnected System

IN EACH example of the modern, interconnected system examined, a similar spatial pattern has appeared. A highly developed core region comprises several huge metropolitan centers which form both the headquarters of political, economic, and, commonly, cultural decision-making and the major centers of manufacturing, trade, and other services. One needs only to recall the Northeastern Seaboard of the United States, the concentration of metropolitan-industrial development along the eastern English Channel in Europe, the Moscow-Leningrad core in the Soviet Union, and the Tokyo-Kobe corridor in Japan.

In each of these examples, the areas outside the central core play peripheral roles, serving most of all as producers of specialized primary production items or of certain types of manufacturing tied to required resources such as minerals, perishable farm products, cheap power, and low-cost labor. Each of the outlying areas also provides a secondary market for products emanating from the metropolitan decision-making core. The major link for each of the outlying regions is the central core, and there is little interaction from one outlying region to another. These regions are more often competitors with each other in the production of goods for the core than complementary traders.

INDUSTRIAL COUNTRIES,
RAW MATERIAL SUPPLYING
NONINDUSTRIALIZED COUNTRIES

When we shift from the national and continental scale of the United States, Europe, the Soviet Union, and Japan to a global view, we find the same phenomena on a much larger scale. As we saw in Chapter 3, the trade of the world is, first of all, between the different industrialized centers and then between those centers and the rest of the world. The amount of movement of either goods or people between one outlying area and another is negligible. Latin-American countries do not trade with one another or with African or Asian countries. Instead they trade with European countries, the United States, Japan, and the Soviet Union. The same holds true for Africa, Asia, and Australia, the parts of the world outside the industrial core regions. The model seems similar to Colonial America when the primary ties of each colony were with the mother country.

The contacts between industrial core regions and peripheral areas in the nonindustrialized parts of the world are between separate cultural groups, just as they are in the Soviet Union or Europe. Each country in the underdeveloped world is a separate culture or is often a variety of distinctive cultures. Each has grown up largely in isolation and, like the feudal kingdoms of the Middle Ages, values its cultural and political independence, even though it is, in fact, economically subservient to the industrial world order.

The Outward Reach of the Interconnected System in Search of Primary Resources

Before turning to the countries of the underdeveloped world, let us examine the outward reach of the modern, interconnected system. Even before Columbus, the primary motive in long-range connection over the world was economic. There have been political and cultural reasons, of course, but the desire for gold, handicrafts, exotic agricultural products, and minerals has increased. Moreover, although there has been migration from the industrial countries to Latin America, Oceania, South Africa, and other places, the desire to obtain products through the adjustment of the native economies has been much more prevalent. The search has been primarily for three sorts of products: minerals, tropical products that cannot be produced in the mid-latitude industrial areas, and staple foods when there is insufficient production at home.

PLANTATIONS AND MINES REPLACE TRADING POSTS

In the early days of the Age of Exploration, European traders were content to depend on the economic organization of the underdeveloped peoples to provide the basic output of desired goods. The ships used in trade could not carry large cargoes, and the demand at home was limited by the scarcity of money. As a result the inefficiencies of native production and the lack of quality control were of little consequence. In those days Europeans were truly "traders." They set up trading posts and forts to barter with the natives for furs, slaves, and regional specialties but did not become involved with production itself.

Soon, however, the demands of the expanding European marketplace became greater than the capacity of a largely subsistence native economy to supply, and Europeans established their own plantations and mines. Colonial territories were established in tropical lands such as islands in the Caribbean and the East Indies where key plantation crops included sugar cane, rice, and indigo.

Essentially the purpose of the plantation system was to get more efficient production from areas that had the potential to produce a specialty crop. Native organization, developed primarily to supply the basic food needs of the local population, was replaced by a system to provide commercial one-crop production designed to provide large-scale output for shipment to markets overseas. This changeover was accomplished by substituting outsiders for native leadership and establishing market control. Thus, local resources were turned from the traditional, locally based, direct support of local people to a specialized output for the overseas market. Wherever possible, not only local resources but also local labor forces were commandeered. Where it was not possible, as in tropical America, the additional labor of African slaves was used.

In early plantation ventures, it was European organization and management that were most important. The Europeans planned the land clearing and crop planting, organized the labor force, established the land transport and port facilities, and handled shipping and financing. The native population, like the land, became a resource utilized by the new organizers. Although basic living patterns of the natives had been broken by European intervention, the plantation organizers gave low priority to the reordering of native lives. All they really wanted was the native input of labor. They disregarded the society they disrupted, leaving the native individual, family, and community to adjust to the new economic system with its political and cultural implications.

Even in colonial times crops were introduced to areas where they had never grown before. Sugar cane was brought from Spain by Columbus, and by 1600 plantations had been established on most Caribbean islands and the northeastern coast of Brazil. The sugar cane industry was probably the largest single industry in the world at the time. In the same way cotton was brought not only to the United States but also to India, Brazil, and other tropical countries. Rubber, first obtained in quantity from wild trees in the Amazon Valley, became the leading plantation crop in Southeast Asia. Coffee was brought from Arabia to Latin America and tea from China to Southeast Asia.

Following the pattern of agriculture, trading

for gold, silver, and precious stones gave way to mining operations which were developed and managed by Europeans. Here, too, cheap native labor was utilized wherever possible, but the primary facet of mining development was the organization of production under efficient European management.

SHIFT TO MASS-PRODUCTION SOURCES AND MASS-PRODUCTION OPERATIONS

Increasing technology and affluence in the modern industrial countries have greatly expanded the interest of these countries in the agricultural and mineral output from the underdeveloped world. Modern ships and railroads have enormously increased man's capacity to move products and to make large-scale enterprises in remote areas economically feasible. But mass production in industrial areas and mass consumption of specialty foods have shifted business interests to those areas that offer the very best prospects for quality and quantity production. Just as the small bog-iron deposits in the industrial countries were abandoned for such as the massive iron-ore fields of Lake Superior when the steel industry shifted to mass production, so overseas production has been localized in the areas that promise the greatest return. There has been a slow sorting out of plantation areas, with greater and greater emphasis on those that can be shifted to mass production. Mineral production has become concentrated in the parts of the world where there are massive deposits of easily accessible, high-quality resources.

LOSS OF COLONIAL CONTROL

One other major facet of the developmental process has changed since colonial days. In the days of widespread European holdings in Latin America, Africa, and Asia, business organizations in the mother countries found it relatively easy to gain permission from their own governments to prospect and develop resources in foreign possessions. The whole philosophy of colonialism called for drawing production from the outlying empires.

Today, however, colonial ties are gone, and in their place stand new countries very jealous of their sovereignty and very conscious of the importance of environmental resources. Like any landowner, these independent countries have control over the land under their domain. For this reason companies in modern industrial countries must not only obtain permission to prospect for minerals or to experiment with crops, they must also negotiate production agreements with the local governments. Although the corporation that invests in developing the mine or plantation and organizes the transportation usually retains basic management control, the rules under which it undertakes development are no longer simple.

The underdeveloped country is the landlord and at any time can foreclose on the leaseholder. Thus, American oil companies were shocked in the 1930s when Mexico nationalized its oil fields, asked the oil companies to get out, and expropriated not only the resource rights but the wells, equipment, and shipping facilities the companies had built. In recent years such actions have become commoner in all sorts of resource developments, sometimes precipitating international incidents and regularly involving long litigation to determine proper compensation to the companies for the facilities.

The World Oil Industry: Case Study in the Complexities of the International Reach for Resources

Perhaps the story of oil, the most important modern energy source in the world, will illus-

trate the complexities of the outreach of the interconnected system into the rest of the world and of the relationships between the corporations, their governments, and the countries in which the resource is found.

A Twentieth-century Industrial Resource

The use of oil and gas is the result of the second stage of the Industrial Revolution and is largely a phenomenon of the twentieth century. The development of the internal-combustion engine and oil and gas burners, all sophisticated engineering developments of advanced industrial technology, triggered this demand. For petroleum, at least, efficient use depends on a complex refining process that breaks the crude oil into a wide variety of derivatives ranging from asphalt to jet fuel. Thus, petroleum and gas are new as natural resources. They may have been formed millions of years ago, but man has perceived a use for them only recently. Moreover, they are resources only to men who have industrial machinery to refine and utilize them.

Major Producing and Consuming Areas and the Changing Geography of Production

Petroleum and natural gas are thus resources today to only those parts of the world that have modern technology. Although one can find jeeps and kerosene lamps and outboard motors in almost every remote corner of the world, the overwhelming concentration of energy-using machinery occurs in the established industrial centers we have investigated—Anglo-America, Europe, the Soviet Union, and Japan. These areas, which account for less than 30 per cent of the population of mankind, consume more than 80 per cent of the petroleum and over 95 per cent of the natural gas used in the world. Eight countries, all familiar to us, have daily consumptions of oil of over 1 million barrels:

the United States, the Soviet Union, Japan, West Germany, Britain, France, Italy, and Canada, in descending order (see Figure 12-1).

Petroleum and natural gas, we will recall, are found in sedimentary rocks, and sedimentary deposits are not found everywhere in the world (Figure 12-2 and Table 12-1).

In 1970 only one major industrialized area—the Soviet Union—was producing a larger share of the world's petroleum than it was consuming (Table 12-2). The United States, by far the leading producing country with almost a quarter of total production, was also the leading consumer with over one-third of total consumption. Western Europe, consuming over 25 per cent, was a major deficit area since it produced only 1 per cent. Japan also produced only a negligible amount of oil. Today all the industrial areas but the Soviet Union depend on major energy sources in the underdeveloped parts of the world (Figure 12-1).

Of the two major oil-producing areas of the underdeveloped world, one is in North Africa and Southwest Asia, which totaled almost 40 per cent of total world output. The second is in Latin America—mostly Venezuela—which produced more than 10 per cent of all production. Neither of these areas is a major consumer, and both are therefore major oil exporters. In fact, their production was developed to provide energy for Europe and the United States.

THE DOMINANCE OF LARGE INTERNATIONAL CORPORATIONS

In the first great period of oil exploration from 1890 to 1950, virtually all effort in the underdeveloped world was carried out by a handful of giant oil companies—British Petroleum, Shell (jointly owned by British and Dutch interests), and five large United States corporations:

Table 12-1 OIL AND NATURAL GAS RESERVES, 1970

REGION	CRUDE OIL (million bbl)	NATURAL GAS (billion cu. ft.)
Anglo-America	47,560	344,162
Latin America	31,143	61,760
Western Europe	6,489	146,375
North Africa–Middle East	381,611	449,803
Subsaharan Africa	6,840	6,400
Far East, Oceania	13,259	48,524
Soviet Bloc	59,460	425,665
World total	546,362	1,482,689

SOURCE: *World Oil: 26th Annual International Outlook Issue,* Aug. 15, 1971, p. 52.

Standard Oil of New Jersey, Mobil, Texaco, Gulf, and Standard Oil of California. Even in the late 1950s these companies controlled 90 per cent of the oil concessions outside of the United States and the Soviet Union.

Oil production in the Soviet Union has been a state-controlled monopoly, while in the United States it has been possible for companies large and small to contract with landowners for drilling rights on particular pieces of property. Thus, oil drilling in the United States has been carried out by large corporations and individual "wildcatters." Since those who drill for oil need not be refiners, the small operator can compete with the large.

Overseas it has been a different story. There a company needs the expertise to deal with the government of a country for a concession to drill for oil. Since in most underdeveloped countries the government controls most of the landholdings, only a single drilling concession

Table 12-2 WORLD OIL DEMAND AND SUPPLY 1970 (thousand bbl per day)

REGION	DEMAND†	PER CENT	SUPPLY‡	PER CENT	EXCESS OR DEFICIT
Anglo-America	16,711	35.7	13,242	27.7	− 3,469
Latin America	2,440	5.2	4,838	10.2	+ 2,398
Western Europe	12,241	26.1	547	1.2	−11,694
North Africa–Middle East	799*	1.7	18,815	39.3	+18,016
Subsaharan Africa	1,183*	2.5	1,457	3.1	+ 274
Far East, Oceania	6,369	13.6	1,398	2.9	− 4,971
Soviet Bloc	7,142	15.2	7,450	15.6	+ 308
World total	46,885	100.0	47,747	100.0	+ 862

*North African demand is included with Subsaharan Africa.
†Oil only.
‡Oil and natural gas.
SOURCE: *World Oil: 26th Annual International Outlook Issue,* Aug. 15, 1971, p. 49.

is commonly offered in a given country or region. The company must prove that it has the financial capacity to mount the tremendous development costs, which often include not only the cost of drilling but also of building roads, pipelines, ports, and storage facilities. The giant international oil companies also own their own tankers and refineries. They have also worked in close association to control the entry of competitors into the field who might threaten international marketing networks.

THE POLITICS OF OVERSEAS PRODUCTION

In the United States oil development and production depend on agreements made with a landowner, providing him with a basic royalty for each barrel of oil produced. In the Middle East and Venezuela the agreements are made with the governments, and production in their major oil pools has been a bonanza for the large oil corporations. These oil fields are among the richest in the world, and since there is no competition in the concession area, wells can be scientifically spaced for most efficient production. As a result production costs in Kuwait have been only 6 cents a barrel, and elsewhere in the Middle East they are no more than 12 cents a barrel.

The major cost has been the royalty payment to the local government. Originally these were modest, consisting mostly of local taxes. But the governments have become more and more aware of the value of their oil resources and of the importance of oil revenues to their countries. Thus, during World War II the first agreements were reached which set the split between government and oil company at 50–50 after taxes. In the 1960s such agreements were further adjusted to 70–30 in favor of the government. Nevertheless, total producing costs in most overseas areas are still far lower than those

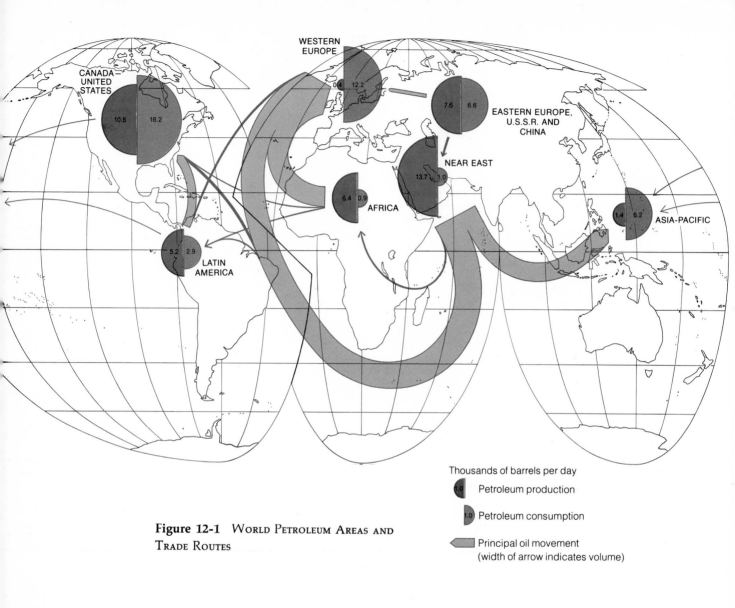

Figure 12-1 WORLD PETROLEUM AREAS AND TRADE ROUTES

CANADA–UNITED STATES

10.8 16.2

WESTERN EUROPE

0.4 12.2

EASTERN EUROPE, U.S.S.R. AND CHINA

7.6 6.6

NEAR EAST

13.7 1.0

AFRICA

6.4 0.9

ASIA-PACIFIC

1.4 6.2

LATIN AMERICA

5.2 2.9

Thousands of barrels per day

1.0 Petroleum production

1.0 Petroleum consumption

Principal oil movement (width of arrow indicates volume)

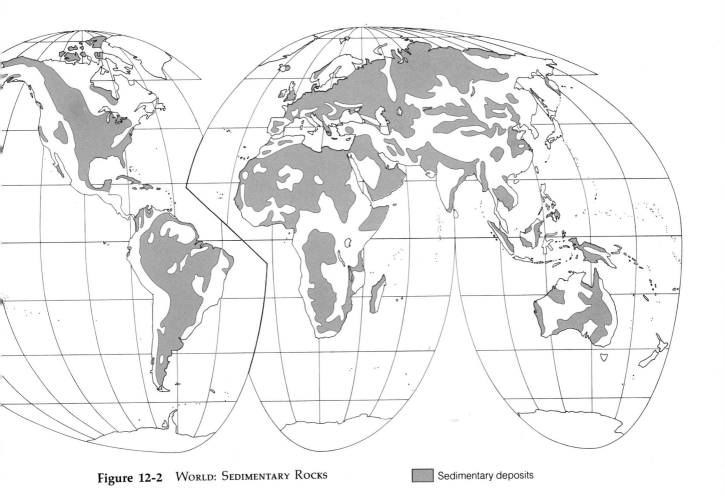

Figure 12-2 WORLD: SEDIMENTARY ROCKS   ▨ Sedimentary deposits

in the United States, and in the 1950s there was a flood of oil imports into the United States. As a result import quotas to protect domestic oil producers quickly followed.

Some countries, however, have not been satisfied with continued outside control of the profitable oil industry. After Mexico expropriated the properties of foreign oil companies in the 1930s, it has operated a nationally owned company. Iran attempted a similar move in the early 1950s but later had to turn operations over to a consortium of foreign oil companies. Recently both Peru and Bolivia, minor producers, have taken over oil operations in their countries. Government seizure is a continuing threat, and oil companies maintain diplomatic staffs to minister to the needs of the governments with which they have concessions.

The Complexities of the World Market

The rise of the major international oil companies with headquarters in the United States, Britain, and the Netherlands has been in part the result of the large investment needed in oil operations. It also indicates the scale of operation needed in the uncertain market of world production and sales. Production and marketing involve not a single product but a great variety of items. Crude oil itself is not a single chemical substance; it contains hundreds of hydrocarbon compounds, plus other minor elements. Some crude oil is thick and asphaltic; some is almost gaseous; some is high in sulfur. Also, the finished refinery products are many and diverse, ranging from asphalt, waxes, and heavy fuel oil to liquefied petroleum gasoline. The composition of products used varies greatly from market to market. In industrialized countries like the United States, gasoline accounts for half the demand for petroleum products, but in non-industrialized ones like India, the demand for kerosene (for lamps) is greater.

The discovery of new oil fields in countries that have not been significant producers has complicated the situation further. Countries seeming to have a near monopoly on world oil exports have found themselves faced with competition from new producing areas, some of which have larger and more productive fields. At the turn of the century, Russia, the United States, the Dutch East Indies, Rumania, and Mexico were the major exporters. After World War I, Venezuela and the first Middle Eastern fields began to dominate the world market. From World War II until 1960, other major fields were discovered in the Middle East, and it became the principal source of world oil exports. Recently major production has developed in the Sahara of North Africa, and now new developments are occurring in Indonesia and southeastern Australia.

But with the control of most oil production in their hands, the large international oil companies have been able to respond quickly to both new markets and new production sources. All have multiple sources of supply, a series of refineries throughout the world, and marketing facilities in many countries. The management problem has been to keep this world-wide system at peak efficiency, shifting crude oil output from field to field in the way best to meet demands. The addition of new flow from Libyan fields, for example, may call for a rethinking of plans to expand production in Kuwait.

The Impact of Oil on the Underdeveloped Oil-producing Countries

While such fluctuating arrangements may be very satisfactory as far as the international oil company is concerned, they are another matter to the producing countries. Libya, Kuwait, Iraq, and Venezuela count on oil revenues for more than 90 per cent of their foreign exchange, and Iran and the Arab countries depend on them almost as much. In an inflationary situation these countries are continually seeking more oil income to provide for needed foreign purchases. If production is cut back, there is heavy pressure to increase their percentage of the oil royalty formula.

Recognizing their precarious position in the oil supply situation, most of the major oil-exporting countries have banded together to form an organization concerned with prices and ratios of return. Equally important, in recent years they have opened up their lands to other oil companies. The Japanese, Italians, and a variety of smaller American oil companies are now in international production operations. When Libya opened its promising Sahara fields after 1955, thirteen different companies applied for concessions.

The development of oil resources in a particular country leads to a conflict of interest. Whereas the company is concerned with fitting the output of that particular producing unit into its global operations and adjusting output to particular market conditions, the country sees oil production in its area as its only source of foreign exchange with which to buy the products of the industrial world. To the underdeveloped country oil is an indirect asset, but the amount of production from its wells is determined by the international company. The government can only exert political pressure to change decisions. It can threaten nationalization of the industry, but this means loss of market because the producing country has neither an internal market to utilize the oil nor the ships needed to transport it to other markets.

In the early 1950s when Iran tried to nationalize its oil production, the large international corporations which controlled almost all the tanker fleets of the world simply boycotted Iranian oil while expanding production in other countries. As a result, the production of the Iranian fields slowly ground to a halt. Even though Iran took over one of the largest oil refineries in the world along with the producing facilities in the fields, and even though it could hire trained technicians to produce and refine the oil, there was no means to ship it. Within a few years Iran was ready to negotiate with a consortium of the major oil companies. Mexico has been able to nationalize oil largely because its growing economy has provided an internal market.

In the 1970s the oil-exporting countries have taken another tack in attempting to gain further control over the production of their own resources. Instead of expropriating the oil-producing companies or demanding further royalty payments, they have been asking to be taken in as part of the investment group in the company. Thus, early in 1972 the Saudi Arabian government reached an agreement with the Arabian-American Oil Company which gave the government a 20 per cent share of the company. Other countries have been pushing for similar agreements. Moreover, demands call for the percentage share to increase with time. Thus, the oil countries are seeking increasing direct participation both in company profits and decision-making, with long-term aims of majority control.

These royalty payments to local government may or may not be made available to the people. Countries such as Kuwait and Saudi Arabia are controlled by absolute monarchs. The money goes to the personal treasuries of the leaders to be used as they wish. Thus, in

THE COMPLEXITIES OF RESOURCE DEVELOPMENT IN AN UNDERDEVELOPED COUNTRY: THE CASE OF COPPER IN WEST IRIAN

The complexities of an international resource development operation can be seen in the recent exploitation of a rich copper deposit in the mountains of West Irian, a remote and little developed part of Indonesia on the island of New Guinea.

The deposit, the largest known to exist in the world, was discovered in 1936 by Dutch explorers when New Guinea was still part of the Dutch East Indies. But the field party's report was not acted upon until 1958 when another Dutch firm discovered it in the files. Lacking adequate financial and research resources, the new company asked Freeport Sulphur, a large American mining-chemical firm, for help.

In 1960 engineers supported by the Dutch government (for it was still nominally in control of West Irian even after the Indonesian republic was established) surveyed the deposit. Two years later the Dutch ceded West Irian to Indonesia. The result was a complete hiatus in development, for the Indonesian government of the time was very nationalistic and strongly anti-Western. For five years no development took place, but when the regime was overthrown in 1967, a new government, much more receptive to foreign resource development, came to power. Within three months an agreement with the government had been reached by Freeport Sulphur, and plans were reactivated. Developmental costs were set at $7.4 million, but it was estimated that another $120 million would be needed to start production.

The result was a complex bit of international money raising. Since American copper smelters had adequate sources of copper, eight Japanese companies contracted to buy two-thirds of the mine's output, and a German company agreed to take the rest. On the basis of these agreements, Freeport Sulphur obtained $20 million from the Japanese firms and $22 million from the West German development bank, guaranteed by the government. The remaining funds were raised in the United States. Seven commercial banks put up $18 million based on Export-Import Bank guarantees. Forty million dollars came from five life insurance companies using Agency for International Development guarantees. It provides risk insurance protecting Freeport Sulphur against war, expropriation, and the inability to convert money from Indonesian accounts.

The actual cost of the on-site undertaking has been tremendous. The company has to build a port that can accomodate 15,000-ton ore ships and cut a 69-mile road through tropical rain forests into the mountains. Then an aerial tramway had to be built to the base of the ore body. The ore will be brought down by conveyor belt, crushed at the base of the mountain, concentrated into slurry, and then pumped through a pipeline to the coast.

Now that a market is assured, the results of this complex process deem it worthwhile. There are an estimated 33 million tons of ore of exceptionally high quality. The assay is 2.5 per cent copper, whereas United States mines are working copper at less than 1 per cent. This means 825,000 pounds of pure copper valued at more than $900 million at 1970 prices. In addition, there are other minerals that can be extracted from the ore as by-products, including an estimated $28 million of gold and $13 million of silver.

In order to encourage the development, the Indonesian government agreed to a three-year tax-free period even though the mine may be in operation. After the first three years, the tax will be 35 per cent for seven years and 41.75 per cent thereafter. But Freeport Sulphur hopes to have exhausted the ore in slightly more than thirteen years.

the early days of oil production in Arabia, the money was used to build palaces and to buy automobiles and other royal luxuries. On the other hand, the emir of Kuwait, with a tiny total population, has chosen to build free schools, hospitals, and dental clinics.

Oil production in underdeveloped countries has not meant new energy sources for national development and only occasionally has it meant improvement for the common people. In all cases it has meant jobs. Workers are needed in construction and on the oil rigs, driving trucks, and working in refineries and in stores and shops associated with the new oil operations, but even in industrialized countries the industry is not a large employer. Far fewer workers are needed in the oil fields than have been traditionally required in coal and iron mining. Refineries are among the most automated of modern industries, and pipelines deliver quantities of oil with far fewer workers than a railroad. Even at the ports tankers can be loaded by attaching a pipe from the storage facilities.

The Situation of Underdeveloped Producing Countries in the Interconnected System

The precarious position of the underdeveloped countries that depend on overseas markets in the great industrial countries can be seen by examining another example of a critical mineral resource: copper. Although it is far less important than petroleum, copper is essential to the modern world and is the foreign-exchange base of several underdeveloped countries.

Copper's two key properties are its ability to conduct electricity and heat and to resist corrosion. Thus, it is most important in the electrical and communications industries and the heating and plumbing industry. It is also vital for heat transfer in engines.

The world's leading copper-producing countries are the large industrial countries—the United States and the Soviet Union. Together they produce over a third of the world's output (Table 12-3). Europe has little copper production, and American consumption is so great that the United States must depend on imports as well as domestic production.

Underdeveloped countries are also major producers of copper and it is of great importance in their export trade position. Zambia's exports in 1968 were 95 per cent copper, and Chile's were 78 per cent. Thousands of workers in each country were employed in mines. Like petroleum, copper has been developed by American and European interests. Fewer than a dozen companies have controlled most of the world's supply. But in recent years there has been strong pressure in the leading copper-producing countries for control of production within their boundaries. Zaire (formerly the Democratic Republic of the Congo) now owns its mines completely; Zambia has 51 per cent of financial control; Chile and Peru have been moving toward complete nationalization.

Control of production facilities has not meant control over the copper market, however, for this depends on consumption in the industrial countries, and copper is notorious for tremendous price fluctuations. Because it is mainly used by industry, its consumption varies sharply with shifts in industrial production. In 1970, at the very time that Freeport Sulphur was planning to develop deposits in West Irian, copper supply was running far ahead of demand. The stockpile in the London Metal Exchange had increased more than four times, and United States inventory supplies were even higher. The price of copper on the London ex-

Table 12-3 MAJOR COPPER PRODUCERS, 1969 (copper in ore in thousands of metric tons)

COUNTRY	METRIC TONS (000)	COUNTRY	METRIC TONS (000)
United States	1,401.2	Peru	206.1
USSR (est.)	900.0	South Africa (est.)	148.0
Zambia	748.2	Philippines	131.4
Chile	697.4	Australia	130.0
Canada	500.2	Japan	120.3
Zaire	362.2	Total	5,345.0
		World total	5,999.0

SOURCE: *Encyclopaedia Britannica, Book of the Year: 1971*, Encyclopaedia Britannica, Inc., Chicago, 1971, pp. 520–521.

change hit a six-year low of 47 cents a pound, down 40 per cent in nine months. The chairman of one of the largest copper companies predicted a surplus of more than 850,000 metric tons by 1975 at present demand rates.

The effects of the slump in copper demand and prices to Zambia, Chile, and Zaire can be readily imagined. Loss of income can be a serious blow to developmental plans and cause extensive unemployment in mining districts. In times of depression, the price of nationalistic control is felt as each country attempts to maintain production in a market it cannot manage.

The producing countries have been meeting to discuss their common problem of the copper price slide and have discussed various solutions, including limits on copper production and a program of massive buying to minimize the price drop. But neither method offers much hope because production cuts would increase unemployment and would work only if all major producing countries agreed. Large-scale stockpile buying to support prices would be possible only if huge loans of international capital could be obtained.

PRODUCTION WITHOUT CONTROL

Variations in the flow and price of copper, oil, or other basic commodities produce serious short-run difficulties for the interconnected system and the industrial countries that are a part of it. But for the underdeveloped countries involved it is little less than disaster. Recent years have seen the rise of political independence, but this has done almost nothing to eliminate the subservient economic position in which most underdeveloped countries find themselves.

It is striking how underdeveloped countries compete with one another for markets for their relatively few major products (Table 12-4). Look at the wide range of countries producing tropical products: coffee, cocoa, peanuts, bananas, rubber, tea, palm products, and tropical woods. The same pattern also exists for subtropical products, sugar, cotton, jute, and rice, and for

Table 12-4 IMPORTANCE OF PARTICULAR COMMODITIES IN EXPORTS OF UNDERDEVELOPED COUNTRIES, 1968 (per cent of country's exports)

SUGAR

Philippines	17
Taiwan	6
Swaziland	18
Mauritius	94
Barbados	81
Cuba	85
Dominican Republic	54
Guyana	35
Haiti	9
Jamaica	20
Mexico	8
Peru	7
Trinidad	5
Malagasy	6
Ecuador	5

COFFEE

Burundi	82
Cameroun	27
Central African Republic	20
Equatorial Guinea	21
Ethiopia	58
Guinea	6
Ivory Coast	34
Kenya	21
Malagasy	31
Rwanda	55
Uganda	54
Tanzania	16
Togo	17
Brazil	41
Colombia	59
Costa Rica	32
Dominican Republic	11
Ecuador	24
El Salvador	47
Guatemala	35
Haiti	38
Honduras	12
Mexico	10
Nicaragua	14
Indonesia	5
Laos	18
Zaire	6

COCOA

Cameroun	23
Ghana	55
Nigeria	25
Equatorial Guinea	44
Togo	24
Dominican Republic	8
Ecuador	15
Western Samoa	38
Ivory Coast	20

COTTON

Afghanistan	12
Burundi	9
Central African Republic	23
Chad	89
Dahomey	10
Upper Volta	19
Tanzania	18
Uganda	22
Sudan	60
Brazil	7
El Salvador	7
Guatemala	16
Mexico	14
Nicaragua	38
Peru	6
Pakistan	14
Syria	37
Turkey	27
United Arab Republic	61
Mali	33

BANANAS

Guinea	10
Western Samoa	5
Somalia	34
Costa Rica	26
Ecuador	44
Haiti	12
Honduras	47
Jamaica	8
Panama	59
Taiwan	6

PEANUTS

Gambia	93
Malawi	11
Mali	8
Niger	77
Nigeria	18
Senegal	77
Upper Volta	9
Sudan	7

RUBBER

Cambodia	25
Ceylon	16
Indonesia	33
Malaysia	33
Singapore	22
Thailand	13
South Vietnam	79
Liberia	17

TEA

Ceylon	57
India	13
Uganda	6
Kenya	18
Malawi	24

TIMBER

Cameroun	6
Congo (Brazzaville)	45
Equatorial Guinea	19
Gabon	5
Ghana	8
Ivory Coast	25
Swaziland	20
Philippines	26
Burma	30
Laos	32
Malaysia	17
Paraguay	17
Honduras	8

PALM OIL OR NUT COPRA

Dahomey	49
Guinea	7
Philippines	28
Togo	6
Ceylon	16
Indonesia	5
Thailand	5
Western Samoa	52
Nigeria	5
Sierra Leone	11

TOBACCO

Malawi	26
Dominican Republic	7
Paraguay	10
Turkey	19
Rhodesia	29

MEAT AND CATTLE

Botswana	90
Chad	6
Kenya	5
Lesotho	29
Argentina	24
Nicaragua	10
Uruguay	34
Lebanon	8
Mongolia	34
Upper Volta	51
Swaziland	5

RICE

Burma	35
Cambodia	42
Guyana	13
Malagasy	11

JUTE

India	15
Pakistan	22

PETROLEUM

Gabon	34
Kenya	12
Colombia	10
Panana	19
Trinidad	78
Algeria	76
Iran	90
Iraq	93
Kuwait	97
Libya	99
Saudi Arabia	94
Southern Yemen	62
Tunisia	17
Venezuela	92
Nigeria	18
Indonesia	35
Singapore	19

COPPER

Zaire	51
Peru	27
Cyprus	20
Zambia	95
Uganda	8
Chile	78

DIAMONDS

Central African Republic	47
Congo (Brazzaville)	6
Zaire	34
Ghana	5
Lesotho	24
Sierra Leone	57
South Africa	13
Tanzania	8

BAUXITE-ALUMINUM

Cameroun	10
Guinea	60
Dominican Republic	8
Guyana	16
Jamaica	49

TIN

Zaire	6
Bolivia	53
Nigeria	6
Rwanda	30
Indonesia	5
Laos	43
Malaysia	20

IRON-ORE

Liberia	75
Brazil	6
Chile	7
Peru	7
Cyprus	8
Venezuela	5
Mauritania	76
Sierra Leone	13
India	7
Swaziland	28

PHOSPHATES

Jordan	29
Morocco	25
Tunisia	27
Senegal	7

SOURCE: *Encyclopaedia Britannica, Book of the Year: 1970,* Encyclopaedia Britannica, Inc., Chicago, 1970; data drawn from inserts of each country.

To understand the problems of underdeveloped countries in managing their own affairs, let us look at an underdeveloped area in the United States. It is a huge portion of the interior of the country, centered roughly around the confluence of the Mississippi and Ohio rivers and extending approximately 600 miles from northeast to southwest and about 300 miles from north to south (Figure 12-3).

This is one of the most productive resource areas in the world. It mines over 90 million tons of coal; only six countries in the world produce more. Less than twenty countries in the world produce as much oil. Iron-ore production is high; and the area is the foremost lead-producing region in the world. There are also major deposits of fluorspar, glass sand, and ceramic clay. This is one of the important cement-producing areas of the United States. Moreover, total agricultural output is greater than such leading agricultural states as Indiana and Minnesota. It includes the leading cotton area in the South, a major rice-producing area, important tobacco fields, and land devoted to growing large quantities of corn and soybeans. As the focal point of the country's largest river network there are unlimited water resources.

Yet in the mid-1960s the area was one of the most economically depressed parts of the country. It had more than 215,000 persons on welfare, 6 per cent of its total population. Median family income in almost every county was less than the national average. In many counties more than half the families had annual incomes below $3,000. More than half of the counties were designated as "depressed" and were eligible for special government assistance.

The basic problem of this area has not been a lack of development. Rather, it has been the nature of the development. Virtually all mineral exploitation has been the result of investment from outside of the area—from St. Louis, Chicago, and New York. Moreover, the development has been designed to carry the raw materials to producing centers outside of the area, generally to metropolitan centers. The coal from one area was developed to serve particular markets. And so on. In most cases different industrial corporations undertook each development.

In agriculture the story is similar. Production has been in the hands of local residents, but the markets have been in various cities. No attempt has ever been made to utilize the primary producing base for an industrial complex within the region. As a result there are only two metropolitan areas, and each of them has a total population of less than 200,000 people. Of the five other communities of over 25,000 people, two are university towns. As might be expected, the transportation system is designed primarily to make connections from the separate productive parts of the area to peripheral cities rather than to center on any point within the area. There is only one main-line railroad.

Internally the people of the area do not think of themselves as part of a single region. They cover parts of six separate states, each separately settled. There are strong local loyalties, and each state sees itself as distinctive within its own borders.

This case illustrates that it is not simply the resources of an area that determine regional success. The people of this region of the United States are a part of the modern, interconnected system. It provides them with employment—in mines, on farms, and in supporting towns. But they receive little share of the profits. If the demand for the coal produced in this area is great, production will be high; if it is low, whole areas lose their productive base. Furthermore, if the mines were mechanized for greater efficiency, nine out of ten miners could find themselves out of work.

Figure 12-3 REGION SURROUNDING THE
CONFLUENCE OF THE MISSISSIPPI AND OHIO RIVERS

Coal mining area

Oil fields

× Lead ore

▫ Iron ore

Eligible redevelopment area

• 10,000 to 25,000 population

• 25,000 to 100,000 population

• Over 100,000 population

petroleum, copper, diamonds, bauxite, tin, iron-ore, and phosphates. Notice, too, how often neighboring countries have the same specialty economies and thus must compete with each other.

The difficult position of the underdeveloped countries in the world-trade structure can also be seen. Not only are they not in control of the market and movement of their products, they are also in the weak posture so characteristic of farmers and other primary producers wherever they are in the world. They produce raw materials which sell at low prices, and need the manufactured goods and services of the industrial world which sell at much higher rates. The foreign exchange gained in the sale of coffee, rubber, or bauxite does not go far when one needs to buy machinery, transportation equipment, and other sophisticated technical supplies from industrial areas.

Moreover, financial control often is not in the hands of the local people, and, as a result, neither is the profit-making component of the transaction. Though payments are made to the government of the country, and though workers in those countries are paid wages, the profit from the operation goes to the foreign company that made the investment. Commonly, total investment in facilities in a country is much less than the profits taken out. This problem is compounded when we recall that in the modern, interconnected system money is made not only in producing but also in financing, managing, transporting, and selling the product. Both New England and Japan succeeded well as primary producers and simple manufacturers because they were able to gain control over the business operations. Soon they were their own entrepreneurs, managing and financing the whole undertaking from the production of raw materials through the sale of the finished prod-

uct. Few underdeveloped countries have managed that.

The "Dual Economies" of Underdeveloped Countries and Problems of National Development

Although the underdeveloped countries are regarded by the industrial countries primarily as sources of raw materials and foods, and although they are dependent on others for the capital for their economic development, the people in most of these countries are not primarily producers or supporting workers of the interconnected system. They belong to a completely different way of life. They are still part of the traditional, locally based society typified by the village of Ramkheri. Moreover, the traditional society is older and has roots deep in the culture of the people. The people not only depend almost entirely upon the resource base of their tiny corner of the world but have their primary cultural and political allegiances within that same local area.

Two Worlds in Juxtaposition

In the underdeveloped world what we really find in each country are "dual economies" in juxtaposition. Mines in Africa depend on workers drawn from the traditional tribal society, and peasant farmers in various parts of the underdeveloped world sell some of their production whenever they can in exchange for a few manufactured products from the outside world. Surely, the two systems are in contact with one another in the local area. Such contacts are more than economic meetings; more often they are also cultural and political. This happens when government officials make efforts to enlist the local population by educating them, taxing them, indoctrinating them, or looking after their health.

TIES WITH INDUSTRIAL COUNTRIES, NOT OTHER DEVELOPING COUNTRIES

Traditionally the people of underdeveloped countries have not had contact with other underdeveloped areas even on their own continent, not to mention other parts of the world. They form separate cultural and economic units. Often these units have been smaller than the countries; the country, much less the continent, does not have a cultural unity.

The ties that have developed are those that have come with the modern, interconnected system. They follow the routes of transportation and communication established by the westerners. Often it is European ties, such as the English or French language, that provide the communication link between one culture and another. Long-distance telephone calls between African countries must be routed through London. Other contacts come when people of different cultures are brought together at a work site, as when workers are recruited from different regions.

At the national level the native leaders of the new countries have been trained in the ways of the modern world through contacts with westerners in western schools, in colleges, or in training programs in the industrial countries. Until the recent entry of the Soviet Union and the Japanese into the field, the contacts native leaders made were generally with the industrial countries that had the closest economic and political ties to their homelands.

It is not surprising that India and the African countries that are members of the Commonwealth of Nations have emerged with governmental structures and even protocol manners very like those of their former colonial masters or that, indeed, they have joined the Commonwealth. Former territories of France, whether in Africa or Asia, tend to show a similar French orientation.

Politically the most important embassies of underdeveloped countries have been those in the capitals of the leading industrial countries. Until recently there has been little basis for discussion with neighboring governments. The United Nations brought diplomatic representatives of the underdeveloped world together and enabled them to discuss their common problems and destinies. In the General Assembly and in the many supporting agencies of the United Nations, the underdeveloped countries have begun to emerge as a special-interest group. As the result of such contacts, world-wide conferences of "underdeveloped countries" have been held.

CULTURAL FRAGMENTATION RESULTING FROM THE IMPACT OF THE MODERN SYSTEM

To gain an appreciation of the problems of superimposing the modern, interconnected system on traditional life in underdeveloped countries, we must remember that new ways add yet another dimension to the already fragmented pattern of separate local cultures. The modern system is not limited to the economic sphere. It is a whole new way of life, offering a different culture with its own values, customs, and life style. Its addition breaks up traditional national patterns and often divides families. Some people in underdeveloped countries, commonly the young people and the city dwellers, heartily accept the modern ways. Others, usually older and more conservative, reject anything new and try to maintain traditional values. Still others are often confused. New values make them question old ways that were once accepted, but they are afraid of anything new. Thus, a society which once had the acceptance of most of its people has become divided into factions. We saw this revealed in the changes that were occurring in Ramkheri. But on the scale of a whole country, the conflict between old and new appears in political as well as cultural form and tears the fundamental fabric of society.

THE POPULATION EXPLOSION AND THE RESOURCE BASE

The introduction of the modern, interconnected system into areas of traditional culture has led to rapid population growth in underdeveloped countries. The population of Africa estimated at 120 million in 1900 was 344 million in 1970, and the pattern holds for Asia and Latin America. Some of this increase, particularly in Latin America, was the result of the in-migration of Europeans and their slaves, but, in large part, the increases represent natural growth by more births than deaths.

Much of the rapid increase can be attributed to the introduction of modern health and sanitation measures. Inoculation against diseases, water purification, higher living standards with better diets resulting, and the use of insecticides have all contributed to lower infant mortality and increased life expectancy. Airlifts of food supplies and medicines have cut the toll taken by famines following floods, droughts, earthquakes, and plagues. Birth rates have not declined, however. In rural areas where children were regarded as economic assets because they can work in the fields, the tradition favoring large families resulted in a population explosion.

In the modern, interconnected system the population explosion means congestion, air-

stifling pollution, and problems of waste disposal. In traditional areas where the population is sparse or the resource base particularly rich, population growth does not appear significant. But in traditional societies where the productivity of the existing economic system is near the upper limits of its capacity for providing food and other basic needs, population growth will produce a new manifestation of the Malthusian problem. Without major improvement in production, sooner or later one of the population controls that Malthus described—famine or war—will appear. Increased productivity depends upon technological change, and that can come only by shifting from the old closed system to longer-range connections.

CHANGES IN THE CLOSED SOCIETY

Of course, there have been changes in the rural world. Some segments of all countries have become productive components of the interconnected system. Moreover, scientific knowledge has spread to the locally based society from the modern world. There have been programs for agricultural education, schemes for improving drainage and irrigation, plans to implement the use of fertilizers, and improvements in simple tools. In the 1960s major breakthroughs were made in the development of "miracle" rice and wheat strains said to increase field yields 50 per cent or better. Even with these improvements the traditional agricultural pattern is still the way of life for most people.

Some romanticists would like to see a return to the "good, simple, primitive life of the past," but this is impossible. In most places population densities are too high for complete support from the local base. Moreover, social and economic institutions have been so modified that societies have become dependent on the modern world and are no longer self-sufficient.

Kikuyu tribesmen of Africa, while still residing in villages and maintaining their traditional social life, now live on poorer lands as a result of the settlement of their areas by Europeans. Money has replaced traditional barter, and cattle and goats are no longer a measure of a man's wealth. Children go to school to read and write, not to become hunters or warriors. The government has encouraged the Kikuyu to abandon their tiny family fields and to work together on larger properties. Coffee and tea are planted for sale. Dairy herds are kept within fenced pastures and no longer allowed to wander wild. The Kikuyu are still terribly primitive by Western standards, but their life has been changed fundamentally. Clearly they could never return to their old way of life.

Governments in underdeveloped countries hoping to improve the standard of living are faced with dual developmental problems. On the one hand, they want to see the modern, interconnected component grow, for with it comes both a rising living standard and increasing political prestige. On the other, they must solve the problem of the population explosion within the traditional segment of the population. In countries where the population pressure on the traditional component is high, the government must often tap the money gained from exports in the modern component to purchase the food necessary to support the overpopulated traditional society.

The problem of development in the modern, interconnected segment of an underdeveloped country centers around control. The industrialized countries of the world have always been interested in development within underdeveloped areas. But they plan for their own ends, not for those of the country in which they in-

vest. Thus, companies from the modern industrial countries are anxious to tap rich, new mining areas and potential plantation regions. They hire local labor to build the new facilities and to help in production. But, as we have shown in the discussions of the oil and copper industries, it is the foreign industrialists who make the basic decisions, control the movement and marketing of products, and reap the profits. Governments of underdeveloped countries get only an indirect share of the operation— royalties and other payments on the amount of product produced.

Underdeveloped countries might like to develop and control their own mines and plantations that export to the world. But struggling countries have little capital to invest in such costly enterprises and far less leadership and management experience than the companies with which they must deal. Nor is there any prospect of raising the necessary capital from taxes. Such poor people have little monetary income and little agricultural surplus to be taken as taxes. Indeed, if population pressure is great, the government will barely be able to provide for its people's needs.

Like the American republic in the eighteenth century, underdeveloped countries commonly have no adequate transportation and communication systems connecting the different regions of the country. The railroads, highways, and ports have been developed by foreign investors to get the products of mines and factories to the coasts, not to provide internal integration for the country. Commonly no transport net connects the capital to all settled parts of the country to offer the necessary base for economic and political integration. Instead, the productive parts of the country are tied to overseas markets, not to the largely subsistence areas elsewhere in their own country.

What to Do with the Closed System

The second problem of underdeveloped countries sometimes looms as that of first priority. With the results of a population explosion, the pressure of the traditional economy on its productive base may be almost critical. The problem may call for short-term programs to provide emergency food while at the same time programs of education try to promote birth control. But in the long run there will have to be an increase in the productivity of the system.

The long-term solution is to increase the output of the subsistence component of the population, whether that increase continues to be used simply to support a locally based system or to begin to provide some cash sales to the subsistence producers. This will require nothing less than an industrial revolution which will result in a complete upheaval.

Again, there is a shortage of both capital and management leadership. People living in a traditional society are conservative with regard to change. Often religious beliefs, built around trying to find meaning in their lives of work and privation, lead them to revere traditional ways.

Too, leadership and power are most often vested in the elders—the most experienced, the most conservative, and the ones with the greatest stake in the status quo. Moreover, living on the brink of starvation makes people extremely reluctant to try something different. Inefficient and unproductive as old ways may be, they have been tested through time, and new ways are regarded as risky. There is need for massive education, not just reading and writing, to foster an appreciation of modern ways.

The dual societies of the underdeveloped areas of Africa, Latin America, and Asia vary greatly. Although each of these countries in various regions of the world stands unique, they are roughly similar. In eastern Asia the problem of population pressure on the land stands out acutely. In the Arab countries and in northern Africa, basically feudal societies have begun to modernize. In tropical Africa some of the newest and smallest countries have recently emerged from colonialism. In Latin America European immigrants are part of the dual economies there. We shall explore each of these submodels in succeeding chapters.

The Problems of the "Dual Economy" in Brazil

The problems of the dual society can be sketched by looking at Brazil, one of the largest of the underdeveloped countries. Brazil has essentially had five different producing areas which feed goods into the interconnected system (Figure 12-4). First, there is the old sugar-producing area of the northeast coast. Second, there are the interior frontier cattle lands in the south. Third are the old rubber lands of the Amazon where, at the close of the nineteenth century, a tremendous rubber boom took place. Fourth, there are the scattered mining communities northwest of Rio de Janeiro where first gold and silver and now iron have been developed. Finally, there is the rich coffee-growing area centered on São Paulo near the coast along the Tropic of Capricorn.

With the exception, perhaps, of the western cattle frontier which had connections with both São Paulo and Rio de Janeiro, each of these different producing regions has had its own transportation system to get products to overseas markets. Ocean ships traveled up the Amazon to the rubber-producing centers, and the ports of Recife and Salvador served the sugar coast. Rio was the focal point for a rail

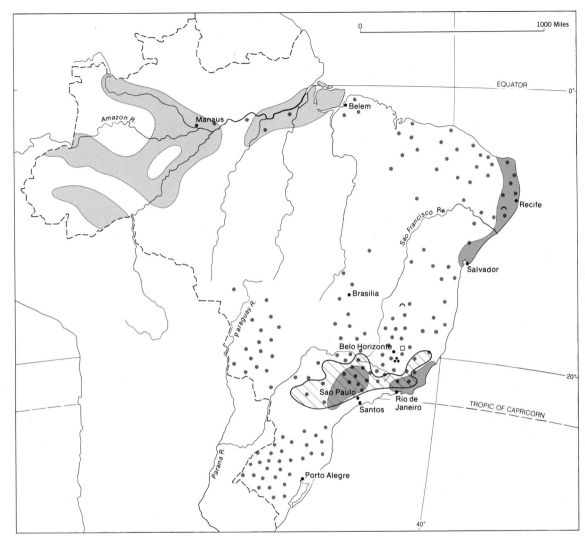

Figure 12-4 BRAZIL: FIVE MAJOR PRODUCING
AREAS

Sugar cane

Coffee

Rubber

Cattle

⌒ Silver

⁂ Gold

▫ Iron ore

network into the mining areas, and São Paulo was the point for another one into the coffee country. Each region has fared differently economically. The booms in rubber and cattle, at least for overseas shipment, are long gone. The sugar coast has never regained the prominent position it held during the colonial period. On the other hand, mining has become profitable, and the São Paulo hinterland has become both the leading coffee-producing district in the world and the economic core of modern-day Brazil.

Outside of these areas and in nonexporting communities within them, the population of Brazil lives in the traditional closed system. Some areas are frontier regions occupied by Indians, traders, and missionaries; others are peasant-farming areas where the local marketplace is the destination of most goods sold.

How does a government put all the disparate and essentially independent components of Brazil into a single functioning country? The question has not yet been answered. If giant Brazil, one of the more advanced countries of the world, has not been able to answer it, how can others, smaller and weaker, accomplish it?

The problem is compounded when one recalls that many underdeveloped countries of the world sprang from European origins, not their own. Their boundaries were set by European powers in colonial days, and those boundaries reflected European interests and compromises, not native interests. Thus, borders cut through the territories of a single native group or surround tribal peoples who have traditionally been rivals. Furthermore, since colonies were organized to exploit the resources of an area, Europeans often set up separate colonial units wherever an individual exporting system focused on a port.

The result is that the new countries are commonly both small and fragmented. Over half of the 102 underdeveloped countries have populations of 10 million people or less, and only 14 have more than 25 million. Eleven countries have less than a million people. Thus, the typical underdeveloped country is in an extremely disadvantageous position in the world when one considers that it must compete economically, politically, and militarily with the industrial world. Yet, even underdeveloped countries with their small populations face the problem of integrating peoples of different local backgrounds, even different languages, into a functioning nation.

Individually these underdeveloped countries and their problems do not loom large in the press and other media of the industrialized countries. But in Latin America, North Africa, and the Middle East, Africa south of the Sahara, and South and East Asia other than Japan is over two-thirds of the people of the world. Thus, their problems are world problems, and their future is of major importance to the whole world. In the next part of this book we will examine three of the most important areas of the developing world: Latin America, Africa south of the Sahara, and South and East Asia, where we will focus on India and China, the two largest countries in the world in terms of population. Each of these areas shares the problem of the dual cultures described in the Brazilian example, but, as with the different industrial areas, each has its own distinctive problems and potential.

SELECTED REFERENCES

Boesch, Hans. *A Geography of World Economy,* Van Nostrand, Princeton, N.J., 1964. A traditional economic geography with emphasis on levels of development and the importance of trade.

Ellsworth, P. T. *The International Economy*, 3rd ed., The Macmillan Company, New York, 1964. A standard text in international economics gives a different view of the interaction of modern and underdeveloped countries.

Lloyd, Peter S. *International Trade: Problems of Small Nations*, Duke U. Press, Durham, N.C., 1968. An economist looks at the position of small underdeveloped countries in a world dominated by the industrial countries.

Rostow, Walter W. *The Stages of Economic Growth*, Harvard U. Press, Cambridge, Mass., 1960. The classic analysis of the problem of economic development.

Thoman, Richard, and Conling, Edgar. *Geography of International Trade*, Prentice-Hall, Englewood Cliffs, N.J., 1967. Another of the small paperback series by this publisher provides a basic treatment of international trade from the viewpoint of a geographer.

PART FOUR   DEVELOPING AREAS: MODERN

AND TRADITIONAL WORLDS

IN CONFLICT

Chapter 13   Latin America: Underdevelopment in an Offshoot of European Culture

LATIN AMERICA has a diverse geographic base. It includes the whole of South America, Central America (the southern portion of North America, i.e., south of Mexico and north of Colombia), Mexico, and the islands of the West Indies (Figure 13-1). It extends across the tropics into the subtropical portions of the northern and southern hemispheres. Its major mountains extend almost the full latitude from north to south, and there are climatic variations in most countries due to changes in altitude.

Yet despite its environmental diversity, the region has been dominated by a single cultural system that affected almost everyone. Latin culture, transported from Spain and Portugal over several hundred years, can be found in modified versions from Argentina and Chile in the south to Mexico in the north. Spain and Portugal dominated the entire region for over 300 years. Although most of the countries have been independent for almost 150 years, vestiges of colonial life are still present in the government, the economy, and the social life of the region. This cultural heritage is probably the most difficult problem facing Latin America today.

Except on a few islands in the Caribbean, Spanish and Portuguese are the dominant languages. Roman Catholicism has been recognized everywhere as the national church, and roots of an essentially feudal economic and political structure are found in every Latin-American country.

The Latin culture exists despite the great ethnic diversity among the people (as we see later in Figure 13-7). In some areas native Indians dominate, particularly in Central America and in the Andes highlands from Colombia down through Peru. In others, such as the West Indies and the northeastern coast of Brazil, descendents of African slaves are a major component of the population. In Argentina, parts of Brazil, and Venezuela, English, German, and

Italian roots are significant. In Guyana are areas that are dominantly East Indian. But while the variations in ethnic character add some diversity to Latin America, these non-Latin peoples have nevertheless found themselves dominated by the Latin-American culture around them. The Indians were overrun by the Spanish and Portuguese, and other ethnic groups were brought in to serve particular functions in an already established system. Despite the centuries that have gone by since the arrival of other groups, the Indians have continued to be a part of the Latin-American system of life.

The cultural unity that we infer from the term *Latin America* is not, in fact, the result of close ties among Latin-American countries or their peoples. A glance at the population map (Figure 6-4) shows that there is little physical contact between the population of one country and another. The centers of population of adjoining countries usually stand as separate islands removed from one another by vast empty stretches. Moreover, the transportation map (Figure 4-7) shows how few surface connections there are between countries. Like the former colonies of Africa and Asia, Latin-American countries have their primary ties to Europe and the United States rather than to their neighbors. Each country stands largely in cultural and economic isolation.

The Dominance of Rural Landlords

From the Latin-American heritage has come the particular submodel of the dual societies typical of the underdeveloped world. This model is similar to pre-industrial Europe. Modern mining, plantation, and industrial developments in Latin America that are a part of the interconnected system have been superimposed on an essentially rural society in which political and economic power has been in the hands of the landowners. At the time of early settlement, land was given out in large tracts to Europeans in favor at the courts of Spain and Portugal or in governmental circles of the colonies themselves. These lands and the wealth they have generated remain largely under the control of the descendents of these first seigniors. The mass of the population serves as the work force: poor, illiterate, and with virtually no political rights.

The basis of Latin-American life has been the hacienda (or estancia) in Spanish lands and the fazenda in Brazil. The hacienda was similar to the plantation in the southern United States. The plantation owner and his family lived in luxury in a "big white house" while the workers were virtually bound to the land in a self-contained community that included not only living quarters but also a store and chapel. The hacienda specialized in a cash crop for sale overseas, but the workers received little return. Nominally free, they were in fact tied to the land by indebtedness to the owner. They were given supplies and food on credit by the hacienda owner, but they could never make enough money to pay off their debts and leave. They paid their obligation by providing labor. They received land and possibly a shack, and tried to eke out a subsistence living on the side after they finished their jobs on the hacienda.

Officially the hacienda has been done away with in many Latin-American countries, but, like caste in India and racial discrimination in the United States, this way of life continues to be of major importance in the culture. Much along the lines of estates of feudal Europe, Japan, or Czarist Russia, the hacienda was part of a system in which the ruling minority lived well and enjoyed power, while the masses who did the work remained both poverty-stricken

Figure 13-1 LAND SURFACE OF LATIN AMERICA

0   1000 Miles

TROPIC OF CANCER

20°

0° EQUATOR

20°

TROPIC OF CAPRICORN

40°

120°   100°   80°   60°   40°   20°

0-1,000 ft.

1,000-3,000 ft.

3,000-10,000 ft.

Over 10,000 ft.

and politically impotent. Governments were drawn from the landed class; sons of the landed gentry joined the Catholic clergy or the army. In this way the wealthy controlled all facets of society.

Today the hacienda has been weakened by the rise of the modern, interconnected system with its emphasis on city life. Many of the landowning families have shifted their interests and become part of a new urban elite consisting of professionals, bankers, and businessmen. The mass of the population in most countries remains as it was in the past: poor, uneducated, and politically weak.

The matter of who shall rule is perhaps the most serious problem of Latin America. With it goes the question of how the wealth will be distributed as the countries develop. In most cases the threat to the ruling clique has been coming not from the rural peasants or the urban poor but from the new middle class of small businessmen, government employees, and professionals who have gained considerable wealth and are now interested in the power that goes with it. Labor movements have also begun among city workers, and their leaders push for urban reforms.

In some countries like Mexico and Chile, the old haciendas have been broken up and parceled out in small units to formerly landless peasants. This is as much a measure of the declining value of the hacienda as a real revolution, however. Plots are often so small that the newly landed peasant has little hope of getting ahead. While the peasants have gained land, the hacienda owner has moved his family and his commercial interests into the city.

Income distribution reflects the continued control of the ruling minority. A recent study by the United Nations Economic Commission for Latin America estimated that the upper 5 per cent of Latin America's income groups re-ceived one-third of the total income, and the upper 20 per cent of the population had almost two-thirds (Figure 13-2). By contrast the lowest 50 per cent, the rural and the poor, had less than 15 per cent of total income, and the lowest 20 per cent received only 3 per cent. Compare these figures with those of the United States. What conclusions can you draw?

Major attempts to change the control of power and money in Latin America have not been widespread. Castro's successful revolution in Cuba is the exception. There a centralized system of state control similar to that of the Soviet Union has been instituted. Both the plans and the goals are in the hands of the government, but, as in the Soviet Union, one of the goals is to redistribute the results of production more evenly among the people than ever before. In Chile a Marxist is pushing land reforms and the nationalization of the country's major industries. In Mexico a real peasant revolt took place early in the twentieth century. Since that time the Mexicans have developed a degree of democracy that is remarkable among Latin-American republics for its effectiveness and stability. There is bureaucratic inefficiency and corruption, but many selfless individuals work within the government for the good of the Mexican people.

Nineteenth-century Independence, but Not Democracy or Industrialization

It is important to recall that the Latin-American countries are not newly emerged from colonialism. Most of them gained independence before 1850. The revolutionary revolts from Spain followed the American Revolution in the United States, but generally they were not attempts by the people as a whole to gain political control democratically. Rather they were revolts by the

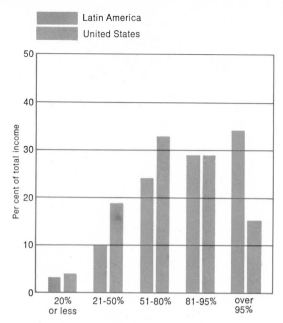

Figure 13-2 Income Distribution in Latin America and the United States

landed gentry who wanted freedom to manage their own lands and to govern themselves. It is this wealthy, powerful minority that has maintained control of most countries ever since. Moreover, it has been this group that has been satisfied with the status quo in economic and social as well as political terms.

The hacienda owner with his vast landholdings and his ready supply of cheap labor has not encouraged technological change. Inefficient as his operation might be, it has usually been on a scale large enough to provide handsomely for himself and his family. It has enabled him to have a luxurious home in the capital city, schooling in Europe for his children, and vacations and buying trips to the United States and Europe. Such a life would hardly interest him in popularly supported political movements such as those under Castro in Cuba, Perón in Argentina, and Allende Gossens in Chile. For the landlord class, "revolution" meant overthrowing the man in power if he did not protect the interests of the landlords. They expected the new leader to come from the same ruling oligarchy that has held power through the years.

If Latin America as a whole, not just the few large cities of Buenos Aires, São Paulo, Rio de Janeiro, Santiago, and Mexico City, is to take its place in the modern world in both economic and political terms, then a fundamental social and political revolution appears necessary, just as it was in Europe at the time of the Industrial Revolution. The political and social stagnation that has pervaded Latin-American countries for so many years must be broken first.

As we look at contemporary Latin America, we can find the full range of life from primitive Indians still in the Stone Age to some of the largest, most modern metropolitan centers in the world. The two different worlds represented by Washington and Ramkheri stand in juxtaposition. No country is all modern or all traditional. Instead, the governments face problems of managing the affairs of two different ways of life, each, as indicated by the illustration of Brazil in Chapter 12, with its own problems.

Stone Age Indians are rather minor problems because there are so few of them left and they occupy the most remote areas. But in Central America and the Andes, more advanced Indian cultures (vestiges of the great Incan, Aztec, and Mayan cultures) make up as much as half the total population. Many still speak their own native dialects as a first language and are not even nominally a part of the Latin culture. Some have been "converted" to Christianity, but many of their traditional religious beliefs have been incorporated in the Christianity they practice. Mostly their lives depend on the produce they can exact from the local region.

Latinos are more widespread than the Indians. They are descendents of European immigrants and may also have Indian or African origins. They were the hacienda workers whose tradition was to work part time for the landlord but to depend on their own plot for much of their subsistence. Some are just squatters on the fringe of settled regions. The landlord lets them stay as long as they just subsist. But if they show signs of developing more products to sell and functioning successfully in the modern system, they may be asked to leave. These people remain centered in the traditional, locally based system almost as much as do the Indians.

Parts of Latin America have been tied to the modern, interconnected system from the first colonial days. Like the British colonies, Latin America's first ties were to Europe. It provided tropical specialties or other products that Europeans wanted. The landlord's share of the hacienda output was designed for export, and so was the produce from the mines and the plantations. Where the modern, interconnected system has been developed on a large scale, as in the coffee lands around São Paulo in Brazil, a full-fledged, regional network of the modern system has developed with its own transportation and communication systems, its own financial organization, and its own specialty food producers supplying fruits, vegetables, meat, and dairy products. São Paulo, the capital of this region, is one of the largest, most modern cities in the world.

But the Brazilian government must deal with the whole spectrum of its populace, including advanced areas of the modern, interconnected system. Some regions are in the interconnected system but are not so successful; in other regions the people are virtually locked within the traditional, locally based system. And this is the case in all Latin-American countries. Elements of both systems exist side by side, often within the same region. Countries with high per capita GNP or energy consumption are closer to the modern, interconnected system; those with low per capita figures are more traditional. Nowhere in Latin America is there a country as wealthy as Japan, Europe, or the other modern industrialized examples we have studied.

The Environmental Base of Latin America

The environmental base in Latin America appears, at first glance, to be similar to that of Anglo-America. Latin America is larger than the United States and Canada combined, and has about four people for every three in Anglo-America. Population density in 1970 was forty persons per square mile compared with thirty persons per square mile in Anglo-America. But the people, culture, technology, and earth environment differ greatly.

MANY POLITICAL ENTITIES

Latin America is divided into twenty-four sovereign countries and several dependent territories (Figure 13-3). Most of these states are small in both area and population (Table 13-1). Seventeen out of the twenty-four independent countries had populations of less than 10 million in 1972—fewer people than the state of Illinois—and twelve were smaller in area as well. Only four countries had populations larger than California. Brazil is exceptional in Latin America. It accounts for 46 per cent of the total land area and 33 per cent of the total population. It is almost 50 per cent larger in area than Mexico and Argentina combined!

CROWDED "POPULATION ISLANDS" IN A VAST SEA OF LAND

The population density of Latin America is very low, but as Figure 13-4 indicates, population distribution is spotty. Most of Latin America has less than two persons per square mile and is, in fact, largely frontier.

The majority of the population lives close to the seacoasts, and even there some areas are sparsely populated. Nowhere in Latin America is there the penetration of population from the coast that one finds in the eastern United States and in the Soviet population triangle. Only in southern Mexico does an important population concentration extend from coast to coast, and there the distance is relatively small.

Each population island forms the core of a different country except in Brazil where there are several nuclei of population along the Atlantic Coast. Population density varies sharply, but the variations are largely a measure of a country's size. Where the country is small and the population island occupies all, or almost all, of the territory as in the West Indies, densities average 200 or more per square mile. Where

the countries are large, the population islands occupy only a part of the total land area, and population densities are low. The highest density in Table 13-1 refers to the tiny island of Barbados with an area of only 200 square miles. Such densities can also be found within other densely settled population islands of Latin America. It just happens that Barbados is such a small nation that it has no empty land to include in the population average.

THE IMPACT OF LATITUDE AND ALTITUDE: A TREMENDOUSLY VARIED ENVIRONMENT

The people of Latin America occupy a tremendously diverse environmental base. The growing potential of equatorial lands in the Amazon lowland is very different from those in northern Mexico and southern Argentina. Still, it should be noted that the greatest share of the Latin-American landmass is within the tropics. Nineteen of the twenty-four independent countries of Latin America, including most of giant Brazil, are tropical. Moreover, Latin America does not extend into high latitudes comparable to those of northern Canada or the Soviet Union. Only southern Argentina and Chile lie beyond the 49th parallel.

This is why most of Latin America fits the designation given to the tropics: "the land where winter never comes." Only in Argentina, southern Chile, and the highlands of northern Mexico are there several months when the danger of frost is present (Figure 6-9). On the edge of the tropics, frosts hit at most only once or twice during the winter.

The environmental base of Latin America is further complicated by the presence of high mountains. Because atmospheric heat decreases rapidly as altitude increases, the mountain areas stand out as environmental anomalies at all latitudes. Even close to the equator there are permanent snows on the high Andes.

Figure 13-3 Political Map of Latin America

Caracas National capital

• Cities over 1 million

Table 13-1 Population Characteristics of Latin-American Countries, 1972

Country*	Area (000 sq. mi.)	Population (000,000)	Current Rate of Growth (per cent)	Population under Age 15 (per cent)	Per Capita GNP (dollars)	Density (per sq. mi.)
Central America						
Mexico	762	54.3	3.3	46	580	71
Guatemala	42	5.4	2.6	46	350	128
El Salvador	8	3.7	3.0	45	290	462
Honduras	43	2.9	3.2	47	260	67
Nicaragua	50	2.2	2.9	48	380	44
Costa Rica	20	1.9	2.7	48	510	95
Panama	29	1.6	2.9	44	660	55
West Indies						
Cuba	44	8.7	1.9	31	280	198
Jamaica	4	2.1	2.1	46	550	525
Haiti	11	5.5	2.4	38	‡	500
Dominican Republic	19	4.6	3.4	47	280	242
Puerto Rico†	3	2.9	1.4	37	1,410	966
Barbados	0.2	0.3	0.8	36	500	1,500
Trinidad and Tobago	2	1.1	1.1	42	890	550
Tropical South America						
Venezuela	352	11.5	3.4	47	1,000	33
Colombia	440	22.9	3.4	47	290	52
Ecuador	109	6.5	3.4	48	240	59
Peru	496	14.5	3.1	45	330	29
Bolivia	424	4.9	2.4	42	160	11
Paraguay	157	2.6	3.4	46	240	17
Brazil	3,286	98.4	2.8	43	270	28
Guyana	83	0.8	2.8	45	340	10
Surinam†	55	0.4	3.2	46	560	7
Middle-latitude South America						
Uruguay	72	3.0	1.2	28	560	42
Argentina	1,072	25.0	1.5	30	1,060	23
Chile	292	10.2	1.9	39	510	35

*Data not available for French Guiana and British Honduras.

†Nonsovereign country.

‡Estimated to be less than $100.

Source: Population Reference Bureau, Inc., *World Population Data Sheet: 1972*, Washington, D.C., 1972. Land area data: Life World Library Series, *Handbook of the Nations and International Organizations*, Time-Life, Inc., New York, 1968.

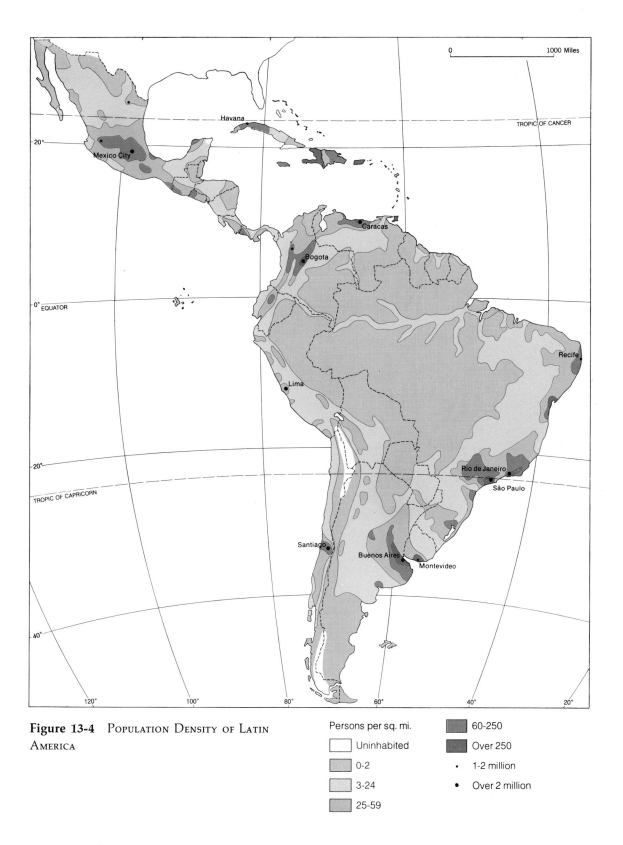

Figure 13-4 POPULATION DENSITY OF LATIN
AMERICA

Persons per sq. mi.

⬜ Uninhabited	◼ 60-250
▨ 0-2	◼ Over 250
▨ 3-24	• 1-2 million
▨ 25-59	● Over 2 million

MEASURING EVAPO-TRANSPIRATION POTENTIALS ON AND AWAY FROM THE EQUATOR

Why do the areas near the Tropic of Capricorn and the Tropic of Cancer have higher evapo-transpiration potential than do the areas near the equator?

1. Remember that the amount of radiant energy received at a place depends on the angle at which the sun's rays strike the earth *and* the length of exposure.

    a. Sun's rays are directly over equator at noon on September 22 and March 21.

    b. Sun's rays are directly over Tropic of Cancer ($23\frac{1}{2}°$ north) at noon on June 21; directly over Tropic of Capricorn ($23\frac{1}{2}°$ south) at noon on December 22.

    c. Length of daylight at equator is 12 hours every day of the year.

    d. Length of daylight at Tropic of Cancer on June 21 is almost $13\frac{1}{2}$ hours; same figure applies to Tropic of Capricorn on December 22.

    e. Between 6° north and 6° south, sun's rays remain almost vertically overhead for 30 consecutive days at the spring and autumn equinoxes.

    f. Between $17\frac{1}{2}°$ and $23\frac{1}{2}°$ latitude, sun's rays are almost vertically overhead at noon for 86 consecutive days during period of the year (June or December) of high sun solstice.

2. In the area near the Tropic of Cancer, the climate is dry, and there is little cloud cover; at the equator, particularly in the Amazon lowland, there is a large amount of cloud cover.

GROWING POTENTIAL

Since the evapo-transpiration potential varies with latitude, tropical areas have much better prospects for plant growth than the mid-latitudes. The tropical areas have up to 50 per cent more radiant heat per year than the southern mid-latitudes of Latin America (Figure 5-4). But we find surprisingly the highest evapo-transpiration potential near the tropical fringes, particularly in the Northern Hemisphere, not at the equator.

LANDFORMS

High mountains extend the full length of Latin America (Figure 13-1). They appear first as extensions of the mountain system of the western United States into Mexico and Central America, and then as the higher Andes Mountains near the western margin of South America, extending all the way to Tierra del Fuego at the southernmost tip. The highest peaks of the Andes are south of Ecuador, and many are higher than the tallest North American peaks.

The rest of Latin America is primarily plain and lowland. As in Anglo-America, a great lowland extends across the South American continent from north to south, but it reaches the Atlantic in the north and south. Uplands similar to the Appalachians occur only along the Venezuelan-Brazilian boundary and in eastern Brazil south of the Amazon River. The Amazon drains a large portion of South American lowland, but it flows from west to east, not north to south like the Mississippi, and so the northern part of the lowland opens on the Atlantic Ocean. Notice that the Amazon closely follows the equator. The south-central portion of the lowland drains away from the equator to form the Río de la Plata where Buenos Aires is situated.

The highland of eastern Brazil rises as a plateau with its steep side facing the Atlantic 2,000 to 8,000 feet above sea level. The plateau drops off gently westward, and most of its drainage is northward into the Amazon and southward into la Plata, rather than eastward over the steep edge of the plateau.

CLIMATIC VARIATIONS WITH ALTITUDE

The presence of mountains in all latitudes adds another variable to growth patterns of Latin America. The rapidly decreasing temperature gradient with altitude causes cool weather and offers less evapo-transpiration potential. (The lowlands, of course, have evapo-transpiration conditions normal to their latitudinal location.)

Latin Americans recognize this change and speak of different living zones with altitude (Figure 13-5). There is the *tierra caliente,* the hot lowland zone from 2,000 to 3,000 feet, where tropical crops such as sugar cane, cacao, and bananas grow; the *tierra templada,* the lower upland zones to 6,000 feet, where it is warm rather than hot, and corn, coffee, and oranges reach the limits of their temperature tolerance; and the *tierra fría,* or the cold country, from 6,000 to about 10,000 feet, where temperatures are cool. Quito, the capital of Ecuador, lies almost on the equator. It is over 9,000 feet above sea level and has an average annual temperature of 56° F. As one would expect at the equator, there is virtually no seasonal variation in temperature.

Climatic variations with altitude are a factor throughout Latin America. More than half of the Latin-American countries experience the changing life zones with altitude—a factor to be taken into account in considerations of their resource potential.

MOISTURE AVAILABILITY AND VEGETATION

Precipitation is greatest along the equator and decreases generally both north and south. There is also heavy precipitation along the eastern side of Central America and along the Pacific Coast of southern South America in latitudes comparable to those of the Pacific Northwest in the United States and Canada. These are the only parts of Latin America to show large moisture surpluses during the wet season. The Amazon lowland, parts of the east coast of Brazil, and southern Chile are the only areas without moisture deficits during the dry periods (see Figure 5-5).

In contrast, more than half of Mexico and parts of Argentina and northern Chile show severe deficits during all seasons. Much of these areas have only scanty desert vegetation similar to the Southwest in the United States.

The wet tropics are forested with dense tropical rain forests. With plenty of moisture and a year-round growing season, the forests have the lushest growth in the world. They are also found along the coast of Brazil and in Central America, but a zone of tropical grassland separates the Brazilian coastal forests from the Amazon forests. There are also grasslands in interior Venezuela along the Orinoco River.

In the middle latitudes of Latin America, the presence of extensive grasslands, scrub forests, and deserts indicates moderate to severe moisture shortages. The pampas of Argentina resemble the grasslands of the interior United States, and the vegetation of central Chile is like that of southern California. In the mountains, vegetation varies with altitude, ranging from lush tropical forests on the lower tropical slopes to arcticlike tundra at the highest levels.

MINERAL RESOURCES

Minerals have been one of the key resources in the development of Latin America since Indian times. The Aztecs and Incas mined gold and silver, and early Spanish explorers sought the sources of the valuable minerals. These minerals are still mined in Mexico and other

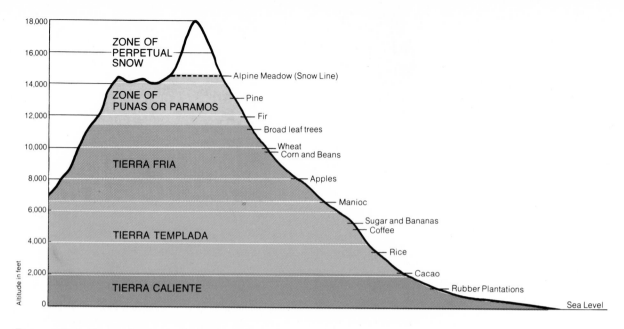

Figure 13-5 LATIN AMERICA: CROSS SECTION OF
TIERRA FRIA, TIERRA TEMPLADA, AND TIERRA CALIENTE

countries, but they have been overshadowed by the modern industrial minerals—petroleum, coal, iron-ore, bauxite, copper, tin, lead, and zinc. Latin America has important deposits of all these, but production is sharply localized. Figure 13-6 shows the four major areas of mineral deposits utilized today. In Venezuela, Colombia, and Trinidad oil is plentiful in the lowlands; iron-ore and bauxite are mined along the high-land fringe on the mainland, iron-ore in Venezuela and bauxite in Guyana and Surinam. The high mountain chain from Mexico to southern South America produces nonferrous metals—copper, tin, lead, and zinc—and some iron-ore. The most important districts for these are in Peru, Chile, Bolivia, and northern Mexico. The Brazilian highlands have important iron-ore deposits and some coal. The West Indies and the northern Caribbean area produce iron and copper, bauxite on the islands, and petroleum in eastern Mexico. Little production has

come from the vast lowland of the Amazon or the east flank of the Andes, but this is still frontier country. Argentina has some petroleum production but little other mineral output.

**Evaluations of the
Latin-American Environment**

The climate, soil, landforms, and minerals in various parts of Latin America have meant something different to traditional farmers and to people living in the interconnected system. Each group has had a different technological base and has perceived the resource possibilities differently. Except for the deserts and high mountains Latin America has productive environments in which plant and animal populations are sufficient to support hunters, gatherers, and, in most places, farmers. For this reason Indians were widely spread over the area at the

Figure 13-6 MAJOR AREAS OF MINERAL DEPOSITS IN LATIN AMERICA

Mineral deposits

■ Iron ore	✕ Lead	◇ Bauxite
▲ Petroleum area	▼ Zinc	⌒ Silver
○ Coal	△ Tin	
▽ Copper	⁘ Gold	

time of Columbus. Even in the areas of advanced Indian culture—in the highlands of Mexico and Peru—where some produce was exchanged with other areas, the economy of the Indians was locally based. Success or failure depended on the growing potential of the local environment as its possibilities were viewed by the local people.

The hot, wet tropical forests were only lightly settled. Although the supply of natural food in such areas was plentiful, there were major problems—particularly of disease from the abundant insect and microbe populations—and difficulties in clearing forest vegetation and maintaining fertility in tropical forest soils. Indians spread over the wetter margins of both tropical and mid-latitude grasslands, but the preferred locations were the highlands, both in the tropics and in the lower mid-latitudes. These highlands were removed from the hot, wet climate of the tropics, the poor soil, and the insects, and their lower slopes were able to produce subtropical crops such as corn, tobacco, and bananas, while the middle valleys were fine grain-growing areas.

Traditional societies suffered not only from their own productive inefficiencies but also from their inability to store the fruits of the environment effectively. This was particularly true in the tropics where, although some products were harvested every month, it was impossible to keep them more than a few weeks because of the rapidness of decay and animal predators.

MODERN, INTERCONNECTED SOCIETY

Modern society has approached Latin America from a rather different point of view. Like the Indians, European settlers preferred the mid-latitude areas of Latin America or the tropical highlands. But commercial interests concerned with the development of particular resources have been able to develop sites almost anywhere. There are mines in mountain valleys over 15,000 feet above sea level and in the driest deserts; there are plantations in some of the hottest, wettest tropical forests. But the modern, interconnected system with its technology and its transportation routes can bring in all necessary support, including workers, food, medical supplies, and even water. Thus, the crucial question is not what the environment is like but rather what its productive possibilities are in terms of a particular crop or mineral. Then the potential for profit must be weighed against the cost of overcoming environmental obstacles.

Migrants, Natives, and Their Offspring in the Population

The characteristics of the population composition in Latin America are very different from those in any of the other underdeveloped areas of the world. Here, as in Anglo-America, the migration of peoples into the region from Europe has been more important than the contribution of the natives. Unlike Africa and Asia, where native people remain dominant in the population, the Indian strain of Latin America has been almost completely fused with European and African immigrants.

Indians and Negroes

The Indian has not been obliterated from Latin America as he has in Anglo-America. The largest Indian populations are in the former centers of advanced Indian civilizations—the highlands of Mexico, Guatemala, and the Andes from Ecuador to northern Chile (Figure 13-7). Indians are still in the majority in Bolivia, Paraguay, and Guatemala and are close to half of the population of Peru. They form the one large group not

Figure 13-7 ETHNIC GROUPS IN LATIN AMERICA

Predominantly Indian

Predominantly European

Predominantly Negro

Mixed European and Indian

Mixed, with large proportion Negro

closely integrated into Latin-American culture. Most still speak native tribal languages.

The black population of Latin America is also considerably greater than it is in Anglo-America. Like the Indians, this group is also sharply localized. The blacks today are descendents of slaves brought from Africa to work plantations in the days before modern machinery. Their distribution correlates highly with the areas of important early sugar plantations along the northeastern coast of Brazil and in the West Indies. Blacks are in the majority in Haiti, in Jamaica, and on most of the small sugar islands. They account for almost half the population of Trinidad and Tobago, and a significant portion of the population of the Dominican Republic, Cuba, Puerto Rico, and Brazil. Mulattos total 60 per cent of the population of the Dominican Republic and constitute a large group in Cuba, Puerto Rico, Trinidad, and Brazil.

Europeans

Although Indians and blacks are sharply localized in the Latin-American population, Europeans are prominent everywhere. In Argentina, Costa Rica, and Uruguay Europeans have not mixed with other groups, but elsewhere they are part of the miscegenation that has produced people of almost every skin color.

The origins of the European population in Latin America are very different from the origins of Europeans in Anglo-America. The settlers of the two Americas came not only from different parts of Europe, but at different times with different purposes.

During the colonial period Latin America seemed to be the prize. The tropical environment offered a source of sugar and other crops that could not be produced in mid-latitude Europe. In addition, the Spanish conquered the two great Indian civilizations, the Aztecs in Mexico and the Inca in Peru, seizing ornate gold and silver art objects.

Latin-American settlement also began much earlier. Europeans explored and settled these areas in the sixteenth century before any permanent colonies were established farther north. Mexico City, Lima, and Havana were major cities when the Jamestown colony was founded. Moreover, European migration to Latin America was more than twice as large as that to North America. An estimated 14 million Europeans entered Latin America from the fifteenth to the eighteenth centuries, whereas only 6 million Europeans migrated to Anglo-America.

But in the nineteenth and early twentieth centuries the situation was reversed. Anglo-America offered productive mid-latitude agricultural lands and a rising industrial economy. It became the land of opportunity for European immigrants. Between 1821 and 1932 an estimated 40 million Europeans entered Anglo-America; only 16 million went to Latin America. Moreover, immigration to the southern areas was sharply localized. Well over half the Europeans went to either Argentina or Brazil.

Since World War II, moderate migration into Latin America has continued, again primarily to Argentina and Brazil but also to Venezuela. To an important degree the destination of these immigrants provides a measure of economic development. It is also true that Latin Americans have emigrated toward the industrialized parts of the world. Large numbers of Cubans and Puerto Ricans have entered the United States, and Jamaicans and other West Indians emigrated to Britain until limitations were established in 1968.

There are striking contrasts between the lives of rich and poor in Latin America.

Most Latin-American countries have been free from colonial domination for more than a century. Even so, today only a minority of the people are a part of the modern interconnected system and a large majority of the people remain bound to traditional ways.

436

The Evolution of the Present Dual Societies

THE COLONIAL PERIOD: SUBSISTENCE AND ESTABLISHMENT OF AN EXTERNALLY ORIENTED ECONOMIC SYSTEM

From the beginning the interest of the Spanish and Portuguese in Latin America was primarily commercial. There is little evidence of settlement groups seeking freedom of religion or trying to establish a utopian society. The expeditions of the sixteenth century were after riches—gold and silver, exotic forest products such as cacao and dyewood. Like Columbus, the first explorers were content to trade with the Indians and attempted no permanent settlements. But the lure of gold brought settlements to the islands of Hispaniola, Cuba, and Puerto Rico in the Caribbean, and these were followed by the conquests of the Aztec Empire of Mexico and the Inca Empire in Peru and Bolivia. These centers of Indian civilization offered accumulated treasures and wealth which could be looted and already determined mineral deposits. They became the focal points of Spanish interest and the chief sources of the wealth of the empire (Figure 13-8).

Mercantilism

The commercial connections between Spain and its Latin-American colonies were closely guarded. All trade had to be with the mother country and, in fact, could only pass through the Spanish ports of Seville and Cádiz. Moreover, all trade in the New World had to move through one of several seaports: Veracruz in Mexico and Portobelo in Panama for goods from the Inca lands; Cartagena in Colombia, San Juan and Santo Domingo for cargoes from Spain to the West Indies. Havana was a shipping port to Spain.

For the Portuguese in Brazil, little wealth was found, for there was no Indian civilization and no gold deposits close to the coast. But sugar plantations were established along the coastal lowland around what are now the port cities of Recife and Salvador, and by 1600 Brazil was the world's chief supplier of sugar. This was the first settlement in Latin America based on plantations and large-scale agricultural exports. Portugal was not able to control the sugar trade, and it was handled by Dutch and English merchants. But soon sugar plantations were developed in the Caribbean islands as well.

Spanish mercantilism attempted not only to restrict all trade to its new colonies but to supply all manufacturing goods as well. But the need for tools, clothing, and buildings resulted in the development of important manufacturing trades in the busy colonial centers, although imports from Spain remained important.

The growth of the export centers also provided a market for other parts of the colonies. Food staples normally were produced in the immediate lands around most commercial centers, but in the case of the miners in the highlands of Bolivia and Peru, food and textiles were carried overland from what is now Argentina, and mules and horses were supplied by livestock operations along the Peruvian coast. In the same way, cattlemen moved into the grasslands of northeastern Brazil from the coast to supply meat to the coastal sugar plantations. Hides and skins from the operation were shipped to Europe.

Labor Supply

Natives, rather than Europeans, were recruited for work on the plantations and in the mines. Those in the mines were commonly slaves, while agricultural labor worked under the encomienda, the antecedent of the hacienda. An encomienda gave the owner not only a vast tract

Figure 13-8 Political Map of Latin America, 1800

■ Spanish lands

□ Portugese lands

of land but a given number of Indians as well. Under the encomienda the Indians were not slaves. Rather they were entrusted to the landowner who was to teach them the Catholic faith and provide various forms of protection and help. There was essentially no payment of wages, and the bulk of the population lived at or near subsistence level. In the sugar lands the supply of native labor proved to be inadequate, and soon African slaves were being imported. In all, 14 or 15 million Africans were brought to these areas, far more than the million brought to the United States.

Developmental Centers

It was the export centers, then, that provided the focus of activity for European settlement in the New World of Latin America. The ports and administrative centers of the commercial system developed into important cities. Mexico City, Lima, and Potosí in the Bolivian mining area were cities of tens of thousands of people before the first English settlement was established in Anglo-America. At the same time, Indians continued living in their traditional subsistence ways in the back country away from these centers.

INDEPENDENCE: POLITICAL INDEPENDENCE BUT INCREASED ECONOMIC DEPENDENCE ON OUTSIDE MARKETS

During the period from independence until World War I, Latin America developed differently from Anglo-America. In the United States this was a time of regional integration and specialization. New transportation and communication developments were used to tie the country together into a series of specialized component parts, while considerable political and cultural unity was building. Latin America responded to the changes brought about by the Industrial Revolution in Europe in a very different way, probably because the former colonies emerged as politically and socially weak entities. Instead of becoming economically independent of European markets, Latin-American countries grew more closely tied to those markets than ever before.

As we have seen, the specialization of European countries in industry, trade, and finance increased their dependence on raw materials and foodstuffs from other parts of the world. Growing affluence and reduced transport costs lowered prices on European markets and increased demand. As a result world trade increased five times between 1820 and 1870, and then repeated that achievement again between 1870 and World War I. The classic pattern of European imports of raw materials and food and exports of manufactured goods was established.

Latin-American population centers, which already had come to depend on European markets for their support during colonial days, reacted by sharply expanding their role as suppliers of raw materials. Colonial restrictions on trade were gone, and the new landlords who ruled the countries were anxious to satisfy their desires for European consumer goods through imports. New plantations and mines and even new producing regions came into being. It was in the latter part of the nineteenth century that the major exporting centers of Latin America were established—the coffee lands near São Paulo, the sugar plantations of Cuba, the banana plantations in Central America, the nitrate and copper mines in northern Chile, and the cattle and grain producing areas of Argentina. Oil production in Venezuela began early in the twentieth century. There developments were on a scale never before seen in Latin America and brought a new demand for labor and capital. But in the nineteenth century slavery was rejected in Latin America as well as in the United States. As slavery was outlawed, a new supply of labor came from European immigrants. In Brazil and Argentina there was a new wave of immigration, not just from Spain and Portugal, but also from Italy, Germany, and other European countries. As in the days of industrial expansion in the United States, European commercial interests invested large amounts of capital in these underdeveloped areas in order to establish vital trade links.

In Brazil and Argentina European investments came not to the plantations and estates that were held by the wealthy landowners. Rather they were made in railroads, public utilities, and other aspects of the infrastructure necessary to get products to ports for shipment overseas. Complete control was established only where major organization was needed, such as in the tropical banana plantations or in the desert wastes of northern Chile where large-scale mining operations were started. In such areas there were no local entrepreneurs who could handle the task.

Like the United States, Latin America was developing regional specialties that fitted the new long-range, interconnected system made possible by the Industrial Revolution. But in Latin America the organizational core of the system remained in Europe. Thus, while local entrepreneurs benefited, often handsomely, by organizing coffee, sugar, and livestock production, and while regional cities like São Paulo, Havana, and Buenos Aires emerged as the business centers of production, the decision-makers—the shippers, financiers, and dealers at commodity markets—remained in Europe.

Moreover, each of the specialized producing regions was in a different country. Where export development was great and local production remained in the hands of the landowners, the population island grew and became pros-

perous. This was the case in the coffee area of Brazil. Such countries emerged as the more advanced in Latin America. But where developments were smaller and all phases remained more in the hands of foreign capitalists, the countries remained weak.

The Ruling Oligarchy and the People

In the nineteenth century the economic and political control of most Latin-American countries was similar to that in Japan at the same time. Power was in the hands of the small landowning class which controlled not only the export-producing units but also the government. The results in Latin America, however, were very different. Unlike the Japanese oligarchy, Latin-American leaders in one country after another were content to be a part of an externally controlled system. The payoff for them was handsome, and they found it a satisfactory arrangement to feed their specialties into the system and to purchase the needed manufactures and luxury goods from overseas.

As a result there was little investment in consumer manufacturing in Latin America. Most of the industry that did develop was concerned with processing export products: refining sugar, processing meat, and smelting and refining metal. Moreover, even this industry was largely in the hands of foreign investors. As in Japan, the mass of the population received little return from the new economic developments anyway. They were Indians, freed slaves, or just subsistence squatters on the land. Or, if they worked on haciendas or plantations, they received little more than a shack and a plot of land. Since they were poor, they provided little market for products of industry. Too, they were poorly educated. The hacienda system lacked the Japanese tradition of industriousness, the same loyalty of the workers to the system, and the strong work incentive found in Japanese feudalism. Hacienda labor had been inefficiently utilized, and there was a tradition of underemployment.

In any case, Latin-American leaders largely chose to ignore the industrial sector of the economy; they were satisfied to depend on exports of raw materials and food specialties. Instead of encouraging industry through financial support and protective tariffs as Japan did, Latin-American countries were content to encourage greater and greater importation of manufactured goods. Thus, while the growth of export markets in most parts of the region during this period provided money which might well have been invested in establishing an industrial base, such an undertaking was not considered. Of course, most countries had much smaller populations and capital bases than Japan, but even in Mexico, Brazil, and Argentina, the industrial sector lagged.

Centers of Economic Activity

The size of export development was reflected in the size of the large cities in Latin America. In 1920 two cities, Buenos Aires in Argentina and Rio de Janeiro in Brazil, had populations of over one million people. Realize that they are not only national capitals but are also headquarters of the two countries most closely tied to the new economic developments in Latin America and to the new immigration. Three other cities had populations over half a million: Mexico City, Santiago, and São Paulo. Other large centers were Montevideo, Uruguay; Havana, Cuba; and Rosario, in the agricultural region of Argentina.

The shift in the major centers of export production from colonial times was striking. Not only did the coffee area of Brazil and the meat and grain sections of Argentina and Uruguay gain major stature, but two major centers of colonial times also declined in importance. In Brazil the sugar-plantation owners were slow to convert to modern milling techniques developed in the nineteenth century, and there was also a shift from primary interest in gold and silver toward industrial minerals. Peru and Bolivia participated in the new mining with production of copper and tin, but these developments never gave the areas the relative importance they had during the colonial period.

THE TWENTIETH CENTURY: THE BEGINNINGS OF INDUSTRIALIZATION

World War I, the depression of the 1930s, and World War II produced major breakdowns in the long-range, interconnected systems of Europe and the United States; this affected Latin-American producers of raw materials and their governments. During the wars, the demand for Latin-American raw materials increased as the European countries undertook all-out war efforts. But because their economies were so preoccupied with war material, there was a breakdown in the supply of imports to Latin America. The depression brought a severe drop in the value of export trade. The total trade of Chile declined 85 per cent in value between 1929 and 1932, and seven other countries in Latin America experienced drops of more than 65 per cent.

The Impact of War and Depression

In World War I the need for domestic supplies of goods normally imported caused some of the larger countries, particularly Argentina and Chile, to embark on increased domestic manufacturing. But prior to World War II, no Latin-American country had a really significant base of heavy industry in steel, metalworking, chemicals, and cement production.

Industrialization

It was the long drought of imports from 1930 through 1945 that changed the thinking of political and economic leaders in Latin America. First, because of the depressed trade of exports during the 1930s and, then, because of the disruption of the manufacturing centers in Europe and the United States during World War II, there were some fifteen years when wealthy groups of Latin-American countries had to cut back on their desires for consumer goods and the essential equipment needed to expand their production facilities.

As a result, since then the governments of the largest Latin-American countries have had programs encouraging the development of modern industry in their countries. Brazil, Argentina, Mexico, Colombia, Chile, and even Peru and Venezuela have their own steel mills and basic metalworking plants. Brazil, Mexico, and Argentina have encouraged makers of automobiles from industrial countries to establish assembly plants and have been pushing to have more and more of the component parts produced in their countries. North American retailers such as Sears Roebuck have been allowed to enter Latin-American national markets with the agreement that an increasing share of the products sold will be made domestically.

It is important to note that the new manufacturing is designed almost exclusively to provide goods for the growing domestic market, not to provide new export products. It will cut down the demand for imports rather than increase exports. The result is that the export trade of Latin-American countries remains essentially as it has been since colonial times—the export of primary products of the resource base, manufactured only to the degree necessary for shipment.

Actually, except for petroleum, total exports from Latin-American countries in the mid-1960s were slightly less than in 1929. Former colonial territories receive preferential treatment both from the British and the Common Market countries. In large measure Latin America has suffered from not having been recently colonial, and African exporters have benefited at the expense of Latin Americans. Also, most Latin-American countries have provided little in the way of research assistance and long-range governmental planning.

Changes in the Trade Structure

As Table 13-2 shows, there has been considerable shifting in the relative importance of particular export commodities that has, in turn, affected particular producing areas. Today in Europe there is less interest in mid-latitude staple foods—wheat, corn, and meat. Europeans have higher per acre yields, depend on more intra-European trade, and draw imports from Canada and Australia. At the same time, with the population increase at home, Argentina now consumes three times as much beef as it exports. Exports of agricultural raw materials, particularly wool and hides, are only half the totals of the early 1930s. Lead, zinc, tin, and nitrate exports are also lower today than 40 years ago.

The greatest gains by far have come in petroleum exports. Crude oil production in Venezuela in 1967 was over ten times that of the early 1930s. Over 90 per cent of Venezuelan oil was exported. Iron-ore exports, though much less in total, also increased sharply in the past 40 years. Over 80 per cent of this output comes from Brazil, Venezuela, and Chile. Chile and Venezuela export virtually all their output even though they have steel mills within their territories. Brazil retains about 10 million tons for domestic consumption.

Table 13-2 Relative Importance of Major Export Commodities of Latin America (per cent)

Commodity	1934–38		1946–51		1963–64*	
Temperate foodstuffs	17.2		10.8		8.2	
Wheat and flour		5.1		4.2		1.7
Maize		6.3		2.0		2.0
Meat		5.4		4.0		4.0
Butter		0.1		0.2		0.1
Cattle		0.3		0.4		0.4
Tropical foodstuffs	21.3		30.0		21.2	
Coffee		12.8		17.4		15.0
Cocoa		1.2		1.6		0.8
Sugar		6.1		10.2		3.3
Bananas		1.2		0.8		2.1
Agricultural raw materials	12.6		11.8		7.6	
Cotton		4.5		4.7		4.3
Wool		4.3		3.7		2.0
Hides		3.5		3.2		0.5
Oils and oilseeds		0.3		0.2		0.8
Forest products	1.0		2.3		1.0	
Timber and manufactures		0.3		1.2		0.8
Quebracho		0.7		1.1		0.2
Fishery products	0.0		0.1		2.4	
Petroleum	18.2		17.3		26.4	
Crude		15.5		14.9		16.3
Refined		2.7		2.4		10.1
Iron and steel	0.0		0.2		2.9	
Ore		0.0		0.1		2.8
Steel		0.0		0.1		0.1
Copper	4.7		3.4		4.9	
Ore		0.3		0.3		0.2
Metal		4.4		3.1		4.7
Lead	1.5		1.3		0.7	
Ore		0.3		0.2		0.3
Metal		1.2		1.1		0.4
Zinc	1.0		0.6		0.6	
Ore		0.8		0.4		0.4
Metal		0.2		0.2		0.2
Tin, predominantly ore	1.6		1.0		0.5	
Nitrates	1.4		0.8		0.3	
Total of above products	80.5		79.6		76.7	

*Excludes Cuba; this affects most the data for sugar.

Source: Grunwald and Musgrove, *Natural Resources in Latin American Development*, Johns Hopkins Press, Baltimore, 1970, p. 21.

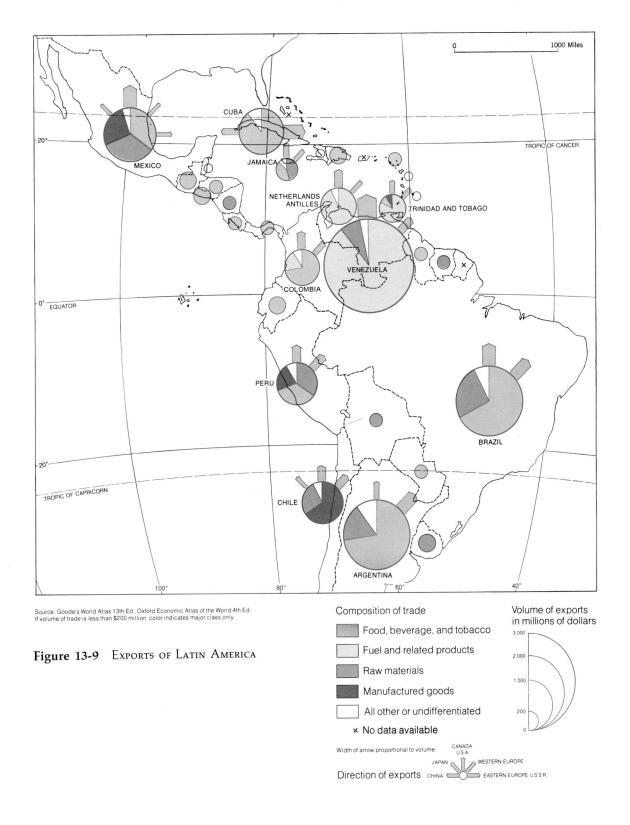

Source: Goode's World Atlas 13th Ed., Oxford Economic Atlas of the World 4th Ed.
If volume of trade is less than $200 million, color indicates major class only

Composition of trade

Food, beverage, and tobacco

Fuel and related products

Raw materials

Manufactured goods

All other or undifferentiated

× No data available

Volume of exports
in millions of dollars

3,000

2,000

1,000

200
0

Width of arrow proportional to volume.

CANADA
U.S.A.

JAPAN   WESTERN EUROPE

Direction of exports   CHINA   EASTERN EUROPE U.S.S.R.

Figure 13-9   EXPORTS OF LATIN AMERICA

One remarkable new export industry in the period since World War II has been the fishmeal and fishoil industry of Peru and, to a lesser degree, Chile. The industry is based on the demand for high-protein livestock feed in the United States and Western Europe and the presence of vast quantities of a single variety of fish, the anchovy, in the cold waters off the Pacific Coast. The anchovy fishing industry and processing plants have risen to serve the export market. Oil is extracted in processing plants along the coast, and it and the resulting fish products are exported, particularly to the Netherlands, West Germany, and the United States.

Through the years the total export of tropical agricultural specialties has stayed about the same. However, coffee and banana exports have increased and sugar shipments declined. Coffee production in Brazil has remained steady, but Colombia has doubled output. Mexico has quadrupled its production. The small Central American producing countries have had increases of 50 to 300 per cent, but the most significant expansion has come in Peru and Ecuador, insignificant coffee producers forty years ago. Brazil, Colombia, and Mexico have important domestic markets, but these account for only a tiny fraction compared to exports. Banana exports, too, have doubled, particularly because Ecuador now provides over one-third of the total Latin-American exports.

Localization of Trade

Actually the major export trade of Latin America is concentrated in only a few centers. Table 13-3 indicates that almost 87 per cent of total exports come from just eight of the twenty major producing countries. Venezuela, a leading world source of petroleum, is by far the most important exporting country with more than 23 per cent of the total for the major producing countries of Latin America. It is followed by Brazil, Argentina, and Mexico, each exporting

approximately one-half to two-thirds the amount of Venezuela. Together, these three countries total 40 per cent of the major countries' exports.

The large Latin-American countries, particularly those in South America, dominate the area's exports. The eight large exporters are the eight largest countries in population, and all but Mexico and Cuba are in South America. But the big eight do not appear in order of their population size. Venezuela, the leading exporter by far, is sixth in population. Argentina, with less than half the population of Mexico, is a larger exporter. Colombia, with more than twice the people of Cuba, Chile, or Venezuela, is the lowest of the eight in exports. All of this indicates the existence of economies with varying degrees of dependence on export trade and, we can anticipate, varying degrees of involvement in the modern, interconnected system. Export trade appears to be much more important to Venezuela and Argentina than to Peru or Colombia, and similar variations can be seen in the smaller countries.

Moreover, most countries in Latin America depend on one, two, or three commodities for virtually all their exports (see Table 13-3). This means that only a few parts of the country are tied into the export system. There are few producing areas with rail or highway links to a port or a handful of ports. This is the case not only with the small countries but also with major exporting countries including Venezuela, Chile, Cuba, and Colombia. Thus, these producing districts and their connecting infrastructure stand as separate islands within each country. The whole producing network is designed not to pull the resources from different parts of the country together. Instead it is designed to deliver the particular specialized products to ports for shipment overseas. Here is the classic example of the commercial export component of the dual economy.

Table 13-3 EXPORT CENTERS OF LATIN AMERICA, 1966 (value of exports in $ millions)

COMMODITY	VENEZUELA	BRAZIL	ARGENTINA	CHILE	MEXICO	PERU	COLOMBIA	ECUADOR	BOLIVIA	URUGUAY	PARAGUAY
Petroleum	**2,508**	‡	14		42	7	47	1	1		
Coffee	17	**764**			82	29	**328**	32			
Copper				641	13	**186**			6		
Sugar		81	7		57	46	8	7			
Meat (beef)		30	393		29					41	14
Cotton		111	4		154	85	2				2
Iron-ore	144	100		78	2	53					
Fish and fishmeal§	2	5	2	34	58	205	1	4			
Bananas		6			1		22	106			
Wheat			280							7	
Forest products§		79		11	2	1	3	2			11
Cocoa	7	72			4			17			
Tin									93		
Zinc					37	34			5		
Lead					28	34			6		
Total all exports†	2,713	1,741	1,593	878	1,228	764	508	171*	126	186	49

COMMODITY	CARIBBEAN ISLANDS			CENTRAL AMERICAN REPUBLICS						TOTALS OF LATIN AMERICA*
	CUBA	DOMINICAN REPUBLIC	HAITI	GUATEMALA	HONDURAS	EL SALVADOR	NICARAGUA	COSTA RICA	PANAMA	
Petroleum									26	2,647
Coffee		21	17	**100**	20	89	24	53	2	1,579
Copper	5									851
Sugar	**480**	76	3			7		9	2	775
Meat (beef)										507
Cotton	4			44	6	24	57			488
Iron-ore										377
Fish and fishmeal	4			1	1	5		1	11	334
Bananas		1		6	70	5	12	29	35	287
Wheat										287
Forest products				2	11			2		124
Cocoa		11						3		109
Tin										93
Zinc					1					78
Lead					2					70
Total all exports†	593	137	34	228	145	189	137	138	89	8,606

*Totals of commodities listed for all Latin America. †Totals of *all* exports for Latin-American countries listes; Ecuador's total is only for commodities listed. ‡All blanks indicate value of less than $1,000,000. §Value in thousands of dollars.
Note: Bold numbers indicate highest value (by commodity), and all values of 100 or more.
SOURCE: Grunwald and Musgrove, *Natural Resources in Latin American Development,* John Hopkins Press, Baltimore, 1970.

Use the figures given in Table 13-4 to determine the relative economic development of Latin-American countries. (Table A-4 in the Appendix gives similar data for more countries.) Determine the countries most integrated into the modern, interconnected system. What countries are least involved? Are the ranges great?

1. We might hypothesize that the higher the relative importance of export trade to a country, the more a part of the modern, interconnected system it is. Does this hypothesis work? For what countries? If it does not work in some countries, which are they? Why might they not fit?

2. If export trade is not a complete measure, what measure would you suggest?

3. Consider the eight largest countries in population and aggregate GNP. Would you consider the modern, interconnected system well developed in each of them? What variations do you see? Do all of them function primarily as exporters to other countries, or do some seem to have internal industrial development?

4. Brazil and Mexico each have over half of their work force engaged in agriculture, yet they are the two leading Latin-American countries in aggregate industrial energy consumption. How do you explain that?

5. Classify the countries as (a) predominantly traditional, (b) important suppliers of exports to other countries, and (c) those with important beginnings of industry of their own.

Table 13-4 INDICES OF ECONOMIC DEVELOPMENT FOR THE LARGEST LATIN-AMERICAN COUNTRIES, 1970

COUNTRY	PER CAPITA VALUE OF EXPORTS (dollars)	PER CAPITA GNP (dollars)	PER CAPITA INDUSTRIAL ENERGY CONSUMPTION (pounds)	PER CENT OF WORK FORCE EMPLOYED IN AGRICULTURE	PER CENT OF NATIONAL INCOME DERIVED FROM AGRICULTURE
High exports					
Venezuela	266	950	5,496	41	7
Cuba	124	310	2,285	42	25†
Chile*	109	480	2,657	28	10
Moderate exports					
Argentina	77	820	3,713	16	16
Peru	76	380	1,340	63	18
Low exports					
Colombia	31	250	1,272	50	29
Brazil	29	250	1,038	52	18
Mexico	28	530	2,651	51	48

*Data for 1969. †Estimate.

SOURCE: *Statistical Yearbook: 1971*, Statistical Office of the United Nations, New York, 1971, pp. 336–337, 383; Population Reference Bureau, Inc., *1971 World Population Data Sheet*, Washington, 1971; *New York Times Encyclopedic Almanac: 1971*, New York, 1971.

Not all countries are tied to one or two export specialties. The exports of Brazil, Mexico, and Peru include a half-dozen or more major types. For Argentina the specific raw-material commodities shown in Table 13-3 make up only a small portion of total exports. In this case there are important exports other than those on the list. In large countries such as these with diverse environments, there are a series of different independent export systems. Brazil is a case in point. The coffee district close to the Tropic of Capricorn is by far the most important, but the sugar, iron, forest-products, and banana areas are all in different parts of the country. The goods from each of these districts move out of the country along separate streams. They are not tied together to contribute to an integrated national economy, as is the case with the industrial countries we have already discussed.

Tourism and the External System

A new sort of dependence on external markets that does not appear in the export totals has emerged in Latin America since World War II. It represents a new evaluation of the environmental resources of the region by developers in the United States and Europe. With increasing affluence, shorter working hours, paid vacations, and jet air transportation, the market for long-range vacation travel has burgeoned. Because it is close to the United States, Latin America has developed as the major tropical center of the travel industry.

Cuba and Mexico have long been destinations of tourists from the United States. Ideological differences between Cuba and the United States have closed off that tourist mecca, but jet travel has opened up the more distant islands of the Caribbean. Not surprisingly, foreign investors have speculated in tropical island real estate and developed hotels and resorts throughout the Caribbean. The influx of large numbers of tourists with their dollars to spend has given an important boost to local economies, but a large share of the profits are siphoned off by external owners of major facilities, making this just one more example of resource-oriented activity controlled by outside interests. But in this case, instead of the products of the environment being transported to other countries, tourists are taking over sections of the Latin-American landscape as their own.

The Dual Societies Today: Differing Degrees of Involvement in the Modern, Interconnected System

Although it is our proposition that each Latin-American country is a combination of both the traditional, locally based society and the modern, interconnected system, we know that there are various degrees of modernization and traditionalism in the region. Some countries are greatly involved in the trade of specialized products from mines and plantations to Europe and Anglo-America; some have populations that are predominantly traditional; and some have even established significant bases of modern industrial production.

EMERGING INDUSTRIAL CENTERS

It would seem that in Mexico and Argentina, perhaps, we have the case of more developed economies, as the emphasis moves from the colonial model of dependence on outside markets to the beginnings of regional integration within the country. Notice in Table A-4 of the Appendix that Mexico has the highest aggregate GNP and energy consumption and that Argentina has almost as much aggregate GNP despite the fact that both are far smaller than Brazil. Both Mexico and Argentina rank high in indices

Table 13-5 INDUSTRIAL PRODUCTION, 1966–1967

Product	Major Industrial Countries								
	MEXICO	BRAZIL	ARGENTINA	VENEZUELA	COLOMBIA	CHILE	PERU	CUBA	URUGUAY
Coal (000 tons)	1,424	1,957	472	34	3,100	1,474	175	99	....
Crude oil (000 tons)	20,773	7,797	17,370	189,204	8,793	1,760	3,453	3,835*	....
Electricity (000,000 kWh)	22,798	34,239	13,504†	9,479	6,350	6,793	4,583	3,709	1,841
Pig iron (000 tons)	1,599	3,373		422	207	442	31	....	....
Steel (000 tons)	3,269	4,436	1,533	667	199	526	62	....	....
Cement (000 tons)	6,152	7,280	4,213	2,278	2,367	....	1,087	800	421
Autos and trucks (000 vehicles)	‡	265	180	‡	....	‡	‡	....	....

Product	Other Countries										
	GUATEMALA	ECUADOR	DOMINICAN REPUBLIC	EL SALVADOR	PANAMA	BOLIVIA	NICARAGUA	COSTA RICA	HONDURAS	PARAGUAY	HAITI
Electricity (000,000 kWh)	526	700	705	550	485	595	335	757	204	205	78
Cement (000 tons)	205	430	310	155	150	65	79	....	....	14	35

*Petroleum products. † Does not include energy generated by industrial plants themselves. ‡ Some automobile assembly.
SOURCE: *Encyclopaedia Britannica, Book of the Year: 1970,* Encyclopaedia Britannica, Inc., Chicago, 1970.

of development. High GNP is a measure of affluence that goes with industrialization, and high-energy consumption is a component of industrial development.

To carry this idea further, notice in Table 13-5 that Mexico, Argentina, Venezuela, Brazil, Colombia, Chile, and Peru have the beginnings of modern industry. They not only have the largest electrical consumption in Latin America, but they also have steel production and, except for Colombia, automobile assembly. This contrasts sharply with other Latin-American countries, which may produce cement and have textile mills and agricultural processing plants, but depend on overseas sources for all the products of modern complex industry.

Notice in Table 13-3 and Table A-4 in the Appendix that after the eight largest exporting countries are considered, the rest of the Latin-American countries, those with populations of less than eight million people, have much lower indices of development.

In a few locations in Latin America, places are tied into the modern, interconnected system, not just as suppliers of resources, but as centers of modern industry, too. Notice that only the largest Latin-American countries have begun to establish a base for industrialization. These are the countries with the threshold size necessary for production.

The aim of internal industrialization is to reduce dependence on overseas sources of manufactured goods. Although theoretically a region might expect to specialize only in the things that it can produce best and to depend on other areas for the rest of its needs, this does not work very well for producers of primary goods. Farmers in the United States have learned this. They sell farm products at low prices and must buy manufactured goods at higher prices. This is the problem of countries such as those in Latin America that export food and raw materials.

Those Latin-American countries that have been able have built up their own industrial base to the point at which they can produce both consumer goods and essentials such as steel and cement for further economic growth face a task that is not easy. It usually calls for establishing tariff walls to protect weak industry from the competition of foreign goods. Anyone wishing to import competing goods must pay duty so that new industrial plants within the country will not be undersold. Large amounts of capital are also needed to build and equip factories. Since most countries of Latin America have had a shortage of available capital, they have encouraged industrial firms from Europe, Anglo-America, and even Japan to build plants in their countries. American automobile companies and Volkswagen, for example, were allowed to build automobile assembly plants in Brazil. But commonly foreign companies are simply not allowed to produce goods within the country and then take the profits out. Usually, new industry is a joint undertaking in which the government of the Latin-American country or some of its own businessmen have an impor-

tant share in the ownership of the corporation. In recent years, the Latin Americans have been asking for an increasing share of the company. In many cases they have the controlling influence which means at least 50 per cent of the stock.

Also, Latin-American countries are no longer satisfied with letting foreign manufacturers import component parts from overseas and then simply assemble automobiles or other products in their countries. Agreements now call for some manufacturing to be done within the country as well. Now automobiles in Brazil are increasingly being made of Brazilian-made components. In the same way manufacturers are being asked to train more and more Latin Americans for supervisory and management jobs as well as for factory jobs.

Some countries have finally taken over all or large parts of the foreign investment within their jurisdiction. The Mexican expropriation of their oil production in the 1930s has already been described. When it took over the government, Castro's regime outlawed all private property in Cuba. In recent years Chile and Peru have nationalized such major foreign industrial developments as oil fields, sugar plantations, and mines. Venezuelans have also been working on agreements with foreign oil companies to turn over control of all operations to their government. Notice in these cases that the industries taken over were those developed primarily to serve export markets. Latin America now prefers to get control of profits rather than to depend on taxes, duties, and royalty payments.

Still with the emphasis on industrialization in Latin America, there has been little attention to integrating the industrial growth of one country with that of another. As in Europe, each country is a separate decision-maker, anxious to enhance its own development. Each has pushed its own industry rather than working cooperatively with other countries to serve their combined markets. Neighboring Peru and Chile, for example, have each developed separate steel industries even though the domestic market in each is marginal. Moreover, the mills are not even complementary to the extent that one might specialize in heavy rods, rails, and constructional components and the other in sheet metal for roofing and siding for buildings. Instead, each produces the same products. As long as countries develop industry in isolation from their neighbors, market size will be the key factor inhibiting Latin-American industrialization. Even in large countries where there is no shortage of people, the incomes of most who are still within the traditional society are very low. For them, the price levels of domestic manufactures are likely to be too high, so they will continue to rely on handmade products or do without altogether. Small plants with inefficient production techniques simply cannot produce goods cheaply enough for such a market.

METROPOLITAN NODES AS INDICES OF THE MODERN SYSTEM

Of the industrial countries we have studied, the large metropolitan centers were foci of the modern, interconnected system. We might logically expect the large urban centers of Latin America to be associated not only with the largest countries, but also with the most advanced foci of industrialization and trade. Evaluate this hypothesis in terms of Figure 13-10. Is it valid?

If the map of large metropolitan areas is a good index of centers of the modern, interconnected system in Latin America, their absence in parts of the region can be taken as a measure of the lack of modern development. Thus, Central America south of Mexico appears

the least developed area, but the islands of the Caribbean (except Cuba) and the South American countries of Paraguay and Bolivia fall into the same category. Considering your examination of the box on economic development, would you consider the above inference to be reasonable?

Even in the larger, more modern countries the distribution of metropolitan areas may be a measure of the uneven development of the country. São Paulo and Rio de Janeiro lie close to one another in one small area of Brazil. This is the coffee region, a very successful but highly localized development. The absence of big centers in most of huge Brazil is striking.

RURAL POPULATION AND SUBSISTENCE LIVING

In most countries of Latin America, the majority of the population is still rural and agricultural. Only in the more highly developed countries of Cuba, Venezuela, Chile, Argentina, and Uruguay is the proportion of people employed in agriculture well below 50 per cent. Even large developing countries like Mexico, Brazil, Peru, and Colombia have more than half of their labor force employed in agriculture. In Haiti, Guatemala, Nicaragua, and Honduras, two-thirds or more of the labor force work the land. Yet despite the large population involved in agriculture, most countries derive less than 25 per cent of their national income from it. This suggests an inefficient use of labor in agriculture and the lack of alternatives. In all countries except, perhaps, Argentina and Chile, a large proportion of the agricultural population is still engaged in subsistence farming. This is particularly true of countries that have high Indian populations. Guatemala, Nicaragua, Mexico, Peru, Bolivia, Colombia, Ecuador, and Paraguay all have large Indian blood strains and domi-

Figure 13-10 LATIN AMERICA:
CENTERS OF THE MODERN INTERCONNECTED SYSTEM

 Raw Materials

▯▯▯ Industrial

— Railway

Metropolitan Population (number)

·   250,000-1,000,000 (35)

•   1,000,000-4,000,000 (11)

●   Over 4,000,000 (4)

Brazil, with much of the Latin-American frontier area, has made tremendous efforts to open the wilderness to settlement. Although its east-west extent is almost as great as the U.S. most people live within 300 miles of the Atlantic Coast. Like the U.S. in the nineteenth century, the Brazilian government has been undertaking projects to encourage its people to move to the frontier.

In 1957 the Brazilian Congress authorized construction of a new capital city in tropical grasslands over 500 miles inland. The magnitude of this decision might be likened to the Continental Congress's deciding to put the new capital, Washington, in western Illinois. But a road was cleared and construction begun (Figure 13-11). Like Washington, Brasilia was designed with the physical plan and buildings created to specification. Already it has over 300,000 people. But rural settlement along the road to Brasilia and in the area around it has been slower than expected. It was hoped that the isolated city market would encourage agricultural settlement nearby.

The Brazilian government has gone one step further. It is building 5,000 miles of highway in the Amazon lowland to provide an east-west overland route across to Peru. The road is to do for Brazil what the transcontinental railway did for the U.S.

We described the problems of agriculture in the hot, wet tropics. What prospects do you see for settlement along the road? There is no current program to prepare farmers. Most clearings (rarely more than 30 acres) follow traditional slash-and-burn methods. The forest cover is burned off before the rainy season; crops are planted after the rains soak the ashes. The larger tree trunks, too troublesome to remove, are left standing. Families support themselves on a semisubsistence basis. It is a problem to get crops to market even in the new cities emerging along the road.

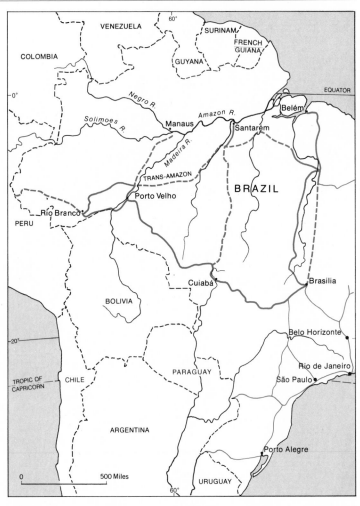

Highways

——— Completed

---- Under Construction

Figure 13-11 Brazil: Highways in the Amazon Lowland

nantly agricultural populations. In Guatemala, Peru, Bolivia, and Ecuador, in particular, large numbers of Indians live essentially as they have for centuries. Their homelands in highland valleys far from the coast were of little interest to Europeans seeking accessible plantation crops, so they were left to continue in semi-isolation. They farm tiny plots of land with a digging stick and hand tools, using most of their production for family consumption, and trade occasional surpluses in the local market for essential goods from other areas. In most of these countries, the Indians live in highland valleys of *tierra fría* above the productive agricultural lands of *tierra templada*. Others live in tropical forests in the lowlands beyond the frontier of European settlement. Speaking native dialects and centering their activities around traditional social and political organizations, these people live essentially within their own separate local worlds.

In the same way, descendents of African slaves and hacienda dwellers with mixed European-native backgrounds live off the land of their local areas. Many are simply squatters on the fringe or part of a landlord's vast property holdings. Farmers in such positions have neither the educational training, the capital, nor the land to operate efficiently.

WILDERNESS AND THE PROBLEM OF FRONTIER SETTLEMENT

Despite the large agricultural populations and vast tracts of land beyond the settlement frontier, in most countries there has been no mass movement to the frontier. In recent years the poor of Latin America, like the poor in rural parts of the United States, have found the lure of city life and its potential riches far more attractive. They flock to the cities without adequate education, capital, or knowledge of urban life, and the results in most countries are squalid slums around most large urban places and a declining rural population.

Latin-American slums are on the edges of the cities rather than in the "inner" city. On the urban fringes rural people have simply settled as squatters on the land and built shacks. Although they do not own the land or pay rent, they look upon their dwellings as their property and constantly strive to improve them. What might start as a cardboard cover on a wooden frame with a bit of tarpaper in time becomes a hut of cement blocks, perhaps with flooring and a corrugated iron roof. Usually there is no water or plumbing, but the inhabitants see it as better than an adobe hut with a thatched roof in a village without hope. These are the pioneers who, not satisfied with what their villages offered, are striving to better themselves economically and socially. To them the city offers much greater possibilities than the rural frontier.

Faced with the problem of trying to develop frontier areas to counteract the urban migration, governments have developed incentive programs to encourage settlement. Particularly active in frontier settlements are the countries with lands in the upper Amazon lowland: Brazil with its remote western lands and Colombia, Ecuador, Bolivia, and Peru with forested lowlands east of the Andes that have never been a functional part of the country. But such development seems to attract more immigrants from other countries than natives. Mennonites have settled land in Paraguay and Mexico, Japanese have gone into the Amazon country, and there have been various land companies and colonies established by North Americans and other outsiders in Peru and Colombia.

In evaluating the laggard development of the rural frontier during the more than 450 years of European settlement in Latin America, it is important to recall that the great frontier areas are in tropical environments, not in the mid-latitudes. The potential for management is very different. Moreover, most of the frontier areas are hot, wet rain forests. The wet tropics provide a satisfactory life with a minimum of effort. Human needs for shelter and clothing are not great. Hunters and gatherers, with almost no capacity for storing perishable goods, find a continuous supply of food, and farmers have learned to manage the tropical environment. They plant a variety of crops together in a clearing and then shift to new clearings when unwanted weeds inhibit growth. They plant root crops, grains, vegetables, and small trees (like bananas) in a seemingly helter-skelter fashion in the clearing. This three-tiered field replicates the great variety of the forest itself and thrives for several seasons before becoming overgrown.

The rain-forest area is uncommonly unsuited to the mid-latitude agricultural system which clears an area, plants one crop, and systematically harvests a crop or two each year. Such a system draws on plant nutrients stored in the soil, nutrients that are replenished regularly by artificially fertilizing the soil. But in the tropics the growth system is dependent on the continuous recycling of nutrients through the plant system rather than on storage in the ground. Tropical soils are storage points for moisture but not for the other essentials for plant growth. Thus, a system that breaks up the variety of plant life in the tropical forest system, that depends on nutrients stored in the soil, and that disrupts the surface soil layer through heavy plowing (all characteristics of mid-latitude agriculture) faces severe problems in the wet tropics.

European settlers have, in general, avoided the forested tropics, and so the major frontier of Latin America in the Amazon lowland of Brazil, Colombia, Ecuador, Peru, and Bolivia remains semiwilderness. Even major schemes by industrial corporations have generally been unsuccessful. An attempt by the Ford Motor Company to establish rubber plantations failed. A scheme to develop 2.4 million acres of timberland by clearing the forest and planting a single species has gotten off to a slow start. Early plantings have been bothered by floods, unusual rains, and the low-spreading growth of the plants.

Frontier growth has been further handicapped by the lack of institutional development in the back country. Development depends on roads and railroads, the availability of land, the incentive of markets. But, except for Brazil's recent efforts in moving the capital and undertaking a new road network into the Amazon, the amount of investment by Latin-American governments in their tropical frontiers has been minimal. Such areas are far from the islands of population that distinguish each country, and in the case of the western countries, they lie across the rugged Andes from population centers. Frontier development has been an almost overwhelming undertaking for the small countries of Latin America faced with the need for further investment by themselves within their population islands. As a result, developments on the frontier have been left in the hands of private companies, such as Ford or the operators of the new timber project, with the understanding that the investor would develop the necessary infrastructure for the enterprise and allow this infrastructure to be used by others moving to the frontier.

ATTEMPTS AT INTERREGIONAL COOPERATION

Despite the continuing separation of the social, political, and economic structures of individual national states in Latin America, there have been attempts in recent years to establish institutions among Latin-American countries similar to the Common Market of Europe. The first formal organization was the Central American Common Market which bound together the five weak Central American republics of Guatemala, Honduras, El Salvador, Nicaragua, and Costa Rica. This pact called for the elimination of tariffs between member countries and established a single set of tariffs for the rest of the world, like the European Common Market. Both external and internal trade within the Common Market has increased, and the availability of a single market for the five has encouraged foreign investment in large-scale enterprises. There was also talk of possible political union, but the organization has been threatened by border disputes and armed conflicts between Honduras and El Salvador. The future of this Common Market depends on the degree to which national governments can overcome traditional self-interest and rivalries with neighbors.

A more ambitious project was the Latin American Free Trade Association which involved all the largest and most advanced Latin-American countries. This organization includes countries which comprise 80 per cent of the population and produce 50 per cent of the output of all of Latin America. LAFTA aimed at reducing trade barriers and at encouraging trade among its members. It has been much less successful than expected, however, probably because it is overambitious in scope at this stage in Latin-American development. Its membership is too widely scattered and diverse, and regional organizations are emerging in its place. The Andean Development Corporation, for example, includes all the Andean countries plus Venezuela, and the Río de la Plata Development Group includes Uruguay, Argentina, Paraguay, Bolivia, and Brazil. The Andean group is working toward the common market goals of internal free trade and a common external set of tariffs by 1990. The Río de la Plata group has emphasized joint exploration and development of the area's resources, but it has been enlarged to include economic integration and development of industries that would complement rather than compete with each other. Most recently small Caribbean countries that were former dependencies of Britain have formed the Caribbean Free Trade Area in order to lower internal tariffs and other barriers between them. Included in this group are Guyana and British Honduras in addition to Jamaica, Trinidad, and various small island units.

Change but Not Solutions

In view of the problems still to be faced, these attempts at economic, political, and social change are moving at an extremely slow pace. Industrialization has just begun, agricultural methods in most places are still primitive, and the entire economy remains basically oriented toward the production of raw materials for industrial Europe and Anglo-America. Political stability has come from power wielded by the military which represents the traditional landed class rather than the people. The mass of Latin Americans have no political voice, live on the brink of starvation, are uneducated, and remain largely ignored by the governments which purport to represent them.

Mexico is cited as the one Latin-American country that has achieved durable political stability and steady economic growth. This is largely the result of the revolution of 1911 which broke up the estates of the oligarchy, gave land to the peasants, and brought social reforms. Mexico is ruled by a benevolent, one-party dictatorship of the Institutional Revolutionary party, and the system works well because it provides some voice within the party for all sectors of the population. But Mexico still has a host of problems. Population is expanding at a rate of 3.5 per cent a year, and the economy is hard pressed to absorb the growing youthful population into the work force. Nearly half the people live in family units earning less than $1,000 annually. Thus, even the most successful of the countries of Latin America has a long way to go.

SELECTED REFERENCES

Bever, Glenn, ed. *The Urban Explosion in Latin America: A Continent in the Process of Modernization,* Cornell U. Press, Ithaca, N.Y., 1967. Experts look at urbanization in Latin America: its causes and its contributions to modernization.

Butland, G. J. *Latin America: A Regional Geography,* 2nd ed., John Wiley, New York, 1966. A standard geographic view of the region of Latin America.

Cole, John P. *Latin America: An Economic and Social Geography,* Butterworth, London, 1965. The same British geographer that did the book on the Soviet Union turns his energies to detailing Latin America.

Furtado, Celso. *Economic Development of Latin America,* Cambridge U. Press, Cambridge, 1970. A treatment of economic development in Latin America since colonial times.

Grunwald, Joseph, and Musgrove, David. *Natural Resources in Latin American Development,* Johns Hopkins, Baltimore, 1970. A "Resources for the Future" analysis of primary resources and their trade. Lots of data.

Nisbet, Charles T., ed. *Latin America: Problems of Economic Development,* Free Press, New York, 1969. Latin America's economic problems.

Urquiri. *The Challenge of Development in Latin America,* Praeger, New York, 1964. A Latin-American economist deals with the question of economic development.

Chapter 14 Subsaharan Africa: The Struggle with a Colonial Heritage

THE PROBLEMS that are being faced in Subsaharan Africa are not new. Every country faced them at one time or another as the following description of England in the seventeenth century suggests:

[The] . . . economy . . . was one in which the methods of production were simple and the units of production were small; in which middlemen . . . were both hated and indispensable; in which agricultural progress was seriously impeded by the perpetuation of communal rights over land. The chronic underemployment of labour was one of its basic problems and, despite moral exhortations, among the mass of the people the propensity to save was low. . . . It was an economy heavily dependent on foreign sources for improved industrial and agricultural methods and to some extent for capital, but in which foreign labour and businessmen were met with bitter hostility. In it ambitious young men often preferred careers in the professions and government service. . . . Men increasingly pinned their hopes on industrialization and economic nationalism to absorb its growing population; but industrialization was slow to come and the blessings of economic nationalism proved to be mixed.[1]

Today, three to four hundred years later, the same situation exists in Africa as the native population emerges from a traditional heritage of tribalism, subsistence living, and many years of outside colonial domination. Most of its people have been living within local traditional societies that formed tightly knit culture groups. Most of the modern economic development in Africa has come as a result of colonial activity designed to exploit the mineral and agricultural resources of the region for external markets in Europe and the United States.

---

[1]From "The Sixteenth and Seventeenth Centuries: The Dark Ages in English Economic History?" *Economica* (London School of Economics), new series XXIV, No. 93 (February 1957), pp. 17–18. Quoted in A. M. Kamarck, *The Economics of African Development*, Praeger, 1967, pp. 47–48.

Despite the years of colonial domination, the African countries emerging today are inhabited by an overwhelming majority of indigenous people. There never was the magnitude of European migration into Africa that there was into the Americas. Probably fewer than five million people in the total population of 258 million are of European origin, most of them in South Africa. Elsewhere no more than two per cent of the population is European, and for the most part the Europeans are temporarily employed only by the government or foreign business interests on a contract basis. Asian immigration has been even smaller, and the Asians in the area total only somewhat over one million persons, half of whom live in South Africa.

INDEPENDENCE FOR TRIBAL PEOPLES

The countries themselves are almost without exception brand-new. As late as 1956 there were only three independent countries in the entire region—Ethiopia, South Africa, and Liberia—but today only three important areas—the Portuguese colonies of Angola and Mozambique, and the territory of South-West Africa—are not independent. Since 1957 more than thirty new states have gained political independence in the region. All the governments of the new countries are in the hands of native peoples except for South Africa and Rhodesia where minorities of European settlers are in control.

We can get some idea of the problems surrounding African independence if we think of European colonization in the Americas and then imagine that the Europeans pulled out, leaving the Indians to rule not their own lands, but the colonial territories established by the Europeans. African life at the time of European colo-nization was not very different from the life of the American Indians when the American colonies were established. The Africans lived in locally based tribes and nations, each with its own cultural heritage, language, customs, political organization, and economic system. Some African tribes lived highly organized lives like the Sioux and Iroquois; a few approached the sophisticated civilizations of the Inca and Aztecs.

European colonization superimposed a new layer of political and economic power on the traditional system. Boundaries were laid out along European lines with little regard for the real patterns of African life. Europeans governed through their own institutions; Christian missionaries introduced new cultural and moral values; and foreign business enterprises established transport facilities to tap valuable resources for their home markets.

The impact of European development on Africa was tremendous, but it was sharply localized. Its effect on most of the rural population was indirect at best. Most people not only continued to support themselves from their traditional local base but also maintained their ancient tribal and village loyalties. The social structure remained that of the closely knit, extended family relationship. In fact, in some places class or caste structures were almost nonexistent; however, in other places they were rigidly enforced.

This pattern has continued after independence. A small fraction of the African population has left the rural communities to be educated by the Europeans or to work in the government or the business community, but most continue their old lives unchanged. New basic crops such as manioc, beans, maize, and peanuts were introduced, and they have been adapted to the existing native diet.

Only a small cadre of Africans, trained and educated by the colonial system, manage the political affairs of the new countries. These new leaders have inherited not only colonial political institutions but also colonial (that is, European) political units. The political states that emerged from colonialism took their identity and boundaries from the colonies. Independence came as one European power after another relinquished control of the colonies under its jurisdiction. The new native leaders inherited countries defined in European terms. These countries are generally too small to be economically viable units in the modern political world, yet they are too large to represent a single tribe or African tribal nation. As it was with the Indian tribes of North America, native Africa has consisted of over one hundred different tribes and nations, numbering from a few thousand to hundreds of thousands.

Africa may seem politically fragmented with more than forty countries (Figure 14-1), but it includes 100 important language variations and a variety of dialects within those groups. The map of African tribes indicates the diversity of the traditional native organization of the continent. Thus, while the map of countries created as a result of European colonization may seem fragmented, it is highly centralized compared to the traditional separation that exists between the locally oriented peoples of Africa.

The European superimposed boundaries inevitably encompassed a number of tribes and nations within a country and sometimes also divided the homelands of groups between two or more different countries. The new countries do not conform at all to the traditional ethnic patterns of Africans, and rivalries among groups pose a serious obstacle to national development. The terrible civil war in Nigeria in 1969 was largely the result of tribal and regional conflict.

Africa's Geographic Base

Africa south of the Sahara is a huge area, 10 per cent larger than Latin America. But its population in 1970 was 9 per cent less than the population of Latin America and only 11 per cent larger than that of Anglo-America. The population density is very low—only 33 persons per square mile, just slightly higher than Latin America's. It is lower than that of the United States and the Soviet Union, neither of which are seriously overpopulated. But population pressure is relative and depends as much on the productivity of the economic system as on density.

The most striking geographic feature of Africa is that it is divided into almost twice as many political units as Latin America. There are forty-one different political units south of the Sahara, and all but four are independent countries (Table 14-1). Yet only six of these have populations of more than 10 million people. Nigeria had an estimated population of more than 55 million people in 1970. Census counts in Africa are very crude, however, and this is little more than an educated guess. Nevertheless, Nigeria clearly has the largest number of people. Ethiopia, the next largest, has less than half as many people as Nigeria. Of the remaining countries and territories, only nine have more than about 5 million people, and four have fewer than half a million. Such populations are equivalent to

Figure 14-1 POLITICAL MAP OF AFRICA

City population

- · 250,000-500,000
- • 500,000-1,000,000
- ● 1,000,000-4,000,000
- ⬤ Over 4,000,000

1: Equatorial Guinea

2: São Tome e Principe (Port.)

Table 14-1 Basic Data on Countries and Territories of Africa South of the Sahara, 1970

Country and Territory	Population (millions)	Population per Square Mile	Rate of Population Growth (per cent)	Per Capita GNP ($)	Energy Consumption (million metric tons of coal equivalent)	Date of Independence
Western Africa						
Dahomey	2.7	63	2.9	90	.08	1960
Gambia	0.4	100	2.0	120	.02	1965
Ghana	9.0	98	2.6	310	1.48	1957
Guinea	3.9	41	2.6	120	.38	1958
Ivory Coast	4.9	39	3.0	310	.98	1960
Liberia	1.5	34	3.0	240	.34	1847
Mali	5.1	11	2.1	70	.10	1960
Mauritania	1.2	3	1.9	140	.11	1960
Niger	4.0	8	2.9	90	.11	1960
Nigeria	55.1	154	2.9	120	2.49	1960
Senegal	3.9	51	2.1	230	.58	1960
Sierra Leone	2.6	93	1.4	190	.26	1961
Togo	1.9	86	2.9	140	.12	1960
Upper Volta	5.4	51	2.1	60	.07	1960
Middle Africa						
Angola	5.5	12	1.3	300	.93	Portugal
Cameroun	5.8	31	2.1	180	.54	1960
Central African Republic	1.5	6	2.6	140	.10	1960
Chad	3.7	7	1.8	80	.08	1960
Congo (Brazzaville)	0.9	7	1.5	300	.19	1960
Zaire	18.8	20	2.8	90	1.57	1960
Equatorial Guinea	0.3	27	1.8	210	.05	1968
Gabon	0.5	5	1.0	630	.41	1960
Portuguese Guinea	0.6	50	0.7	250	.04	Portugal
Eastern Africa						
Burundi	3.6	327	2.0	60	.03	1962
Ethiopia	25.0	53	2.2	80	.79	1941
Kenya	11.2	50	3.1	150	1.72	1963
Malagasy	7.3	32	2.6	130	. . . .	1960
Malawi	4.4	98	2.6	80	.20	1964
Mozambique	7.7	26	1.9	240	1.12	Portugal
Rwanda	3.6	360	3.0	60	.04	1962
Somalia	2.8	11	2.4	70	.10	1960
Rhodesia	5.3	34	3.3	280	2.76	1965
Tanzania	13.2	36	2.5	100	.92	1961
Uganda	9.8	89	2.7	130	.70	1962
Zambia	4.1	14	2.5	400	2.32	1964
Mauritius	0.8	1,248	2.4	240	0.15	1968
Southern Africa						
Botswana	0.6	3	2.0	110	. . . .	1966
Lesotho	1.0	83	2.0	90	. . . .	1966
South Africa	22.1	46	3.0	760	63.25	1910
Swaziland	0.4	57	2.9	180	. . . .	1968
South-West Africa	0.4	1	2.0	. . . .	. . . .	South Africa

Source: *World Bank Atlas: Population, Per Capita Product and Growth Rates,* International Bank for Reconstruction and Development, Washington, 1972; *Statistical Yearbook: 1971,* Statistical Office of the United Nations, New York, 1971; *Handbook of the Nations and International Organizations,* Life World Library Series, Time-Life Books, New York, 1968.

Table 14-2 Comparative Per Capita GNP, 1970 (number of political units in each category)

Region	Over $1,000	$750–$1,000	$500–$750	$250–$500	$100–$250	Under $100
Africa south of the Sahara	0	1	1	7	19	12
Latin America	2	3	8	11	2	0
North Africa–Southwest Asia	3	0	3	8	5	2
South and East Asia	1	2	0	5	13	4

Source: *World Bank Atlas: Population, Per Capita Product and Growth Rates,*
International Bank for Reconstruction and Development, Washington, 1972.

those of American states ranging from Michigan down to the least populated: Alaska, Wyoming, and Vermont.

The Meaning of GNP
for Underdeveloped Areas

Table 14-1 shows the very low living standard of most people of this region. While South Africa is above the world average in per capita GNP, 31 of the other countries have per capita GNP below $250 a year (Table 14-2). This is in striking contrast to the per capita GNP for the United States in 1970 of $3,670.

We have been using per capita GNP as a measure of wealth in the modern, interconnected world, but it is really a very different sort of measure in underdeveloped areas. GNP is a monetary measure, designed as a useful index for a money economy. But much of the production of traditional society is never translated into monetary terms. A family may support itself almost entirely from the subsistence production from its own lands. It consumes almost all it produces and has little surplus to sell. Its total commercial sales in the market

place during the year may be $100 or less, and much of that may be transacted by bartering without using money at all.

But the question of measuring subsistence production remains. If the living standard of the subsistence family is very low, its total assets, or its total production in money terms, may be very low when compared to a city worker's annual wages. But if the family continues to live from year to year on a given standard, life is supported, and the threshold necessary for life has been reached, regardless of the level of actual production. One could ask whether a threshold should be considered in money terms, or whether it should be measured in terms of hours of contentment, units of family togetherness, or degrees of personal satisfaction. There is no meaningful way to "cost out" the input and output of a subsistence farm as one might cost out the production of a factory worker.

But whereas GNP cannot be used as a measure of the standard of living of people in underdeveloped areas such as Africa, it can serve as a measure of the degree to which the country is a part of the money economy or outside it. GNP stands as a measure of the relative impor-

Table 14-3 Comparative Per Capita Consumption of Industrial Energy Sources, 1969 (number of political units in each category)

Region	Million Metric Tons of Coal Equivalent					
	Above 5	2.5–5	1–2.5	0.5–1	0.1–0.5	Under 0.1
Africa south of the Sahara	1	1	6	6	11	12
Latin America	9	2	5	6	4	2
North Africa–Southwest Asia	8	2	4	4	3	3
South and East Asia	9	1	2	0	3	0

Source: *Statistical Yearbook: 1971*, Statistical Office of the United Nations, New York, 1971.

tance of the modern, interconnected (money) economy as compared to the traditional, locally based (nonmoney) economy. If a country has a GNP of $200 a year, one can assume that a much larger segment of its population is in the modern, interconnected system than would be the case in a country with a GNP of $50 per capita. In the latter case we would presume that most of the population is in the traditional system.

If GNP is used in this way, it is obvious that Subsaharan Africa is one of the least developed parts of the world in terms of the modern system. The modern, interconnected system is prominent in certain areas that produce goods for the rest of the world. But in most African countries this type of production is very small when compared to the total population.

Affluence and Energy Use

Figures on the consumption of modern energy sources can be used in the same way to support the thesis developed with GNP data. The greater the use of modern industrial energy sources, the greater a country's role in the modern, interconnected system. Conversely, countries that rely on animate energy are presumably still traditional societies.

When one considers the lack of modern energy consumption (Table 14-3), the lack of involvement in the money economy, and the small population of most African states, one can understand their extremely weak position in the modern world. If they are to participate fully in the modern world, they would seem to need a major technological revolution and the development of at least a considerable degree of interstate economic cooperation.

Widely Spread Rural Population and a Few Urban Nodes

As we can see from the world map of population (Figure 6-4), the population distribution in Subsaharan Africa is very different from that in Latin America. There are fewer areas with very low densities in Africa, indicating less wilderness area beyond the frontier of settlement. The wider spread of population also suggests that African population distribution is much more the result of indigenous traditional settlement than outside immigration. There are concentrations of population well back from the coast, but as in Latin America, high densities generally stand as separate islands.

In Table 14-1 we can see that few of the countries of Africa south of the Sahara have dense populations. The highest densities of over three hundred per square mile occur in some of the smallest countries, but most African countries have densities lower than those of Latin-American countries and far lower than those of Europe. In measuring the meaning of these densities, we must consider three factors: (1) the degree to which population is dependent on its local resource base rather than outside supplies, (2) the degree to which the people can use the resource base, and (3) the degree to which the base can support the economy. As an underdeveloped area, Africa is still dependent on its own resource base. It has not had the technical capability to use the environment fully, and in many respects the environment of tropical Africa presents major obstacles to any development.

The population distribution of Africa today has resulted from an interpretation of the uses of that environment throughout Africa's long history. Europeans have interpreted the African environment differently, however, and their use of the resources has superimposed a different population pattern on the traditional one. They measured the region in terms of resources that could be exploited for export, developed production centers where they could find raw materials, and established political capitals and seaports where they were accessible to outside transport routes. Many Africans have left their traditional homelands to migrate to the jobs that these centers offer. The large concentrations of population in South Africa, in the copper belt of Zaire, Zambia, and Rhodesia, and in the interior of east Africa are examples of such European-inspired development.

In general, African economic systems use the land extensively, and population densities remain low. Traditionally most African cultures have been dependent on shifting cultivation, herding, hunting, fishing, and gathering. These activities depend primarily on harvesting what is growing wild in the natural environment rather than on expending extra labor to make the land produce particular products. Tribes dependent only on gathering, fishing, and hunting in rain forests and those that are primarily livestock herders in drier areas require many square miles to support a small population. An economy based on shifting cultivation can support a slightly larger population, but it still requires a large area. Even in the few places where Africans have developed more intensive agriculture, the land is not nearly as densely settled as it is in the agricultural areas of Asia.

Mining and even plantation agriculture in the modern system call for intensive use of manpower to extract, process, and ship the products to market. Size of population is proportional not only to the intensity of labor use but also to the amount of the external demand for the product. Large-scale mining operations, such as those in the copper belt or the diamond-mining district of Witwatersrand in South Africa, have resulted in the immigration of large numbers of Africans from far away. The concentrations of production and population stand at focal points in transportation systems designed to get the resources to external markets. In turn these areas draw on a wide network of agricultural producing regions for the food supply necessary to support the working population. Agricultural food producers who supply the mining and plantation developments no longer farm their land on a subsistence basis. Instead they use it to produce cash crops for the African urban market.

Rapid Population Growth after Centuries of Stagnation

One other aspect of Africa's population map should be noted. Although the population of most African areas is growing very rapidly, growth is largely a recent phenomenon. Between 1650 and 1850, the subsistence economy was tapped by slave traders who carried off an important segment of the youthful population. This exodus, coupled with a high death rate, caused a decline in Africa's total population. Later, Europeans exploring and occupying land not only exposed the local populations to European diseases but also spread native ones. Stanley on his explorations of 1887 carried the sleeping sickness across Africa. The Italians invading Ethiopia in the 1890s are thought to have brought in cattle with rinderpest. Within a few years, the disease swept east Africa, killing millions of cattle, the support of herding tribes. As a result, large grazing areas returned to bush, the native environment of the tsetse fly that carries the sleeping sickness. The effect of these diseases was that until the 1930s most local governments were concerned with depopulation, not population growth.

Today public health measures have eliminated much of the danger of disease and death rates have dropped sharply. With no out-migration from the region, populations are growing rapidly and are expected to continue to rise because such a large proportion of the population is young. Nearly half of the African population is under the age of fifteen, as compared with less than one-third of the population in most industrial countries. The large proportion of young people means a large population coming into the childbearing ages in the future and promises explosive population growth unless birth control is encouraged. Moreover, the need to provide necessary schools, hospitals, and jobs for the young is a tremendous burden on the new African governments.

A TROPICAL ENVIRONMENT

The African environment is almost exclusively tropical. The continent extends southward only to 32° south latitude, and only South Africa is primarily in the mid-latitudes. The Cape of Good Hope at the tip of South Africa is closer to the equator than is Savannah, Georgia. In the north, only small portions of Mauritania and Mali lie beyond the Tropic of Cancer.

Obstacles to Agriculture

Contrary to the impressions of early explorers, who viewed only the wetter coastal fringes of the continent, tropical rain forests cover less land area in Africa than in either Latin America or Asia. Most of tropical Africa is covered by a natural variety of grasses and open woodlands. In wetter areas the woodlands predominate; in drier places trees are less numerous and grasses are most prominent. But in most of Africa the combination of grasses and trees produces a parklike landscape different from either the forests or the grasslands found in the middle latitudes.

The presence of open woodlands and tree-strewn grasslands rather than forests is an indication of an insufficient moisture supply for full forest growth. Indeed, the problem of soil-moisture shortage is chronic to most of tropical Africa. While there are year-round temperatures conducive to rapid plant growth over most of tropical Africa, the areas of open forests and grasses suffer from pronounced dry seasons during the year and from great fluctuations in rainfall from year to year. The greatest potential evapo-transpiration occurs along the fringes of the tropics rather than at the equator, and the drier areas, where there is less cloudiness, have greater potential evapo-transpiration than the rain forests. But all areas, except a small portion of the rain forest on the equator, have water deficits during a dry season and the deficiency increases in the tropics. At the same time the only moisture surpluses during the wet season occur in the forest areas of west Africa near the equator.

The moisture deficiencies of most of tropical Africa indicate that farmers without irrigation will be unable to obtain the full potential for growth that temperature conditions would permit. Most of the continent has relatively low agricultural productivity, particularly for the traditional native subsistence farmer or herder who does not have scientific knowledge or modern capital. Over the centuries the natives have worked out a functioning system by trial and error. But even the Europeans who have worked with this land scientifically find it most difficult to farm.

Soils present problems to Africans as they do in other tropical areas. Even in the dry tropical areas most soils are thin and low in organic matter. When land is cleared of natural vegetation, the heat of the sun kills essential soil bacteria, and heavy rainstorms destroy the vital soil structure and leach away mineral components necessary for plant growth. Native agriculture has adjusted to these problems by planting a profusion of different crops in clearings to provide soil cover and by abandoning clearings when productivity decreases. Only the best soils have supported continuous cultivation year after year, since native agriculture generally uses no fertilization and depletes the fertility of the soil.

Most of Africa south of the Sahara stands 800

feet or more above the sea and is separated from the lowland by a sharp rise. This is plateau rather than plain, for once the rise from the coastal lowland is made, the land is level or rolling, not hilly or mountainous (see Figure 14-2). There are no mountains comparable to Latin America's Andes in either altitude or extent. There are probably fifty peaks in the Andes higher than Mount Kilimanjaro, which is Africa's highest mountain, and there are only nine mountains in all of Africa which rise to heights of over 13,000 feet.

Two Different Areas of Subsaharan Africa

When climate is considered along with land surface, Subsaharan Africa can be divided between the higher, drier, and cooler eastern and southern portions, and the lower, wetter, and hotter western and northern areas. This division has been most prominent in the pattern of settlement by Europeans. No significant settlement by them has occurred in west Africa, but east and south Africa have been the sites of settlement in agricultural as well as in urban and mining areas.

Rain Forests: The Wet World of Africa

In west Africa and along the coastal lowlands there are lush tropical forests similar to those in Latin America. Here, too much moisture—not too little—is the problem. Moisture in an area with high temperatures throughout the year results in rapid vegetative growth. The hothouse conditions produce a tremendous variety of plants, but they also result in the rapid proliferation of insects and bacteria which foster rapid decay of plants and cause disease. Malaria and yellow fever are carried by mosquitoes, and hookworm, amoebic dysentery, and schistosomiasis are also problems. The greatest

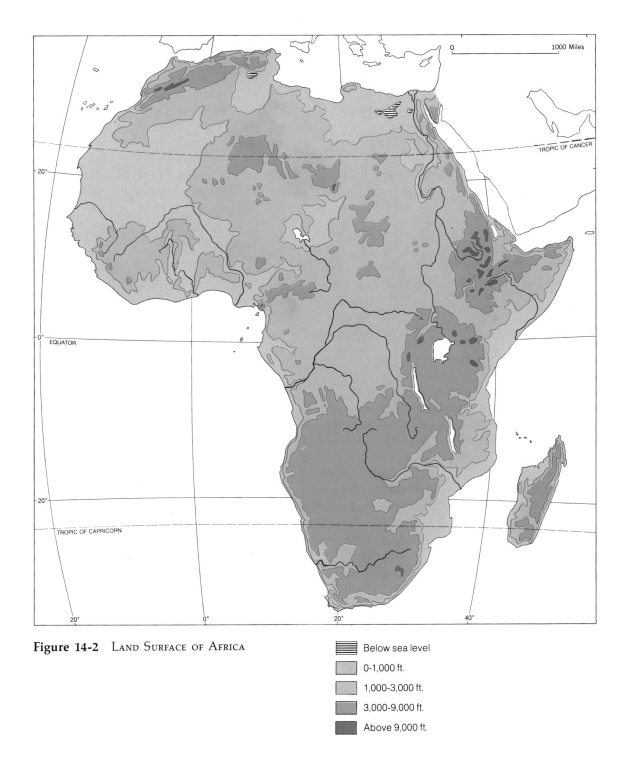

Figure 14-2 LAND SURFACE OF AFRICA

Below sea level
0-1,000 ft.
1,000-3,000 ft.
3,000-9,000 ft.
Above 9,000 ft.

FARMING IN A RAIN FOREST: A PROBLEM NOT YET SOLVED

In most underdeveloped countries one of the great problems is the increasing pressure of population on the traditional, locally based economies. Great pressure of increased numbers of people is seen on the traditional system of shifting cultivation in the tropical forests. Under that system farmers essentially rotate the fields that they cultivate rather than rotate crops in continuously cultivated fields as is common in the mid-latitudes.

This system, common among peoples in the tropical forests where both degree of technology and population density are low, depends on "slash and burn clearing" done with machetes and on "bush fallow" by which fields are used for two or three years and then are allowed to return to wild vegetation for six to ten years. When the fields are planted, a great variety of crops are intermixed. This is often called "three-story agriculture" because some crops such as yams and cassava are root crops, others such as grains and vegetables grow near the surface, and others such as plantains are treelike.

Farmers depending on traditional trial-and-error experience and using only hand tools such as the hoe and digging stick make as much use of the environmental systems as possible. Clearing is done during the dry season, and the debris is burned off to leave the ash for important plant nutrients. The variety of crops planted not only keeps rapid-growing weeds from taking over but has the advantage of providing the farm family with the essential variety of food needed during the year. Different crops mature at different times, so there is some harvesting through most of the year. Root crops also can be left as long as two years in the ground before they deteriorate; thus there is a natural storage system. Most important, the new clearing depends on the natural soil regeneration process; green fertilizer is provided by abandoning the field to forest regrowth for the years between the reclearing of the area.

But in Ghana, Nigeria, and other African countries there is increasing pressure on the rain-forest lands for use in agriculture. Population growth in these countries is as much as 3 per cent a year. Moreover, forest lands are in demand for commercial plantation production of tree crops such as cacao, rubber, and oil palm. The result of such pressure is that the bush-fallow time between use and reuse of a clearing has been reduced to two or three years with the result being inadequate refertilization from fallowing and inadequate yields in the recleared areas.

Given this problem, what solutions might be tried?

Two possibilities are apparent: (1) the addition of fertilizer to the farming system to reduce or eliminate the fallow period needed and (2) the planting of special high-nutrient fallow cover or of fallow crops that would have some food value.

But in either case the present agricultural system calls for fundamental changes in the perspective and work practices of the people using shifting cultivation. The use of fertilizer requires educating farmers in the quantity and time of its application. Moreover, such farmers will either have to be subsidized by the government for fertilizer purchases, or they will have to enter further into the cash economy. In the latter case there needs to be a guaranteed market for cash-crop sales, some price guarantees, and adequate storage and shipping facilities—in short, a whole new structure for agriculture.

In the case of more valuable fallow growth, there is the problem of planting desired crops where the present growth is natural and almost inevitable. Clearing land for fallow planting does not fit into present work practices. Moreover, the brush in the fallow areas presently is used as a major source of wood supply for cooking and other fuel needs. One possibility would be the addition of mechanized clearing and planting equipment, but this would probably mean government subsidization of the program.

In the case of any of the solutions, there is the problem that individual clearings are small and each farmer works several fields interspersed among his neighbor's fields. The use of either fertilizer or mechanization depends almost surely on the redistribution and consolidation of land which in turn calls for fundamentally altering the present farming system. But the existing traditional system is no longer capable of supporting the population. Something must be done. What?

scourge of all is the sleeping sickness spread by the native bloodsucking tsetse flies. In its different forms sleeping sickness is fatal to both humans and domestic livestock.

Since the rapid-decay cycle goes on throughout the year, it is difficult to store produce without modern controlled storage facilities. But because the rain forest continues to produce fruits throughout the year, it will support a limited population. The people who live in these areas have acquired some immunity to the worst diseases.

From the start Europeans considered the climate unhealthy. West Africa was known as the "white man's grave," and the wet tropical areas of west Africa have never received any appreciable influx of Europeans. Government and company officials and their families have served tours of duty, but most have viewed the assignments as temporary. West Africa has the lowest percentage of Europeans of any part of Africa.

Africa: Almost Unknown to the Outside World until the Last Century

The African coastline was explored by Europeans before much was known of the coast of the Americas. The ancient empires of Ghana and Mali are supposed to have been the chief sources of gold for Europe before the discovery of America. Nevertheless, most of Africa beyond the coast was out of contact with the rest of the world until less than a hundred years ago.

The first important contacts were across the Sahara, and the most sophisticated African societies in political and economic terms were those of Ghana and Mali along the northern desert fringe in west Africa. There was a steady trade between the contrasting lands on either side of the Sahara. In many ways these cultures were similar to early societies in Europe and Japan. Political control was built upon an economic system that included both traditional subsistence agriculture and trade connections with other areas. Most of the population depended on the local agricultural base, but the wealth of the empire, in the hands of the nobility, came from the profitable cross-desert trade. Merchants shipped gold, ivory, spices, hides, forest products, and slaves northward across the desert and received salt, copper, cloth, and dried fruits in return.

This trade represented an earlier model of the interconnected system, and nodal points were necessary in the transport and business management of such trade. Trading cities such as Timbuktu were centers of commerce and learning, and political kingdoms rose above the level of the subsistence tribe. Over the trade routes came the Moslem religion, and much of the population of the desert fringe remains Moslem to this day.

By the time Europe established trade contacts with Africa, much of the trans-Saharan trade had broken down, and the prosperous empires declined. Invasions by nomadic peoples with firearms disrupted political control, and they took over the trade. They brought herds of livestock that broke up the delicately balanced agricultural system of the area.

For centuries afterward the Sahara desert proved a formidable barrier to overland trade, and the coastal zone, particularly the wet tropical lowland along the Atlantic, was forbidding to Europeans. There are few good harbors along the African coasts, and for centuries ocean-going vessels had been forced to anchor offshore and transfer cargo by means of small boats. Only in the twentieth century have artificial harbors been built to allow ships to reach dockside.

Because vast stretches of the African coast are backed by dense tangles of mangrove swamp, Europeans were content to carry on trade with the natives at points along the coast and let native merchants bring wanted goods out to the coast for trade. The same was true, in general, of the east coast where Arab and Indian traders had contacts before the arrival of the Europeans. The natives controlled the trading system and brought gold, ivory, and slaves to the coast from great distances, but they did not develop a real empire. Instead, they tapped each local trading system in the tribal village or at markets and periodic fairs. They also benefited from the booty taken as a result of tribal warfare. Routes of trade were commonly paths rather than roads, and human carriers were used for transport. The connections were from tribe to tribe and nation to nation, and the sum of these local interchanges added up to regional systems that brought goods to the coast.

Despite such contacts traditional society in most of Africa remained within the tribal group and the local area. The success of life in any locality depended on both the productiveness of the environment and the ability of the group to perceive its usefulness. Thus, there was a multitude of different local variations in the subsistence economy throughout Africa. Variations were twofold. The range in environments from one part of the country to another resulted in general regional patterns. In the drier areas nomadic herding predominated; in areas with somewhat better moisture conditions livestock herding was combined with slash-and-burn cultivation. In wetter areas, particularly near the equator, raising livestock was impossible because of the tsetse fly, and shifting agriculture was combined with forest gathering, hunting, and fishing. In areas with better soil and moisture conditions or where population pressure was great, communities maintained permanent fields. Some farmers terraced hillsides with stone walls, rotated crops, kept cattle for manure, collected weeds to make compost, and used wood ash or night soil to maintain the fertility of their fields.

The Slave Trade: First Exploitation of an African Resource

The presence of European traders along the coasts and on the desert fringe brought some changes to Africa. The traders were interested in African goods, but since transport was difficult, trade was limited to commodities of great value and little bulk—gold, ivory, pepper, and gum—and to those that were mobile, for example, slaves. Sections of the west African coast were known as the Gold Coast, Ivory Coast, and Slave Coast, but trade in these commodities took place at trading outlets throughout all Africa.

European and Arab traders established trading posts at or near the coasts and depended largely on native-controlled trade to gather what they wanted from the continent. Sales allowed the Africans to purchase European guns, powder, rum, cloth, and hardware. For the most part there was little surplus from their subsistence life with which to trade. Slaves provided the one ready surplus commodity, and tribe fought with tribe for hostages. Within the tribal structure the chief might sell his subjects, or in dire circumstances the parent a child. Personal security was lost over much of Africa, and the area fell into anarchy. Rulers along the coast built up kingdoms as middlemen in the slave trade, and their power blocked any plans of European traders to move inland.

While there is no good estimate of the total number of slaves taken from Africa, it is assumed to have been at least 10 and perhaps 15 million. West Africans went to the West Indies and North America, East Africans to Arabia and the Ottoman Empire. Whatever the number, it is clear that the slave trade drew off the strong young people, a resource that any region could ill afford to lose.

Colonialism: Product of Europe's New Industrialization

European colonialism burgeoned in the latter part of the nineteenth century. New industries brought an increasing demand for raw materials, and the developing urban society and its increasing wealth opened a new market for tropical specialty products. At the same time, steam-powered, steel-hulled ships made possible the movement of vast quantities of goods by sea, and railroads could reach the interior of continental regions.

Europeans also had political and cultural interests in Africa. There was political prestige in claiming new lands, and explorers were folk heroes to European society. The tales that they brought back concerning the primitive conditions of the native life they had seen inspired a wave of missionaries to convert the heathen and to better the lives of the downtrodden.

Whereas in 1884 European claims in Africa consisted only of coastal footholds plus South Africa (Figure 14-3), by 1895 virtually the whole of Africa south of the Sahara was controlled by European powers (Figure 14-4). At first, development had been turned over to private trading companies, and soon these companies were raising capital to build railroads and ports.

The first important commodity from the tropical African coast was palm oil for soap, candles, and lubricants for machinery. Lever Brothers, the well-known soap company, has trading interests in west Africa that date from this period.

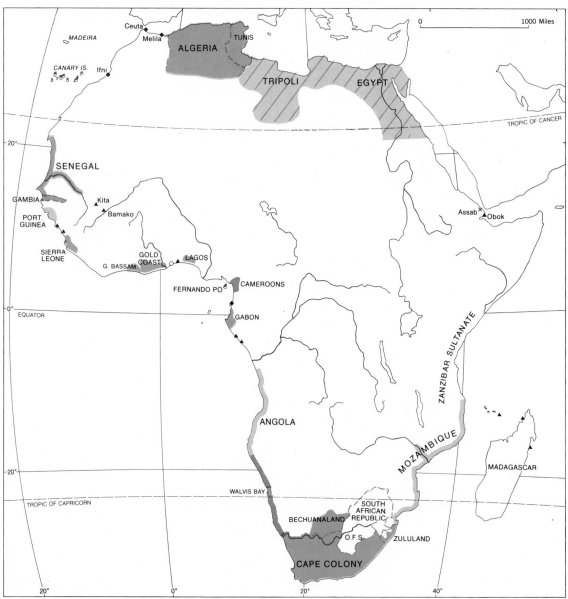

Figure 14-3 Political Map of Africa, 1884

▨	British (Br)
▲	French (Fr)
◉	German (G)
◆	Spanish (Sp)
▨	Portuguese (P)
▨	Turkish vilayets
☐	Independent
×	Italian (It)
▲◆○×	Footholds or trading stations

Map labels: MADEIRA, Ceuta, Melila, CANARY IS., Ifni, ALGERIA, TUNIS, TRIPOLI, EGYPT, TROPIC OF CANCER, SENEGAL, GAMBIA, Kita, Bamako, Assab, Obok, PORT. GUINEA, SIERRA LEONE, G. BASSAM, GOLD COAST, LAGOS, FERNANDO PO, CAMEROONS, GABON, EQUATOR, ZANZIBAR SULTANATE, ANGOLA, MOZAMBIQUE, MADAGASCAR, WALVIS BAY, TROPIC OF CAPRICORN, BECHUANALAND, SOUTH AFRICAN REPUBLIC, O.F.S., ZULULAND, CAPE COLONY

1000 Miles

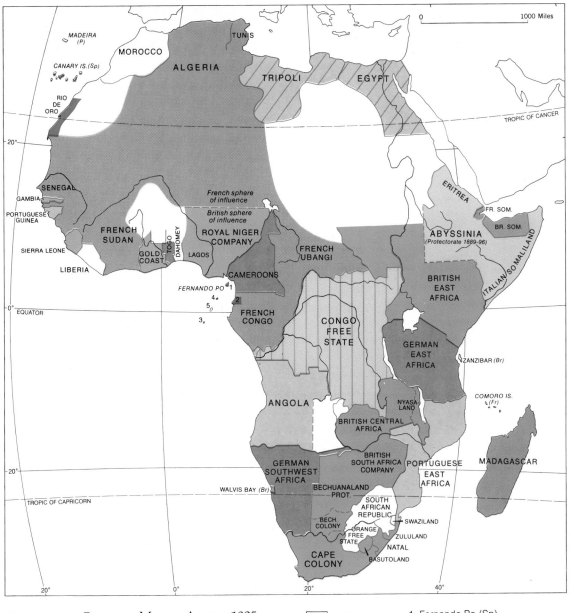

Figure 14-4 POLITICAL MAP OF AFRICA, 1895

British (Br)

French (Fr)

German (G)

Belgian (Be)

Italian (It)

Spanish (Sp)

Portuguese (P)

Turkish vilayets

Independent

1. Fernando Po (Sp)
2. Rio Muni (Sp)   3. Annobon (Sp)
4. Principe (P)   5. Sao Tome (P)

475

The trading companies were not profitable, however, and soon African development was largely in the hands of European governments. Although there were several heated struggles by native peoples to remain free of colonial control, in most cases land was taken by treaties containing promises of trade, protection, or both. Boundaries were arbitrarily established often by compromises between European powers and usually with no relation to traditional boundaries. The land of the Somali was divided five ways; there were British, French, and Italian Somalilands, a Somali enclave in Kenya, and another in Ethiopia. The Somali are still disputing with the latter states.

European investment in Africa was made both by colonial governments and private investors interested in resource development projects. By 1936, about $6 billion had been invested in Africa, slightly less than half of it by the colonial governments and the rest by private investors. Most government money went to build railroads, ports, and public works, whereas the greatest part of private investment was used to develop the mines.

DEVELOPMENT AS COUNTERPOINT TO LATIN-AMERICAN TROPICAL TRADE

Originally tropical Africa was seen by Europeans in terms of its potential for producing agricultural specialties, but much more attention came to be put on the development of mineral resources. Perhaps because Africa did not come into the functioning European colonial system on a large scale until the twentieth century, agricultural specialization was less concentrated than it was in Latin America. By the time African areas were under European control, Latin-American plantations had already been well organized to produce cotton, coffee, sugar, and bananas, and Asian colonies were producing rubber, tea, and jute. Africa has pro-

duced all of these crops, but it is not a primary source of supply for any of them.

It is, however, Africa south of the Sahara and not Latin America that has emerged as a treasure house of precious metals. Africa produces more than two-thirds of the world's gold, most of the world's diamonds, and more than one-fifth of its copper. A very large share of European investment went into production facilities for these three minerals alone.

Thus, although European political control in colonial times covered almost the whole of Africa, economic investment by Europeans was not only centered around three major minerals, but around the few localities where they were produced. By far the largest amount of economic investment went into South Africa where first gold and then diamonds were discovered. Most of the rest of European investment went into the copper-mining districts of Zambia, Zaire, and Rhodesia. Investment in other colonies was minor by comparison, and because much of it was in plantation agriculture, it was widely scattered.

ISLANDS OF EUROPEAN DEVELOPMENT IN A SEA OF RURAL AFRICA

With investment in mines and plantations, the European colonial powers and the favored private investors established localized "islands" of the modern, interconnected system within the vast native subsistence system. The nature of this localized system can readily be seen by examining Figure 14-5. Notice that areas of commercial production, particularly those that are intensively developed, do stand out as scattered islands. Railroads reach back from ports to make connection with these productive areas. They can move products to export markets but do not form an extensive rail network within any country, much less within this part of Africa.

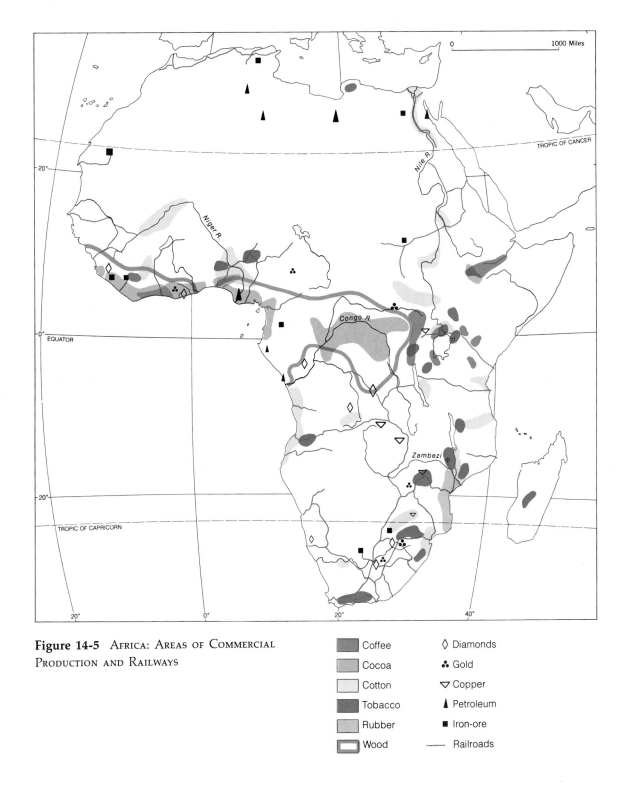

Figure 14-5 Africa: Areas of Commercial Production and Railways

Coffee
Cocoa
Cotton
Tobacco
Rubber
Wood

◊ Diamonds
⚜ Gold
▽ Copper
▲ Petroleum
■ Iron-ore
— Railroads

Migratory Labor: A Life Style Bridging the Dual Economies

The commercial islands and their transport links have formed new centers of employment for indigenous Africans. The mines and plantations need workers, and since slavery has long been outlawed, most work is done by men recruited from rural tribal units. Such individuals often travel hundreds of miles to live in workers' dormitories and work in the mine. Typically the worker's family remains behind, continuing in its traditional life style. This is possible because much of the work in agriculture has always been handled by women. Males were traditionally hunters and warriors and only incidentally concerned with farming. They cleared new fields and watched after livestock but did not participate in the daily farm routine.

The men who now work in the commercial mines are transients in the interconnected system. Neither they nor the companies see them as permanent miners. They are recruited and assigned for a contract period of three years. A worker stays until he feels that he has acquired enough money to meet his family's needs or until some pressing tribal or family matter calls him back. Since he is essentially a laborer as far as the mining company is concerned, another young man from the countryside can be hired to take his place if he does not return for another contract period.

These migratory workers, more than any other people we have seen, are men of both systems. On the average they spend over 60 per cent of their working life from age sixteen to forty-seven outside their tribal area. It is not an ideal system. The mining companies believe that the transient nature of the work force is inefficient, and it is hard to imagine that the workers benefit from it either.

Remarkably, the migratory workers cross national boundaries regularly as they move between work and village. The low densities in rural areas make it necessary to draw workers from a very large area. Mines in South Africa utilize workers from Lesotho, Botswana, Swaziland, and Malawi. Malawians also work in Rhodesia, Rwandans work in Uganda, and people from Niger and Chad work in the cotton fields in the Sudan. Some people from Nigeria and Upper Volta are employed in Ghana and the Ivory Coast. In this way the nodes of production in the modern, interconnected system serve as growth points that pump money into widespread rural areas.

People from countries such as Malawi depend on employment in other countries for money. To a degree, the emerging modern, interconnected system of one country depends on the expansion of resource production in another. This is a particularly difficult situation for African countries whose citizens work in South Africa and Rhodesia where the governments are controlled by European minorities. How can such countries enforce economic sanctions against apartheid policies when their citizens depend on jobs in racist countries?

Obviously, the presence of persons who move between one cultural-economic system and another results in modifications in each system. The workers of mines and plantations come from various rural areas and bring their own customs and languages with them. Training programs and living facilities must take into account such ethnic diversity. Workers are exposed to a whole set of new standards and values, and they must learn to work under foremen who speak a different language. And, when these men return to their villages, they inevitably transmit some of this new culture to their families.

Gertrude Stein once said that she found the
ancient masks and sculpture of Africa too sophisticated for her taste.

New African nations are faced with reaching
a resolution between the old and the new. They
have pride in their cultural heritage, but strive
to take their place in the modern world of politics and economics. The result is a pastiche of
old and new ways.

484

LARGE CITIES: INDICES OF THE MODERN INTERCONNECTED SYSTEM

The system of transient workers can only operate well where masses of unskilled workers are needed. It is not an answer to the industrial and business needs of mechanized mines and modern cities. There is a growing need for people trained to handle complex tasks, and a decreasing need for masses of unskilled workers. Industry calls for a continuous, stable work force committed to continuing to work within the modern, interconnected system. In South Africa, where industrialization has moved well ahead with production of steel, machinery, automobiles, and appliances, migratory labor has been largely replaced by a permanent population of indigenous Africans who have given up tribal life. Thus, in large cities throughout Africa, a trend of accelerated urban growth and rural depopulation has begun, similar to ones in the industrial countries and Latin America, but on a much smaller scale.

Nevertheless, Africa south of the Sahara remains the most rural of all regions of the world. The rural population of most African nations exceeds 80 per cent of the population. South of the Sahara there are just two metropolitan areas with over one million persons: Johannesburg, center of mining and industrial development in South Africa, and Kinshasa, capital of Zaire. Only seven other cities have over half a million people, and two of them are in South Africa (Figure 14-1). A dozen African countries have capital cities with populations of less than 100,000 persons, and seven of these are below 50,000. As a rule, the capital of an underdeveloped country is the largest city. Only South Africa and Nigeria have more than two cities of 100,000 or more people.

The size and number of large cities in a country correlate more closely with their amount of commercial development than with total population size. South Africa, with less than half the population of Nigeria, has eleven cities of over 100,000 as compared to seven in Nigeria. Ethiopia, which is second in terms of population, has only one city of more than 100,000.

POLITICAL AS WELL AS ECONOMIC LINKS

Large cities are an indicator of the degree to which a country has become involved in the modern, interconnected system. The capitals are not only the political centers around which the internal affairs of the country develop but are also the links with the modern, international political scene that bring all countries together. Even the smallest, least developed country has links with the United Nations and with the major political powers of today's world. Such contacts are vital not only in political terms but also for economic and cultural programs. Through such contacts come most of the investment capital for economic development. Africa receives assistance from individual industrial countries, the United Nations, and the World Bank, and private investments in transportation and resource development. These sources also provide a stream of technical assistance for development programs and offer ties to advanced institutions of higher education through which the new leadership of African countries can be trained.

Political Capitals: Heritages of Colonialism

Almost without exception these cities were created by the European colonial governments. They built streets, government buildings, and residences of officials after the style of their homelands. Governmental form also reflected European views. And since the leaders of new native governments were trained in the industrial countries, the form of both the physical city and its institutions still follow Western culture even after independence. It should not be surprising that business cores of almost all African cities appear very familiar to visitors from industrial countries. Moreover, these cities are very modern, since they have grown tremendously in the last few years.

Capital cities may be the center of life in African countries, but they also impose major burdens on the new governments. Paternalistic colonial governments developed elaborate, free public services for cities including free schools, free medical services, free roads, free water, free sewerage, free or subsidized electricity, and subsidized housing, often at the expense of the agricultural population in the rest of the country. Today, weak, underfinanced national governments believe that they must not only continue such services but that they must also expand them to make their capital a worthy show place of national life. And, city populations are very vocal in their demand for such services. Thus, in a country with very limited financial resources, money spent to subsidize city life leaves less for other essential national investment.

Governments as an Economic Force

The dominance of government in the economic affairs of the country was typical of colonial regimes and continues today. The government is the single largest enterprise, and its expenditures usually amount to about 20 per cent of gross national expenditures. If publicly owned railways, electricity, authorities, ports, and other basic industries are included, governmental expenditures amount to one-third of the total, and the private-investment sector of the economy may be less than that.[1]

---

[1] For comparison, in the United States private-business production accounts for over 80 per cent of total production, whereas the government total is about 11 per cent.

The government is also the chief employer. It hires from one-third to half of the wage earners of the country, and most government employees live in the capital city. But we must remember that the number of paid workers in the labor force of the largely subsistence African countries is very low, usually from 5 to 10 per cent.

Government jobs traditionally carry the highest prestige and pay the highest wages. Most educated Africans aspire to such positions. But even the relatively massive government system cannot absorb all the Africans looking for clerical and other low-level government jobs, and large numbers of urban Africans with school educations are unemployed. Once their families have made sacrifices to send them to school, they are regarded as failures if they become farmers like the uneducated, so they do not take laboring jobs.

As the locus of government, the big city is the decision-making center for the entire country. It is the city-based government that sets fiscal priorities, decides on rural land programs, and underwrites the development of transportation and public utilities for the whole country. It approves resource concessions for foreign investors and trade programs, and also determines who will be educated, what the cultural image of the country will be, and even what the national language will be.

THE PROBLEMS OF A COLONIAL PAST

Each of the European colonial powers had its own style of colonial development with varying degrees of political independence, social concern, and economic development. Naturally the newly emergent independent states reflected the differing styles of the colonial governments. Zaire, for example, was known as the Belgian Free State under the direct control of King

Leopold and was operated as a gigantic hacienda with an absentee landlord. When the territory was taken over by the government of Belgium in 1908, the natives were regarded as childlike wards to be managed by paternalism. Profit was of first importance to the king, who had little concern for the people. The natives had duties, and some rights; their "place" was determined by the Belgians, and they were given no preparation for self-government. None of the people were trained to be leaders, and when independence came there was virtually no one to take charge.

France, which ruled its colonies in a very authoritarian way at first, later espoused the principle that all French territories were a part of France and all residents were Frenchmen. They made an attempt to educate as many of the natives as possible to become responsible and intelligent French citizens.

The Portuguese, who continue to hold colonies, also strove for assimilation of all colonies and all colonial peoples as Portuguese. Interbreeding and intermarriage between natives and Europeans is encouraged. Theoretically, natives who can read Portuguese and who will accept Christianity can acquire full Portuguese citizenship, but only if they put aside tribal loyalties, prove that they are of good character, and demonstrate that they can earn enough to support a family. Obviously few natives can achieve that status. The remaining natives are subject to another set of rules and can be sentenced to forced labor by authorities.

British colonial rule varied greatly from colony to colony, partly depending on the interest of Europeans in settlement. In hot, wet Africa where they showed little interest, they were not allowed to own land, but in cooler, drier east Africa European immigrants not only owned land but forced natives from their traditional homelands to make way for their settlements. Because of this, independence in British colonies brought different problems in different places. In some the question was how to arrange native rule; in others it was whether the native majority or a white minority should rule.

Where European populations were present in former British colonies, they have largely taken control. This is the case in the Republic of South Africa where less than 20 per cent of the population is of European descent. A strictly enforced system of apartheid keeps blacks from having any power. They must live in their own compounds adjacent to the major industrial areas or in areas designated as native reserves. Basically the blacks in South Africa have no political rights. The government determines what work they may do, where they may live, and whether or not their families can live with them. They are temporary migrants in most of the country and must produce an identification card on demand.

In Rhodesia the European minority comprises less than 6 per cent of the total population but is in control of the government. In Kenya the black population controls the government despite the fact that a European population of less than 1 per cent remained dominant in business and commercial agriculture after independence. Those Europeans who have stayed in Kenya have retained their economic position.

Most other African countries are now largely in the hands of native peoples, although their leaders have been trained in European institutions. Government forms follow European models, and the boundaries and cultural patterns established by Europeans also remain. There is a great gap between the life of the urban African who lives in the modern, interconnected system and the majority of Africans who still live in traditional societies.

Most governments are strongly nationalistic and proud of their native cultural heritage. They wish to govern their own affairs and develop their own institutions. Yet, in fact, their governmental structure is based on colonial patterns, and in almost half of the new countries either English or French is the official language. Africans apparently believe that to function in the modern political and economic world, government and business leaders must speak a European language. Of course, they inherited a colonial system designed to work in a European language. The rural tribesman who speaks only his local dialect is cut off from the mainstream of national life and is at a disadvantage in the modern, interconnected system in his own country.

But the largest metropolitan centers are more than places of government. They are the major commercial centers and the headquarters for the modern business organizations in the country as well. Here are found the national management functions of the large international companies that are developing the resources of the country as part of the modern, interconnected system centered in industrial areas outside of Africa. For this system, begun in colonial times, remains important to the foreign investor and essential to the new government as a source of foreign exchange, taxes, and native jobs.

It is logical to have the business management operations of a company located in the largest city. The local corporate management in the capital has direct air and communications connections to the corporate headquarters elsewhere, while at the same time it is at the node of the national transport and communications net and at the seat of government where economic decisions are being made. Thus, in the capitals of the proud, new countries are found the management operations of the international corporations whose chief concern is the export of the country's resources for use elsewhere. The impact of the modern, interconnected world is very great, and to an important degree capital size is proportional to the importance of the country in the external, interconnected system. In the largest capitals private office buildings housing local headquarters of these corporations and trading companies are becoming as prominent as government buildings.

Major Exports Localized in a Few Countries

Most of the export trade from Subsaharan Africa originates in just a few countries (Table 14-4). The Republic of South Africa accounts for more than one-third of all exports, and the top six countries account for more than two-thirds of the total. In twelve countries the leading export is a mineral, and in South Africa gold, diamonds, copper, and manufactured products lead the list. It is the one industrialized country in Africa.

Per Capita Exports as a Measure of Modernization

Among the top six countries, only Zambia and South Africa have exports of very high per capita value. They amount to $172 and $160 per person, respectively, figures that are higher than those of any Latin-American country except Venezuela and Chile. In none of the others does the per capita figure reach $100, but Liberia, Gabon, and Swaziland rank favorably with the leading trading countries in Latin America. In Ghana, Zaire, and Nigeria the per capita figures are much lower, and Nigeria's total of only $11 is even lower than that of Brazil. Neither of the most populous countries in Latin America shows a high ratio of exports to population.

The relatively high per capita value of exports in Africa as compared to Latin America would suggest considerable potential purchasing power in the struggle by the African countries

Table 14-4 VALUE OF EXPORTS OF AFRICAN COUNTRIES SOUTH OF THE SAHARA, 1967

COUNTRY	TOTAL VALUE ($ millions)	PER CAPITA VALUE, ($)	LEADING COMMODITIES (Per cent of country's total exports)	
Western Africa				
Dahomey	15	6	Palm oil	21
Gambia	16	40	Groundnuts	43
Ghana	308	39	Cocoa beans	58
Guinea	54	15	Bauxite	63
Ivory Coast	325	83	Coffee	32
Liberia	159	144	Iron-ore	75
Mali	17	4	Cotton	30
Mauritania	70	65	Iron-ore	92
Niger	33	10	Groundnuts	69
Nigeria	680	11	Petroleum	30
			Cocoa beans	23
			Groundnuts	18
Senegal	139	38	Groundnut oil	43
Sierra Leone	70	29	Diamonds	65
Togo	32	19	Phosphates	38
Upper Volta	18	4	Animals	51
Total	1,936			
Central Africa				
Angola	238	46	Coffee	52
Cameroun	158	30	Coffee	28
			Cocoa	28
Central African Republic	28	20	Diamonds	47
Chad	27	8	Cotton	83
Congo (Brazzaville)	47	62	Diamonds	34
Zaire	446	28	Copper	60
Gabon	121	242	Petroleum	30
Total	1,065			
Eastern Africa				
Burundi	16	5	Coffee	83
Ethiopia	101	5	Coffee	56
Kenya	167	17	Coffee	30
Malagasy	104	16	Coffee	32
Malawi	57	14	Tea	27
Mauritius	64	84	Sugar	94
Mozambique	133	19	Cotton	18
Rwanda	14	4	Coffee	55
Somalia	28	11	Hides and skins	42
Rhodesia	281	64	Tobacco	20
Tanzania	222	19	Cotton	16
Uganda	184	23	Coffee	54
Zambia	658	172	Copper	93
Total	2,029			
Southern Africa				
Botswana	7	12	Meat	....
Lesotho	7	8	Cattle	....
South Africa	3,026	160	Gold	35
			Copper	4
			Diamonds	5
Swaziland	42	168	Sugar	12
Total	3,082			

SOURCE: Kamarck, *The Economics of African Development*, Praeger, New York, 1971, pp. 112–115.

Table 14-5 Comparison of Electricity Production in Leading African Countries South of the Sahara and Latin America, 1970

Country	Aggregate Electricity (000,000 kWh)	Population (millions)	Per Capita Electricity (000 kWh)
South Africa*	48.6	22.1	2,199
Brazil†	41.6	92.7	449
Mexico	28.5	50.6	563
Argentina	16.8	23.2	724
Venezuela†	11.0	10.3	1,068
Chile	7.4	9.7	763
Colombia†	7.3	21.6	338
Rhodesia	6.0	5.3	1,132
Cuba‡	4.7	8.3	566
Zaire	2.9	18.8	154
Ghana	2.9	8.6	337
Nigeria	1.5	55.0	27
Zambia	0.7	4.1	171

*Includes South-West Africa (Namibia). †Data for 1969. ‡Data for 1968.
SOURCE: *Encyclopaedia Britannica, Book of the Year: 1972,* Encyclopaedia Britannica, Inc., Chicago, 1972; *World Bank Atlas: Population, Per Capita Product and Growth Rates,* International Bank for Reconstruction and Development, Washington, 1972.

to modernize. However, there are a number of problems, particularly the small size of the populations of most African countries. While the per capita figure may appear significant, the aggregate total remains small. Moreover, there is the question of how the income is distributed. Money may be spent on governmental institutions, national symbols, or national defense, rather than on economic development.

In Latin America we used per capita exports as an initial index of the degree of industrialization. Using this measure, we would expect Zambia to be the most highly industrialized country south of the Sahara, with Rhodesia and South Africa well behind and with Ghana, Zaire, and Nigeria lagging badly. But, in fact,

only South Africa has any significant steel production, and if electrical energy production is used as a measure, all the other African exporting countries show up poorly in comparison with Latin-American exporters, as we can see in Table 14-5. Apparently, except for South Africa, there is very little development of home-based modern industry. The small, newly emergent African countries still depend on imports of manufactures for their needs.

MOST OF AFRICA REMAINS UNDERDEVELOPED

In per capita terms, South Africa shows up far ahead of any of the other countries, with export figures comparable to those of many European countries. In terms of this measure of industrialization, South African development is greater per capita than any of the Latin-American countries.

Aside from South Africa which might well have been considered an example of the modern, industrialized world in Part Three and which is run by its European immigrant population rather than by indigenous Africans, no country south of the Sahara has taken more than the first steps leading to industrialization. The major manufacturing countries are those processing export products—the copper smelters and refineries in Zambia, Rhodesia, and Zaire, and the aluminum plant in Ghana.

The Beginnings of Industry

Still, most of these countries have food-processing plants for local markets and produce simple building materials, textiles, and other consumer goods that require little capital investment or skilled labor. Some automobile parts are produced, tires are retreaded, and trucks are made from parts shipped from European countries.

Almost every one of these countries has an oil refinery built by one of the international oil companies to give the company a dominant position in the local energy market. There are even small iron and steel mills in Rhodesia, Kenya, Uganda, and Nigeria. But this is just a beginning. In tiny African countries with such small proportions of their population in the money economy, there is commonly neither the market that would attract foreign investors nor the local capital available for native initiative. Only where the government sees a particular industry as part of its development plan or of particular prestige value is an industry established. Outside of South Africa, Subsaharan Africa has no industrial developments comparable to those in parts of Mexico, Argentina, and Brazil.

In per capita terms two small African nations, tiny Gabon and Liberia, place great emphasis on exports (see Table 14-4). For their size they rank among the highest of any nonindustrialized countries in the world with regard to their involvement in foreign trade. At the other extreme Ethiopia, Dahomey, Mali, and Upper Volta are among the lowest countries in the world with regard to per capita exports. They appear only marginally connected into the world-wide economic system. In fact, Mali and Upper Volta's exports are not primarily geared to non-African markets; they provide livestock to neighboring countries, much in the way that Ireland sells food to Britain. Thus, they are in an extremely weak position because they depend on an African market that, in turn, depends on markets external to Africa. In the same way, newly independent Lesotho and Swaziland, part of a customs union with South Africa, serve that country with food and raw materials.

Rapid Growth in Exports in Recent Years

In contrast to Latin America where exports are about the same as they were in 1929, Africa increased its exports almost tenfold between 1938 and 1968 (Table 14-6). Most of this growth has come since World War II. In 1950 the value of trade was almost three times what it had been in 1938, but between 1950 and 1968 exports increased more than three times the base figure.

Remarkably the growth has occurred in almost every country. Except in Zaire, Guinea, and Dahomey, much of the growth has come since 1960, the period after independence. As a result of the general growth everywhere, the major export countries have not changed much in terms of relative importance. Kenya and Malagasy are comparatively less important today, and

Table 14-6 GROWTH OF EXPORT TRADE IN COUNTRIES OF AFRICA SOUTH OF THE SAHARA: 1938, 1950, 1960, 1968 ($ millions)

COUNTRY	1938	1950	1960	1968
South Africa	496	946	2,018	3,235
Nigeria	47	253	475	587
Zambia	49	140	362	759
Rhodesia	52	135	193	275*
Zaire	64	273	501	516
Ghana	76	216	325	336
Ivory Coast	11	79	151	425
Angola	15	75	124	276
Tanzania	22	82	174	241
Uganda	4	81	120	186
Sierra Leone	12	22	83	96
Kenya	40	58	113	176
Liberia	2	28	83	169
Cameroun	7	47	97	189
Senegal	22	72	113	151
Mozambique	9	37	78	154
Gabon	....	....	48	125
Central African Republic	⎫	⎫	14	36
Chad	⎬ 7	⎬ 43	13	28
Congo (Brazzaville)	⎭	⎭	18	49
Ethiopia	....	28	77	106
Malagasy	24	69	75	116
Mauritania	....	....	....	72
Guinea	3	11	55	55
Swaziland	....	....	....	42
Malawi	....	21	24	58
Somalia	5	6*	25*	30
Togo	2	9	15	39
Niger	1	3	13	38
Mali	....	....	....	17
Dahomey	3	13	18	22
Gambia	2	6	8	15
Upper Volta	†	†	4	21
Lesotho	....	....	....	7
Total	966	2,753	5,417	8,647

*Estimate· †Included with Ivory Coast.

SOURCE: Kamarck, *The Economics of African Development*, Praeger, New York, 1971, pp. 115–116.

leaders such as South Africa, Nigeria, Zambia, Zaire, Ivory Coast, and Ghana have grown rapidly in absolute terms, but there has been little change in their order.

A COMPETITOR OF LATIN AMERICA

In considering the recent growth of Africa as a source of exports, it is useful to compare it with trade in South America, another largely tropical area. Recall that, in contrast to the rise in Africa's exports, Latin-American exports have not increased significantly in the past forty years. In this comparison, it is important to recognize that colonial development occurred in Latin America more than a hundred years earlier than it did in Africa, and commercial development in Africa is largely a phenomenon of the twentieth century. Because they have similar resources, Africa really could offer little that was not already being developed in Latin America. As a result the early exploitation of Africa emphasized gold, diamonds, and palm oil, products that are unique to Africa.

Mostly the Same Exports

For the most part, exports from Subsaharan Africa compete with products from Latin America and elsewhere. They include the basic tropical and subtropical export products: coffee, cocoa, cotton, tobacco, copper, petroleum, iron-ore, and timber. Of these, only copper and cocoa are produced in greater quantity in Africa than in Latin America (Table 14-7).

Of the tropical agricultural export products, only palm oil is indigenous to Africa. All the other products were introduced by Europeans during colonial days to provide revenue for the new colonies and to compete with products from the independent American countries. Except for the cocoa market, however, Africa has not been able to make up for its late start, and

it trails Latin America in terms of every other major export.

Localization of Export Centers

Few African countries have more than two products going into export. Most individual countries are dependent for their exports on only a handful of commodities (see Table 14-8). But production of most commodities within the continent is usually sharply localized. Copper is produced in Zambia, Zaire, Uganda, and Rhodesia (Table 14-8). Except for the Ivory Coast, most coffee comes from the highlands of east Africa; cocoa comes from the lowlands of west Africa; groundnuts grow on the drier grasslands just south of the Sahara; tobacco and cotton are raised in the north and east; diamonds are found in South Africa, Angola, Zaire, and the Congo (Brazzaville); and wood comes from the tropical rain forests of the west.

Increasing Demand, Protected Markets, and an Available Environment

The recent increase in African production reflects the growing demand for its products during the last half-century, particularly since World War II. The new affluence of the industrial countries, particularly in Europe, which has been Africa's traditional market, has been a vital factor in improving the African export base. World consumption of copper increased almost two and one-half times between the 1930s and the 1960s; coffee and cocoa consumption almost doubled; and cotton use rose over 40 per cent.

African countries have benefited from old colonial trade ties with European countries which were continued after independence. Former British colonies that became members of the Commonwealth have had free access to markets in Britain, and former French colonies were given associate status in the Common Market with preferential treatment. These ad-

Table 14-7 COMPARISON BETWEEN AFRICAN AND LATIN-AMERICAN
PRODUCTION OF EXPORT PRODUCTS, 1925–1967

	Avg. 1925–29	Avg. 1940–44	1950	1960	1967
Copper (000 metric tons)					
Africa	120.7	429.2	519.1	975.8	1,128.3
Latin America	379.5	564.2	481.5	796.9	922.1
Petroleum (000 cubic meters)					
Africa (Nigeria)	237	1,331	2,656	16,744	181,370
Latin America	28,327	47,232	110,669	210,401	277,217
Iron-ore (000 metric tons)					
Africa (incl. North Africa)	3,751	2,666	7,004	15,464	43,001
Latin America	2,183	1,526	5,628	43,643	62,890

	1929/30 1933/34	1939/40 1943/44	1950/51	1960	1967
Coffee (000 metric tons)					
Africa	72.9	172.0	266.5	817.5	1,145.0
Latin America	1,988.7	1,623.6	1,784.0	2,945.0	2,750.0

	1930–34	1940–44	1950	1960	1967
Cocoa (000 metric tons)					
Africa	376.0	407.2	485.4	661.3	968.2
Latin America	162.1	208.9	260.0	367.4	337.1

	1925–29	1940–44	1950/51	1960/61	1966/67
Cotton (000 metric tons)					
Africa	438	479	685	904	1,078
Latin America	259	740	864	1,305	1,462

	1925–29	1946–49	1950	1960	1966
Forest products (000 cubic meters roundwood)					
Africa	. . . .	90,550	89,780	190,073	238,378
Latin America	. . . .	177,910	191,978	209,347	274,268

SOURCE: Grunwald and Musgrove, *Natural Resources in Latin American Development*, Johns Hopkins Press, Baltimore, 1970.

Table 14-8 LEADING EXPORT COMMODITIES OF AFRICAN COUNTRIES SOUTH OF THE SAHARA, 1967 ($ millions)

COMMODITY	VALUE	COMMODITY	VALUE
Copper		Senegal	90
Zaire	261	Niger	22
Rhodesia	8	Gambia	12
Uganda	17	Other	210
Zambia	612		456
	898	Petroleum	
Coffee		Nigeria	204
Ivory Coast	104	Gabon	36
Angola	124		240
Malagasy	33	Cotton	
Ethiopia	57	Uganda	44
Uganda	99	Tanzania	36
Kenya	50	Chad	22
Cameroun	44	Mozambique	22
Tanzania	33	Central African Republic	7
Zaire	26	Other	11
Togo	4		142
Central African Republic	6	Tobacco	
	580	Rhodesia	53
Diamonds		Malawi	15
South Africa	162	Zambia	7
Sierra Leone	46		75
Angola	43	Wood	
Zaire	22	Ivory Coast	84
Congo (Brazzaville)	16	Gabon	30
Central African Republic	14	Congo (Brazzaville)	17
Other	31	Ghana	28
	320	Swaziland	10
Cocoa		Other	7
Ghana	192		176
Nigeria	156	Iron-ore	
Ivory Coast	55	Mauritania	64
Cameroun	44	Liberia	119
Togo	10	Sierra Leone	14
	457	Swaziland	6
Groundnuts		Guinea	3
Nigeria	122		206

SOURCE: Kamarck, *The Economics of African Development*, Praeger, New York, 1971, pp. 112–114.

vantages over the independent Latin-American countries have helped Africa's international position tremendously.

Africans have also benefited from attempts by other producing areas to improve the price for raw materials. Coffee production in Africa was encouraged by efforts of Latin-American countries to establish a floor on coffee prices. With a guaranteed minimum price in world markets, much of the risk of entering coffee production is gone, and the tropical highland areas of Africa are excellent coffee-growing land. Likewise the efforts of the United States, the world's leading cotton exporter, to support domestic prices through acreage control and price subsidy on domestic sales have raised cotton prices throughout the world and encouraged other countries to increase production.

The African Way of Life: Two Veneers of Modernization on a Traditional Base

Subsaharan Africa clearly lives with a dual economy. Modern Africa is the result of the superimposition of two layers of the modern, interconnected system over the traditional, locally based way of life. The first of these layers was colonialism, which began when Europe exploited African resources and then carved political realms out of the continent, ignoring the cultural groupings of the native peoples. Independence was not granted to Africa as a whole. Instead each of the political units the Europeans had created was treated separately. Colonialism had produced a cadre of Western-trained Africans who took over the governments of the new countries and perpetuated European political and economic institutions. As a result of this heritage, the new African states still follow traditional European models to a great extent, despite their independence.

The second layer of the interconnected system to affect Africa has been that of the economic system powered by foreign investors who have been attracted to Africa's resources. They have required an infrastructure of transportation and communication routes, cities and ports in order to get products to the coast for shipment overseas. Africa has almost no interlacing infrastructure that unites even the modern, interconnected system of the region into a functioning whole. Instead each commodity development has its own individual transport corridor to the coast and, aside from air travel and radio communication, one African country has little direct interaction with its neighbors. Nor has this economic intervention ended with independence. Virtually all resource development in Africa is still under the auspices of the large international firms.

Up to now, the interconnected system has significantly touched only a tiny fraction of Africa's population. It reaches a small, modernized elite, the particular export-producing areas, and the big cities. There has been some development of commercial agriculture in traditional, tribal subsistence areas, but such development depends on transport and marketing facilities that are simply not yet available in most of Africa. Tribal peoples affected by migratory labor opportunities are only partly involved in the modern system. Working in the large mines provides a way to pay taxes and buy desired staples of the modern system, but it offers little incentive to break with traditional social and political patterns.

COUNTRIES BOTH TOO LARGE
AND TOO SMALL

The political divisions of Africa, made arbitrarily by Europeans, are both too large and too small. The countries are larger than local cultural groups, yet most people regard themselves

We have now examined two underdeveloped areas: Latin America and Africa—each with dominantly tropical environments, each with the juxtaposition of traditional and modern societies. The countries of each area have problems in taking their place in today's political and economic world. The people of each area are faced with making choices between old ways and new.

Do you think that the two areas are fundamentally alike in terms of their developmental problems? Do they have the same potential for development? What do you see as the major obstacles to the development of each?

Both areas consist in large part of countries with very small populations. What value would there be in developing closer economic ties with their neighbors or with countries in other culture realms?

The underdeveloped areas are often spoken of as "the third world." Do you see a commonalty in social or political terms?

Underdeveloped peoples are very jealous of their independence. To what degree are the countries of these areas free? Do you see different degrees of independence in the two areas?

Do you see any analogies between the developmental problems of Africa and Latin America today and those of the United States as it emerged from colonialism? What prospects do these areas have for developing along the lines of the United States or Japan? Should this be their aim?

first as members of their own tiny local world and only secondarily as citizens of the new country. Often large groups still regard themselves as subject peoples. This was the case when the Ibo of Nigeria revolted in the late 1960s.

Viewed from an external economic perspective, however, most of the countries of Africa are far too small in population. Nigeria with its 55 million people is the only country with a population size even comparable to the European nations which, in turn, see the need for economic integration to provide a modern-scale market. Most African countries have less than 5 million people—numbers that hardly provide an adequate base for modern industry and commerce, assuming that most of the people were part of a modern, interconnected system. With most of the population still primarily at the subsistence level, such countries have little hope of generating either the capital or the labor force needed for industrialization. Most African countries will continue to be dependent on external sources for capital and technical help for a long time to come. Moreover, plans for the political and economic integration of Africa are much less developed than in Latin America. Four small former French territories in west Africa have joined in a customs and economic union to work out common external tariffs and internal economic policies. The East African Community also has been formed by Uganda, Kenya, and Tanzania, countries that share common rail, post, and telecommunications systems. But these are only tiny beginnings, and even they are having difficulty surviving.

Problems of Redeveloping a Fragile Environment

Too little is known of the African environment to be able to guarantee the success of any economic development plans that call for radical changes in farming and herding practices or for the establishment of major facilities that produce change in a local environment. There is also much still to learn about the impact of agriculture on the fragile tropical environment with its tremendous evapo-transpiration potential. Scientific stations have been set up, but as Africans begin to learn about the environment, they are still confronted with the problem of translating ideas into practice on land that is now mostly held by very conservative subsistence farmers.

Modern facilities often produce environmental changes that result in unplanned side effects. In 1971 a huge new dam was completed in the Ivory Coast. It formed a lake which is 1,700 square kilometers in area and provides enough water to generate about a fifth more electric power than the country had. In twenty years it should be able to irrigate 25,000 acres of land, but it also drowned 178 villages, acres of farmland, and giant **teak** forests, and displaced great numbers of wildlife. Nearly 100,000 people had to be moved into new settlements. The greatest danger is that the new dam may result in a serious threat of disease. A much smaller one in 1964 touched off raging epidemics of river blindness and sleeping sickness as insect carriers found the new environments around the lake to their liking.

In an important sense the experts from advanced countries as well as the African leaders do not really know how best to plan for African development. The initial stages have only recently been undertaken and the area has only recently begun to be scientifically studied. One can wonder whether it will be possible to produce the dramatic changes in African life that have occurred, for example, in England over the past three hundred years. One can also question whether Africa can afford to wait three hundred years to join the modern interconnected system.

SELECTED REFERENCES

Arkhurst, Frederick S. *Africa in the 70s and 80s: Issues in Development,* Praeger, New York, 1967. Results of a symposium of African experts dealing with future economic, political, and social prospects.

Davidson, Basil. *The African Genius,* Atlantic Monthly Press, New York, 1969. This volume for popular readers is called "an introduction to cultural and social history."

Grove, A. T. *Africa, South of the Sahara,* 2nd ed., Oxford U. Press, New York, 1970. A basic geographic treatment of this culture region.

Hance, William A. *Africa: Economic Development,* Praeger, New York, 1967. An economic geography of Africa.

Herskovitz, M. *The Human Factor in Changing Africa,* Knopf, New York, 1962. An anthropologist who has specialized in Africa gives his insights into African development.

Lloyd, Peter. *Africa in Social Change,* Praeger, New York, 1968. This volume uses West Africa. as a case study of social change in Africa.

Prothero, R. Mansell. *People and Land in Africa South of the Sahara,* Oxford U. Press, New York, 1972. A new geography by a British scholar.

Chapter 15 India and China: Two Approaches to the Population Problem

THE PROBLEMS of Asia seem at first glance to be the same as those that we have examined in Latin America and Africa: underdevelopment, dual societies, low per capita gross national product, high population growth, and a large proportion of young people both coming into the labor force and into the childbearing years. Population densities are higher in most Asian countries than we have seen in either Latin America or Subsaharan Africa, and are reminiscent of those in Europe (Table 15-1). But here populations of 100, 300, 500, and 800 per square mile must be supported by largely traditional economic systems rather than by a highly specialized, long-range interconnection.

Still, most Asian countries have total populations of manageable size, 10, 20, or 30 million people—less than the largest European countries or Brazil or Nigeria. While they face major developmental problems, the scope of these challenges seems within the range of known solutions even though there are difficulties in implementation.

But China and India are something else again. Their huge populations are unique in the world, and possible solutions are beyond the scope of any developmental programs ever conceived anywhere. China has more than 780 million people, over three and a half times the population of the United States. China's growth rate is small, but even with a rate of 1.7 per cent the 1973 increase was estimated as 13.4 million people. India has more than 580 million people but a higher growth rate. Its population increased by 15.3 million people in 1972, more people than the total population of most countries of the world.

These two nations face issues of development and industrialization on a scale never before encountered. Europe at the beginning of the Industrial Revolution had less than one-seventh of the present population of China and less than one-fifth that of India today. The United States

Table 15-1 Population Data on Selected Countries of Asia, 1972

	Population 1972 (000,000)	Population per Square Mile	Annual Population Growth (per cent)	Population under 15 Years (per cent)	Per capita GNP (dollars)
Pakistan*	146.5	366	3.3	45	110
India	584.8	477	2.5	42	110
Bhutan	0.9	49	2.2	†	60
Nepal	11.8	217	2.2	40	80
Ceylon	13.2	521	2.3	41	190
Burma	29.1	111	2.3	40	70
Thailand	38.6	194	3.3	43	160
Singapore	2.2	9,821	2.2	39	800
Malaysia	11.4	89	2.8	44	340
Indonesia	128.7	223	2.9	44	100
South Vietnam	18.7	283	2.1	†	140
North Vietnam	22.0	358	2.1	†	90
Laos	3.1	33	2.5	†	110
Cambodia	7.6	109	3.0	44	130
Philippines	40.8	352	3.3	47	210
Taiwan	14.7	1,058	2.3	43	300
Hong Kong‡	4.4	11,055	2.4	38	850
China	786.1	213	1.7	†	90
North Korea	14.7	316	2.8	†	280
South Korea	33.7	887	2.0	40	210
Mongolia	1.4	2	3.1	31	460
Japan	106.0	743	1.2	24	1,430

*Pakistan population = 66.9 million; Bangladesh population = 79.6 million.   †Unavailable or unreliable.   ‡Nonsoverign political unit.

Source: Population Reference Bureau, Inc., *World Population Data Sheet: 1972*, Washington, D.C. 1972; *Handbook of the Nations and International Organizations*, Life World Library Series, Time-Life Books, New York, 1968.

had less than 50 million people when it began industrializing. Economists and developmental specialists can present blueprints for expansion to underdeveloped countries, but no one can say with assurance how the problems of India and China are to be solved. Yet the problem of their development must be faced. One-third of the world's population cannot be ignored.

Population Densities near Malthusian Limits

The populations of India and China are near Malthusian limits. In times of less than normal agricultural output there is the danger of famine, and the countries must rely on outside sources of food. In 1943 1.5 million people died as a result of the famine. During the winter of 1960–61 only grain importation from Canada and Australia and emergency rationing stopped widespread famine in China.

Traditional agricultural methods which have supported the Chinese and Indian nations for so long are no longer adequate. Population has expanded within a predominantly static economic system. Not only must further growth be slowed, if not stopped, but there must be a change in the basic economic system. Otherwise the rest of the world will have to subsidize India and China. But such a plan in itself would require trade on an unprecedented scale. Moreover, even if it were feasible, it is unlikely for social and political reasons.

Site of Two of the World's Great Civilizations

Despite their present problems, India and China represent two of the world's most ancient and advanced civilizations. The roots of Indian and Chinese civilization date back beyond 2000 B.C. When Europeans first established contact with them in the thirteenth century, these civilizations were more sophisticated than the Europeans in artistic, scientific, and political developments. The explorers who first reached the Americas were, in fact, searching for new sea routes to Asia to tap the lucrative trade in spices, porcelains, and silks of these wealthy societies. As late as 1793 when an envoy of the British king was received at the court of the Chinese emperor, he was brushed aside with the following comment, "I have already noted your respectful spirit of submission . . . . I do not forget the lonely remoteness of your island cut off from the world by intervening wastes of sea," but "our Celestial Empire possesses all things in prolific abundance and lacks no product within its borders. We do not need to import the manufactures of outside barbarians in exchange for our own produce."

PRIMARILY AGRICULTURALLY
BASED CULTURES

Despite the wealthy and literate upper class and skilled and productive craftsmen, both China and India were agriculturally based civilizations. Farm practices today are essentially those developed by the civilizations thousands of years ago, using simple tools and human and animal power.

The major crop wherever it can be planted is rice. It is the most important food crop in the world because it is the primary support of these huge populations. The most productive varieties grow in standing water and call for irrigation. Hence, farmers in the wet tropical and mid-latitude lands of Asia have created a man-made water control system on a scale unknown anywhere else. Over thousands of years, the lowland river valleys have been transformed by peasant farmers. Dams, level diked fields which allow water to flow on and off the land, and terraced hillsides are evidence of the tremendous effort man has made, assisted only by animals and simple tools. This farming system is the most intensive in the world. The lowlands were turned into giant garden patches, carefully manicured by millions of peasant families whose way of life has passed almost unchanged from generation to generation. Even today, the lowlands of China and India are among the most productive areas in the world.

For hundreds of years 80 to 90 per cent of the huge population of Asia has been agricultural, living in small farm villages. Farm properties are small. What is more, an individual farm is fragmented into as many as five separate pieces. In Asia farming is a family enterprise; children help transplant seedlings and harvest the crops. This system produces almost 80 per cent of the world's rice and large quantities of grains and vegetables. Yet without modern capital and scientific know-how, it appears to offer little hope for expanding output fast enough to keep up with population growth.

A look at the frost-free period (see Figure 6-9) shows that much of this part of Asia has favorable temperatures for plant growth throughout the year. Given adequate rainfall, it is possible to grow two and sometimes three crops a year in India and in peninsular Asia. Two crops can be grown on the better land of southern China. Northern China and the highlands of Tibet, however, have very limited frost-free seasons.

Metropolitan Areas in Basically Rural Countries

Life in this part of the world centers around the local village. Nevertheless, there have been and continue to be large cities in both India and China. In 1800 China had eight cities with more than 250,000 population; Calcutta in India had 600,000 people at that time. Today, China has eighteen metropolitan areas of over one million people, and India has eight. Most of the large metropolitan areas of Asia continue to be political centers. Eleven of the eighteen Chinese cities and seven of the eight Indian centers are capitals of provinces. However, the largest metropolitan areas in the region—Shanghai, Bombay, and Calcutta—are not capitals of their countries. Rather they are ports and were centers of foreign trade in the nineteenth century. In contrast to the United States or Europe, only 5 per cent of the Chinese population lives in the largest metropolitan areas; in India only 4 per cent.

Mature, Advanced Cultures

In both China and India the traditional economy was almost feudal. Agriculture was the basic occupation of the populace, and most of the production was consumed locally. Landlords held political and economic power in particular parts of the country, and shares of the crops went to them or on to regional centers to support the upper classes and the emperor's court. Local craftsmen in farming villages depended on payment in food for their services.

THE IMPORTANCE OF THE FAMILY AND THE VILLAGE

In both countries, the family was the basic social unit, and the village the fundamental sphere of life. Villages were governed by the heads of leading families; contact with the central government was to be avoided if at all possible.

Traditional upper-class culture was also based on the extended family. The wealthy avoided manual labor and regarded literacy and education, ownership of land, and, in China, public service as the highest social goals and symbols of status. In India the caste system evolved based on the role of each family in the economic structure. Farmers formed the dominant caste; above them were the warriors and aristocrats, and the priests and intellectuals; below them were the menial laborers. These four castes in turn were divided into 3,000 subcastes, each with its own status, traditions, rules, and regulations. In general, one was born into a particular caste and could not marry outside it or have social contact with persons outside. Outside the caste system altogether were the untouchables whose very shadow may defile a caste Indian. They comprise about 10 per cent of the population. In China there was no caste system, but in reality there was little social mobility either.

DIFFERENT PERSPECTIVES OF LIFE

India and China differ sharply in their traditional view of the basis of life. In India otherworldly religion has dominated the traditional culture, while in China human relations and the organization of society have taken precedence over the otherworldly. Although the Indian is concerned with the fate of the individual soul, the ultimate rewards are not in this world but in a better birth in a subsequent reincarnation. The Hindu believes world progress is both improbable and temporary. His religion tells him that the universe, like man, dies and is reincarnated. In China the power of the state was based on the belief that the emperor's presence was the manifestation of Heaven. The state came first, with religion as its support. The civil service bureaucracy selected by examinations and promoted by merit ratings supervised social conduct, not the priests.

CHINESE UNITY AND INDIAN DIVERSITY

Through history China has had greater unity than India. Although there are a multitude of local dialects in China, a single writing system was developed. In the ninth century a form of printing was invented. This provided the scholarly upper class throughout the Empire with a common cultural base, although the complex writing system put literacy out of reach of the average man. It is estimated that until the eighteenth century more literature had been printed in Chinese than in all other languages of the world combined.

In India the lack of a single spoken or written language is a reflection of the many contrasting political and cultural forces in its history. Lying at the crossroads of land routes from the west and northwest, India has been the scene of conquest by various external groups since the invasion of Alexander the Great. Huns have come from central Asia. The wave of Moslems came from the Middle East. Today, in addition to the multitude of language differences there is great religious diversity. Although India is nominally Hindu, the importance of the Moslem presence was seen in the partition of British India into essentially Moslem Pakistan and Hindu India. Today Hindu India includes some Moslems, Buddhists, Christians, and members of such minor sects as the Jains and Parsees.

COLONIAL INDIA AND SUBJUGATED CHINA

The early industrialization of Europe enabled it to dominate the rest of the world. Even these two great Asian civilizations suffered the humiliation and divisive force of European power. When Britain began trading in India, political control within the Indian civilization was in disarray. The Moslem Mogul Empire occupied the northern area, tiny principalities bickered and fought in central India, and the old Hindu Empire had fallen in the south. Thus, in a period of about 130 years ending in 1885, the British gradually gained political as well as economic control of the entire area.

The Chinese, whose emperor had spoken so haughtily to the British envoy late in the eighteenth century, faced the humiliation of granting special trading privileges to the British less than half a century later. Until then all trade with China was carried on by "factories" of European traders in the city of Canton in south China. There the traders were forced to live in a foreign quarter and were prohibited from entering Canton proper. All trade had to be through a government-controlled monopoly. Chinese would not even teach their language to the foreign traders.

During the Opium Wars (1839–1842) modern British military power proved the Chinese Empire to be almost defenseless. As a result the island of Hong Kong was ceded to the British, and five Chinese ports were opened to trade with permission for aliens to live there outside the jurisdiction of Chinese courts. Christian missionaries were granted permission to enter China after years of prohibition. Soon other industrial nations obtained the same privileges in other ports. By the end of the nineteenth century, China had been divided into spheres of influence, and the British, German, French, Russian, and Japanese were all granted sweeping and exclusive economic rights in particular parts of China.

The impact of the Europeans on India and China was very different from their impact on Africa. European investment in export production tended to dominate the sparsely populated, tribal areas of Africa. By contrast, in the densely populated, highly developed socio-economic systems of India and China, the changes were of a different order. There the emphasis on trade with Europe was minor in relation to the mass of the population still living on a traditional subsistence basis.

The Minor Importance of Plantations and Mines

Moreover, the Europeans found developing trade with India and China to be an operation very different from that in Africa or Latin America. Elsewhere commerce had begun with trading posts that drew off a share of the production of the traditional native economies, but it soon shifted to specialized output from European-established plantations and mines. In China and India this shift to European-controlled production rarely took place. Production of silk, cotton, jute, and tobacco continued to come from native farms rather than from European-created plantations. Tea plantations in the Assam Hills of northeast India and small, localized developments of coffee and rubber elsewhere in India are the exceptions. In most cases export crops represented a tiny

fraction of either the land or output of native farms. Thus, the traditional agricultural systems of India and China added the new function of producing the export commodities that the Europeans wanted.

The mass of traditional Indian and Chinese agriculture was, however, simply too much to be affected by the small amount of European trade. Some production from primarily subsistence farms had always gone to supply the wealthy and their retainers in the urban courts. The new European trade was simply an expansion of that sort of demand, and the primary role of agriculture on any farm remained the support of the people of that area. Moreover, European traders found very little opportunity for development of their own plantations. They were confronted by an intense subsistence agricultural system, not migratory agricultural tribes.

Asia as a European Market

It was markets created by the huge populations rather than the lure of raw materials that stirred European trading activity. In the nineteenth century the new factories of Europe were turning out vast quantities of textiles, hardware, and small metal products, and India and China offered a larger market than all the other colonial areas combined. Even though subsistence farmers can buy very little, a few purchases by each of the families in these vast populations created an important market. As a result a flood of European goods moved into the ports of India and China and gradually filtered into the marketplaces of most villages.

The new European-organized trade had several pronounced effects even though its total impact on the traditional agricultural society seemed small. Factory goods could be sold for less than the products of local artisans, and large numbers of the skilled workers who had been a part of the village community abandoned their crafts. In the new European port cities, textile and clothing mills and factories making various metal products and processed foods also appeared. Local labor could be employed for just a few cents a day.

Europeans also built modern railroads and telegraph lines, and steamboats were introduced on China's main rivers. Before long, a continuing flow of European traders, scientists, bureaucrats, and missionaries were bringing new ideas and new ways of life to all parts of the country. Since railroads required fuel for engines and metal for rails, there was a need for local sources of supply, and new coal mines and rudimentary steel plants were started.

All of this European-dominated activity involved native people. Merchants and shopkeepers emerged to provide outlets in the villages and among the native population of the cities. Workers were employed on the railroads, in the warehouses and factories, and at the ports. In India the government hired many natives as clerks, inspectors, tax collectors, and soldiers. These workers learned English, came in daily contact with European customs, and were often trained in European schools. They were far from the traditional rural society of most of their countrymen. This group, caught in the conflict of old and new values, formed the basis for discontent and change. It led the nationalistic movements for independence and tried to find a means of blending the traditional cultural heritage with the modern world.

THE EUROPEAN-INDUCED POPULATION EXPLOSION

Perhaps the most significant change in India and China as a result of the coming of the Europeans has been the "population explosion." Estimates suggest that the population of Asia as a whole increased more than five times in the last three hundred years, and most of this increase occurred in India and China.

This growth was almost entirely the result of a natural increase in the population caused by an excess of births over deaths; there was virtually no immigration into either area. The explosiveness of the increase was due in large part to changes brought about by European science and technology. Birthrates changed little during this period. Children provided an extra supply of energy for farming in a rural society, and the family was highly revered. By contrast, the death rate fell markedly when Europeans introduced sophisticated medical care and new discoveries about sanitation and public health. New food crops such as potatoes and maize were introduced to improve the food supply, and the skeletal network of railroads and roads enabled food supplies to be transported to areas where natural disaster struck. Finally, Europeans enforced peace in both countries.

Population Growth in an Essentially Static Economy
The most important aspect of the tremendous population growth in India and China is that it took place within an essentially static economic system. Changes in the system—the introduction of commercial crops, new factories, and modern transportation—increased productivity, but the scale of these changes has been very small when measured against the tremendous populations. In 1950 the per capita value of all industrial output in India was estimated at only $7, and it was probably less in war-decimated China. In 1950 as in 1650, these countries were essentially nonindustrial in terms of the tremendous population to be served. Thus, it is not hard to see how the great Chinese and Indian cultures, the equal of any in the world three hundred years ago, have had their material existence deteriorate not only by

Western standards but also in terms of their own earlier histories.

The problem of scarcity is unevenly felt, however. In India a minority of the population has entered the commercial system brought by the Europeans. They and the economic and political institutions they represent absorb most of the commercial-industrial output of the country. It is the rural village that has suffered the most, for it is in the agrarian communities that the traditional system has remained static. Each generation has had more mouths to feed. The presence of Europeans and modern political control has also brought changes in the patterns of land ownership and increased the taxes to be paid from the produce of the land. Where there was once some surplus to be sold, there is no longer enough to go around. Hence, the living standard and the basic diet of the population have declined.

It is not surprising, therefore, to find that people from rural areas are lured to the cities by tales of prosperity. But, like the Puerto Ricans who move to New York, they are untrained for urban jobs and for urban life. Most Indians find no jobs in the city. Nor can the Indian government afford welfare payments. Western visitors who may never get into the country to see an Indian village are struck by the abject poverty of cities like Calcutta. The problem has been accentuated by millions of refugees uprooted from rural areas after the establishment of East Pakistan in 1949 and again during the war with Pakistan in 1971.

In China the government has more control over the movement of people into the cities. It has embarked on a campaign of rural reform and propaganda glorifying rural life. Even professionals and students from the cities are pressed into service in rural communities, and every effort is made to minimize the lure of the city.

Two Different Models of the Development Theme

In the late 1940s new governments in both China and India inherited the problem of a national economic system unable to supply its population properly—the Indian government of Mahatma Gandhi achieved independence from Britain in 1946 and the Chinese regime of Mao Tse-tung defeated the Chinese Nationalists in 1949 (Figures 15-1 and 15-2).

Each government has had the same aim: to build its rural economic base so that it can adequately support the population. At the same time they want to develop a modern economy that will provide industrial and military strength. But their development plans are very different. India, though strongly nationalistic, has built its new political-economic system firmly on the British model. It has an elected parliament and a socialist economic system that relies heavily on private enterprise. The government of India makes financial plans and sets economic priorities, but a large sector of the economy remains in private hands, including farmlands that support rural India and many of the major industrial and business concerns that support city life. As a result the government has no direct control of total economic development. It tries to present goals, provide loans, offer assistance, levy taxes, and make regulations to direct the private sector in following its plans.

China, by contrast, has chosen the communist model first pioneered by the Soviet Union. The central government not only maintains control, but makes all basic developmental decisions and then sees that they are carried out. It sets priorities and has the power and investment support to push them forward.

The result of these two different forms of development programs has been very different

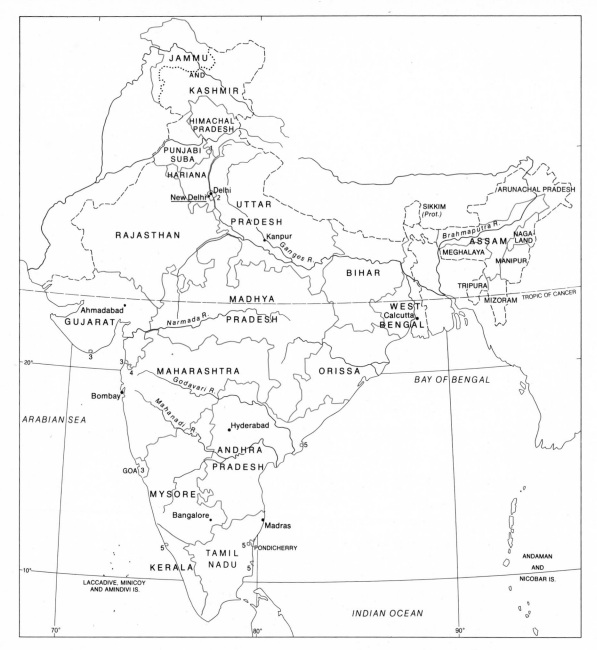

Figure 15-1 POLITICAL MAP OF INDIA

1. Chandigarh   2. Delhi   3. Goa, Daman and Diu
4. Dadra and Nagar Haveli   5. Pondicherry

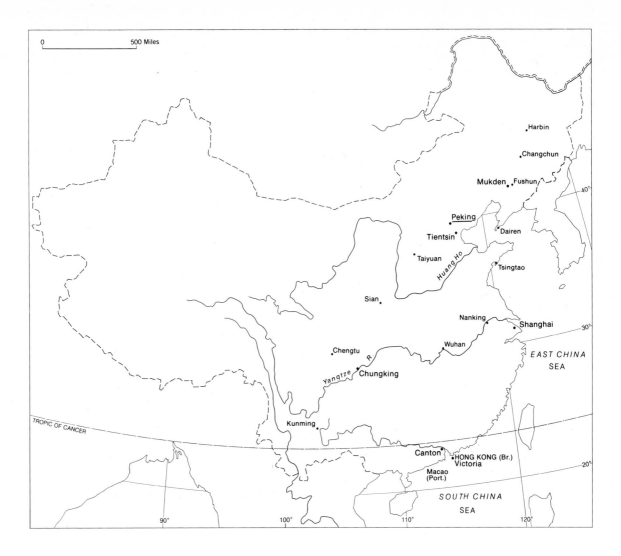

Figure 15-2 Political Map of China

degrees of support from the industrial countries that are in a position to provide economic and technical assistance. India has received important amounts of assistance from the United Nations, Britain, the United States, West Germany, and the Soviet Union.

China, on the other hand, was refused admission to the United Nations until 1971 because the United States continued to recognize the Nationalist regime of Taiwan as the legitimate government for China. The primary concern of Mao's regime has been with internal development. There has been virtually no trade with capitalistic industrial countries. For more than a dozen years the Soviet Union provided both capital goods and technical assistance, but after an ideological break in the 1960s, China refused all further aid from them. Thus, the Chinese have been developing their economy with a minimum of outside assistance and foreign trade with the industrial countries. It has only been since its admission to the United Nations and President Nixon's visit in 1972 that normal trade relations between China and other industrial countries have begun.

The push of each government toward economic development must also be seen in terms of the great cultural heritages of these countries. Each government sees itself re-establishing the proud position the country once held in pre-colonial times. In particular, much of China's appeal to its people is the pledge, presented continually in government propaganda, to re-establish China as the world's greatest cultural and political power.

THE TWO-FOLD DEVELOPMENT TASK

For both India and China, the developmental problem is much the same. Like every country with ambitions to be a first-rate political power, industrialization must provide an arsenal of modern military weapons including jet aircraft, rocket launchers, atomic weaponry, and the scientific skill necessary to maintain the arsenal. Japan proved that such industrialization could bring national recognition quickly.

For India and China these problems are in many ways similar to those of the United States and the Soviet Union rather than to those of Japan. Although the population density of China and India is much higher than that of the United States and the Soviet Union, it is much less than that of Japan or the small Western European countries. China and India are huge by world standards and have relatively large resource bases both in terms of agricultural and mineral resources. Their problem is not to develop specialty products for export but to develop their resource bases to serve their own people. They must change a largely locally based system into one that will function on a national basis.

How can a nation become industrialized while it struggles to support a population of two or three times that of the U.S.?

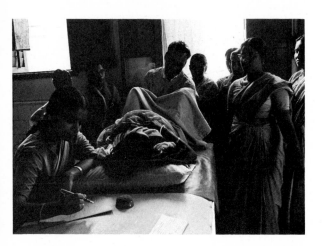

Because of their tremendous populations, India and China must face the problems of industrialization on a larger scale than any country has experienced thus far. They must continue to increase their food production by traditional methods and at the same time build the foundations of modern industry: dams, power plants, steel mills, and railroads. Yet, while the world watches, India and China are following very different strategies toward development.

Modifications on the Soviet and British Models

At first glance, the two countries seem to have attacked their problem in the same way: The governments have embarked on a series of long-range economic plans, like the five-year plans of the Soviet Union. In China the central government has become the major decision-maker and manager of the whole economy. Central planners determine what the country's needs are in terms of industrial development and the supporting food base. Priorities are set among all the myriad facets of the economy of this huge country and goals established in terms of the availability of production, manpower, and supporting facilities such as transportation. Production quotas are set for rice production, steel output, and for every other commodity needed in the economy, and then the needed resources are allocated to achieve those goals. Moreover, national quotas are allocated regionally over the country until each county, commune, and factory has been presented with its expected share of the national plan.

Once allotments have been made to the local producing units, whether factory or commune, it is up to the local working unit to plan its own ways of achieving its quota. Local councils then muster their manpower, resources, and capital to do the task assigned. In this structure the whole economy from the largest to the smallest item is under the review of the central government.

India has followed a plan of socialization very much like that in Britain. The state finances essential new projects—hydroelectric irrigation works, transportation, communication, and even new industries such as steel mills. Moreover, it has taken over from private firms banking and major insurance companies. Public investment is planned to total more than half of each five-year plan, but private investment is still strongly supported. There are many privately owned industries including the country's largest steel operation. Moreover, agriculture remains in private hands. The Indian government worked hard to carry out land reform. The vast hereditary estates of the upper castes and those created under the British regime were broken up. Land was given to previous village tenant farmers. But not all land was distributed in this manner, and large landowners are being encouraged to prosper by the government. They tend to be the most efficient farmers, and the Indian government recognizes that the best way to cope with the continuing pressure for farm production is to focus resources on the most successful farms. So these farms get subsidized credit, fertilizer, and the new high-yielding "miracle" seeds. Thus, India with its socialistic goals is creating a group of successful capitalistic farmers. Big farm operations often return $100 per acre on farms of two hundred acres. In a country where the average income is less than $100 per year, that is big money.

Different Environmental Bases

Although the problem of industrialization is essentially the same for India and China, the pressure of population on the environment is very different (Figures 15-8 and 15-10). China is three times the size of India in area but has almost twice the population. Thus, while the Chinese population is incredible to Western minds, India's pressure on the environmental base is even more overwhelming. India is about a third the size of the United States and has a population slightly smaller than the United States and Western Europe combined. Figures 15-3 and 15-4 show India and China compared in area to the United States.

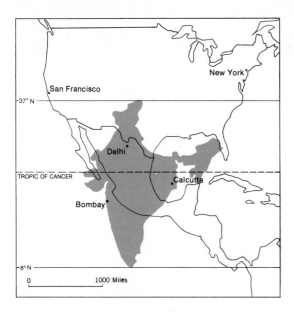

Figure 15-3 India Compared to the United States in Area and Latitude

INDIA AND CHINA IN CLOSER PERSPECTIVE

In our preliminary examination of India and China, we have emphasized comparisons of the two. But there are also important differences in the human and environmental resources on which they draw and in the way that each is undertaking its developmental programs.

India: Building in the Midst of Diversity

In area India is one of the ten largest countries in the world. One would expect to find important environmental variations within its borders that present different production possibilities. And, since India's problem, first of all, is feeding its tremendous population, the variations in agricultural production from place to place are a key to the country's problems. Most Indians are farmers who depend on production from their own local area to provide the essentials of life. Variations in agricultural output suggest differences in the environmental base from one part of India to another and, as we will see later, differences in population density.

Since grain rather than meat is the staple Indian diet, the distribution of its production (Figures 15-5a, b, and c) provides a measure of growing potential. Rice is the favored grain. Wheat is grown only where rice cannot grow; it usually does not do well in tropical environments. Millet and sorghum are raised only where other grains are not possible.

Notice that the rice lands of India are concentrated in the lower Ganges River valley in the northeast and along the coastal fringes of the Indian peninsula. While rice will grow anywhere in the tropics and subtropics, the major varieties are wet crops requiring standing water on the field during much of the growing period.

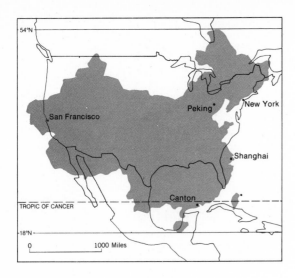

Figure 15-4 CHINA COMPARED TO THE UNITED STATES IN AREA AND LATITUDE

Hence, rice is grown in areas with abundant rain or access to irrigation water.

Over most of the peninsula millet and sorghum are the dominant grains. The fact that these rougher grains are grown reflects environmental conditions more than choice. Millet and sorghum are often planted in the same fields. If summer rains are good, the millet dominates, but if the year is dry, sorghum is the chief harvest. As we saw in Ramkheri, both can be used to produce a porridge or ground into a meal for bread or cakes.

The relative importance of these different grain regions can be seen by comparing their areas of production (Figures 15-5a, b, and c) and population densities (Figure 15-8). Since most of India's rural population still lives like the villagers of Ramkheri, very little grain moves from one rural agricultural area to another. There is little surplus to sell and most villagers would not have the money to buy it anyway. Though some grain is sold to markets in cities, major movements occur only during times of crop failure, when emergency arrangements are made to redistribute surpluses to areas in need. Most cities must depend on imported grain, which accounts in part for the fact that India is a leading importer. The primary function of India's agricultural system is simply to feed the people of the country, and the agricultural program will succeed or fail on that basis. India cannot, however, afford to shift areas that are producing export or important industrial crops into food production.

In contrast to Latin America or Africa, agricultural exports are a very secondary matter to India. Cotton and other crops are used as raw materials in the growing industrial programs. Exports of tea, jute, and cotton provide important foreign exchange, and increasing amounts of jute and cotton are being manufactured within the country, not only to clothe Indians

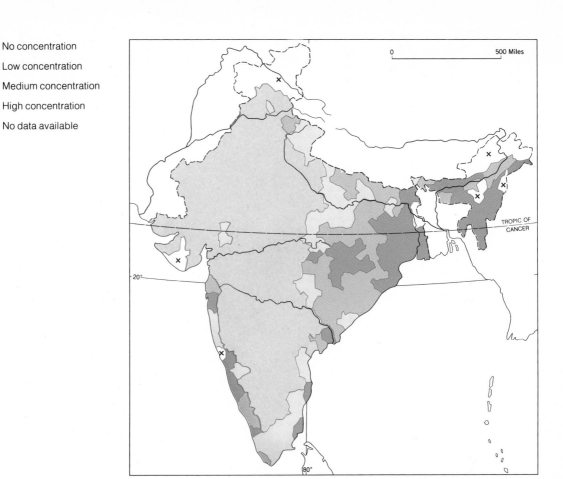

No concentration
Low concentration
Medium concentration
High concentration
x   No data available

Figure 15-5a   CONCENTRATION OF RICE IN INDIA

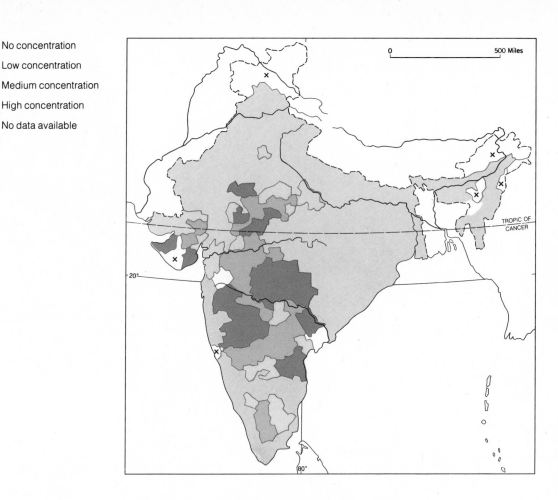

Figure 15-5b Concentration of Sorghum in India

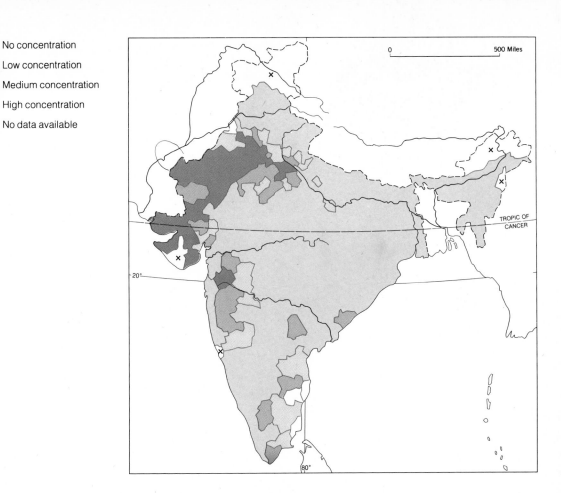

Figure 15-5c CONCENTRATION OF MILLET IN INDIA

but also for export. Cotton manufacture is one of India's largest industries, and tea accounts for 40 per cent of agricultural exports.

Agricultural Problems

The map of major cereal regions indicates one of India's leading agricultural problems: a shortage of moisture over much of the country. Except along coasts in rice-growing areas and along the southern flank of the Himalayas in the north, the deficits in the dry season are comparable to those in the Great Plains and desert areas of the United States (Figure 5-5). Over most of the cereal-producing area there are no surpluses after the rainy season.

Dryness

Since water shortages are endemic to most Indian agricultural areas, better water supplies could improve productivity over much of the country. Most Indian wells have been dug by hand methods and are limited to tapping ground water just below the surface. But modern well-drilling equipment can reach hundreds, even thousands, of feet into the ground. The Indian government has undertaken well-drilling programs in many Indian villages. Equally important has been the construction of storage ponds, called "tanks," to collect run-off during the wet season for use during the dry months. Just as the Department of Agriculture's program of pond construction has been an important factor in improving farming in some parts of the United States, the tank digging program of the Indian government is being counted on to improve vital water supplies in areas of the Indian peninsula where there is a shortage of water for all purposes during the dry season.

One of the major efforts of India's agricultural program under the five-year plans has been to provide irrigation. Large-scale projects using the run-off from the mountains in the northwest have been undertaken. Three major dams were under construction in 1970. These schemes are modeled after the huge multi-purpose projects of the United States and the Soviet Union and require great capital investment. They will open new acreage in dry areas which have hitherto been impossible to farm, but they are likely to be little more than stopgap measures. With modern irrigation, productivity will be increased, but the number of acres being added to India's total agricultural land is very tiny. Moreover, soil and water must be closely managed to keep soil salinization from occurring.

Underutilization of the World's Largest Cattle Population

A second problem of Indian agriculture is well known: the presence of the largest cattle population in the world. In 1969–1970, this population was estimated at 176 million plus 54 million water buffalo (Figure 15-6). In most countries such cattle numbers would be looked upon as a major agricultural asset. In predominantly Hindu India, however, cattle are considered sacred. The bull is associated with the Hindu god Shiva, and the cow is recognized as the representative of "mother earth." All cattle are inviolate, and even those castes that eat meat do not eat beef. This huge cattle population is not a source of direct food supply except for dairy products. Instead it competes for the scarce food supply.

The cattle have important uses in the agricultural system, however. They and the buffalo are the chief source of draft power. They plow fields, power threshing machines, and pull carts. Moreover, since there are few trees in most of India, animal dung mixed with straw and dried in the sun is the chief fuel for cooking and

Each dot represents 500,000 head

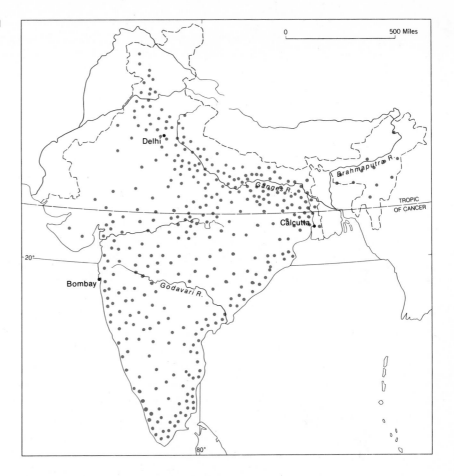

Figure 15-6 CONCENTRATION OF CATTLE IN INDIA

heating. Nearly half the cattle manure that might be used as fertilizer on the land is used as fuel.

Since cattle in India are essentially beasts of burden, they are most populous in areas where crop production is greatest. But the Indian agricultural system is designed to produce food crops for humans, not animals. There is little room in the system for growing fodder, and pasture land is limited. Work animals are fed "hay" of rice and other grain stalks that are left after the harvest. Without seeds it is not very nutritious, and the supply is woefully inadequate. Such animals are too undernourished to provide good power, dairy products, fuel, or fertilizer.

The biggest problem is the indiscriminate, largely uncontrolled breeding of cattle. Government plans call for gradual reduction in the number of cattle and improvement in breeding. But implementing such plans are very difficult in Hindu India. It is very difficult to implement a program unless the Indians change their attitudes about livestock management.

Low Agricultural Productivity

In contrast with other parts of Asia where the intensive agricultural system produced some of the highest per acre crop yields in the world even before the introduction of modern scientific techniques, India's yields are among the lowest. An important component in India's problem is the drier environment, but much of the problem is the result of Indian farming practices. The land has been worked continuously for thousands of years without adequate nutrient replenishment. In most areas the people were generally ignorant of proper methods. Little fertilizer was used, tools and cultivation practices were poor, and seed selection was bad.

Overcoming India's agricultural problem will require more than new wells, water tanks, and better breeds of cattle. The attitudes of farmers will have to be refashioned to accept innovation. A new infrastructure of roads and markets must be developed that will bring India's farmers into the cash-crop economy. There will need to be a better system of credit and a readjustment of farm prices. There will have to be better storage facilities and more processing plants for farm produce.

Plans for Agricultural Development

Indian agricultural plans still rest largely on expanding the output on existing farmlands. There is no significant frontier land that can be developed. The productivity of India's farmers must be increased; resources must be better utilized. India needs to modernize agriculture, not in order to release workers to industry in cities, but to increase productivity of the land. Thus, it is not mechanization that is called for, but increased productivity within the traditional farming system based on man and animal power. The essentials needed are education, markets for surplus production, credit on reasonable terms, and agricultural research. But since none of these factors are independent of the others, all of these changes are needed simultaneously. The farmer who learns about new methods must be assured that he can market his products and that he will have the money to undertake necessary changes whether they be purchases of new seeds or fertilizer. All this requires remodeling the rural way of life, a task which, as we saw in Ramkheri, cannot be quickly accomplished.

The goals for India seem modest, but the task has been overwhelming. There are more than 750,000 Indian villages and about 100,000,000 farm families whom the new program must reach. With the help of agricultural specialists and community development experts, the Indians have been working on the problem for more than twenty years. Early strategies called for schools and programs to train one influential leader from a given village with the idea that he would return and spread the word. Farmers have been encouraged to organize into cooperatives, to provide credit, and to share know-how. Village councils were started both to educate and to make people feel a part of change.

Yet even when the new ideas are presented, it is hard to change tradition, particularly tradition that has been much the same as far back as anyone in the village can remember. And the Indian government, in contrast to the Chinese or the Soviets, is trying to work with individuals. There are no collective farms or communes that manage the productive lands. Land is in private working units that average less than ten acres in size. The owner may be illiterate and in debt. He has no ready access to additional working capital. Under these circumstances India faces the biggest educational and developmental program in the world. Moreover, the Indian government itself is far from efficient. It has an elaborate bureaucracy burdened with a mass of directives and cumbersome communications channels.

The greatest obstacle, however, has been the scale of the problem. The Indian government found that, in attempting to carry the program to its five-year plans, whatever money was budgeted was swallowed up by bureaucracy. A government attempting a development program could not follow formulas used in Japan or in smaller developing areas.

In recent years the concentration of development efforts has affected just a few areas over the country. Great effort has been expended to find the key places where agricultural investment can be expected to yield high returns. Areas with good soil and water and with adequate connections to regional markets are chosen, and planners concentrate on these areas. But making these lands productive requires a prodigious amount of organization, cooperation, and patience. Farmers are provided with supplies, storage facilities, transport, credit, and price incentives. The most significant need is probably fertilizer. Indian land has been worked for two or three thousand years, and the biological cycle for restoring soil fertility is inadequate. Chemical fertilizers, plus some organic material to maintain the water-holding capacity of the soil, are vital if the food output is to increase fast enough to feed the people. In response to this situation, India has expanded fertilizer production.

From 1962 to 1968 fertilizer consumption increased sixfold, and by 1975 plans call for increasing that output four times. Increased use can be seen as an index of progress made in intensifying agriculture. But it does not pay to use heavy doses of fertilizer unless other aspects are also improved, particularly water availability and the quality of seed. Remarkably India has been keeping fertilizer demand, supply, and distribution roughly coordinated over this vast agricultural country.

But India's agricultural effort has come at severe cost. In 1968 alone, over a third of India's $1.4 billion foreign-exchange deficit resulted from imports of food, fertilizer, phosphate, and petroleum, all primarily involved in the agricultural program.

By some measures, India's agricultural program has had remarkable success. Rice output in 1967 was 40 per cent greater than in the early years of Indian independence. Wheat production for the same period was up 35 per cent. By 1970, with the addition of miracle grain varieties, the agricultural problem was shifting to questions of adequate storage, and shipment and delivery before spoiling. Restrictions on interstate movement of grain were lifted, and with the help of imports there was a reserve of more than 6 million tons. The 1969–1974 five-year plan calls for self-sufficiency in food production by 1974.

Yet, in spite of remarkable progress, India's central problem remains one of improving the food-to-population ratio. Cutting imports is one thing; improving the diet of the people is another. The average daily diet contains less than 2,000 calories, 400 less than the generally recognized requirement.

For success there will have to be stabilization of India's population growth, or any gains will be lost to the Malthusian doctrine. The government has been making the greatest effort to reduce the rate of population growth in absolute terms. An elaborate birth control educational campaign has been instituted. Vasectomies have been legalized, and a government program encourages men to have the operation. But the vastness of the population problem—in terms of capital, educational efforts, and expert manpower needed to execute the program—is overwhelming. Even if the effort was multiplied many times, the rate of growth would continue to rise in the immediate future because the death rate will almost certainly continue to drop. India has worked successfully during the past twenty years to eliminate epidemic and endemic diseases. Millions of children carried through their early years by new disease-control measures are now reaching childbearing age. Imagine what it would be like without the large-scale birth control efforts now underway.

DEVELOPING INDUSTRY

When the Indian government achieved independence from Britain, it inherited the beginnings of modern industry. In the nineteenth century the British had thought of industry mostly in terms of export products. British cap-italists took advantage of India's agricultural exports and cheap labor. Jute mills developed in and around Calcutta, cotton mills sprang up in the Bombay area, and the large Indian population constituted a ready market for domestic textiles. The large ports were connected to all major parts of the country by a road and railway network. Originally for military and political purposes, the network now enables goods from the large port cities to be marketed over most of the country. Not surprisingly, the widespread marketing of factory-made goods has hurt traditional, small-scale handicraft industries located in the villages.

Even the basic ingredients for modern industry were established under the British. Not only was the railway and road network laid out, but coal was produced for use in factories, and a steel mill was built near Calcutta, close to coal and iron deposits. Between the two world wars this plant became the largest steel-producing unit in the British Empire and one of the world's lowest-cost producers of steel.

India's Push for Industrialization

India's industry at the time of independence was still insufficient in the quantity of goods produced and in the sophistication of production. But because of its other problems, India could not afford to spend the money necessary to import its industrial needs. Then, as now, the basic need is to build up domestic industry to supply the national market.

Industrialization has had a top priority in Indian economic plans. The government has used duties on imports to protect newly developing industries and to limit foreign exchange. It has encouraged foreign investors to enter the market in partnership with Indian investors if they are willing to train Indians to do supervisory as well as menial jobs. Like any plan for industrialization, emphasis falls on heavy industry—iron, steel, aluminum, and finished metal products such as bicycles, sewing machines, automobiles, and trucks. The second emphasis is on fertilizers and other chemicals needed to serve the Indian market.

First results appear significant. The establishment of three government-owned plants built with foreign assistance from Britain, the Soviet Union, and West Germany helped increase steel production three and one-half times between 1950 and 1969. Coal output, too, increased almost two and one-half times in twenty years. India makes the trucks and buses that are important to the country's political and cultural pride as well as to its economy. It has its own growing chemical industry, exports textiles, and makes its own telephone equipment. It even makes a variety of machine tools of sufficient quality and quantity to export to industrialized countries.

India is certainly among the ten leading industrial countries in the world, but its output in 1969 totaled less than 2 per cent of the world's industrial output. That places it roughly at the level of Sweden, a country with less than 10 million people, or Yugoslavia, one of the least developed of European countries. But India has 15 per cent of the world's population! Industrial production has risen from 9 pounds per person in 1950 to more than 20 pounds per person, but the United States produces more than half a ton per person and Japan almost one ton per person. India's industrial growth is still below the world average. In the past twenty years Japan's industrial production has more than doubled, whereas India's has increased less than 30 per cent. Thus, although India's industry is proceeding a bit more rapidly than its population growth, it is not keeping pace with the industrial powers of the world.

The Problem of Scale Again

India's problem is less one of a shortage of resources than a human problem of organization adequate to cope with the scale of development needed. India lacks certain key resources such as petroleum, timber, and some of the nonferrous metals, but it has adequate supplies of both coal and iron-ore and excellent possibilities for hydroelectric development. Although it has far better resources than Japan, India lacks the experienced leadership and organizational skills of the Japanese. The imbalance is between the physical industrial plant and the human component needed to make the plant operate at full capacity and efficiency. Indian management and workers are reluctant to accept modern industrialization fully.

Large-scale Industry

As did most modern industrial countries, India located its factory-industry complex in the large metropolitan areas that are traditional centers of skilled workers and capital and in new urban centers built near basic industrial resources. Seven of the eight metropolitan areas with one million or more people are major manufacturing centers (Figure 15-7). By far the most important centers are the two largest cities, Calcutta and Bombay, ports originally developed by the British as junctions of empire trade. Each is surrounded by supporting industrial towns. Calcutta has an iron and steel district to its west. Bombay is near both an old complex of textile mills and agricultural processing towns, and a new area for chemical, engineering, and modern light industry. The third and fourth industrial districts are centered around the cities of Ahmadabad, north of Bombay, and Bangalore in the southern interior. New Delhi and Madras, two of the largest cities in the

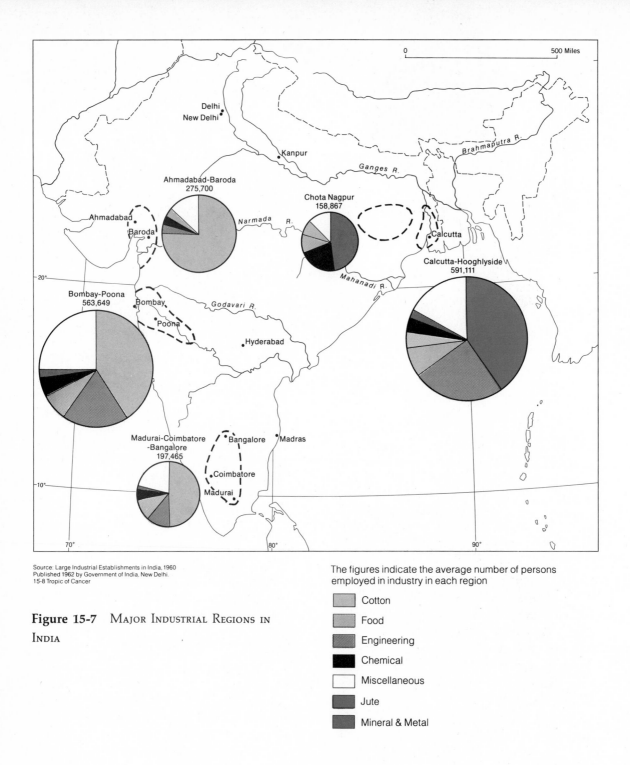

Source: Large Industrial Establishments in India, 1960
Published 1962 by Government of India, New Delhi.
15-8 Tropic of Cancer

The figures indicate the average number of persons
employed in industry in each region

- Cotton
- Food
- Engineering
- Chemical
- Miscellaneous
- Jute
- Mineral & Metal

Figure 15-7 MAJOR INDUSTRIAL REGIONS IN
INDIA

country, are major industrial centers in their own right. Both have shifted from traditional textile mills to more modern technical industry. Madras, a port, concentrates on heavy industry, while New Delhi, the political center, shows a greater orientation toward consumer goods.

Industry in Domestic Development

The first five-year plans gave top priority to industrial development, but increasingly the agricultural problem has become India's first order of business. In the long run industrial development in India may depend most on the agricultural changes that the government is trying to implement. For, today, the Indian domestic industry is perhaps handicapped most by the poverty of the hundreds of millions of largely subsistence villages. Most of these people have cash incomes of less than $100 a year. Their needs are simple: a few condiments, pots, some kerosene, a few hand tools, and a simple plow. Moreover, Indian policy since independence has encouraged local village craftsmen and cottage industry, which uses the home of the craftsman as opposed to a factory or even a shop. Dependent on manpower rather than modern energy sources, cottage industry is very ineffective when compared to a factory assembly line. But, as we saw in Ramkheri, it has important advantages to the villager. The craftsman can repair or design products to fit individual needs, and he can be paid with a portion of the farmer's crop rather than cash. He also usually offers easy credit terms until the harvest. Indian planners had hoped cottage industry would expand to take care of the demands of villagers so that early industrial development could concentrate on heavy industry.

Indian industry also suffers from problems common to other developing countries. The country has always had a trained intellectual elite, but there is a severe shortage of mana-gerial leadership. Economic controls can cause terrible delays. Labor unions have developed along political, not just economic, lines, and there is also the usual shortage of particular raw materials and small parts.

THE SPATIAL PATTERN OF FUNCTIONING INDIA

As the population map indicates the core area of India is along the Ganges Valley in the north (Figure 15-8). The states from Delhi on the west to Calcutta on the east account for 38 per cent of India's population. In that zone the state of Uttar Pradesh has a population of more than 70 million which would make it one of the ten largest counties in the world, and the state of Bihar has almost as many people. The other densely populated areas appear along the coasts of the peninsula. It is the interior of the peninsula, the desertlike northwest, and the northeast frontier that are least densely settled.

Yet even the Ganges population center does not function as an entity. To its west the people speak Hindi, and around Calcutta in the east the language is Bengali. The people in the west are wheat cultivators; those in the east are rice growers. The center of the east is Calcutta; the center of the west is Delhi.

And, while the Ganges area does have both major agricultural output and some of the most industrialized centers in the country, it is much less a focus than the cores of major industrial countries are. Calcutta, the largest and most important metropolitan center of India, anchors one end of the Ganges core, and New Delhi, capital of the country, anchors the other. But, Bombay, the second most important commercial-industrial metropolitan area, is not included. Moreover, all but one of the other metropolitan centers with a million or more population are outside the Ganges area. The

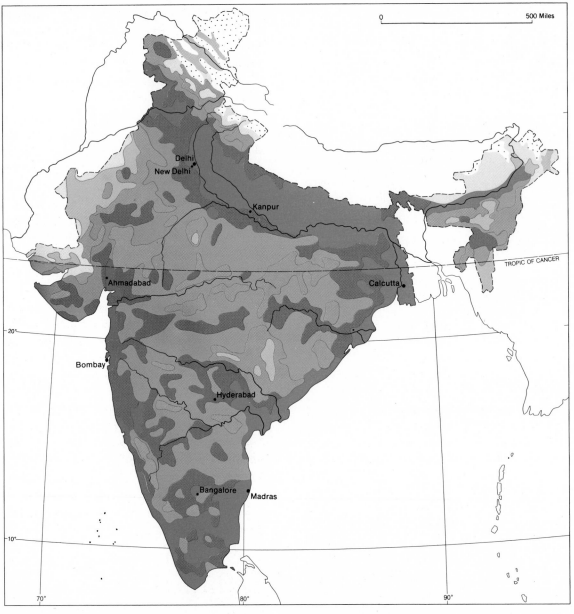

Figure 15-8 POPULATION DENSITY OF INDIA

Persons per sq. mi.

- ░ Uninhabited
- ▫ Less than 2
- ▫ 2 - 25
- ▫ 26 - 125
- ▫ 126 - 250
- ▫ 250 - 520
- ▫ More than 520

Metropolitan area population

- • 1 - 2 million
- • Over 2 million

INDIA: AN AGRICULTURAL AND ECONOMIC ANALYSIS

Table 15-2 provides some indices of economic conditions in the fifteen member states of India.

1. Assume that each state is essentially self-sufficient in terms of agricultural production. Would you expect diets to vary sharply from one part of India to another? Are variations most closely associated with population density, agricultural output, or grain production? Why would you think this relationship holds?

Do you think the Indian economy depends on regional self-sufficiency in agriculture? Why?

Do the areas of high per capita net agricultural output occur in the areas of rice production?

2. A modern, interconnected system depends on manufacturing, modern energy sources, and modern transportation systems. Do you see any correlation between these factors and per capita income as revealed in the table?

We consider education to be a component of the modern, interconnected system. Is there a close correlation between literacy as a measure of education and per capita income in Indian states?

3. In the United States and other modern areas, we have seen core areas that can be identified by their high development and affluence. Are there any such areas among the Indian states?

first-order movement between large centers in India moves between the three largest centers: Calcutta, Bombay, and New Delhi. The second-order connections are between those places and other metropolitan centers scattered throughout the country.

Despite the local orientation both economically and culturally of the majority of its people, India does function as a national entity more than most of the underdeveloped areas we have examined thus far. Despite its diversity of languages and local cultures, India is a political entity. Its member states are larger in size and population than most of the underdeveloped countries. Most, in fact, do have nationalistic movements. But, because Britain organized India into a single colonial entity instead of a series of separate colonies, most Indian leaders have regarded the entire subcontinent as an entity. Gandhi, who led the fight for independence, pushed for one India, and it was only the fear of the Moslem minority that resulted in the creation of Pakistan.

To govern so large and diverse an area is not easy, but it is held together by a nationally oriented and organized civil service administration—a Western-educated, strongly nationalistic elite—spread over the country and by a modern transportation and communication system that is probably the best of any underdeveloped country.

INDIA: LEFT TO SOLVE ITS OWN PROBLEMS

Development in India since independence has not been spectacularly successful, and the one-sixth of the human race that lives within this one country has reaped little benefit from the modern world. Yet the industrial countries have shown surprisingly little concern for India's problems. United States policy has been tuned to the strategic "hot spots" of the world,

Table 15-2 INDICES OF REGIONAL DEVELOPMENT IN INDIA, 1960s

REGIONS AND STATES	POPULATION DENSITY PER SQUARE MILE, 1961	PER CAPITA INCOME, 1961 (rupees)	PER CAPITA NET AGRICULTURAL OUTPUT, 1960–61 (rupees)	ACRES OF FOOD GRAIN PER 100 ACRES OF FARMLAND, 1960–61	PER CAPITA VALUE ADDED BY MANU-FACTURING, 1964 (rupees)	PER CAPITA ELECTRICITY CONSUMPTION, 1964–65 (kWh)	ROAD LENGTH PER 100 SQUARE MILES, 1964	PER CENT OF POPULATION LITERATE, 1961
Northwest								
Jammu and Kashmir	41	248	130	226	7	27	9	11
Punjab	430	467	193	114	23	111	61	24
Rajasthan	152	328	127	74	8	27	20	15
Ganges Valley and northeast								
Uttar Pradesh	649	245	141	163	11	25	63	18
Bihar	691	211	97	193	21	56	49	18
West Bengal	1,032	386	147	264	89	113	77	29
Assam	252	323	183	267	19	7	46	27
Central								
Gujarat	286	369	162	175	53	75	26	31
Madhya Pradesh	189	324	149	86	14	31	20	17
Orissa	292	227	157	144	16	73	43	22
Maharashtra	333	483	156	130	90	105	31	30
South								
Mysore	318	284	153	137	25	53	61	25
Andhra Pradesh	339	326	130	160	12	28	39	21
Madras	669	368	146	282	36	38	66	31
Kerala	1,127	338	148	829	16	38	191	47
India (including Union territories)	348	371	144	156	32	59	40	24

SOURCE: B. K. Narayan, *Southern States through Plans: A Regional Study,* Sterling Publishers, Ltd., Delhi, India, pp. 81, 144.

particularly in Latin America and the smaller countries of Southeast Asia. The largest annual loan from the United States to India was $435 million in 1966, equivalent to about 87 cents per capita. The size of loans has declined in recent years. The Common Market countries have been more concerned with African development, and Britain has been in no financial position to underwrite development on a large scale. But the Soviet Union has provided India with increasing economic and military support in recent years. It was the Soviet Union that most strongly backed India's support of the independence movement of Bangladesh.

But India has been of interest to the large industrial corporations. They are lured by the huge internal market and the need of the Indian government for private investment to complement its own expansion of transportation and key industries. The government prefers to have Indian participation in the control of foreign industrial operations in the country and calls for the use of Indian nationals in management roles. But these rules have been flexible for industries the government badly needs. Foreign investment is greatest in oil, chemical, and metal production.

China: Centralized Control over the World's Largest Development Job

Under the Communist regime China centralized control and decision-making in the national government. The political rulers in Peking demand control over economic and moral life of the entire country. Such control, which seems characteristic of a Communist-controlled society, is also a tradition of China's culture. Throughout Chinese history a highly developed and complex civil service bureaucracy carried out the power of the emperor and established the rules of life for the people. These rules included a code that placed service to the state above family loyalty and family loyalty above personal gain.

These ideals of life have been recaptured and augmented by the Chinese Communist regime. The individual's obligation to the state, not enlightened self-interest, is the highest goal for the Chinese, according to the plan. Thus, the People's Republic of China will succeed as a result of the prodigious effort of people working for goals set by the central government. In place of financial reward, the people will get satisfaction from being part of the national program. The sayings of chairman Mao Tse-tung, like the words of Confucius in earlier days, set the goals for the state and its people and describe proper conduct. In China today there is a strict code of morality. Discipline is very important, just as it was in Japan.

Little distinction is made between the sexes in Chinese society. Women as well as men are expected to contribute their full effort to the programs of the new China. Women hold high-ranking positions in the government and in professions, and work alongside men in all occupations. They are measured in terms of their contributions to society rather than in terms of physical appearance. Sex is a personal thing between man and wife, and it is unimportant to the individual's contributions to the state. To free women for regular work, day-care is an essential part of Chinese life. It is provided by the state.

Making the centralized Chinese system work in the world's most populated country is a tremendous undertaking, and such an attempt requires audacious self-confidence on the part of the government. Centralization is a great asset when the system functions and the central government makes correct decisions, but it can produce horrendous disasters if anything goes wrong. The Chinese found this during the Great Leap Forward in the late 1950s and early 1960s. At that time the government's emphasis on industrial priorities and the collectivization of agriculture disrupted agricultural production, and the problem was exacerbated by bad weather. The result was a severe food crisis for the country that required emergency imports and forced a re-evaluation of national economic priorities.

THE ENVIRONMENT MOST LIKE THE UNITED STATES

In terms of environmental setting, China is probably more like the United States than any other country in the world. It is almost exactly the same size and has the same mid-latitude position. Eastern China has somewhat greater latitudinal range than the continental United States (see Figure 15-4). Both countries have ocean frontage along their entire eastern extremity and extend almost the same distance from east to west. Of course, China, as part of the huge Eurasian landmass, has no western seacoast.

Climatic Diversity

Like the United States, China has great climatic diversity. South China has year-round growing temperatures, whereas most of north China has less than half of that, and northern Manchuria has only ninety days (Figure 6-9). Most of southeastern China gets more than 60 inches of rainfall a year, 50 to 70 per cent more than most of the southeastern United States (Figure 6-5). This rainfall, coupled with year-round growing temperatures, offers great growing potential. At the same time, most of northeastern China receives less than 40 inches of precipitation annually, less than that of the northeastern United States. To the west more than

half of China is arid like the western United States. Rainfall ranges from less than 20 inches annually to fewer than 10 inches in the Gobi desert of Sinkiang-Uigur and in parts of Tibet. Even in the mid-latitudes such rainfall is inadequate for agriculture without irrigation.

Landforms

Although China has climatic conditions comparable to those of the United States, the land surface is very different (Figure 15-9). China has more mountains and highlands. Most of south China is hill land similar to Appalachia. Western China is similar to the mountainous area of the western United States, but on a grander scale. The highest mountains rim the country along the frontiers of the Soviet Union and India, while Tibet is mostly high plateau about 10,000 feet above sea level.

Although there are significant differences between the earth environment of China and the United States, growing potentials are about the same (Figures 5-4 and 5-5). But the Chinese culture has made very different evaluations of the rather similar environment. The swamplands are highly prized by the Chinese for rice culture, but such land in the United States was about the last occupied by European settlers. By contrast, grasslands, not too different from those of the Great Plains, were among the last lands developed by the Chinese. While there are distinctive environmental differences, the distinctions between Chinese and American life seem to be more cultural than environmental.

Crowded, but Not Exceptionally So

In rough outline, Chinese population distribution appears similar to that of the United States. The eastern portions of the country are heavily populated, while the west is almost empty (Figure 15-10). In China the distinctions between east and west are even more pronounced,

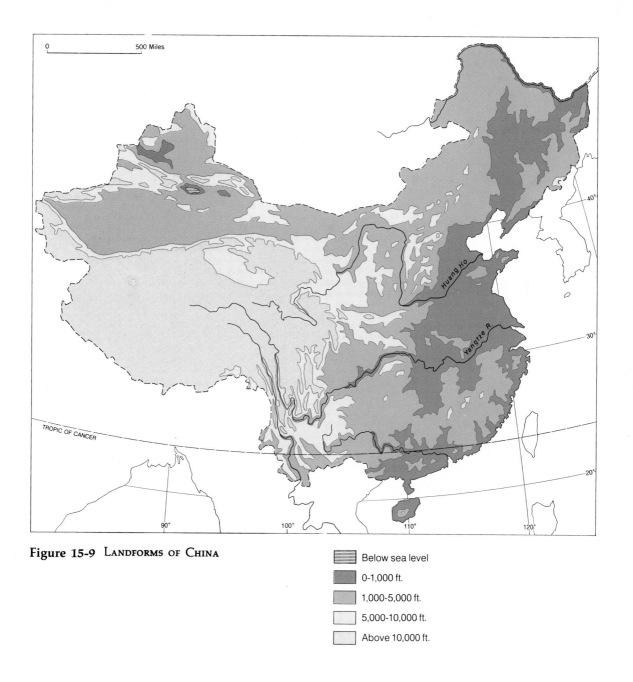

Figure 15-9 LANDFORMS OF CHINA

0 ——— 500 Miles

Below sea level
0-1,000 ft.
1,000-5,000 ft.
5,000-10,000 ft.
Above 10,000 ft.

Huang Ho

Yangtze R.

TROPIC OF CANCER

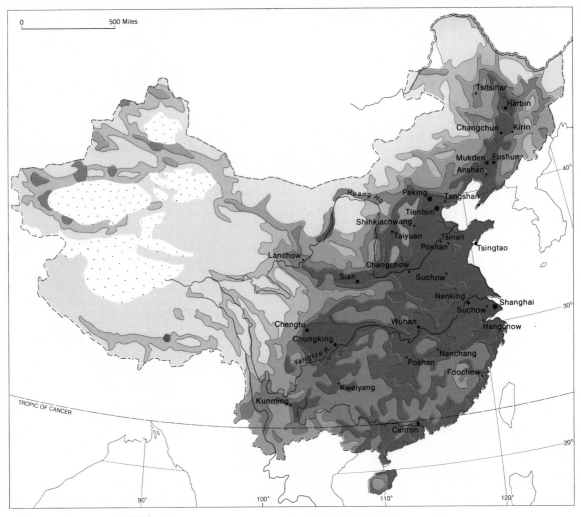

0 500 Miles

TROPIC OF CANCER

Source: C.I.A. Communist China map folio 1967

Figure 15-10 POPULATION DENSITY OF CHINA

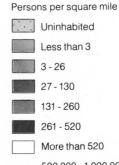

Persons per square mile

Uninhabited

Less than 3

3 - 26

27 - 130

131 - 260

261 - 520

More than 520

· 500,000 - 1,000,000 population

• 1,000,000 - 3,000,000 population

● More than 3,000,000 population

however. Over three-fourths of the Chinese population is found in the east on less than one-third of the land. In terms of distance from the seacoast, this would be the same as having three-fourths of the United States population living between Omaha, Nebraska, and the Atlantic Ocean. But in China the total population within that eastern area is two and a half times the total population of the United States.

Population density in the crowded areas is about 400 persons per square mile, and more than two-thirds of these people live directly from agriculture. Population per square mile of cultivated land is more than 1,600. Such figures are not exceptional; they are comparable to those of Japan, India, the Netherlands, and Belgium. But they are exceptional when applied to a country the size of China, and they indicate the tremendous burden placed on the agricultural system.

Separate, Local Agricultural Systems

China's regional crop specialization provides a measure of environmental growing conditions perceived by the traditional agricultural system. Here, as in India, the population depends on grain as the staple food, and the different crop possibilities result in varied diets from place to place as well. In China the order of grain preference is the same as in India. Rice is the first choice, wheat is second, and millet and sorghum are grown on poorer lands (Figure 15-11).

China's humid tropical and subtropical south is well suited to rice, and two crops a year are possible in many places. As a result production in terms of calories per acre is very high, and the highest rural densities are supported. In the Yangtze valley of central China, rice can still be grown as the chief summer crop, but winter wheat replaces the second rice crop each year.

In northern China in most places conditions are too dry in summer for rice or spring wheat,

and kaoliang, a giant sorghum, is the summer crop instead. They are supplemented with soybeans, grown not only for oil but also as a bread substitute. On the drier western margin of populated China, the crop choice shows adaptation to more arid conditions. North of the Yangtze, winter wheat is the leading crop, and farther northwest where winters are more severe, millet and spring wheat dominate.

Root crops such as yams and potatoes provide the basic supporting crops. Cotton, the major industrial crop, is grown in northern China rather than in the south. In all of these areas livestock are raised either to serve as beasts of burden in the largely nonmechanized agricultural system or as scavengers which provide an important additional food source. Cultivation is intensive, and in most of China the land is worked as a giant garden patch.

Through the centuries this farming system primarily has fed the farm population, particularly the individual family and the cluster of families that occupy the myriad of farm villages of rural China. In the process it supplied the cities and towns, too. Although China has eighteen metropolitan areas whose populations total 40 million people, this figure is only slightly more than 5 per cent of the total of the country. That population scattered through the metropolitan areas of the country has not represented a major burden on either production or transportation. Almost every farm contributed a small amount of production in this commercial system and, in turn, received some cash payment.

COMMUNISTS INHERIT INDUSTRY AND THE RESULT OF TWENTY YEARS OF WAR

With its cities China even in pre-Communist times had an industrial base. But to a large degree foreign investors built that industry

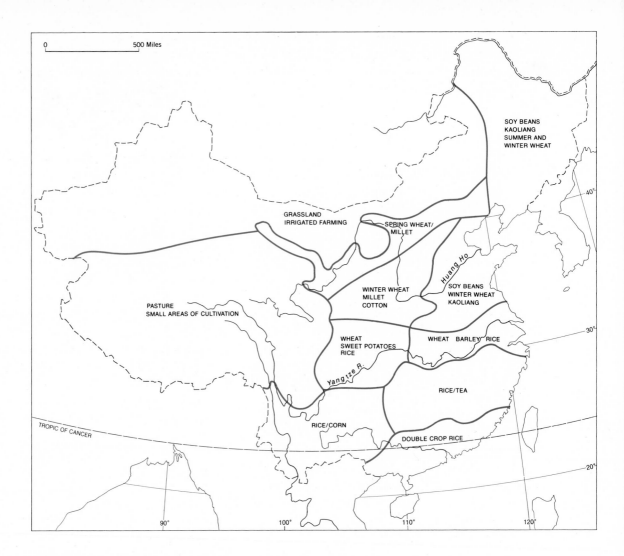

Figure 15-11 Agricultural Regions of China

either to process Chinese raw materials or to provide clothing and supplies to the population, particularly the city workers most involved in the monetary economy. Manufacturing was concentrated in the major ports such as Shanghai, Canton, and Tientsin. During the 1930s and early 1940s, Japan had built a heavy industrial base in Manchuria, relying on local supplies of coal and iron, but its output was destined to supply the Japanese Empire, not China.

This was the system that the Chinese Communists inherited when they came to power in 1949. It was a system that had been suffering through almost twenty years of warfare, first against the Japanese and then against the Chinese Nationalists. Except in Manchuria, investment in industry had been neglected during the period of war. The modest railroad system had suffered from war damage, and national control had been lost in most regions of the country. Local landlords dominated both the political and the economic scene in many areas, and tenants were forced to make exorbitant payments for the use of land. In the postwar period the Soviets who occupied industrial Manchuria systematically "liberated" machinery and railroad cars to rebuild their own war-devastated economy.

The Communist Development Plan

Once in power the Chinese Communists began to consolidate the nation within the framework of the Communist political and economic philosophy of the Soviet Union. They promised agrarian reform to the peasantry, and in the early 1950s the holdings of the landlords were redistributed among the landless families in the villages. But the major economic thrust of the government followed the pattern of the Soviet Union.

Initial emphasis and major investment priority was placed on modern heavy industry as a base for expanding industrial activity and military power. The first five-year plan (1952–1957) assigned almost half of the state's investment to industry, over 80 per cent of this amount for heavy industry. Only 3 per cent of the investment went directly to agriculture.

The Soviets provided the basic industrial support for the new Communist regime by offering trade, foreign assistance, technical advisors, and the opportunity for Chinese students to train in Soviet universities.

Heavy Industry

Chinese strategy called for developing a full, modern industrial base to provide independence from foreign sources of supply along the Soviet model. The program has resulted in successful completion of modern metalworking, chemical, and petroleum plants. The design and construction of jet aircraft, the detonation of nuclear bombs, and the launching of space vehicles attest to Chinese progress.

To accomplish this in a country beleaguered by a shortage of modern overland transportation, the Chinese have relied primarily on the expansion of industry in the traditional industrial centers of the major cities (Figure 15-12). Most important has been the heavy industrial complex in Manchuria centered in Mukden, Fushun, and Anshan. Second have been the great port of Shanghai and the lower Yangtze centers of Hangchow and Nanking. Third has been the Peking-Tientsin capital area. All other major metropolitan areas are centers of regional industry.

In addition, new centers have emerged at sites of new raw-material development such as the iron and steel industry on the upper Yellow River and the Yangtze, and oil refining and petrochemicals at Lanchow on the newly built railroad to Sinkiang-Uigur. The development of new sites takes tremendous amounts of capital,

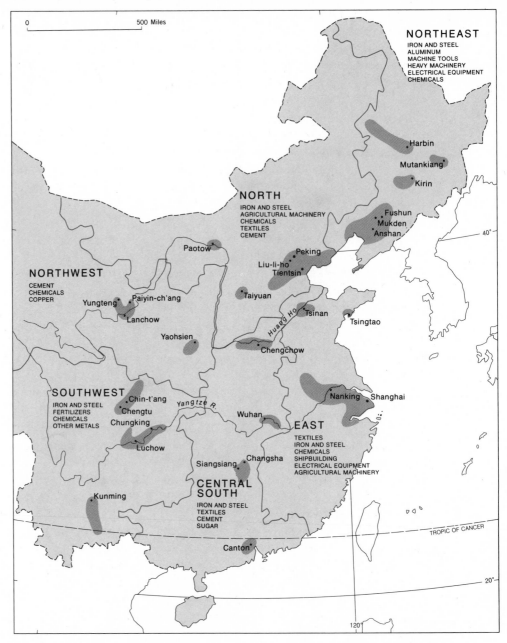

0   500 Miles

NORTHEAST
IRON AND STEEL
ALUMINUM
MACHINE TOOLS
HEAVY MACHINERY
ELECTRICAL EQUIPMENT
CHEMICALS

Harbin

Mutankiang

Kirin

Fushun
Mukden
Anshan

NORTH
IRON AND STEEL
AGRICULTURAL MACHINERY
CHEMICALS
TEXTILES
CEMENT

Paotow

Peking
Liu-li-ho
Tientsin

NORTHWEST
CEMENT
CHEMICALS
COPPER

Yungteng   Paiyin-ch'ang

Lanchow

Taiyuan

Yaohsien

Huang Ho

Tsinan

Tsingtao

Chengchow

SOUTHWEST
IRON AND STEEL
FERTILIZERS
CHEMICALS
OTHER METALS

Chin-t'ang
Chengtu

Chungking

Luchow

Yangtze R.

Wuhan

Nanking   Shanghai

EAST
TEXTILES
IRON AND STEEL
CHEMICALS
SHIPBUILDING
ELECTRICAL EQUIPMENT
AGRICULTURAL MACHINERY

Siangsiang   Changsha

CENTRAL SOUTH
IRON AND STEEL
TEXTILES
CEMENT
SUGAR

Kunming

Canton

40°

20°

TROPIC OF CANCER

120°

Figure 15-12 INDUSTRIAL AREAS OF CHINA

—— Economic-region boundary

Major industrial area

however, and such projects must be carefully planned in advance.

This modern industrial development must be seen, in the short run at least, as serving the government's long-range goals rather than the direct needs of the people. The government determines what will be produced in the factories; priorities, like those of any emerging nation, are toward heavy industry including steel, chemicals, cement, machinery, and electrical equipment. Included are the military needs for gaining political status and security. All the demands of a modernizing state are given first priority over consumer goods. The state's priorities may include fertilizer for agriculture, farm tools, and transportation facilities to get crops to market. Clothing appears on the priority list of necessity, as do bicycles, but cosmetics and automobiles do not.

The new industrialization of China is based on resources from the agricultural segment of the economy. Both workers and working capital are drawn from the rural areas. The private landowners of China have been replaced by the government. As in the Soviet Union, normally, each producing unit receives a quota before harvesting takes place, and the result, when totals from the whole country are added together, is a massive supply of agricultural produce to be turned over to the government. Most of the government's share is sold to city workers, and some may be exported. In either case it produces money that can be used in the industrial development program.

THE SAME AGRICULTURAL SYSTEM, BUT A STRUCTURAL AGRICULTURAL REVOLUTION

All agricultural growth has been achieved without fundamentally changing production from the intensive garden-patch agricultural system. In fact, the methods of farming look the same today as they have for centuries. The change has been in control. As we have seen, the Chinese Communists promised land reform, and in the early days of their control accomplished it. By 1956 all farmers were being encouraged or coerced into participating in producer cooperatives where they pooled their land and worked the combined lands in work teams. Returns were a share of the harvest proportionate to the amount of land contributed.

Gradually, the basis for sharing in the cooperative changed to the amount of work done by members of the cooperative. Soon "advanced cooperatives" were established in which all farmers turned over titles of their private land to the cooperative, and the basis of compensation became the time a man worked for the cooperative. In the process this changeover eliminated the advantages of the large- and middle-sized landowner and put the landless and small owner on the same basis. Gradually the cooperatives became collectives which held land rights, equipment, and livestock collectively rather than individually, and where the designated leaders of the collective made management decisions.

In 1958 the Central government moved even one step further than the Soviets. In what was called the "Great Leap Forward," they formed communes by consolidating collective groups into units with 18,000 to 20,000 people and brought the inhabitants of numerous nearby villages together. These communes were planned to be far more than agricultural entities. They were to be social and cultural units with supporting educational and social services. They were also to function as the local units of government. And finally they were to undertake industrial projects, build roads and dams, and carry out such developmental operations as reforestation and land reclamation. Under the system even traditional family groups were reorganized with separate dormitories for men and women and communal dining halls. Special boarding schools for children were established. Workers received free food, clothing, and a small salary. Women were expected to do much of the field work in agriculture, in order to release men for work on irrigation facilities, roads, and factories. This increased the available work force.

But such a program was far too great a departure from traditional Chinese life to be fully accepted by the people, and the plan was too hastily implemented to function properly. Many of the rural people resisted these changes. Moreover, there were droughts in some areas and floods in others. Shortages required China to import wheat and rice.

Faced with the failures of Great Leap Forward, as in the Soviet Union, pragmatism replaced idealism, and the government was forced to backtrack on its ambitious schemes. It abandoned communal living and allowed families to have private plots of vegetables, other cash crops, and livestock. Traditional trade fairs were revived for selling the surplus. Much of the working management shifted from the larger commune level to more local production brigades and teams that really represented the old collective and cooperative organizations. Ownership was returned to the cooperative level, and each team was guaranteed the fruits of its production. Thus, there could be differences in living standards among the individual production teams within the same commune.

The Commune as the Primary Institution in the New China

Still, the objectives of the multipurpose commune are fundamental to Chinese Communism. Long-term aims call for elimination of individ-

ual incentives and encouraging people to work for the goals of the state and commune. Chinese Communism aims to help those people without political power, without a high level of education, and without material wealth. These people are the rural majority of China.

In contrast to every other industrial country, the cities of China are not seen as the high points of civilization. The Chinese are striving to minimize the level of urbanism and to move many essentially urban functions including industrialization and government control to rural areas. The commune is designed to serve as the basic organizational unit.

In carrying out this rural program, China, unlike India, has worked to gear its price structure to the needs of the rural people rather than the city population. The creation of locally based commune industry is an attempt to provide inexpensive industrial goods to rural areas. Communes are to be as independent of city factories as possible. Local industries are seen as a part of the policy of integration of industry and agriculture and of town and countryside, and communes are expected to supply their own basic nonagricultural needs as well as food. (City factories supply national needs.)

Communes are recognized as functioning entities of local government. Moreover, they are social and cultural units operating their own educational and medical facilities and their own militias. Management cadres work one-half day in the fields as well as in the office, so as not to lose touch with the masses. Everyone is encouraged to work for group achievement, not individual success.

Thus, each local area at the level of counties, not provinces, is to build from the commune base and become a self-contained economic, cultural, and political unit. Ideally, it should provide for most of its own economic needs and most of its own management and decision-making—within the strongly centralized guidelines and authority of the central government, of course.

Counties are encouraged to establish integrated local industrial systems with their own iron and steel plant, collieries or other power sources, agricultural processing facilities, metalworking and cement plants, fertilizer works, and insecticide and consumer-goods factories. Many of these will supplement commune-run operations, and others will depend on commune workers as sources of labor.

Since the Cultural Revolution in the late 1960s, the central government has also been trying to replace elite education with universal education of the people. People trained in universities and technical schools are expected to return to rural areas to work. The government wants to end the traditional Chinese reverence for scholarship and professional pursuits. The emphasis formerly placed on higher education is now placed on practical skills, not on classical studies. The Chinese tradition of an intellectual elite is being reoriented to encourage functional education for all. Communes have public schools for all children, and more and more students in higher education come from the communes. All of this fits the plan to modernize the life of the largely rural population without a massive movement of people to urban areas.

Thus, the Maoist cultural revolution aims at modernizing China not according to Western models but within the value structure of traditional Chinese culture. Chinese leaders plan not just to bring China into the modern world but to provide an alternative form of modernization to the world. The Chinese experiment becomes significant not only to other underdeveloped areas but to the industrialized countries of the world as well.

INCREASING AGRICULTURAL OUTPUT

The traditional Chinese agricultural system, with its intensive methods concentrated on the best land resources of the country, has been a productive one. Yields have always been much higher than in India. Still, the Chinese Communists have increased agricultural productivity appreciably. A small amount of additional land on the drier margins of agricultural China has been put into cultivation by using more mechanized methods and by irrigation and other land reclamation projects. But essentially the increase has come through the application of the same methods used in India: planting improved crop strains, carrying out scientifically approved farm practices, and developing adequate storage facilities.

But the Chinese control the basic resources of the country, especially the agricultural land, whereas in India each farm property is in private hands. The commune system cuts down sharply on the number of contacts that the government has to make in carrying out development plans. In addition, communes are managed by leaders who are commonly not only accepting of new methods but are also trained by the government. The government can be sure that developmental programs will be carried out on each commune to the maximum ability of the commune leadership. The variables are the talent and manpower of the local commune and the particular environmental conditions of the year; but they do not include the individual farmer's conservatism or his individual poverty as in India.

As in India, the problem is not so much knowing what to do as how to carry it out on a vast scale, but China has major cultural and historical advantages. Chinese culture has placed a premium on service to the state and

SMALL-SCALE PRODUCTION
ON THE BYWAYS OF CHINA

Plans for the rural areas of China call for integrated regional economies on the local level rather than at the provincial level. Since agriculture has been dominant and has always provided local support to its own population, the problem of regional integration is to provide essential industrial raw materials, fuel sources, and manufacturing facilities.

Fuel for the local economies comes from two primary sources: local coal mines and small hydroelectric plants. In Yunnan Province in southern China in 1970 there were an estimated 2,100 small coal mines with a total production of 1.8 million tons, and Chekiang Province south of Shanghai had over 1,200 small mines. Several Chinese provinces each claim over 2,000 small hydroelectric power stations, with Kansu at the headwaters of the Yellow River having more than 4,400. Electric lights have replaced the pine and bamboo torches used for lighting in remote rural areas, and radio networks tie together out-of-the-way villages in the highlands.

Manufacturing is centered around four small industries essential to local needs: iron and steel manufacture, machine building (especially farm machinery), chemicals (particularly fertilizers and insecticides), and cement works. Other small-scale industries include agricultural and timber processing, textile and clothing manufacture, papermaking, and pharmaceuticals. The idea is to use local materials, manpower, and funds.

Emphasis is on small but complete local industrial systems. In Chiyuan County, Honan, such a system includes an iron and steel plant, collieries, sulfur mills, machinery plants, a metallurgical plant, and plants or mills producing refractory materials, cement, chemicals, ceramics, knitwear, and paper among twenty county-run plants and mines and seventeen commune-run factories. In addition production brigades on communes have their own farm-tool shops, small coal pits, and workshops processing agricultural and farm by-products. In Fencheng County, Kiangsi, a system comprising iron and coal mines, a coke oven, iron-smelting furnace, steel-making furnaces, and rolling mills produces several hundred tons of rolled steel. In Kiangsu Province two-thirds of the counties have small nitrogenous fertilizer plants, and half of them have small phosphate works.

Output for the county in 1969 included 4,000 tons of pig iron, 99 tons of steel, 6,500 pieces of farm machinery, and some 1,300,000 small farm tools.

Given the wide range of environmental conditions over China, what problems would you anticipate in developing real local economic self-sufficiency?

Are there problems of small-scale production? Can steel be made as effectively in small plants as in large? If so, what are the advantages of the large-scale steel plants that have developed into some of the largest industrial mass-producing facilities in countries such as the United States, the Soviet Union, and Japan? What problems will China have to face as it moves in this direction?

on the development of the organizational and communications systems toward management by the government of the country. The government can take advantage of a largely homogeneous culture and a common written language, plus great national pride in China's heritage and destiny. The Communist regime seems to have capitalized on these most valuable assets.

The Chinese system operates on two different levels, each loosely, but importantly, interconnected. The lives and support of most of the people are on a local level, that of the county, the commune, or even the work team. The national governmental system with its implications for the economy and the social and cultural life of the people works from Peking, the capital, to the twenty-two provinces, the five autonomous regions, and the municipal districts of the three largest cities—Shanghai, Peking, and Tientsin. Plans for the political, economic, and social functioning of the entire Chinese state are made in Peking. Shanghai and Tientsin play special roles, too. Shanghai, the largest port, is China's most important international center and the major focus of manufacturing and internal commerce. It functions somewhat as New York City does in relation to Washington, D.C. It is the largest metropolitan area in the country. Tientsin is the combined port and industrial base closest to Peking. It is the country's leading chemical center and has a wide range of both heavy and light industry. Most of the government programs filter down through the capital cities of the provinces and the five autonomous regions.

Peking's ties to the major provincial political capitals are provided by the country's transportation and communications links. The most important of these are within the populous areas of China. Thus, in contrast with the United States, the busiest links are those on a north-south axis rather than east-west. The first-order of interaction is between Peking and centers within heavily populated eastern China. Next connections are made from Peking to the outlying areas in the west—Singkiang, Tibet, and the others.

At the local level traditional life within China has been centered on the individual family, the village, and the local market. But this has been in terms of a dominantly agricultural system. As we have seen, the Communists have made conscious efforts to change agricultural villages into collectives with industrial, trading, and social services. Communes and counties within provinces have largely self-contained economies and political structures. Each area is to contribute food and industrial resources to the country as a whole and, of course, each political unit must follow national policies.

Unlike India the Chinese communist plan for the rural areas of the country has been to revolutionize not only agricultural production but the whole of rural life. Ideally the communes were to provide the basic industry, services, and amenities of urban life. Moreover, traditional symbols and styles of living—the family and the village—were to be replaced. Patterns of social and economic activity drawn from the modern, secular world were imposed on the traditional society with its emphasis on the sacred and the status quo. To provide direct ties with the modern world, doctors, technicians, and scientists trained in the cities are expected to take assignments in the communes.

With such revolutionary goals, it is not surprising that the Great Leap Forward and the Cultural Revolution resulted in opposition and turbulent unrest. The Communists were trying to accomplish in a few years what the industrialized nations took years to complete. Moreover, they aimed to resolve contradictions between rural area and city, agriculture and industry, that even now exist in other modern countries. It is remarkable that a country which faces as many problems as China does would try experiments not tried elsewhere.

Going It Alone

For the present, China seems to have solved the population problem. There is no famine or serious food shortage. Birth control clinics in the communes appear to have the rate of population growth under greater control than in India. But still the population growth is tremendous in absolute terms—almost 14 million more people each year. As in India, the absence of famine and starvation is one thing; raising the living standard of such a huge population is another. This is the ultimate test that will determine the success of the Chinese communist model of development. Still we must recognize that the government is making very conscious and significant efforts to equalize the living standards between the rural and urban population and between rich and poor.

In terms of development, China more than India has depended on its own resources with regard to both natural resources and sources of investment capital. Since the United States supported the anti-Communist Nationalist regime during the civil war that followed World War II, it still has not formally recognized the Chinese Communist government and for many years blocked China's entry to the United Nations. Moreover, during the Great Leap Forward and the Cultural Revolution, China made an ideological break with the Soviet Union. Thus, China is cut off in terms of economic aid and technical assistance from the two greatest powers of the world today. While most European countries and underdeveloped countries recognize China and carry on trade and diplomatic relations, they have contributed little to the nation's vast capital and technical needs, and China has been forced into becoming self-sufficient.

However, through trade and aid with underdeveloped areas, particularly African countries, China worked to develop its own sphere of influence in the world. It is building a 1,100-mile railroad for Tanzania and Zambia. But the greatest trade is with Japan, its nearby industrial neighbor. China exports textiles and clothing, metallic ores, rice, tea, and coal, and imports grains and machinery.

In the 1970s the isolation of China from the power circles of the world began to break down. In 1971 China was admitted to the United Nations as a permanent member of the Security Council where it sits along with the United States, the Soviet Union, France, and Britain. In 1972 the President of the United States visited China in an effort to open up channels for communications and trade.

In the past ten years China has achieved a position of political prestige in the world far above that of India. China's political influence will probably continue to grow. In economic terms China has also made considerable progress since the end of World War II, but the road to modernization is difficult and many challenges still lie ahead. In the next twenty to fifty years, the stories of India and China may be the great ones of the world. Their success or failure may well determine the future of all mankind, simply because more than a third of the human race will be involved.

SELECTED REFERENCES

India

Brown, Joe D. *India*, rev. ed., Life World Library Series, Time-Life, New York, 1967. A general introduction to India.

Lewis, John P. *Quiet Crisis in India,* Brookings
Institution, Washington, 1962. India's economic problems are highlighted in a scholarly
study.

Neale, Walter C. *India, the Search for Unity, Democracy and Progress,* Van Nostrand, Princeton,
N.J., 1965. An introduction to India and its
problems.

Spate, O. H. K., and Learmonth, A. T. A. *India
and Pakistan,* Methuen, London (Barnes and
Noble, New York), 1967. A tremendous compendium of material on the geography of
these countries and on their regions.

Whyte, R. O. *Land, Livestock and Human Nutrition
in India,* Praeger, New York, 1968. Focus on
the problem of feeding India's population.

China

Buchanan, Keith. *Transformation of the Chinese
Earth,* Praeger, New York, 1970. A China
scholar looks at the changes under the Chinese Communists.

Donnithorne, A. *China's Economic System,* Praeger,
New York, 1967. The economics of Communist China.

Shabad, Theodore. *China's Changing Map, National and Regional Development 1941–1971,*
Praeger, New York, 1972. A geographer
known for his detailed regional studies ex-
amines the growth of the Chinese economy.

Traeger, T. R. *A Geography of China,* Bell, London, 1966. A basic geography of China.

Tuan, Yi-Fu. *China,* Aldine, Chicago, 1965. A
humanistic and cultural geographer looks at
China.

Chapter 16 Epilogue:

Implications for the Future

HAVING EXAMINED the present world from a geographic perspective, we are left with the question of what the future holds as these different systems of spatial organization continue to develop, interact, and modify each other.

The Challenge of the Environment

Throughout history, technology has been concerned with finding new ways to manipulate the environment. The creation of the modern, interconnected system which is not tied to a geographic base and which can protect itself against most natural disasters attests to man's ingenuity. Only recently have the problems of managing the environment been realized. The future may indeed suggest that the dangers inherent in the interconnected system may be too great and that the nonindustrialized countries of the world should not follow the same road.

In the traditional society a man's ability to alter the physical environment is limited. Even in India and China where dense human populations have re-engineered almost the entire landscape, life is still directly affected by the exigencies of nature. A severe drought, an insect plague, or a typhoon will cause suffering and disaster by diminishing the local reserves of life's essentials.

Throughout the traditional system, lack of technology extremely limits man's effectiveness in managing environments. With only man-power, animal power, and simple hand tools, production is an unsure undertaking. Planting and harvesting are so slow that only small fields can be harvested. Rarely are there large surpluses of agricultural products that can be saved from one growing season to the next. Moreover, without modern storage facilities, surpluses that do not rot are often destroyed by predators.

Under such circumstances man depends a great deal on the environment for his material existence.

Culture also has adjusted to, or developed around, environmental conditions. As in Ramkheri, festivals commonly come at harvest time, weddings and other important events such as pilgrimages take place during dry seasons or winter, and special religious ceremonies occur at planting time.

Because traditional societies are limited in their ability to manipulate the environment, they are less likely to damage the ecological balance than technologically advanced societies are. Wastes from the activities of man are largely organic in origin, and the environment is usually capable of decomposing and recycling them. Durable items, such as houses, are difficult to produce and are kept for generations. When buildings are destroyed, their materials may be reused in other structures. People often make multiple use of a stream—for bathing, drinking, watering domestic animals, and waste disposal. If the contamination is not too extreme the population may build up enough immunity to continue using the water. Where populations are dense, such multiple use of water supplies can, of course, cause severe epidemics.

Among modern societies oscillations in the performance of the physical environment are minimized over seasonal or annual periods. The technological sophistication of these societies in improving plant growth, controlling pests, reshaping the land, and relocating water has greatly increased productivity of the environment for human use, though often at the expense of other life. The capacity of storing reserves also gives these societies security against famine. This security is even greater now that men can share the resources of environments of the entire world and need not rely on only their local environments.

The cities exemplify man's greatest insulation from the environment. Only the severest snowstorm disrupts the routine of the city, and even then it is usually only for a few hours. Where winters are extreme and where there are great amounts of snow, cities generally anticipate such conditions. Only the highly unexpected brings chaos to the urban scene. Man has created his own climate over most of the city. He has remodeled the landscape with bulldozers, filled in vast tracts of low-lying land, and dug tunnels for subways and utilities.

When a breakdown occurs in the city, it often is the result of inadequate planning. Lack of sufficient electricity can be a problem during hours of peak consumption and faulty equipment or an accident can cause a break in the system. These breakdowns are not merely the result of technological failures; they are partly the result of priorities. A society that can go to the moon could, for example, produce more energy if it wanted to.

POLLUTION AND RECYCLING

Although the environment has only an indirect impact on man in the urban world of cities, urban man is an increasing threat to the environment. The problem is the result of two factors: the crowding of vast numbers of people in city and metropolitan areas, and the tremendous material consumption of these metropolitan systems. In fact, the modern system consumes resources at such a rate that some people fear the basic resources of the earth's crust will be exhausted.

More recently the problem has taken on a new dimension: tremendous consumption has resulted in great quantities of refuse and abandoned structures in metropolitan areas. These wastes accumulate faster than the physical environment can decompose them, increasing the problem of disposal.

Perhaps the greatest problem is the waste that

occurs in the production of energy. Automobiles, jet airplanes, and thermal electric plants are outstanding examples of energy generators. Energy transformation creates waste and heat which are discharged in high concentrations into the atmosphere or bodies of water. People generally assume that the atmosphere and water systems will absorb and diffuse the wastes. Recently the automobile exhaust smogs over major cities and the pollution of water has made people increasingly aware that the concentrated loads of wastes going into the air and the waters of our cities have reached such high levels that the cleansing action of these systems is not adequate to do the job. The consequences if these conditions continue are not inviting. The inputs have been so great that changes in the character of the earth's atmospheric and oceanic systems have been observed, changes that might make the earth uninhabitable.

The tremendous consumption of material resources in metropolitan centers is also a severe problem. Vast quantities of other consumer, industrial, and business wastes must be disposed of. Unlike the wastes of traditional societies, much of this is not organic and will not decompose readily. For the most part the solutions of modern man have been the time-honored ones of burning, burying, and abandoning, each of which creates additional problems of pollution. Gradually it is being recognized that such a use-and-discard philosophy depletes resources and compounds environmental pollution.

An alternative view is to regard the wastes of metropolitan centers as growing stocks of reusable material resources that will have to be the mines of the future. Think of all the steel that rests in junked cars, machinery, and the other metallic refuse of cities. As the rich, easily accessible deposits of minerals are depleted through mining, the dumps of metropolitan areas may become the richest metal lodes in the world. The organic waste of cities could be recycled as fertilizer, enormous amounts of paper could become the source of raw material for new paper manufacture, and used glass could become a large resource for making new glass.

Recycling might well supplant primary mining and pulp production as they exist today. Vast recycling operations might spring up around major metropolitan centers. These would, however, require a basic reorientation of modern mineral- and paper-processing facilities. In the United States, for example, the paper industry has a basic part of its production system located near forests for pulp production, but such locations would be unsuitable if waste paper replaced pulp as the basic raw material, as we saw in the potlatch industries example. Neither New York nor New England has any sizable primary steel plants, yet they are primary resource locations for scrap steel. Also, the fertilizer industry is closely tied to the availability of mineral fertilizers and to the farm market. Each of these industries would require a reallocation of capital resources.

There is no question that economic logic and the goal of waste management provide attractive arguments for recycling solid wastes since large urban centers are simultaneously the location of the resources and the market. But the consequences of reorganizing the locations of such industries as paper, steel, and glass as part of recycling systems to handle urban wastes would have economic, social, and political ramifications that should be carefully considered. Imagine the repercussions of relocating the paper-processing industry. This decrease in employment opportunities in rural areas would also lower the tax base, increase unemployment, and encourage young people to move to the cities. Such a switch would also intensify the concentration of activity around cities and result in the premature abandonment of rural facilities. It is questionable whether the economic and environmental advantages would be sufficient to justify the social and political problems likely to arise if more emphasis is placed on city living.

The industrial gain for the urban centers would be accompanied by still greater crowding, more traffic congestion, further reduction of open space, and greater sprawl. From a geographic context, the concentration of facilities for recycling of wastes, despite its great economic attractiveness for industry, could well create other, more serious economic and social consequences for the public. Perhaps the alternative of relocating the solid wastes rather than the processing industries would make better sense, even though it is not the most economical solution. Another alternative is to cut down drastically on the total consumption of materials that create solid wastes. Imagine the possible ramifications of such a solution!

From this example it is clear that in considering any change in the environment, we must anticipate the consequences of the alternative courses of action. The problems of pollution, resource management, industrial development, and the management of urban environments cannot be viewed as isolated factors independent of one another.

Two Systems and the Destiny of Man

A major difference between the traditional society and the modern system is in the degree to which men have control over their destinies. In the traditional closed society a place may be dependent on the whims of nature, but decisions on how to live with the environment and further develop the place lie with the group that occupies it. Since outside contacts are of secondary importance, the local group must rely on its own capabilities and knowledge in deciding how to use the environment. Often the

range of choice is limited by technological inadequacies and a narrow cultural perspective, but the local group does, in fact, chart its future course.

In the modern, interconnected system people have lost primary control over their destiny. Decisions are made by the central government and by large corporations with headquarters far away. Decisions concerning the development of mineral resources, the establishment of manufacturing plants, and the development of resort areas rarely are made by people of the locality. Scientific survey and extra capital are often beyond the means of the local people. They are dependent on governments and corporations. The future of a community depends on someone else's decision to build a new plant or close an old one, to develop or abandon an existing mine, to create a beach resort or ski slope. Many communities rise and fall as outside decision-makers dictate.

In the United States it is possible to find communities in the same region that show very different degrees of development. In one place, outside interests have made major investments in production or recreational facilities, and the community is booming. In the other, an old investment has been abandoned, and the community is a ghost of its former self. Some local people may have contributed to the decisions. An active chamber of commerce may have made an effective presentation which attracted a new plant to the booming community. The people in the declining town may have been less aggressive, or their town may have been less attractive because it was built to support another sort of activity. In any case the local people play minor roles compared to the decision-makers from outside the area. This is also the case in state-controlled systems, like those of China and the Soviet Union. Restricting the number of decision-makers theoretically allows decisions to be made more efficiently with regard to production and the social welfare of people. But here again the voice that people have in their destinies is limited.

In the interconnected world local communities do not determine their own cultural ways either. Fashions are established by the manufacturers of ready-made clothing. Styles in a country the size of the United States and in most industrialized countries are becoming homogeneous. People in every community are exposed to almost the same set of cultural values. Regional variations in dialects, cuisines, and even moral values have been affected by network television, nationwide supermarket chains, and popular magazines. With the use of communications satellites, people with access to television and radio throughout the world can view the coronation of a king, the funeral of a president, and an international soccer match. Music, news, films, and commercials all set standards. Even the values taught in schools and churches are largely determined by hierarchies at distant locations.

Modern society has also managed to protect itself against environmental dangers. The modern, interconnected system operates on such a grand scale and has such an extensive communications system that any community struck by a calamity can be supplied with the necessities of life almost immediately.

Environmental matters are not man's only concern, however. He is now threatened by breakdowns in the systems he has created. Strikes, wars, equipment failures, human error—these have become major threats to the modern world. Man's ability to manipulate the environment is also a problem. Some changes have been so great that they threaten the existence of life. Too much carbon dioxide is entering the atmosphere, life in streams is being destroyed by nitrogen and phosphorus, and animal life is being seriously harmed by the use of insecticides. The modern complex system provides one form of security, but we have that security at great cost.

The Establishment and the Individual

The irony of the modern system is that it requires more conformity and also provides more freedom of choice. In the traditional system alternatives are limited by the knowledge of the group. Generally individuals are not literate and cannot afford radios. The range of personal contacts is small, occupational choices are few, and, more important, the group puts a high price on conformity. There is little possibility for dissent because major attention must be placed on simply surviving. Work occupies virtually all the time of the able-bodied individuals in the community. Little time is left for philosophy, and should a new idea emerge, the group's conservatism would question the risk of trying something new. A farmer in Ramkheri may be told that a new strain of rice or a different set of practices holds the promise of much greater yield. But can he be sure? He knows the old variety or practice does not produce any surplus, but it does produce. If the new variety fails, he and his family will go hungry. There is no room for error, since there is no supply of grain laid away.

Yet the individual in the traditional society still retains much of the decision-making for his own welfare. Each farm family is essentially a separate enterprise, and the head of the household is the entrepreneur. In the same way the local craftsman fashions unique items, but each item is essentially the same because he has limited capital for experimentation and is producing for a limited market.

In the modern, interconnected system many workers perform repetitive tasks. Company officers make all but routine decisions, and factory and office work is so subdivided that the individual worker has almost no opportunity for individual action, much less individual creativity. The situation is typified by assembly-line work which calls for workers to operate or monitor machines that perform the same individual task minute after minute, hour after hour, day after day, year after year. A punch-press operator's task is to turn out each day as many uniform objects as time-and-motion experts and labor representatives have agreed is reasonable to expect. There is no variation; the operator has become part of the machine. It is the same with a typist, a punch-card operator, and a mailroom clerk. It is not uncommon for a given worker to spend thirty or forty years doing essentially the same thing each working day. Gone, in this modern system, is the variety that the native craftsman finds in taking a product from raw materials to finished form. It is also difficult for the regimented workers to have pride in their accomplishments.

But the modern worker may work only eight hours a day for five days a week or, increasingly, ten hours a day for four days a week. For two or three days each week, on increasingly more holidays, and for two to four weeks of vacation, he is free to do as he pleases. Moreover, he has four to eight hours a day between work and sleeping. Such a worker is well educated by world standards—he can read and write. He is affluent when compared with traditional workers. He has the luxury of time to recreate in the way of his choice and is free to think whatever he wishes. He may be conditioned to sit in front of television absorbing presorted thoughts, but even there he is exposed to a greater variety of ideas than is the individual in the traditional system. Network news gives him some idea of what is happening in other places, and if he is interested, he can read more about it. He hears discussions of war, economic measures, new taxes, and proposed laws.

REACTING TO CHANGE

In both traditional and modern systems accomplishing change is difficult. There is little question that innovation does not come easy in traditional society. It is only when national leadership decides that changes should be made, as in Japan in the nineteenth century, that significant ones take place within traditional society.

We often assume, however, that change is indigenous to the modern system and that technological progress is indeed welcome. Social and political changes, on the other hand, are not. In the United States the political boundaries of most cities remain unchanged through the decades, even though metropolitan areas— the geographic cities—have expanded far beyond boundaries created in the nineteenth century. County and township governments designed for rural areas in days of horse-drawn transportation remain unchanged in the jet age. An important segment of the voting population of the United States speaks of returning to the democracy set down by the Constitution for a rural society of almost two hundred years ago. In Europe the need for political change to accomodate the Common Market is confronted by nationalistic traditions within existing political states.

Not only are the traditionalists within modern society often in the positions of power, but the institutions themselves also have deep cultural roots. Such institutions, operating under the burden of cultures heaped with tradition, markedly limit the efficiency of modern technology. As a result new technology produces more and more problems. One of the ironies of the current world is that we can articulate many of the basic solutions to today's problems, but we cannot implement these solutions because of our cultural and institutional conservatism. Theoreticians tell us that cities can be made livable, that pollution can be controlled, and that world peace is possible, but we cannot solve the human problems thwarting the execution of such plans. It is far easier to carry out new technological missions such as putting a man on the moon than to undertake basic changes in existing human structures on earth.

How Are Choices Made?

In traditional society knowledge of the potential of the environment is extremely limited. Moreover, with existence dependent on animate sources of transportation, only local resources can be used. Commonly the labor force is underemployed in terms of what it might accomplish if it were highly organized. It is this sort of underemployment that the Chinese have been attempting to reduce by their reorganization of communes. There is also a problem of labor management in the modern system. Normally an employer has numerous applicants for a job, and he might well have more if the full labor force could be mobilized. All economies in the modern system, even those closely controlled, have unemployment. And automated machines offer ready alternatives to human labor.

Also there are usually many alternate sites where a given economic activity might take place. Rich oil deposits, for example, have been found in many parts of the world, and production must be allocated among them. A manufacturer of textiles or electronics seeking a new plant site where there is cheap labor has hundreds of small towns in the South from which to choose as well as locations in other countries such as Hong Kong and Korea. Steel mills, automobile plants, and aluminum plants can also locate at a variety of sites. Although the specifications for giant installations narrow the range of possibilities, and although plants are located on the basis of precise industrial location surveys, the number of optional locations is still large. Often the actual decision depends on noneconomic factors such as the impressions gathered by decision-making company executives as they play golf with the president of a local chamber of commerce. However they are made, such decisions have a tremendous effect, both on the areas chosen and on the areas by-passed.

What Next?

For many years the "progress" that created the modern, interconnected system was believed to be the solution of every problem. Now it is clear that the interconnected system has created larger problems than ever existed, and many people wish to withdraw from the modern world. They want to re-create the "good old days" when people lived a rural life, ate "organic" foods, and loved one another. Our examination of traditional society indicates that such a life has major problems, too, and there are few truly untouched traditional communities left. Virtually all places have some tie to the modern system. Of the people living in a traditional society, some have a small amount of produce to sell, others receive money from children who leave to work in cities, and others receive care from a mobile medical team. In many cases these ties have broken down the system in such a way that the native population cannot return to the former life of the entirely closed system. Often there are too many people to support, or the land has been overworked as a result of commercial production. Even in the most isolated places people have developed dependence on the outside world for certain products.

JUXTAPOSITION OF TWO WORLDS

In virtually all parts of the world, both of the systems can be found. We saw them in the United States, where despite the dominance of the modern system, there are regions still tied to a traditional economy—even within the inner portions of the largest metropolitan centers. This is also true in other industrial countries.

In the underdeveloped regions of the world, there are dual economies wherever the modern world has imposed itself. Islands of the modern system abound in the midst of vast areas of largely traditional, often tribal, society. In large nonwestern cities the two systems exist side by side to an even greater degree than they do in the modern industrial countries because the city slums in the underdeveloped world are populated by people who are not supported by the modern system through welfare payments. Instead they support themselves "off the land" as they did in their rural homelands by growing garden patches, keeping scavenging animals, and gathering the wastes thrown away by the urban society.

It is in large metropolitan areas that the two systems are found in closest physical juxtaposition. But, despite their physical proximity, the two worlds may remain essentially apart. The gap between those in the slums and the wealthy is a tremendous one whether the poor live in the inner city and the rich in the suburbs as in the United States, or whether the slums are on the periphery and the fine homes close to the center of town as in most underdeveloped countries.

In any metropolitan area, however, more particularly in the cities of the underdeveloped world, one can find individuals and families at every stage between the extremes of the traditional and modern societies. Often the lot of these "in-betweens" is the most difficult of all. They are dissatisfied with traditional life and often refuse to have further contact with it, yet they are on the outer fringes of the modern system and are not fully at home there either. This group includes the Indians who have come to Latin-American cities, the European immigrants and the poor blacks from rural areas who have menial jobs in American cities, and the West Indians and Pakistanis in British cities. In many cases these unfortunate individuals are not really a part of either system. They cannot return to their old ways, nor is there much prospect for them in the urban world. Their chief hope is that their children, raised in the city and educated to its needs, will rise to full membership in the modern society in the next generation. This goal was accomplished by many European immigrants in the past century. But the modern system of the twentieth century has far more individuals trying to work within it than are essential to it.

The Universal Culture

The new culture is becoming world-wide. It is present not only in the developed industrial countries of North America and Europe but also in the cities of every continent, virtually in every country. But unlike the traditional, the modern culture has not spread out from a single center over the landscape to encompass a given region of the world. Rather, using ships, airplanes, and telecommunications, it has followed a hop-skip pattern over the world, by-passing vast areas in the process. This new culture is emerging right in the midst of each of the major culture realms of the world. At the present time

What are the limits of the earth environment supporting mankind? It is obvious that the planet Earth is finite and contains only so much material for use. But at present there really is no way to measure its natural resources because components of the earth environment become resources only as man finds them useful. Thus, the limits of the finite earth environment are only part of the limits question.

Equally important is the question of the limitations of man's knowledge of the resource possibilities. How nearly complete is man's knowledge of the earth and its possible resources? How completely developed is man's technology for using the earth? And to what extent will man be able to tap the resources of the earth environment or even the environment of the universe? At present we have reached the moon, but have only been able to drill six miles along the earth's 4,000-mile radius. We really have little comprehension of the earth's finite limits.

it is causing considerable friction between old and new in many places, but in the long run it may provide a basis for greater understanding among nations.

For the time being the bridgeheads of the new culture are the large metropolitan areas of the world. The transportation and communications networks are oriented toward the cities. The big metropolitan areas are most closely linked with one another, and it may be that Washingtonians have more in common with people in Peking than they do with people in rural Virginia.

Culture diffuses from one node in the network to the others, finally embracing business and financial transactions, political affairs, clothing fads, music, and social and moral values as well. The inputs to modern fashion that spread from metropolitan area to metropolitan area are often drawn from the varied traditional cultures of the world, but they are put together by modern designers in ways different from those in which they originally appeared, the better to appeal to modern cultural tastes. The contrast in dress between city dwellers and people in rural areas is marked in all areas of the world. In Latin America and Africa a person's place within the modern system can be judged not only by his ability to speak European languages, but also by his clothing. In the United States the ghetto dweller quickly adopts modern dress to prove that he is part of the urban scene. Links between major cities are stronger than ties between some cities and more remote service areas.

The implications for the world today and the future of the modern, interconnected system are basic to life on earth. The culmination of the traditional system with its limited transport and communications possibilities was the nation-state. This state spread governmental control over a single contiguous area, commonly representing the territory of a given culture group.

Today the nation-state is perhaps the most fundamental symbol of the cultural and economic unity of a region. The areas that have emerged from colonialism since World War II have established themselves as nation-states in the mode of European states in the nineteenth century. In the past thirty years, more than sixty new nation-states have appeared in the world.

Yet, from the point of view of the modern culture, the nation-state may be outmoded. It certainly seems to be archaic in economic terms. The modern, interconnected system operates on a world basis, not a national one. Goods, capital, and people regularly move from metropolitan centers in one country to those in another, from one continent to another. In so doing they must face obstacles to movement such as quotas, duties, passports, visas, and changes in currency.

But governments of nation-states are maintaining national rivalries that necessitate armies and military preparedness. This activity supports a large component of modern industry—the military-industrial complex, as it is called in the United States—and diverts capital to essentially nonproductive investment. One irony of the modern world is that new nation-states in Africa and Asia, jealous of their sovereignty, support armies using modern military equipment, yet they have no production facilities for modern armaments themselves. They must purchase such goods or receive them as "aid" from modern industrial countries. Moreover, it is trouble in underdeveloped areas such as Vietnam and the Middle East that has produced some of the great crises of the past twenty-five years. In industrial countries militarism has caused world wars and led to development of nuclear devices and intercontinental missiles capable of destroying all life.

The modern industrial system has adjusted to the existence of political nation-states. The leading industrial corporations and financial enterprises have capital resources and manpower expertise greater than most countries. Those that function on an international basis take advantage of the opportunities provided by the nation-state structure: protected markets in one country, low wages in another, low taxes here, special incorporation laws there.

The future of the world involves the juxtaposition of old and new systems. Traditional values that have been passed on for thousands of years have come in conflict with new ideas that have emerged with the rapid technological changes of recent centuries. The economies of industrialized and nonindustrialized societies have become dependent on each other and the possibility of a world political system is increasing. Television and other rapid forms of communication are breaking down cultural barriers and creating a world culture as well. It is the proposition of this book that the future as well as the present is wrapped up in the interaction and competition between the two worlds of man.

Appendix

Table A-1 WORLD POPULATION, ENERGY CONSUMPTION, AND GNP, BY COUNTRY

	Country	Population, Midyear 1970 (millions)	Total Energy Consumption, 1969 (million metric tons of coal equivalent)	Per Capita Energy Consumption, 1969 (kilograms)	Total GNP, 1968 ($ millions)	Per Capita GNP, 1968 ($)
1.	China	759.6	392.16	505	65,520*	90*
2.	India	554.6	103.76	193	43,000**	84**
3.	USSR	242.6	1,010.20	4,199	265,290*	1,110*
4.	United States	205.2	2,189.45	10,774	880,100	4,375
5.	Pakistan (including Bangladesh)	136.9	11.80	93	15,287	140
6.	Indonesia	121.2	11.43	98	10,868	96
7.	Japan	103.5	289.40	2,828	141,882	1,404
8.	Brazil	93.0	43.74	481	29,717	337
9.	West Germany	58.6	295.06	4,850	134,625	2,238
10.	United Kingdom	56.0	286.01	5,139	103,139	1,866
11.	Nigeria	55.1	1.88	29	4,559¶	76¶
12.	Italy	53.7	129.24	2,431	75,414	1,430
13.	France	51.1	177.17	3,518	127,304	2,550
14.	Mexico	50.7	51.08	1,044	26,744	566
15.	Philippines	38.1	9.68	261	10,990	306
16.	Thailand	36.2	6.85	197	4,964	147
17.	Turkey	35.6	15.85	461	12,750	380
18.	United Arab Republic	33.9	7.18	221	6,090	102
19.	Spain	33.2	44.85	1,354	25,784	790
20.	Poland	33.0	131.92	4,052	28,424*	880*
21.	South Korea	32.1	19.96	641	5,900	194
22.	Iran	28.4	15.66	562	7,960	295
23.	Burma	27.7	1.55	58	2,057	78
24.	Ethiopia	25.0	.57	23	1,480**	63**
25.	Argentina	24.3	37.03	1,544	17,458	739
26.	Canada	21.4	185.45	8,794	62,254	2,997
27.	Colombia	21.4	12.10	591	7,118	359
28.	North Vietnam	21.2	....	....	1,872*	90*
29.	Yugoslavia	20.6	25.29	1,243	10,302*	510*
30.	Rumania	20.3	52.58	2,628	15,132*	780*
31.	South Africa	20.1	60.86	2,746	14,176	654
32.	South Vietnam	18.0	6.30	353	3,022**	178**
33.	Zaire	17.4	1.39	81	874§	79§
34.	Afghanistan	17.0	.43	26	1,203	60
35.	East Germany	16.2	96.94	5,697	24,453*	1,430*
36.	Sudan	15.8	1.52	100	1,560**	109**
37.	Morocco	15.7	2.93	195	3,036	208
38.	Czechoslovakia	14.7	88.24	6,120	17,856*	1,240*
39.	Algeria	14.0	6.28	470	2,773§	248§
40.	Taiwan	14.0	12.06	874	4,199	312
41.	North Korea	13.9	....	....	3,250*	250*
42.	Peru	13.6	8.20	623	3,718	291
43.	Tanzania	13.2	.74	58	708§	67§
44.	Netherlands	13.0	60.00	4,661	25,378	1,994
45.	Ceylon	12.6	1.44	118	1,805	151
46.	Australia	12.5	63.94	5,200	29,786	2,476
47.	Nepal	11.2	.12	11	801	75

Country	Population, Midyear 1970 (millions)	Total Energy Consumption, 1969 (million metric tons of coal equivalent)	Per Capita Energy Consumption, 1969 (kilograms)	Total GNP, 1968 ($ millions)	Per Capita GNP, 1968 ($)
48. Kenya	10.9	1.56	148	1,324	130
49. Malaysia	10.8	4.56	452	3,057¶	314¶
50. Venezuela	10.8	21.60	2,153	9,146	944
51. Hungary	10.3	29.73	2,888	9,996*	980*
52. Chile	9.8	11.58	1,210	5,316	569
53. Belgium	9.7	54.21	5,429	20,738	2,156
54. Iraq	9.7	5.83	623	2,520	279
55. Portugal	9.6	5.77	603	5,009	529
56. Ghana	9.0	1.33	155	1,987	237
57. Greece	8.9	10.16	1,150	7,554	858
58. Uganda	8.6	.50	53	992	107
59. Bulgaria	8.5	29.45	3,491	6,468*	770*
60. Cuba	8.4	8.69	1,053	2,542*	310*
61. Sweden	8.0	46.02	5,768	26,250	3,315
62. Saudi Arabia	7.7	5.44	755	2,452**	351**
63. Mozambique	7.7	1.24	169	482§	71§
64. Austria	7.4	22.08	2,995	11,350	1,544
65. Cambodia	7.1	.28	42	920¶	147¶
66. Malagasy	6.9	.42	64	778	120
67. Switzerland	6.3	19.81	3,172	16,972	2,761
68. Syria	6.2	2.80	477	1,160	214
69. Ecuador	6.1	1.59	270	1,311	230
70. Cameroun	5.8	.47	83	936	165
71. Angola†	5.7	.78	144	358§	71§
72. Yemen	5.7	.07	14	240‡	48‡
73. Upper Volta	5.4	.05	10	245¶	49¶
74. Haiti	5.2	.15	31	423	91
75. Guatemala	5.1	1.17	234	1,533	315
76. Mali	5.1	.11	22	405¶	88¶
77. Tunisia	5.1	1.25	248	1,048	213
78. Rhodesia	5.0	2.87	564	1,139	230
79. Denmark	4.9	25.15	5,142	12,294	2,527
80. Finland	4.7	16.82	3,576	8,065	1,720
81. Bolivia	4.6	1.05	218	837	179
82. Malawi	4.4	.18	41	246	58
83. Ivory Coast	4.3	.76	180	1,248	304
84. Dominican Republic	4.3	.97	232	1,169	290
85. Zambia	4.3	2.14	509	1,288	316
86. Hong Kong†	4.2	3.71	931	1,550§	442§
87. Norway	3.9	17.06	4,430	9,044§	2,368
88. Guinea	3.9	.37	96	333§	99§
89. Senegal	3.9	.56	148	830	225
90. Niger	3.8	.06	15	319¶	88¶
91. Chad	3.7	.06	18	220**	64**
92. Rwanda	3.6	.04	10	135¶	42¶
93. Burundi	3.6	.03	9	141§	46§
94. El Salvador	3.4	.53	156	910	279
95. Ireland	3.0	8.63	2,953	3,043	1,046
96. Laos	3.0	.16	54	168§	67§
97. New Zealand	2.9	7.28	2,623	4,861	1,767

Table A-1 (*continued*)

Country	Population, Midyear 1970 (millions)	Total Energy Consumption, 1969 (million metric tons of coal equivalent)	Per Capita Energy Consumption, 1969 (kilograms)	Total GNP, 1968 ($ millions)	Per Capita GNP, 1968 ($)
98. Uruguay	2.9	2.61	914	1,833	650
99. Israel	2.9	6.08	2,154	4,084	1,488
100. Somalia	2.8	.07	27	160§	69§
101. Puerto Rico†	2.8	8.45	3,068	4,093	1,503
102. Lebanon	2.8	1.82	689	1,238**	491**
103. Dahomey	2.7	.09	33	194¶	81¶
104. Honduras	2.7	.57	230	618	256
105. Sierra Leone	2.6	.18	70	373**	153**
106. Paraguay	2.4	.30	131	509	228
107. Jordan	2.3	.68	308	553	263
108. Albania	2.2	1.26	608	800*	400*
109. Singapore	2.1	1.54	762	1,495	752
110. Jamaica	2.0	2.01	1,032	969	507
111. Nicaragua	2.0	.63	328	687	373
112. Libya	1.9	.90	482	2,545	1,412
113. Togo	1.9	.12	67	209¶	124¶
114. Costa Rica	1.8	.56	331	745	456
115. Central African Republic	1.5	.05	36	141	108
116. Panama	1.5	1.91	1,348	836	609
117. Mongolia	1.3	....	....	516*	430*
118. Southern Yemen	1.3	.59	484	179§	167§
119. Liberia	1.2	.35	307	254	225
120. Mauritania	1.2	.11	95	166¶	155¶
121. Trinidad and Tobago	1.1	4.65	4,468	748	733
122. Lesotho	1.0	....	....	75¶	88¶
123. Congo (Brazzaville)	.9	.18	202	153§	188§
124. Mauritius	.9	.13	161	198**	256**
125. Bhutan	.8	....	....	45§	62§
126. Kuwait	.7	7.17	12,588	2,221	4,113
127. Guyana	.7	.66	896	215	298
128. Cyprus	.6	.79	1,252	438	704
129. South-West Africa (Namibia)	.6	....	....	....	....
130. Botswana	.6	....	....	55¶	96¶
131. Gabon	.5	.22	458	267**	565**
132. Surinam†	.4	.84	2,165	118	375
133. Luxembourg	.4	With Belgium	With Belgium	775	2,305
134. Swaziland	.4	....	....	64**	167**
135. Barbados	.3	.17	668	111	438
136. Malta	.3	.32	991	183	575
137. Equatorial Guinea	.3	.05	161	72	240
138. Iceland	.2	.70	3,453	416	2,072

*Per capita GNP taken from *1971 World Population Data Sheet,* population for 1968 taken from *1968 World Population Data Sheet,* and total GNP calculated from these two figures.

†Nonsovereign nation.

‡1958 data.

§1963 data.

¶1966 data.

**1967 data.

Source: Population Reference Bureau, Inc., *1970 World Population Data Sheet,* Washington, D.C., 1970; *Statistical Yearbook, 1970,* Statistical Office of the United Nations, New York, 1970, pp. 356–359, 603–605.

Table A-2 Leading Metropolitan Areas Where Headquarters of Major Companies Are Located—Ranked by Total Assets, 1969 ($ 000,000)

	Industrial Corporations	Unclassified Corporations	Banks	Life Insurance Companies	Retail Companies	Transportation Companies	Public Utilities	Total
New York	54,072*	5,438	89,676	56,343	7,293	7,951	54,416	377,189
	(206)	(20)	(11)	(7)	(14)	(9)	(11)	(278)
Chicago	25,558	629	17,294	2,026	10,601	7,757	3,768	67,613
	(94)	(3)	(4)	(2)	(7)	(8)	(2)	(120)
San Francisco	11,776	503	36,036	....	906	3,424	3,815	56,460
	(23)	(1)	(4)		(2)	(4)	(1)	(35)
Detroit	31,686	....	9,246	....	904	....	1,529	43,365
	(22)		(4)		(3)		(1)	(30)
Los Angeles	17,377	510	12,637	2,352	535	729	3,698	37,868
	(45)	(3)	(3)	(2)	(2)	(2)	(2)	(59)
Newark	2,524	476	....	29,033	120	281	2,152	34,589
	(13)	(1)		(2)	(1)	(1)	(1)	(19)
Philadelphia	8,274	....	7,701	3,970	766	6,875	1,554	29,120
	(28)		(4)	(3)	(3)	(2)	(1)	(41)
Pittsburgh	18,069	138	6,090	....	146	....	....	24,443
	(29)	(1)	(2)		(1)			(33)
Hartford	2,600	....	....	19,259	....	....	1,133	22,992
	(4)			(5)			(1)	(10)
Boston	5,059	541	3,836	12,627	439	78	939	21,639
	(20)	(2)	(1)	(2)	(3)	(1)	(1)	(30)
Cleveland	9,236	56	5,122	....	142	2,719	....	17,275
	(32)	(1)	(3)		(2)	(2)		(40)
Dallas	4,852	50	3,890	1,430	198	538	1,371	12,787
	(17)	(2)	(2)	(2)	(1)	(2)	(1)	(27)
Minneapolis–St. Paul	5,059	79	1,196	1,146	955	3,093	1,011	12,539
	(19)	(1)	(1)	(2)	(2)	(3)	(1)	(29)
St. Louis	6,672	27	1,291	465	826	1,775	1,135	12,191
	(19)	(1)	(1)	(1)	(1)	(2)	(1)	(26)
Houston	5,131	339	....	....	....	....	5,627	11,097
	(9)	(1)					(5)	(15)
Milwaukee	2,758	....	....	5,720	....	....	824	9,302
	(18)			(1)			(1)	(20)
Cincinnati	3,932	....	....	2,577	1,561	....	....	8,070
	(9)			(2)	(2)			(13)
Akron	6,260	....	....	....	....	82	856	7,198
	(6)					(1)	(1)	(8)
Portland, Ore.	1,825	....	3,548	....	....	....	786	6,159
	(15)		(2)				(1)	(18)
Atlanta	465	240	1,568	....	114	558	2,487	5,432
	(4)	(3)	(1)		(1)	(1)	(1)	(11)

* 54,072 is the total assets in millions of dollars; 206 is the number of industrial corporations.
Source: "Directory of 1,000 Leading Corporations," *Fortune*, May 1970.

Table A-3 CITIES WITH HEADQUARTERS OF THE 500 LEADING
INDUSTRIAL CORPORATIONS, 1970

CITIES	RANK	ASSETS OF CORPORATIONS ($ 000,000)	PER CENT OF TOTAL ASSETS	NUMBER OF CORPORATION HEADQUARTERS
New York*	1	158,121.224	39.4	131
Detroit*	2	33,281.244	8.3	13
Chicago*	3	25,053.388	6.2	49
Pittsburgh*	4	18,589.526	4.6	14
Los Angeles*	5	17,066.990	4.2	20
San Francisco*	6	10,974.768	2.7	10
Cleveland	7	9,384.356	2.3	17
Philadelphia*	8	7,693.939	1.9	14
Akron	9	7,061.305	1.8	5
St. Louis*	10	6,636.532	1.7	11
St. Paul–Minneapolis*	11	5,627.752	1.4	11
Stamford–Greenwich, Conn.	12	5,046.261	1.3	7
Dallas*	13	4,843.540	1.2	7
Wilmington, Del.	14	4,512.965	1.1	3
Houston*	15	4,198.435	1.0	2
Paterson et al., N.J.*	16	3,804.037	0.9	7
Allentown–Bethlehem, Pa.	17	3,585.394	0.9	2
Greensboro et al., N.C.	18	3,345.364	0.8	4
Bartlesville, Okla.	19	3,102.280	0.8	1
Rochester, N.Y.	20	3,100.285	0.8	2
Toledo	21	3,004.869	0.7	7
Boston*	22	2,858.099	0.7	9
Seattle, Wash.*	23	2,816.243	0.7	3
Dayton, Ohio	24	2,633.708	0.7	4
Midland, Mich.	25	2,619.776	0.7	1
Milwaukee*	26	2,382.590	0.6	11
Richmond, Va.	27	2,379.693	0.6	2
Cincinnati*	28	2,235.494	0.6	4
Portland, Oreg.*	29	2,143.848	0.5	3
Newark, N.J.*	30	1,984.982	0.5	7
Boise, Idaho	31	1,984.048	0.5	1
Peoria, Ill.	32	1,982.530	0.5	3
Hartford, Conn.	33	1,888.849	0.5	3
Middletown, Ohio	34	1,846.144	0.5	1
Tacoma, Wash.	35	1,643.662	0.4	1
San Jose, Calif.*	36	1,622.409	0.4	4
Moline, Ill.	37	1,404.847	0.3	1
New Orleans*	38	1,401.838	0.3	1
Findlay, Ohio	39	1,299.606	0.3	1
Neenah, Wisc.	40	904.314	0.2	1

Table A-3 (*continued*)

Cities	Rank	Assets of Corporations ($ 000,000)	Per Cent of Total Assets	Number of Corporation Headquarters
Providence, R.I.	41	895.124	0.2	1
Ashland, Ky.	42	846.412	0.2	1
Indianapolis*	43	843.164	0.2	3
Tampa*	44	696.584	0.2	1
Syracuse, N.Y.	45	671.820	0.2	2
Oklahoma City	46	667.940	0.2	1
Canton, Ohio	47	638.051	0.2	2
Fort Wayne, Ind.	48	576.872	0.1	3
Benton Harper, Mich.	49	573.929	0.1	1
Corning, N.Y.	50	548.274	0.1	1
Nashville, Tenn.	51	533.731	0.1	1
Atlanta*	52	524.179	0.1	3
Buffalo, N.Y.*	53	513.905	0.1	2
Baltimore*	54	504.956	0.1	3
Corpus Christi, Tex.	55	490.190	0.1	1
Lancaster, Pa.	56	487.725	0.1	1
New Brunswick, N.J.	57	481.640	0.1	1
Buchanan, Mich.	58	444.246	0.1	1
Honolulu, Hawaii	59	415.440	0.1	1
Kansas City, Mo.*	60	411.087	0.1	2
Columbus, Ohio	61	386.903	0.09	1
Harrisburg, Pa.	62	373.123	0.08	2
Kalamazoo, Mich.	63	360.807	0.08	1
Rockford, Ill.	64	344.615	0.08	1
Eldorado, Ark.	65	343.914	0.08	1
Fort Mills, S.C.	66	334.671	0.08	1
Battle Creek, Mich.	67	317.117	0.07	1
Erie, Pa.	68	306.012	0.07	1
Wichita, Kans.	69	299.603	0.07	2
Decatur, Ill.	70	297.006	0.07	2
Worcester, Mass.	71	292.591	0.07	1
Louisville, Ky.	72	283.771	0.07	1
Greenville, S.C.	73	282.408	0.07	1
Elkhart, Ind.	74	281.955	0.07	2
San Juan, P.R.	75	278.497	0.07	1
Niagara Falls, N.Y.	76	270.085	0.06	1
West Point, Ga.	77	269.859	0.06	1
Waterbury, Conn.	78	269.505	0.06	1
Enka, N.C.	79	266.869	0.06	1
Bridgeport, Conn.	80	262.702	0.06	2
Reading, Pa.	81	259.131	0.06	2

Table A-3 (*continued*)

Cities	Rank	Assets of Corporations ($ 000,000)	Per Cent of Total Assets	Number of Corporation Headquarters
Kannapolis, N.C.	82	252.144	0.06	1
Denver, Colo.*	83	221.835	0.05	1
Amsterdam, N.Y.	84	215.153	0.05	1
Camp Hill, Pa.	85	211.257	0.05	1
Meriden, Conn.	86	210.727	0.05	1
Omaha, Nebr.	87	207.210	0.05	3
North Bergen, N.J.	88	188.561	0.04	1
Birmingham, Ala.	89	187.282	0.04	1
Lancaster, Ohio	90	186.332	0.04	1
New Britain, Conn.	91	184.995	0.04	1
West Haven, Conn.	92	182.695	0.04	1
La Crosse, Wisc.	93	180.403	0.04	1
Youngstown, Ohio	94	177.158	0.04	1
Chula Vista, Calif.	95	168.415	0.04	1
Quincy, Ill.	96	161.851	0.04	1
Edison, N.J.	97	153.941	0.03	1
Port Edwards, Wisc.	98	152.795	0.03	1
South Nashua, N.H.	99	147.293	0.03	1
Eden, N.C.	100	145.736	0.03	1
Le Sueur, Minn.	101	144.887	0.03	1
Madison, Wisc.	102	141.776	0.03	1
Tecumseh, Mich.	103	141.438	0.03	1
Brockway, Pa.	104	130.964	0.03	1
Kankakee, Ill.	105	122.360	0.03	1
Sioux City, Iowa	106	117.065	0.02	2
Racine, Wisc.	107	117.060	0.02	1
Austin, Minn.	108	115.788	0.02	1
Fremont, Mich.	109	115.623	0.02	1
Norwalk, Conn.	110	111.252	0.02	1
Newton, Iowa	111	99.910	0.02	1
Memphis, Tenn.	112	99.773	0.02	1
Saline, Mich.	113	84.118	0.02	1
Phoenix, Ariz.	114	72.830	0.02	1
Eau Claire, Wisc.	115	60.548	0.01	1
Greeley, Colo.	116	49.561	0.01	1
Waterloo, Iowa	117	36.170	0.009	1
Spencer, Iowa	118	33.068	0.009	1
Rock Point, Mo.	119	19.854	0.008	1

*Standard Metropolitan Statistical Areas (SMSAs) of 1 million or more population in 1970.
Four SMSAs are the headquarters of any of the 500 leading corporations: Washington, D.C.; Anaheim, Calif.; Miami, Fla.; and San Bernardino, Calif.
Source: "Directory of 1,000 Leading Corporations," *Fortune*, May 1970.

Table A-4 INDICES OF LATIN-AMERICAN DEVELOPMENT

COUNTRY	POPULATION, 1970 (millions)	AGGREGATE GNP, 1970 ($ millions)	PER CAPITA GNP, 1970 ($)	INDUSTRIAL ENERGY CONSUMPTION, 1967 AGGREGATE (million metric tons)	INDUSTRIAL ENERGY CONSUMPTION, 1967 PER CAPITA (kilograms)	TOTAL EXPORTS, 1967 AGGREGATE ($ millions)	TOTAL EXPORTS, 1967 PER CAPITA ($)	IMPORTANCE OF AGRICULTURE,* AS A PER CENT OF NATIONAL EMPLOYMENT	IMPORTANCE OF AGRICULTURE,* AS A PER CENT OF NATIONAL INCOME
Mexico	50.7	26,871	530	48.99	1,073	1,145	25	51	48
Brazil	93.0	23,250	250	33.61	392	1,654	20	41	18
Argentina	23.2	19,024	820	32.08	1,380	1,465	64	16	16
Venezuela	10.4	9,880	950	20.76	2,220	2,885	320	41	7
Colombia	21.6	5,400	250	10.80	521	510	27	47	28
Chile	9.8	4,704	480	10.63	1,163	913	103	25	5
Peru	13.6	5,168	380	7.68	620	774	64	63	18
Cuba	8.4	2,604	310	8.20	1,033	715	90	34	16
Uruguay	2.9	1,508	520	2.36	847	159	59	18	14
Guatemala	5.2	1,664	320	1.03	218	199	44	65	26
Ecuador	6.1	1,342	220	1.21	219	200	38	53	31
Dominican Republic	4.1	1,189	290	.69	178	157	42	58	24
Jamaica	1.9	874	460	1.78	951	222	123	49	11
El Salvador	3.5	980	280	.54	170	207	69	58	28
Panama	1.5	870	580	1.66	1,249	92	70	44	23
Trinidad and Tobago	1.0	870	870	4.26	4,215	443	443	20	12
Bolivia	4.9	735	150	.89	234	145	39	48	21
Nicaragua	2.0	740	370	.48	271	147	86	56	28
Costa Rica	1.7	765	450	.52	327	143	95	44	25
Honduras	2.5	650	260	.41	169	156	66	70	40
Paraguay	2.4	552	230	.27	127	48	23	52	33
Haiti	4.9	343	70	.15	33	33	7	83	45
Guyana	0.8	272	340	.66	964	113	161	33	25

*From *The New York Times Encyclopedic Almanac 1972,* The New York Times, New York, 1972.

SOURCE: Population Reference Bureau, Inc., *1971 World Population Data Sheet,* Washington, D.C., 1971; *Statistical Yearbook: 1968,* Statistical Office of the United Nations, New York, 1968; *World Bank Atlas: Population, Per Capita Product and Growth Rates,* International Bank for Reconstruction and Development, Washington, D.C., 1972.

Table A-5 COMPARISON OF POPULATION DENSITIES AND
LAND AREA, 1959

AREA	POPULATION (per cent)	LAND AREA (per cent)	DENSITY (persons per sq. mi.)
EUROPEAN SETTLEMENT			
Traditional Slavic homelands			
Northwest	3	1	84
Center	21	4	117
Byelorussian SSR	4	1	101
Ukrainian-Moldavian SSRs	21	3	183
Subtotal	49	9	485
Other European homelands			
Baltic	3	1	90
More recent European settlement			
Volga	5	2	58
Urals	7	3	57
Western Siberia	6	5	25
Krasnoyarsk–Irkutsk	2	7	8
North Kazakhstan	3	7	10
North Caucasus	6	2	85
Subtotal	29	26	243
Total	81	36	818
TRADITIONAL NON-EUROPEAN SETTLEMENT			
North (Europe and western Siberia)	3	13	6
Caucasus	5	1	132
Central Asia	8	11	18
Eastern Siberia	1	25	1
Far East	2	14	4
Total	19	64	161
Grand total	100	100	

SOURCE: Cole and German, *A Geography of the U.S.S.R.*, Butterworth & Company, Ltd., London, 1961.

Index

China (continued)

agriculture of, traditional, 502–505, 539, 543, 545

ancient civilization of, 502–504

central planning in, 518, 535–536, 541, 543–545, 547

climate of, 536

communes in, 543–545, 556

    as economic units, 544–545

    as socio-political units, 544–545

compared to the United States, map, 520

core, 547

Cultural Revolution in, 545, 547–548

culture of, 503–504, 545, 547

decision-making in, 518, 535–536, 541, 543–545, 547–548

development problems of, 500–502, 510

economic system of, 518, 546–548

environmental base of, 548

    similarity to the United States, 536, 539

European spheres of influence in, 504–505

five-year plans of, 541

foreign aid to, 510

foreign relations with, 548

government in, 507, 510, 518, 535–536

    control of, 507, 525, 535–536, 543, 545

Great Leap Forward, 536, 543, 546–548

growing potential in, maps, 110–111; 536

growing season of, map, 148–149; 536

industry in, 539, 541; map, 542; 543

    heavy, 541; map, 542

    local, 545; box, 546

land surface of, 536; map, 537

metropolitan areas of, 157, 503, 539, 541

missionaries to, 504–505

and Nationalist China, 510, 548

political map of, 509

population

    distribution of, 536; map, 538; 539

    growth of, 500

    problem of, 501–502, 518

    size of, 500

rainfall in, map, 142–143; 536

references, 549

rules of conduct in, 535

rural reorganization in, 547

Soviet assistance to, 507, 510, 541, 548

China and India

development problems of, photo essay, 511–517

European impact

    on investment, 505

    on trade, 504–505

    on village manufacturing, 505–506

five-year plans of, 518

as a market for European goods, 505–506

mines in, 505

plantations in, 505

population explosion in, 506

similarities to the United States and Soviet Union, 510

villages in, 503

Cuba, 419–420, 422, 427, 430, 440–443, 446; box, 448; 450–452

cultivated land of the world, by country, map, 155

culture, 50, 56–57, 62–96, 218–219, 245, 265, 270, 308–309; tables, 309, 311; 407–409, 416–417, 478, 503–504, 545, 547

decision-making, 51–54, 168–169, 236, 241–243, 245–254, 256, 303, 312–314, 316–317, 321–322, 330, 342–345, 349–351, 367–368, 378–379, 381, 394–395, 398–399, 401–402, 406, 410, 417, 419–420, 487, 507, 510, 518, 535–536, 541, 543–545, 547–548, 553–556

deserts, 109, 116, 471–472

diet, 161–162; map, 163; 164

dual economies, 51–55, 59, 91, 406–411, 413, 417, 420–421, 438–443, 446, 449–452, 455–458, 460–461, 476, 478, 505, 510, 557

earth environment, 98–132; photo essay, 119–125

ecological balance, 551

economy, 52–54, 178, 265, 267, 269, 301; table, 304; 313–314, 316–317, 322, 325, 330, 342–345, 349–350, 354, 417, 420–432, 438–443, 446; table, 448; 449, 464–465, 518, 547–548

employment, 155; map, 158; 401, 478–479

energy consumption, 80; table, 82; 82–83; table, 84; 159; map, 160; 465, table and map, 491; 552

Photo Credits

Stock, Boston; *right*, Patricia Gross/Stock, Boston; *bottom*, Ken Heyman; **250** and **251**, Pictograms by Shigeo Fukuda and Yoshiro Yamashita-Masaru Katzume, art director; **252**, Group for Environmental Education; **253** *top left*, Franklin Wing/Stock, Boston; *right*, Daniel Brody/Stock, Boston; *bottom*, Paul Arnaud.

Page 271, Frank Siteman; **272** *top left*, Aero-Camera/Rotterdam; *right*, Lufthansa Archiv; *center right*, Boro Facilte Eirenn/Dublin; *bottom left*, Mastboom Uliegbedrijf/Rotterdam; *right*, French Government Tourist Office; **273** *top left*, Management James Sawders; *right*, Max Jacoby; *bottom left*, Norwegian Embassy; *center right*, Norwegian National Tourist Office; *bottom right*, French Government Tourist Office; **274**, French Government Tourist Office; **275**, Patricia Gross/Stock, Boston.

Page 333, Novosti; **334**, Henri-Cartier Bresson/Magnum; **335**, Novosti through World Health Organization; **336** *top left*, Annan Photo Features; *right*, Novosti; *bottom left*, Annan Photo Features; *right*, Novosti; **337**, Novosti; **338**, Mark Riboud/Magnum; **339**, Eve Arnold/Magnum.

Page 369, Mark Riboud/Magnum; **370**, United Nations; **371**, Standard Oil; **372**, Associated Press through Wide World; **373** *top*, Consulate General of Japan; *bottom*, Japan Air Lines; **374** *top*, Henri-Cartier Bresson/Magnum; *bottom*, Asuka-Eu Nora/Japan; **375**, Geographical Survey Institute/Tokyo.

Page 433, Paul Conklin; **434** *top*, Standard Oil of N.J. collection; *bottom*, Paul Conklin; **435** *top left*, Standard Oil; *right*, United Nations; *bottom left*, United Fruit; *right*, Standard Oil; **436**, Lupe/Stock, Boston; **437**, Franklin Wing/Stock, Boston.

Page 479, courtesy of UNESCO; **480,** Owen Franklin/Stock, Boston; **481,** United Nations; **482** *top left*, Mobil Oil; *right*, Food and Agricultural Organization; *bottom left*, United Nations; *right*, UNESCO; **483** *top left* and *right*, George Rodger/Magnum; *bottom left*, George Rodger/Magnum; *right*, United Nations;

484, Owen Frankin/Stock, Boston; **485,** Owen Frankin/Stock, Boston.

Page 509 *top*, Emil Schulthess/Black Star; *bottom*, Food and Agricultural Organization; **510** *top left*, Annan Photo Features; *right*, Marc Riboud/Magnum; *bottom left*, United Nations; *right*, Indian Consulate; **511** *top*, Marc Riboud/Magnum; *bottom*, Care photo; **512** *top*, Marc Riboud/Magnum; *bottom*, United Nations; **513** *top left*, Annan Photo Features; *right*, Marc Riboud/Magnum; *bottom left*, Marilyn Silverstone/Magnum; *right*, Indian Consulate; **514**, Rene Burri/Magnum; **515**, Ernie Bischof/Magnum.

Chapter opener photos: chapter 1, Paul Conklin; chapter 2, Food and Agricultural Organization; chapter 3, NASA; chapter 4, United Nations; chapter 5, Peter Menzel/Stock, Boston; chapter 6, John Shelton; chapter 7, U.S. Department of Transportation; chapter 8, Harvard Laboratory for Computer Graphics; chapter 9, Boro Facilte Eirenn/Dublin; chapter 10, Novosti; chapter 11, United Nations; chapter 12, Peter Menzel/Stock, Boston; chapter 13, Standard Oil of N.J. collection; chapter 14, Food and Agricultural Organization; chapter 15, Food and Agricultural Organization; chapter 16, The Martin Co.

Part opener photos by J. Seeley. Globe by Replogle.

map- p. 170